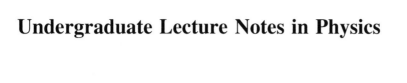

Undergraduate Lecture Notes in Physics

For further volumes:
http://www.springer.com/series/8917

Undergraduate Lecture Notes in Physics (ULNP) publishes authoritative texts covering topics throughout pure and applied physics. Each title in the series is suitable as a basis for undergraduate instruction, typically containing practice problems, worked examples, chapter summaries, and suggestions for further reading.

ULNP titles must provide at least one of the following:

- An exceptionally clear and concise treatment of a standard undergraduate subject.
- A solid undergraduate-level introduction to a graduate, advanced, or non-standard subject.
- A novel perspective or an unusual approach to teaching a subject.

ULNP especially encourages new, original, and idiosyncratic approaches to physics teaching at the undergraduate level.

The purpose of ULNP is to provide intriguing, absorbing books that will continue to be the reader's preferred reference throughout their academic career.

Series Editors

Neil Ashby
Professor, Professor Emeritus, University of Colorado, Boulder, CO, USA

William Brantley
Professor, Furman University, Greenville, SC, USA

Michael Fowler
Professor, University of Virginia, Charlottesville, VA, USA

Michael Inglis
Professor, SUNY Suffolk County Community College, Selden, NY, USA

Elena Sassi
Professor, University of Naples Federico II, Naples, Italy

Helmy Sherif
Professor Emeritus, University of Alberta, Edmonton, AB, Canada

Bruce Cameron Reed

The History and Science
of the Manhattan Project

 Springer

Bruce Cameron Reed
Department of Physics
Alma College
Alma, MI
USA

ISSN 2192-4791   ISSN 2192-4805   (electronic)
ISBN 978-3-662-50914-2   ISBN 978-3-642-40297-5   (eBook)
DOI 10.1007/978-3-642-40297-5
Springer Heidelberg New York Dordrecht London

Printed on acid-free paper

Springer is part of Springer Science+Business Media (www.springer.com)

This work is dedicated to Laurie
You will always be The One

Preface

In August, 1945, two United States Army Air Force B-29 bombers each dropped single bombs on the Japanese cities of Hiroshima and Nagasaki. These new "atomic" bombs, colloquially known as *Little Boy* and *Fat Man*, each exploded with energies equivalent to over 10,000 tons of conventional explosive, the normal payload of 1,000 such bombers deployed simultaneously. Hiroshima and Nagasaki were both devastated. A few days later, Japan surrendered, bringing an end to World War II. In a speech to his people on August 15, Emperor Hirohito specifically referred to "a new and most cruel bomb" as one of the reasons for accepting surrender terms that had been laid out by the Allied powers. A later analysis by the United States Strategic Bombing Survey estimated the total number of people killed in the bombings to be about 125,000, with a further 130,000–160,000 injured.

While historians continue to debate whether the bombs can be credited with directly ending the war or simply helped to hasten its end, it is irrefutable that the development and use of nuclear weapons was a watershed event of human history. In 1999, the *Newseum* organization of Washington, D.C., conducted a survey of journalists and the public regarding the top 100 news stories of the twentieth century. Number one on the list for both groups was the bombings of Hiroshima and Nagasaki and the end of World War II. Journalists ranked the July, 1945, test of an atomic bomb in the desert of southern New Mexico as number 48, and the Manhattan Project itself, the U.S. Army's effort under which the bombs were developed, as number 64. The Manhattan Project was the most complex and costly national-level research and development project to its time, and its legacy is enormous: America's postwar military and political power, the cold war and the nuclear arms race, the thousands of nuclear weapons still held in the arsenals of various countries, the possibility of their proliferation to other states, the threat of nuclear terrorism, and public apprehension with radiation and nuclear energy all originated with the Project. These legacies will remain with us for decades to come.

The development of nuclear weapons is the subject of literally thousands of books and articles, many of them carefully researched and well-written. Why, then, do I believe that the world needs one more volume on a topic that has been so exhaustively explored?

Source material on the Manhattan Project can be classed into four very broad categories. First, there are many synoptic semi-popular histories. This genre began with William Laurence's *Dawn over Zero* (1946) and Stephane Groueff's 1967 *Manhattan Project: The Untold Story of the Making of the Atomic Bomb*. The current outstanding example of this type of work is Richard Rhodes' *The Making of the Atomic Bomb* (1986); references to a number of others appear in the "Resource Letters" by myself cited in the Further Reading list at the end of this section. Second are works prepared as official government and military histories, primarily for academic scholars. The original source along this line was Henry DeWolf Smyth's *Atomic Energy for Military Purposes*, which was written under War Department auspices and released just after the bombings of Hiroshima and Nagasaki. More extensive later exemplars are Hewlett and Anderson's *A History of the United States Atomic Energy Commission*, and Vincent Jones' *United States Army in World War II: Special Studies—Manhattan: The Army and the Atomic Bomb*. Third are the numerous biographies on the leading personalities of the Project, particularly some of the scientists involved. Well over a dozen biographies have been published on Robert Oppenheimer alone. Finally, there are specialized technical publications which require readers to be armed with some upper-undergraduate or graduate-level physics and allied sciences to appreciate fully.

Synoptic volumes are accessible to a broad audience, but tend to be limited in the extent of their technical coverage. Interesting as they are, one can read the same stories only so many times; eventually a curious reader must yearn for deeper knowledge: Why can only uranium or plutonium be used to make a fission weapon? How does one compute a critical mass? How was plutonium, which does not occur naturally, created? Official histories are superbly well-documented, but also tend to be non-technical; they are not meant to serve as student texts or popularly-accessible treatments. Biographies are not usually written to address technical matters, but here a different issue can creep in. While many biographies are responsible treatments of the life and work of the individual concerned, others devolve into questionable psychological or sociological analyses of events and motivations now decades in the past, where, not inconveniently, the principals have no opportunity to respond. Some of the synoptic-level treatments fall prey to this affliction as well.

The bottom line is that after many years of teaching a college-level general-education course on the Manhattan Project, I came to the conclusion that a need exists for a broadly-comprehensible overview of the Project prepared by a physicist familiar with both its science and history. My goal has been to try to find a middle ground by preparing a volume that can serve as a text for a college-level science course at a basic-algebra level, while also being accessible to non-students and non-specialists who wish to learn about the Project. To this end, most chapters in this volume comprise a mixture of descriptive and technical material. For technically-oriented readers, exercises are included at the ends of some chapters. For readers who prefer to skip over mathematical treatments of technical details, the text clearly indicates where descriptive passages resume.

Another motivation for taking on this project is that, over time, access to sensitive information regarding historically important events inevitably becomes more open. At this writing, almost 70 years have elapsed since the Smyth report, 50 have passed since the publication of Hewlett and Anderson's *New World*, and over 25 since Rhodes' *Making of the Atomic Bomb*. In the meantime, a considerable number of technical and non-technical publications on the Project have appeared, and many more original documents are readily available than was the case when those authors were preparing their works. From both a personal professional perspective and an access-to-information viewpoint, the time seemed right to prepare this volume.

Writing about decades-old events is a double-edged sword. Because we know how the story played out, hindsight can be perfect. We know which theories and experiments worked, and which did not. The flip side of this is that it becomes far too easy to overlook false starts and blind alleys, and set out the story in a linear this-then-that sequence that gives it all a sense of predetermined inevitability. But this would not give a due sense of the challenges faced by the people involved with the Project, so many aspects of which were so chancy that the entire effort could just as well have played *no* role in ending the war. After the discovery of nuclear fission, it took some of the leading research personalities of the time well over a year to appreciate how the subtleties of nuclear reactions might be exploited to make a weapon or a reactor. Even after theoretical arguments and experimental data began to become clear, technological barriers to practical realization of nuclear energy looked so overwhelming as to make the idea of a nuclear weapon seem more appropriate to the realm of science-fiction than to real-world engineering. Physicist and Nobel Laureate Niels Bohr was of the opinion that "it can never be done unless you turn the United States into one huge factory." To some extent, that is exactly what was done. Again, my goal has been to seek a middle ground which gives readers some sense of the details and evolution of events, but without being overwhelming.

The scale of the Manhattan Project was so great that no single-volume history of it can ever hope to be fully comprehensive. After the Project came under Army auspices in mid-1942, it split into a number of parallel components which subsequently proceeded to the end of the war. This parallelism obviates a strict chronological telling of the story; each main component deserves its own chapter. Thousands of other publications on this topic exist precisely because many of those components are worthy of detailed analyses in their own right. Thus, the present volume should be thought of as a gateway to an intricate, compelling story, after which an interested reader can explore any number of fascinating sub-plots in more depth.

It is my sincere hope that you will enjoy, learn from, and seriously reflect upon the science and history that unfold on the following pages. I hope also that they whet your appetite for more. Sources of information on the Project are so extensive that a single individual can hope to look at but a few percent of it all; I have devoted well over a decade of my professional career to studying the Manhattan Project, and know that I still have much to learn. The future will need more scientists and historians to serve as Manhattan Project scholars. To students reading these words, I invite you to consider making such work a part of your own career.

A Note on Sources

As much as possible, I have drawn the information in this book from primary documentary sources (see below), from works whose authors enjoyed access to classified information (Smyth, Hewlett and Anderson, Jones, and Hoddeson, et al.), from memoirs and scholarly biographies of individuals who were present at the events related, and from technical papers published in peer-reviewed scientific journals. A list of references appears at the end of each chapter under the heading Further Reading; a detailed list of citations can be found at www.manhattanphysics.com.

One will occasionally find that even very credible sources report the details of events slightly differently: dates may vary somewhat, numerical quantities and funding amounts differ, lists of individuals involved may be more or less consistent, and so forth. Written records of meetings were often deliberately left incomplete. The Manhattan Project's commander, General Leslie Groves, frequently preferred to issue only verbal orders, and some information still remains classified. As a result, full understanding of some aspects of the Project is simply not possible. To fill in the gaps, it is necessary to extrapolate from what is known to have happened, and to work from the potentially fallible recollections of individuals involved. In such cases, I have tried to work with the most authoritative sources available to me, but I do not doubt that some errors and inconsistencies have crept in. For these I apologize to my readers in advance.

The primary source of Manhattan Project documentary material is four sets of documents available on microfilm from the National Archives and Records Administration (NARA) of the United States. These comprise a total of 42 rolls of film, and readers who are motivated to explore them need to be aware that information on a given topic can be spread over multiple rolls within each of the four sets, and that documents on a given topic within a given roll by no means always appear in chronological order. Some documents that are still classified are deleted from the films. The four sets and their NARA catalog numbers are as follows:

A1218: Manhattan District History (14 rolls). This massive multi-volume document was prepared as an official history of the Project after the war by Gavin Hadden, an aide to General Groves. Known to historians and researchers as the MDH, these documents are a fundamental source of information on the Project.[1]

---

[1] As this book was going to press, the Department of Energy began posting the MDH online at <https://www.osti.gov/opennet/manhattan_district.jsp>. In particular, previously redacted material on the K-25 plant (Sect. 5.4) is now being made available.

M1108: Harrison-Bundy Files Relating to the Development of the Atomic Bomb, 1942–1946 (Records of the Office of the Chief of Engineers; Record Group 77; 9 rolls).

M1109: Correspondence ("Top Secret") of the Manhattan Engineer District, 1942–1946 (Records of the Office of the Chief of Engineers; Record Group 77; 5 rolls).

M1392: Bush-Conant File Relating to the Development of the Atomic Bomb, 1940–1945 (Records of the Office of Scientific Research and Development, Record Group 227; 14 rolls).

An index for each set can be viewed by searching its catalog number on the NARA ordering website (select "Microfilm" from the tabs at the top of the page): https://eservices.archives.gov/orderonline/start.swe?SWECmd=Start&SWEHo= eservices.archives.gov.

Acknowledgments

I have benefited from spoken and electronic conversations, correspondence, suggestions, willingness to read and comment on draft material, and general encouragement from John Abelson, Joseph-James Ahern, Dana Aspinall, Albert Bartlett, Jeremy Bernstein, Alan Carr, David Cassidy, John Coster-Mullen, Steve Croft, Gene Deci, Eric Erpelding, Patricia Ezzell, Ed Gerjuoy, Chris Gould, Robert Hayward, Dave Hafemeister, William Lanouette, Irving Lerch, Harry Lustig, Jeffrey Marque, Albert Menard, Tony Murphy, Robert S. Norris, Klaus Rohe, Frank Settle, Ruth Sime, D. Ray Smith, Roger Stuewer, Michael Traynor, Alex Wellerstein, Bill Wilcox, John Yates, and Pete Zimmerman. If I have forgotten anybody, I apologize; know that you are in this list in spirit. Alma College interlibrary loan specialist Susan Cross has never failed to dig up any obscure document which I have requested; she is a true professional. Angela Lahee and her colleagues at Springer deserve a big nod of thanks for believing in, committing to, and seeing through this project.

In addition to the individuals listed above, the fingerprints of a lifetime's worth of family members, teachers, classmates, professors, mentors, colleagues, students, collaborators, and friends are all over these pages; a work like this is never accomplished alone. I thank them all. Over the course of several years, Alma College has generously awarded me a number of Faculty Small Grants which have gone toward developing this work, as well as a sabbatical leave during which this book was completed.

Most of all I thank Laurie, who bore with it.

Further Reading

Books, Journal Articles, and Reports

S. Groueff, Manhattan Project: *The Untold Story of the Making of the Atomic Bomb*. (Little, Brown and Company, New York, 1967). Reissued by Authors Guild Backprint.com (2000)

R.G. Hewlett, O.E. Anderson Jr., *A History of the United States Atomic Energy Commission, Vol. 1: The New World, 1939/1946*. (Pennsylvania State University Press, University Park, PA 1962)

L. Hoddeson, P.W.Henriksen, R.A. Meade, C.Westfall,*Critical Assembly: A Technical History of Los Alamos during the Oppenheimer Years*, (Cambridge University Press, Cambridge 1993)

V.C. Jones, *United States Army in World War II: Special Studies—Manhattan: The Army and the Atomic Bomb* (Center of Military History, United States Army, Washington, 1985)

W.L. Laurence, *Dawn over Zero: The Story of the Atomic Bomb*, Knopf, (New York 1946)

B.C. Reed, *Resource Letter MP-1: The Manhattan Project and related nuclear research*. Am. J .Phys. **73**(9), 805–811 (2005)

B.C. Reed, *Resource Letter MP-2: The Manhattan Project and related nuclear research*. Am. J. Phys. **79**(2), 151–163 (2011)

R. Rhodes, *The Making of the Atomic Bomb*, (Simon and Schuster, New York,1986)

H.D. Smyth, *Atomic Energy for Military Purposes: The Official Report on the Development of the Atomic Bomb* Under the Auspices of the United States Government, 1940–1945. (Princeton University Press, Princeton 1945). Available electronically from a number of sources, for example, <http://www.atomicarchive.com/Docs/SmythReport/index.shtml>.

Websites and Web-based Documents

Newseum story on top 100 news stories of twentieth century:
<http://www.newseum.org/century/>

Contents

Chapter 1
Introduction and Overview

The official military designation of the program conducted by the United States Army to develop atomic bombs—which are more correctly termed "nuclear weapons"—was "Manhattan Engineer District" (MED). In official documents, this designation was often contracted to Manhattan District, and in postwar vernacular became "The Manhattan Project." Despite its size and complexity, this effort was carried out with remarkable secrecy: some 150,000 people were employed in the project, most of whom labored in complete ignorance as to what they were producing. Indeed, it has been estimated that perhaps only a dozen individuals were familiar with the overall program. By August, 1945, the cost of the Project had reached $1.9 billion out of a total cost to the United States for the entire war of about $300 billion. Two billion dollars for any one element of the war was a monumental amount; the Manhattan Project was an organizational, engineering, and intellectual undertaking that had no precedent.

This book offers an overview of the science and history of the Manhattan Project. To help orient readers, this chapter is devoted to a description of the general nature of the Project and how this book is organized.

1.1 Chapters 2 and 3: The Physics

Chapters 2 and 3 are devoted to the background scientific discoveries that led up to the Manhattan Project. Since this is really a lesson in what is now well-established physics, I forgo blind alleys for the sake of brevity. Even then, these chapters are lengthy.

Nuclear physics as a scientific discipline can be said to have started with the discovery of natural radioactivity in 1896, almost a half-century before Hiroshima. In the following decades, experimental and theoretical work by various researchers, located mostly in England, France, Germany, Denmark, and Italy, unraveled the nature of radioactivity and the inner structure of atoms. By the early 1930s, our now-common high-school image of an atom comprising a nucleus of protons and neutrons being "orbited" by a cloud of whizzing electrons was largely established.

B. C. Reed, *The History and Science of the Manhattan Project*,
Undergraduate Lecture Notes in Physics, DOI: 10.1007/978-3-642-40297-5_1,
© Springer-Verlag Berlin Heidelberg 2014

During the middle years of that decade, it also came to be realized that radioactivity could be induced artificially through experimental conditions set up by human beings, as opposed to waiting for the phenomenon to happen through random natural processes. Millions of people who have been treated with radiation therapy are beneficiaries of that discovery. Chapter 2 covers the history of nuclear physics from the discovery of radioactivity to the mid-1930s.

Late 1938 witnessed the stunning discovery that nuclei of uranium atoms could be split apart when bombarded by neutrons. In this process, a split uranium nucleus loses a small amount of mass, but this mass corresponds to a fantastic amount of energy via Albert Einstein's famous $E = mc^2$ equation. The amount of energy released per reaction in such cases is *millions* of times that liberated in any known chemical reaction. This process, soon termed *nuclear fission*, lies at the heart of how nuclear weapons function. Only a few weeks after the discovery of fission, it was verified (and had been anticipated in some quarters) that a by-product of each fission was the liberation of two or three neutrons. This feature is what makes a *chain reaction* possible (Fig. 1.1). These "secondary" neutrons, if they do not escape the mass of uranium, can go on to fission other nuclei. Once this process is started, it can *in principle* continue until all of the uranium is fissioned. Of course, practice always proves more difficult than theory (phenomenally so in the case of the Manhattan Project), but this is the fundamental idea behind nuclear reactors and bombs. Chapter 3 deals with the discovery and interpretation of nuclear fission.

Any interesting scientific discovery always opens more questions than it resolves. Ergo: Could any other elements undergo fission? Why or why not? Was there a minimum amount of uranium that would have to be arranged in one place to have any hope of realizing a chain reaction? If so, could the process be controlled by human intervention to give the possibility of an energy source, or would the result be an uncontrolled explosion? Or did the fact that the uranium ores of the Earth had not spontaneously fissioned themselves into oblivion millennia ago

Fig. 1.1 Schematic illustration of the start of a chain reaction. The secondary neutrons may go on to strike other target nuclei

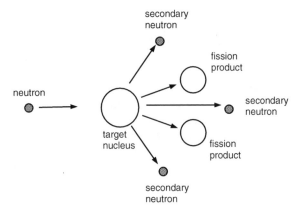

mitigate against any such possibilities?[1] As the world advanced toward global conflict in the early months of 1939, nuclear physicists investigated these questions.

By the time of the outbreak of World War II in September, 1939, the pieces of an overall picture were starting to come together. Only the very heavy elements uranium and thorium looked to be fissile. For reasons that are elucidated in Chap. 3, thorium ended up playing no role in the Manhattan Project. This left uranium as the only possibility as a source of energy or as an explosive, but the prospects looked bleak. As it occurs naturally in the Earth's crust, uranium consists predominantly of two isotopes, U-235 and U-238.[2] (If you are unfamiliar with the concept of isotopes, read the footnote below. Definitions of a number of technical terms appear in the Glossary at the end of this book.) But these isotopes occur in far-from-equal proportions: only about 0.7 % of naturally-occurring uranium is of the U-235 variety, while the other 99.3 % is U-238. By early 1940, it was understood that only nuclei of the rare isotope U-235 had a useful likelihood of fissioning when bombarded by neutrons, whereas those of U-238 would primarily tend to somewhat slow down and subsequently absorb incoming neutrons without fissioning. Given the overwhelming preponderance of U-238 in natural uranium, this absorption effect promised to poison the prospect for a chain reaction.

To obtain a chain reaction, it appeared necessary to isolate a sample of U-235 from its sister isotope, or at least process uranium in some way so as to increase the percentage of U-235. Such a manipulation of the isotopic abundance ratio is now known as *enrichment*. Enrichment is always a difficult business. Since isotopes of any element behave identically so far as their chemical properties are concerned, no chemical separation technique can be employed to achieve enrichment. Only a technique that depends on the slight mass difference (~ 1 %) between the two isotopes could be a possibility. To this end, the prospects in 1940 were limited: centrifugation, mass spectrometry, and diffusion were the only techniques known. Unfortunately, they had been applied successfully only in cases involving light elements such as chlorine, where the percentage differences between isotopic masses is much greater. It is no wonder that Niels Bohr, who was the most significant contributor to the understanding of the roles of different isotopes in the process of fission, was skeptical of any prospect for harnessing "atomic energy."

---

[1] It was subsequently discovered (1972) that naturally-occurring chain reactions in deposits of uranium ores in Africa did occur about two billion years ago.

[2] In any atom, the number of orbiting electrons normally equals the number of protons in the nucleus of the atom. This number, the so-called *atomic number*, usually designated by Z, is the same for all atoms of the same element, and dictates the chemical properties of the atoms of the element. For oxygen atoms, $Z = 8$; for uranium atoms, $Z = 92$. Different *isotopes* of the same element have differing numbers of neutrons in their nuclei, but —because they have the same Z – have the same chemical properties. In particular, the nuclei of U-235 and U-238 atoms both contain 92 protons, but respectively contain 143 and 146 neutrons; for each of these isotopes, the numbers following the "U-" gives the total number of neutrons plus protons in their nuclei. There is a third isotope of uranium, U-234, which contains 142 neutrons, but its natural-occurrence level is only 0.005 %; it plays no role whatever in our story.

By the middle of 1940, however, understanding of the differing responses of the two uranium isotopes to bombarding neutrons had become more refined, and an important new concept emerged. This was that it might be possible to achieve a *controlled* (not explosive) chain reaction using natural uranium *without enrichment*. The key lay in how nuclei react to bombarding neutrons. When a nucleus is struck by a neutron, various reactions are possible: the nucleus might fission, it might absorb the neutron without fissioning, or it might simply scatter the neutron as a billiard ball would deflect an incoming marble. Each process has some probability of occurring, and these probabilities depend on the *speed* of the incoming neutrons. Neutrons released in fission reactions are extremely energetic, emerging with average speeds of about 20 million meters per second. For obvious reasons, such neutrons are termed "fast" by nuclear physicists. As remarked above, U-238 tends to ultimately absorb the fast neutrons emitted in fissions of U-235 nuclei. However, when a nucleus of U-238 is struck by a *very slow* neutron—one traveling at a mere couple thousand meters per second—it behaves as a much more benign target, with scattering preferred to absorption by odds of somewhat better than three-to-one. But—*and this is the key point*—for such slow neutrons, U-235 turns out to have an enormous fission probability: over 200 times greater than the capture probability for U-238. This factor is large enough to compensate for the small natural abundance of U-235, and renders a chain reaction possible. *Slowing fission-liberated neutrons is effectively equivalent to enriching the abundance percentage of U-235.* This is such an important point that it is worth repeating: If neutrons emitted in fissions can be slowed, then they have a good chance of going on to fission other U-235 nuclei before being lost to capture by U-238 nuclei. In actuality, both processes will proceed simultaneously. Counterintuitively, neutron capture by U-238 nuclei actually turns out to be indirectly advantageous for bomb-makers, as is explained in the third paragraph following this one.

How can one slow a neutron during the very brief time interval between when it is born in a fission and when it strikes another nucleus? The trick is to work not with a single large lump of uranium, but rather to disperse it as small chunks throughout a surrounding medium which slows neutrons without absorbing them. Such a medium is known as a *moderator*, and the entire assemblage is a *reactor*. During the war, the synonymous term "pile" was used in the literal sense of a "heap" of metallic uranium slugs and moderating material. Ordinary water can serve as a moderator, but, at the time, graphite (crystallized carbon) proved easier to employ for various reasons. By introducing moveable rods of neutron-absorbing material into the pile and adjusting their positions as necessary, the reaction can be controlled. It is in this way that natural-abundance uranium proved capable of sustaining a *controlled* nuclear reaction, but not an *explosive* one. Reactor engineering has advanced phenomenally since 1945, but modern power-producing reactors still operate via chain-reactions mediated by moderated neutrons.

To be clear, a reactor cannot be made into a bomb: the reaction is far too slow, and even if the control rods are rendered inoperative, the reactor will melt itself long before blowing up—as seen at Fukushima, Japan. But reactor meltdowns are a digression from the main story of the Manhattan Project. In early 1940, it still appeared that to make a chain-reaction mediated by *fast* neutrons—a bomb—it would be necessary to isolate pure U-235.

In May, 1940, however, an insight gleaned from theoretical physics opened the door to a possible route to making a fission bomb without having to deal with the challenge of enrichment. It was mentioned above that neutron absorption by U-238 nuclei will still continue in a reactor. On absorbing a neutron, a U-238 nucleus becomes a U-239 nucleus. Based on extrapolating from some experimentally established patterns regarding the stability of nuclei, it was predicted that U-239 nuclei might decay within a short time to nuclei of atomic number 94, the element now known as plutonium. It was further predicted that element 94 might be very similar in its fissility properties to U-235. If this proved to be the case, then a reactor, sustaining a controlled chain reaction via U-235 fissions, could be used to "breed" plutonium from U-238. The plutonium could be separated from the mass of parent uranium fuel by conventional chemical means, and used to construct a bomb; this is what obviates the need to develop enrichment facilities. Within months, these predictions were partially confirmed on a laboratory scale by creating a small sample of plutonium via moderated-neutron bombardment of uranium.

By the time of the Japanese attack on Pearl Harbor in December, 1941, it appeared that there were two *possible* routes to developing a nuclear explosive: (1) isolate tens of kilograms of U-235, or (2) develop reactors with which to breed plutonium. Each method had potential advantages and drawbacks. U-235 was considered almost certain to make an excellent nuclear explosive; the prospect looked as solid as any untested theory could be. But those tens of kilograms would have to be separated *atom by atom* from a parent mass of uranium ore: Bohr's national-scale factory. As for plutonium, the anticipated separation techniques were well understood by chemical engineers, but nobody had ever built a reactor. Even if such a new technology could be developed and mastered, might the new element prove to have some property that obviated its value as an explosive?

Motivated by the existential threat of the war and the prospect that German scientists were likely thinking along the same lines, the scientific and military leaders of the Manhattan Project made the only decision that they could in such circumstances: both methods would be pursued. In the end, both worked: the Hiroshima bomb utilized uranium, while the one dropped on Nagasaki used plutonium.

1.2 Chapter 4: Organization

A project involving 150,000 people, dozens of contractors and universities, and a budget of nearly $2 billion (over $20 billion in 2013 dollars) could be an organizational nightmare in the best of circumstances. The possibilities for waste and mismanagement were tremendous, particularly given that it was all to be done in secrecy with little outside oversight. How was such a monumental effort effectively initiated and administered?

Such an undertaking cannot spring up fully-formed overnight. Chapter 4 explores how possible military applications of nuclear fission were first brought to the attention of the President of the United States in the fall of 1939, and how government and military support for the endeavor came to be organized. Between 1939 and 1942, this support was under the authority of various civilian branches of the government, although it was being conducted in secrecy. Various committees funded and oversaw the work, which also benefitted from critically timed interventions on the parts of a few key individuals who felt that the pace of efforts was not active enough. By mid-1942, various lines of research in both Britain and America led to the conclusion that both nuclear reactors and weapons could be feasible, but that they would require engineering efforts far beyond the experience of a university research department or even the budget and resources of a single large industry. The only organization capable of mounting such an effort with the requisite secrecy was the United States Army. Given the scale of construction involved, the Project was assigned, in the fall of 1942, to the Army's Corps of Engineers. Chapter 4 describes the administrative history of the Project to early 1943, by which time the Corps was firmly in command. Administrative aspects subsequent to this time are more conveniently discussed in relevant individual Project-component chapters.

1.3 Chapters 5–7: Uranium, Plutonium, and Bomb Design and Delivery

Two major production facilities and a highly-secret bomb-design laboratory were established to advance the work of the Manhattan Project. These facilities are the subjects of Chaps. 5, 6, 7. The production facilities were located in the states of Tennessee and Washington, and were respectively devoted to obtaining nearly pure U-235 and breeding plutonium. These facilities are discussed in Chaps. 5 and 6. The bomb design laboratory was located at Los Alamos, New Mexico, and is the subject of Chap. 7, which also describes some of the training of air crews selected to deliver the bombs to their targets.

The uranium facility located in Tennessee was designated by the name Clinton Engineer Works (CEW), after the small town near Knoxville where it was located. Spread over a roughly 90-square mile military reservation were three separate enrichment facilities, plus a pilot-scale nuclear reactor, supporting shops, chemical processing laboratories, electrical and water utilities, and food services, housing, hospitals, schools, shopping centers, and other amenities for the workers. Overall, the CEW consumed nearly $1.2 billion in construction and operating costs. The three uranium-enrichment facilities were code named Y-12, K-25, and S-50. As described in the following paragraphs, each utilized a different method of enriching uranium.

Y-12: This facility comprised over 200 buildings, and enriched uranium by the process of *electromagnetic mass spectroscopy*. The fundamental principle involved here is that when an ionized atom or molecule is directed into a magnetic field, it will follow a trajectory whose path depends, among other things, on the mass of the atom or molecule. (An atom is said to be ionized when one or more of its outer orbital electrons have been removed, leaving the atom with a net positive charge.) To put this into practice, a uranium compound was heated until it became vaporized. The vapor was then ionized, and directed as a narrow stream into a vacuum tank sandwiched between the coils of a huge electromagnet. Atoms of the two different isotopes will then follow slightly different trajectories, and can be collected separately. Ideally, the ions move in circular trajectories. For the strength of the magnetic field that was employed, the separation of the ion streams was at most only about one centimeter for trajectories of diameter 3 meters. To get a sensible rate of production of bomb-grade uranium, over 900 magnet coils and nearly 1,200 vacuum tanks were put into operation. The design of these "electromagnetic separators" was based on a particle-accelerating "cyclotron" developed by Ernest Lawrence of the University of California; the CEW cyclotrons were known as "calutrons," after *Cal*ifornia *U*niversity Cyclo*tron*. In practice, this process tends to be difficult to control and the efficiency can be low, but every atom of U-235 in the Hiroshima *Little Boy* bomb eventually passed at least once through Lawrence's calutrons. Ground was broken for the first Y-12 building in February, 1943, and operations began in November of that year. Some 5,000 operating and maintenance personnel kept Y-12 running. The bill for this facility ran to some $477 million in construction and operating costs.

K-25: At over $500 million in construction and operating costs, this was the single most expensive facility of the entire Manhattan Project. Imagine a large U-shaped factory, four stories high (one underground), half a mile long, and about 1,000 feet wide (Fig. 1.2).

This enormous structure housed the *gaseous diffusion* plant of the Project; this process was also known as *barrier diffusion*. The premise of this technique is that if a gas of atoms of mixed isotopic composition is pumped against a thin, porous

Fig. 1.2 An aerial view of the K-25 plant. *Source* http://commons.wikimedia.org/wiki/File:K-25_Aerial.jpg

metal barrier containing millions of microscopic holes, atoms of lower mass will pass through the barrier slightly more readily than those of higher mass. The result is a very minute level of enrichment of the gas in the lighter isotope on the other side of the barrier. The term "microscopic" is meant literally here: the holes need to be about 100 Ångstroms in diameter, or about three billionths of an inch. A characteristic of this process, however, is that only a very small level of enrichment can be achieved by passing the gas through the barrier on any one occasion. To achieve a level of enrichment corresponding to "bomb-grade" material (90 % U-235), the process has to be repeated sequentially thousands of times. K-25 incorporated nearly 3,000 enrichment stages, and was the largest construction project in the history of the world to that time. Construction began in June, 1943, but difficulties in securing suitable barrier material delayed the start of operations until January, 1945. Some 12,000 people were employed to operate the plant.

S-50: The S-50 plant enriched uranium by a second diffusion-based process known as *liquid thermal diffusion*, which is often simply termed *thermal diffusion* to differentiate it from the *gaseous* process employed in the K-25 plant. While liquid diffusion is rather inefficient, it is relatively simple. Imagine a vertical arrangement of three concentric pipes (Fig. 1.3). The innermost one is heated by high-temperature steam pumped through its center. A second, intermediate, pipe closely surrounds the innermost one, with a clearance of only a quarter of a millimeter between the two. Liquefied uranium hexafluoride is then fed under pressure into the annulus between the two pipes. The intermediate pipe is surrounded by the third pipe, through which cold water is pumped to chill the outside surface of the intermediate pipe.

The hexafluoride thus experiences a dramatic thermal gradient across its quarter-millimeter width. The result is that liquid containing the lighter isotope moves toward the hotter pipe, while heavier-isotope material collects toward the cooler one. The hotter material rises by convection while the cooler descends, leading to an accumulation of material slightly enriched in the lighter isotope at the top of the column. From there, the lighter-isotope-enriched material can be harvested and sent on to another stage. The S-50 plant utilized 2,142 such three-pipe columns, each 48 feet high. Due to political wrangling, the decision to proceed with the S-50 plant was not made until June, 1944, but construction proceeded so rapidly that preliminary operation of the plant was begun in September of that year; full construction was essentially complete by January, 1945.

Originally, these various enrichment methods were thought of as individual horses competing in a race to see which one could start with natural uranium and most efficiently produce bomb-grade material in one process. But as they were put into operation it became clear that they worked better as a team; some proved more efficient at various stages of enrichment than at others. In the end, uranium began its journey by being processed through the S-50 plant to receive a slight level of enrichment (to 0.86 % U-235), from where it went on to the K-25 plant (to 7 %), and thence through one or two separate stages of the Y-12 calutrons to get to 90 % U-235.

Fig. 1.3 Sectional view of a thermal diffusion process column in the S-50 plant. Uranium hexafluoride consisting of a mixture of light (U-235) and heavy (U-238) isotopes is driven into the narrow annular space between the nickel and copper pipes. The desired lighter-isotope material is harvested from the top of the column. From Reed (2011)

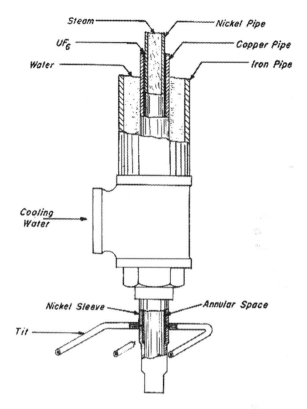

Under Project auspices, the world's first nuclear reactor, an experimental graphite-moderated pile, achieved a self-sustaining chain reaction in early December, 1942. Located at the University of Chicago, this pile, code-named CP-1, operated at an estimated power output of one-half of a Watt. CP-1 was strictly experimental; its purpose was to demonstrate that a chain reaction could be created and controlled. The rate of formation of plutonium in a reactor depends on the reactor's power output, and the power level of CP-1 was a far cry from the *millions* of Watts (MW; megawatts) estimated to be required to breed plutonium rapidly enough to produce a bomb in a sensible length of time. Later plutonium-production reactors were designed to operate at 250 MW, and three were built.

Engineers were naturally dubious of scaling a new technology from Watts to hundreds of millions of Watts, so it was decided to build an intermediate-stage "pilot" reactor to test cooling and control systems, and to create a few hundred grams of plutonium for research purposes. Known as X-10, initial plans were to locate this pile outside Chicago, but it was instead built at the Clinton site for reasons of safety and centralization of operations. X-10 was a forced-air-cooled, graphite-moderated reactor designed to operate at a power level of 1 MW, although later improvements in the cooling system permitted operation at 4 MW. X-10 began operation in November, 1943.

Chapter 6 describes the 250-MW plutonium production reactors. Initially, these were also to be sited in Tennessee as well, but the prospect that a catastrophic accident could doom all of the Project's production facilities led to a decision to locate them at a remote site in south-central Washington, where they could be cooled with water drawn from the Columbia river. The Hanford Engineer Works (HEW, or just Hanford) occupied an enormous area, over 600 square miles altogether. To provide a margin of safety, the reactors were situated six miles apart along the banks of the Columbia. Ten miles south of them were located facilities for chemical separation of the plutonium, and over ten miles yet further away were facilities for fabricating the slugs of uranium fuel fed to the reactors, as well as a constructed-from-scratch village for housing plant personnel and their families. Ground was broken at the Hanford site in April, 1943. The first reactor achieved criticality in late September, 1944, but unanticipated problems caused a three-month shut-down while modifications were effected. Ultimately, the Hanford reactors produced the kilograms of plutonium necessary for the *Trinity* test and the Nagasaki bomb.

Perhaps the most famous Manhattan Project facility was the Los Alamos Laboratory, which is the subject of Chap. 7. Established in the spring of 1943 in the high desert of northern New Mexico and directed by physicist J. Robert Oppenheimer of the University of California, the task of this secret installation was to design the weapons that would be powered by the uranium and plutonium being produced in Tennessee and Washington. In theory, the work facing Los Alamos scientists seemed straightforward. Fissile elements like U-235 or Pu-239 possess a so-called *critical mass*, a minimum mass necessary to sustain a chain reaction. The precise value of the critical mass depends on factors such as the density of the material, its probability for undergoing a fission reaction, and the number of neutrons liberated per fission. Much of the experimental work at Los Alamos involved obtaining accurate measurements of these quantities. With these numbers in hand, the critical mass can be calculated via mathematical relationships from an area of physics known as diffusion theory, which was a well-established science long before 1943.

For sake of argument, suppose that the critical mass for some material is 50 kg (which is not far off the mark for U-235). It turns out that you will get a more efficient explosion if you have more material available than just one critical mass, so imagine that you have 70 kg. To make your bomb, form your 70 kg into two pieces, say each of mass 35 kg, and simply arrange to bring them together when you are ready to detonate your device. In effect, this is exactly what was done in the uranium-based Hiroshima bomb. Inside a long cylindrical bomb casing was mounted the barrel of a naval artillery gun. One piece of the uranium, the "target" piece, was mounted at the far (nose) end of the barrel, while the second piece, the "projectile," was loaded into the breech (tail) end (Fig. 1.4). When radar and barometric sensors indicated that the bomb had fallen to a pre-programmed detonation height, a conventional powder charge was ignited to propel the projectile piece into the mating target piece. There are ancillary considerations such as providing for a source of neutrons to initiate the chain reaction at the desired moment, but this is the basic idea of how a so-called fission "gun bomb" operates.

The Hiroshima bomb contained about 60 kilograms of U-235, but, overall, the bomb weighed nearly 5 t (this book will often abbreviate "ton" or "tons" as "t"; the term may be used as a weight or TNT energy equivalent). Much of this was the weight of the artillery cannon, but there was another significant contributor: the target end of the cannon was surrounded by a steel tamper of mass several hundred kgs. The tamper serves two functions. First, by briefly retarding the expansion of the bomb core as it detonates, one buys a bit more time (microseconds) over which the chain reaction can operate. Second, if the tamper is made of a material which reflects escaping neutrons back into the core, it gives them another chance at causing fissions. Both effects enhance weapon efficiency. An efficiency increase of a factor of ten over an untamped device is quite possible, so it is certainly worth going to the effort of providing a tamper. The presence of a tamper complicates the calculations, but Los Alamos physicists, aided by early electronic computers, became very adept at such work as they balanced issues of fission physics, electronics, ordnance, neutron initiators, and the payload limit of a B-29 bomber. Remarkably, despite its destructive power, the Hiroshima bomb had an overall efficiency of only about 1 %.

The plutonium bomb, however, was a very different matter. Reactor-produced plutonium proved to exhibit a fairly high level of *spontaneous fission*—a natural, completely uncontrollable process. Because of this, it was predicted that if one tried to make a gun-type bomb using plutonium, the nuclear explosion would start itself spontaneously before the target and projectile pieces were fully mated. The result would be an expensive but very low-efficiency explosion, a so-called "fizzle." Two possible approaches to avoiding this problem were evident: find a way to use less fissile material (lower spontaneous-fission rate), and/or assemble the sub-critical pieces more rapidly than could be achieved with the gun mechanism. Both approaches were utilized. The critical mass of a fissile material depends on its density; greater density means a lower critical mass. Thus, if you have a mass of material that would be *subcritical* at *normal* density, it can be made critical by crushing it to a higher density; the result is that you can get away with using less material than would "normally" be required. This led to the idea of an

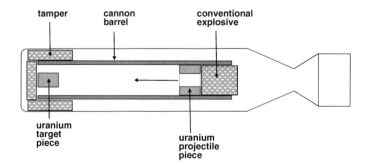

Fig. 1.4 Schematic illustration of a gun-type fission weapon. The uranium projectile is fired toward a mating target piece in the nose

implosion weapon, wherein a subcritical core (and hence one with a low rate of spontaneous fission) is surrounded with explosive material configured to very rapidly detonate *inwards*. By using a very fast-burning explosive to achieve the crushing, the "assembly time" can also be reduced. The hard part is that the implosion has to be essentially perfectly symmetric, with all of the pieces of surrounding explosive detonating within about a microsecond of each other. Los Alamos scientists and engineers devoted enormous effort to perfecting this never-before-tried technique. Questions as to the feasibility of implosion were so serious that it was decided to use some of the precious Hanford plutonium in a full-scale test of the method. This was the *Trinity* test of July 16, 1945, the world's first nuclear explosion (Fig. 1.5). The test was a complete success, and just three weeks later the method was put to use in the Nagasaki bomb.

With the above descriptions, you now have an idea why the Hiroshima U-235 *Little Boy* uranium bomb was a long, cylindrically-shaped mechanism, while the Nagasaki Pu-239 *Fat Man* plutonium bomb was a bulbous, nearly spherical arrangement (Fig. 1.6).

Fig. 1.5 *Left* The *Trinity* fireball 25 ms after detonation (*Source* http://commons.wikimedia.org/wiki/File:Trinity_Test_Fireball_25ms.jpg). *Right* The fireball a few seconds later (*Courtesy of the Los Alamos National Laboratory Archives*)

Fig. 1.6 *Left Little Boy* in its loading pit. *Right* The *Fat Man* bomb. Note signatures on tail. *Sources* http://commons.wikimedia.org/wiki/File:Atombombe_Little_Boy_2.jpg; http://commons.wikimedia.org/wiki/File:Fat_Man_on_Trailer.jpg

At its peak, the Los Alamos Laboratory employed only about 2,500 people, but without their efforts the work of tens of thousands of others at Clinton and Hanford would have been for naught. The accomplishments of the scientists and engineers at Los Alamos are now legendary in the physics community.

A bomb needs to be transported to a target, and this was the task of the so-called 509th Composite Group, the Army Air Force unit specifically formed to deliver the products of the Manhattan Project. The selection, training, and deployment of this group are described in Chaps. 7 and 8.

1.4   Chapters 8 and 9: Hiroshima, Nagasaki, and the Legacy of Manhattan

The Manhattan Project culminated with the bombings of Hiroshima and Nagasaki. Chapter 8 examines the bombings: selection of target cities, political consider-ations as to actual use of the bombs versus a demonstration shot, the aircraft, the flight crews and their training, the missions themselves, the explosive energies of the bombs, and their effects on people and structures. To provide some context, casualty rates for some of the Pacific island-hopping campaigns of the war are discussed briefly, as are the plans and expected casualty rates for the proposed invasion of Japan which was scheduled for November, 1945. Planning for postwar administration of nuclear energy is also discussed.

Chapter 9 examines the some of the legacies of the Manhattan Project. In postwar years, improvements in the design of nuclear weapons accumulated rap-idly, culminating with fusion-based devices known popularly as hydrogen bombs. The number of countries possessing nuclear weapons grew to five by the mid-1960s and now stands at about twice that number. The number of nuclear war-heads held by these countries peaked at over 70,000 in the mid-1980s (a figure which astonishes most people), and the world is now almost literally awash in bomb-grade uranium and plutonium. While the number of warheads has since declined significantly due to various arms-control treaties, thousands of nuclear weapons are still deployed and will remain so for years to come. As these issues are not really directly germane to the Manhattan Project, the discussion in this Chapter is intended to give readers only a brief outline of postwar developments as an epilogue to the main story.

By now you should appreciate the validity of the assertion in the Preface that no one volume can hope to cover every aspect of the Manhattan Project. Given this, it is important to mention what topics this book addresses only briefly or not at all. As this book is devoted primarily to historical and technical aspects of the Project, I offer very little in the way of personality profiles. That a small number of Soviet agents working at Los Alamos transmitted information back to Moscow despite a widespread counterintelligence effort is well-known, but as my focus is the science and organization of the Project, I forgo any detailed analysis of this matter. I also do not examine the German wartime program to investigate possible applications of atomic fission (which never proceeded even to the stage of an operating reactor), the infamous Bohr/Heisenberg meeting of 1943, the interning of German scientists at Farm Hall in Britain after the war, or the American effort to discover what German scientists were up to in the nuclear field, the so-called *Alsos* mission. All of these events are discussed in many excellent books and articles. Some aspects of the Project involved efforts which ended up not being put into large-scale operation (such as the use of centrifuges for enrichment), and so are mentioned only tangentially where appropriate. Postwar developments such as arguments for and against the development of fusion weapons and Robert Oppenheimer's scandalous 1954 security hearing lie well outside the scope of this book, and are not discussed at all.

Further Reading

Books, Journal Articles, and Reports

R.G. Hewlett, O.E Anderson, Jr, *A History of the United States Atomic Energy Commission, Vol. 1: The New World, 1939/1946* (Pennsylvania State University Press, University Park, 1962)

V.C. Jones, *United States Army in World War II: Special Studies—Manhattan: The Army and the Atomic Bomb* (Center of Military History, United States Army, Washington, 1985)

G. Parshall, Shock Wave. U. S. News and World Report **119**(5), 44–59 (31 July 1995)

B.C. Reed, Liquid Thermal Diffusion during the Manhattan Project. Phys. Perspect. **13**(2), 161–188 (2011)

R. Rhodes, *The Making of the Atomic Bomb* (Simon and Schuster, New York, 1986)

Chapter 2
A Short History of Nuclear Physics to the Mid-1930s

Until the late 1930s, the study of radioactivity and nuclear physics were relatively low-profile academic research fields whose applications were limited primarily to medical treatments such as radiation therapies for cancers. But within a very few months between late 1938 and mid-1939, some investigators began to realize that one of the quieter provinces of pure research could become a geopolitical game-changer of immense military potential. How did this transformation come to be?

To set the stage for a description of the development of nuclear weapons, it is helpful to understand a lengthy progression of underlying background discoveries. Our understanding that atoms comprise protons, neutrons, and electrons; that nuclei of various elements occur in different isotopic forms; that some elements are radioactive; and that nuclear weapons can be made are now facts of common knowledge: not many people know, or ever wonder how, their predecessors divined such knowledge. The purpose of this chapter is to give an overview of how the scientific community came to these understandings.

In developing such an overview there are many facts to be considered, and they interconnect so tightly that it can be difficult to decide how to order the presentation. I use a largely chronological ordering, but slavishly maintaining a chronological description would be awkward in that full understanding of some phenomena took decades to develop. A dramatic example of this is that neutrons were not discovered until 1932, a full twenty years after the existence of nuclei had been established (Fig. 2.1). Consequently, there are points where I abandon the chronological approach for the sake of coherence or to digress on some background material. Also, there are instances where a concept introduced in one section is revisited in a later one for fuller elaboration. Readers are urged to consider the sections of this chapter as linked units; treat them as a whole, and re-read them as necessary.

The discovery of the neutron in 1932 is considered such a pivotal event that nuclear physicists divide the history of their discipline into two eras: that time before awareness of the neutron, and that time after. In keeping with this, the first section of this chapter covers, in a number of subsections, important developments from the discovery of radioactivity in 1896 up to 1931. Sections 2.2 through 2.4

B. C. Reed, *The History and Science of the Manhattan Project*,
Undergraduate Lecture Notes in Physics, DOI: 10.1007/978-3-642-40297-5_2,
© Springer-Verlag Berlin Heidelberg 2014

take the story from 1932 through the discoveries of the neutron and artificially-induced radioactivity, and Enrico Fermi's neutron-bombardment experiments of the mid-1930s. Section 2.5, which can be considered optional, fills in some technical details that are skirted in Sect. 2.1.4.

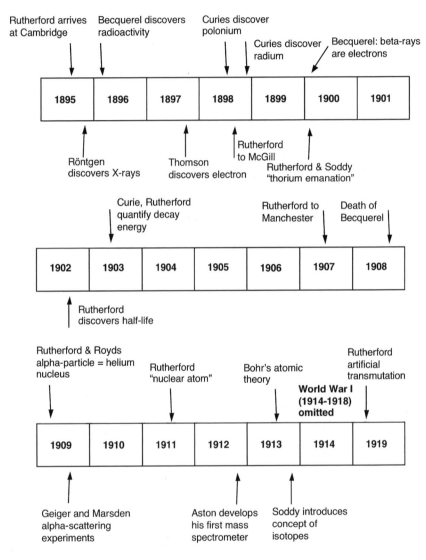

Fig. 2.1 Chronology of early nuclear physics

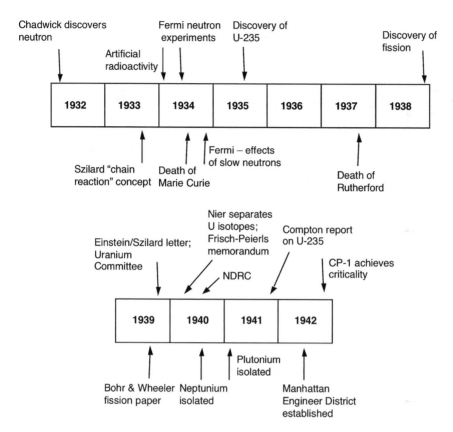

Fig. 2.1   continued

2.1 Radioactivity, Nuclei, and Transmutations: Developments to 1932

The era of "modern" physics is usually considered to have begun in late 1895, when Wilhelm Conrad Röntgen, working in Germany, accidentally discovered X-rays. Röntgen discovered that not only could his mysterious rays pass through objects such as his hand, but they also ionized air when they passed through it; this was the first known example of what we now call "ionizing radiation". A part of Röntgen's discovery involved X-rays illuminating a phosphorescent screen, a fact which caught the attention of Antoine Henri Becquerel, who lived in France. Becquerel was an expert in the phenomenon of phosphorescence, where a material emits light in response to illumination by light of another color. Becquerel wondered if phosphorescent materials such as uranium salts might be induced to emit X-rays if they were exposed to sunlight. While this supposition was wrong, investigating it led Becquerel, in February 1896, to the accidental discovery of

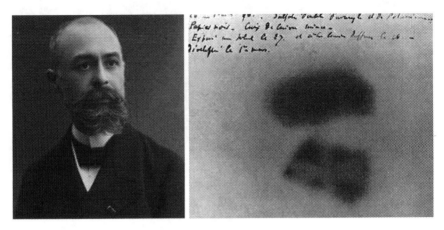

Fig. 2.2 Henri Becquerel (1852–1908) and the first image created by "Becquerel rays" emitted by uranium salts placed on a wrapped photographic plate. In the lower part of the plate a Maltese cross was placed between the plate and the lump of uranium ore. *Sources* http://upload.wikimedia.org/wikipedia/commons/a/a3/Henri_Becquerel.jpg; http://upload.wikimedia.org/wikipedia/commons/1/1e/Becquerel_plate.jpg

radioactivity. Becquerel observed that samples of uranium ores left on top of wrapped photographic plates would expose the plates even in the *absence* of any external illumination; that is, the exposures seemed to be created by the uranium itself. When the plates were unwrapped and developed, an image of the samples would be apparent (Fig. 2.2). Nuclear physics as a scientific discipline originated with this discovery.

We now attribute these exposures to the action of so-called "alpha" and "beta" particles emitted by nuclei of uranium and other heavy elements as they decay naturally to more stable elements; the nature of these particles is elaborated in the sections that follow. Some of the decay timescales are fleeting, perhaps only minutes, while others are inconceivably long, hundreds of millions or billions of years. In the latter event it is only because there are so many trillions upon trillions of individual atoms in even a small lump of ore that enough are likely to decay within a span of seconds or minutes to leave an image on a film or trigger a Geiger counter. (The operation of a Geiger counter is described at the end of Sect. 2.1.5.) A third form of such emission, "gamma rays," was discovered by French chemist Paul Villard in 1900. Gamma-rays are photons, just like those entering your eyes as you read this, but of energies about a million times greater than visible-light photons.

By the time of Becquerel's death in 1908, the field of research he had opened was producing developments which would lead, by about 1920, to humanity's first true scientific understanding of the structure of the most fundamental constituents of matter: atoms and their nuclei. Remarkably, just less than a half-century would pass between the discovery of radioactivity and the development of nuclear weapons.

2.1.1 Marie Curie: Polonium, Radium, and Radioactivity

Becquerel's work came to the attention of Marie Sklodowski, a native of Poland who had graduated from the Sorbonne (part of the University of Paris) with a degree in physical science in 1893; the following year she would add a degree in mathematics. In 1895 she married Pierre Curie, a physicist at the Paris School of Physics and Chemistry (Fig. 2.3). Seeking a subject for a doctoral thesis, Marie turned to Becquerel's work, a subject about which not a great deal had been published. She set up a laboratory in her husband's School, and began work in late 1897.

Becquerel had reported that the energetic "rays" emitted by uranium could ionize air as they passed through it; in modern parlance the rays collide with molecules in the air and cause them to lose electrons. Pierre Curie and his brother had developed a device known as an electrometer for detecting minute electrical currents. Making use of this device, Marie found that the amount of electricity generated was directly proportional to the amount of uranium in a sample. Testing other materials, she found that the heavy element thorium also emitted Becquerel rays (a fact discovered independently by Gerhard Schmidt in Germany), although not as many per gram per second as did uranium. Further work, however, revealed that samples of pitchblende ore, a blackish material rich in uranium oxides, emitted more Becquerel rays than could be accounted for solely by the quantity of uranium that they contained. Drawing the conclusion that there must be some other "active element" present in pitchblende, Curie began the laborious task of chemically isolating it from the tons of ore she had available. By this time, Pierre had abandoned his own research on the properties of crystals in order to join Marie in her work.

Spectroscopic analysis of the active substance proved that it was a new, previously unknown element. Christening their find "polonium" in honor of Marie's

Fig. 2.3 Marie (1867–1934) and Pierre (1859–1906) Curie; Right Irène (1897–1956) and Frédéric Joliot-Curie (1900–1958) in 1935. *Sources* http://commons.wikimedia.org/wiki/File:Mariecurie. jpg; http://commons.wikimedia.org/wiki/File:PierreCurie.jpg; http://commons.wikimedia.org/ wiki/File:Irène_et_Frédéric_Joliot-Curie_1935.jpg

native country, they published their discovery in July, 1898, in the weekly proceedings of the French Academy of Sciences. That paper introduced two new words to the scientific community: "radioactivity" to designate whatever process deep within atoms was giving rise to Becquerel's ionizing rays, and "radioelement" to any element that possessed the property of doing so. The term "radioisotope" is now more commonly used in place of "radioelement," as not all of the individual isotopes of elements that exhibit radioactivity are themselves radioactive.

In December, 1898, the Curies announced that they had found a second radioactive substance, which they dubbed "radium." By the spring of 1902, after starting with ten tons of pitchblende ore, they had isolated a mere tenth of a gram of radium, which was enough for definite spectroscopic confirmation of its status as a new element. In the summer of 1903 Marie defended her thesis, "Researches on Radioactive Substances," and received her doctorate from the Sorbonne. In the fall of that year the Curies would be awarded half of the 1903 Nobel Prize for Physics; Henri Becquerel received the other half.

2.1.2 Ernest Rutherford: Alpha, Beta, and Half-Life

In the fall of 1895, Ernest Rutherford (Fig. 2.4), a New Zealand native, arrived at the Cavendish Laboratory of Cambridge University in England on a postgraduate scholarship. The Director of the Laboratory was Joseph John "J. J." Thomson, who in the fall of 1897 was credited with discovering the electron, the fundamental, negatively-charged particles of matter which account for the volumes of atoms. It is rearrangements of the outermost electrons of atoms which cause the chemical reactions by which, for example, we digest meals to provide ourselves

Fig. 2.4 Left Ernest Rutherford (1871–1937) about 1910. Right Seated left to right in this 1921 photo are J. J. Thomson (1856–1940), Rutherford, and Francis Aston (1877–1945), inventor of the mass spectrograph (Sect. 2.1.4). *Sources* http://commons.wikimedia.org/wiki/File:Ernest_Rutherford.jpg; AIP Emilio Segre Visual Archives, Gift of C.J. Peterson

with the energy we use to do useful work such as the preparation of book manuscripts.

Rutherford's intrinsic intelligence, capacity for sheer hard work, and unparalleled physical insight combined with propitious timing to set him on a path to become one of history's great nuclear pioneers. Soon after Rutherford arrived in Cambridge, Röntgen discovered X-rays. As a student, Rutherford had developed considerable experience with electrical devices, and the Cavendish Laboratory was well-equipped with Thomson's "cathode ray tubes," the core apparatus for generating X-rays. Rutherford soon began studying their ionizing properties. When the discovery of radioactivity was announced, it was natural for him to turn his attention to this new ionizing phenomenon.

Rutherford discovered that he could attenuate some of the uranium activity by wrapping the samples in thin aluminum foils; adding more layers of foil decreased the activity. Rutherford deduced that there appeared to be two types of radiation present, which he termed "alpha" and "beta." Alpha-rays could be stopped easily by a thin layer of foil or a few sheets of paper, but beta-rays were more penetrating. Henri Becquerel later showed that both types could be deflected by a magnetic field, but in opposite directions and by differing amounts. This meant that the rays must be electrically charged; alphas proved to be positive, and betas negative. Becquerel also later proved that beta rays were identical to electrons. Alpha-rays were much less affected by a magnet, which meant that they must be much more massive than electrons (further details on this point appear in Sect. 2.1.4).

In the fall of 1898, Rutherford completed his studies at Cambridge, and moved to McGill University in Montreal, Canada, where he had been appointed as the McDonald Professor of Physics. Over the next three decades he continued his radioactivity research, both at McGill and later back in England. This research would contribute to a series of groundbreaking discoveries in the area of atomic structure, and would earn him the 1908 Nobel Prize for Chemistry.

Rutherford's first major discovery at McGill occurred in 1900, when he found that, upon emitting its radiation, thorium simultaneously emitted a product which he termed "emanation." The emanation was also radioactive, and, when isolated, its radioactivity was observed to decline in a geometrical progression with time. Specifically, the activity decreased by a factor of one-half for every minute of time that elapsed. Rutherford had discovered the property of *radioactive half-life*, the quintessential natural exponential decay process.

As an example, suppose that at "time zero" you have 1,000 atoms of some isotope that has a half-life of 10 days. You can then state that 500 of them will have decayed after 10 days. You cannot predict *which* of the 500 will have decayed, however. Over the following 10 days a further 250 of the original remaining atoms will decay, and so on. Remarkably, the probability that a given atom will decay in some specified interval of time is completely independent of how long it has managed to avoid decaying; in the subatomic world, age is *not* a factor in the probability of continued longevity.

The following paragraphs examine the mathematics of half-life. Readers who wish to skip this material should proceed to the paragraph following Eq. (2.8).

If the number of nuclei of some radioactive species at an arbitrarily-designated starting time $t = 0$ is N_O, then the number that remain after time t has elapsed can be written as

$$N(t) = N_O e^{-\lambda t}, \tag{2.1}$$

where λ is the so-called decay constant of the species. If $t_{1/2}$ is the half-life of the species for some mode of decay (alpha, beta, ...), λ is given by

$$\lambda = \frac{\ln 2}{t_{1/2}}. \tag{2.2}$$

Since half-lives run from tiny fractions of a second to billions of years, there is no preferred unit for them; one must be careful to put t and $t_{1/2}$ in the same units when doing calculations.

What is measured in the outside world is the rate of decays R, which is determined by taking the derivative of (2.1):

$$R(t) = \frac{dN(t)}{dt} = -\lambda N_O e^{-\lambda t} = -\lambda N(t). \tag{2.3}$$

The meaning of the negative sign is that the number of nuclei of the original species steadily decreases as time goes on; what is customarily quoted is the *absolute value* of $R(t)$.

Since one is more likely to know the mass of material m that one is working with rather than the number of atoms, it is helpful to have a way of relating these two quantities. This is given by

$$N = \frac{m N_A}{A}, \tag{2.4}$$

where N_A is Avogadro's number and A is the molecular weight of the species. Tradition is to quote A in grams per mole, which means that m must be expressed in grams.

Marie and Pierre Curie adopted the rate of decay of a freshly-isolated one-gram sample of radium-226 as a standard for comparing radioactivity rates of different substances. This isotope, which has a half-life of 1,599 years, is a rather prodigious emitter of alpha-particles. With $A = 226.025$ g/mol,

$$N_O = \frac{m N_A}{A} = \frac{(1\ \text{g})(6.022 \times 10^{23} \text{nuclei}/mol)}{(226.025\ \text{g}/mol)} = 2.664 \times 10^{21}\ \text{nuclei.} \tag{2.5}$$

To compute the decay rate in nuclei per second, convert 1,599 years to seconds; 1 year $= 3.156 \times 10^7$ s. Hence 1,599 years $= 5.046 \times 10^{10}$ s, and the initial decay rate will be

$$R_O = \lambda N_O = \frac{(\ln 2)(2.664 \times 10^{21}\ \text{nuclei})}{(5.046 \times 10^{10}\ \text{s})} = 3.66 \times 10^{10}\ \text{nuclei}/s. \tag{2.6}$$

This rate of activity, slightly rounded, is now known as the *Curie*, abbreviated *Ci*:

$$1\,Ci = 3.7 \times 10^{10} \text{decay/s}. \tag{2.7}$$

This is a large number, but a gram of radium contains some 10^{21} atoms and so will maintain its activity for a long time; even after several years it will possess essentially the same rate of activity it began with.

In many situations, a Curie is too large a unit of activity for practical use, so in technical papers one often encounters *millicuries* (one-thousandth of a Curie, *mCi*) or *microcuries* (one-millionth of a Curie; *μCi*). Household smoke detectors contain about 1 *μCi* of radioactive material (37,000 decays per second) which ionizes a small volume of air around a sensor in order to aid in the detection of smoke particles. A more modern unit of activity is the *Becquerel* (*Bq*); one *Bq* is equal to one decay per second. In this unit, a smoke detector would be rated as having an activity of 37 kiloBecquerels (*kBq*). If you would like to try a quick exercise, imagine that you have 1 *kilogram* of plutonium-239, which has $A = 239.05$ g/mol and a half-life for alpha-decay of 24,100 years. You should be able to prove that the decay rate would be 62 *Ci*, a substantial number. We will see in Chap. 7 that decay rates are an important consideration in nuclear weapons engineering.

To better understand the comment above about decay probability being independent of age, consider the following argument. If the number of undecayed nuclei at some time is $N(t)$, then (2.3) tells us that in the subsequent dt seconds the number that will decay is $dN = \lambda\, N(t)\, dt$. The probability $P(t, dt)$ that any given nucleus will decay during these dt seconds is therefore the number that do decay, divided by the number that was available at the start of the interval:

$$P(t, dt) = \frac{\lambda N(t)\, dt}{N(t)} = \lambda\, dt. \tag{2.8}$$

As claimed, $P(t, dt)$ is independent of the time t that had elapsed before the interval considered.

Rutherford sought to identify what element the thorium "emanation" actually was, and to do so teamed up with Frederick Soddy, a young Demonstrator in Chemistry (Fig. 2.5). They had expected the emanation to be some form of thorium, but Soddy's analysis revealed that it behaved like an inert gas. This suggested that the thorium was spontaneously transmuting itself into another element, a conclusion that would prove to be one of the pivotal discoveries of twentieth-century physics. Soddy initially thought that the emanation was argon (element 18), but it would later come to be recognized as radon (element 86). Various isotopes of thorium, radium, and actinium decay to various isotopes of radon, which themselves subsequently decay.

In the 1920s half-life came to be understood as a quantum–mechanical process that is a manifestation of the wave-nature of particles at the atomic level; it is a purely probabilistic phenomenon. This more sophisticated understanding has no bearing on the issues discussed in this book, however: for our purposes, we can

Fig. 2.5 Frederick Soddy (1877–1956). *Source* http://commons.wikimedia.org/wiki/File:Frederick_Soddy_(Nobel_1922).png

regard radioactive decay as an empirical phenomenon described by the mathematics developed above.

2.1.3 Units of Energy in Nuclear Physics and the Energy of Radioactive Decay

In this and the following section we break with chronological progression to fill in some background physics on the units of energy conventionally used in nuclear physics, and the history of the discovery of isotopes. We will return to Rutherford in Sect. 2.1.5.

In the rare circumstances when people consider the quantities of energy that they consume or produce, the unit of measure involved will likely be something such as the kilowatt-hours that appear on an electric bill or the food-calories on a nutrition label. Science students will be familiar with units such as Joules and physical calories (1 cal = 4.187 J). The *food* calorie appearing on nutrition labels is equivalent to 1,000 physical calories, a so-called *kilocalorie*. The food calorie was introduced because the physical calorie used by physicists and chemists is inconveniently small for everyday use.

The words *energy* and *power* are often confused in common usage. Power is the *rate* at when energy is created or used. For physicists, the standard unit of power is the Watt, which is equivalent to producing (or consuming) one Joule of energy per second. A kilowatt (kW) is 1,000 W, or 1,000 J/s. A kilowatt-hour (kWh) is 1,000 W times one hour, that is, 1,000 J/s times 3,600 s, or 3.6 million Joules. A 60-W bulb left on for one hour will consume (60 J/s)(3,600 s) = 216,000 J, or 0.06 kWh. If electricity costs 10 cents per kWh, your bill for that hour will be six-tenths of one cent. The cost would still be a bargain at twice the price, so you can keep reading.

When dealing with processes that happen at the level of individual atoms, however, calories, Joules, and Watts are all far too large to be convenient; one would be dealing with exceedingly tiny fractions of them in even very energetic reactions. To address this, physicists who study atomic processes developed a handier unit of energy: the so-called *electron-Volt*. One electron-Volt is equivalent to a mere 1.602×10^{-19} J. This oddly-named quantity, abbreviated eV, actually has a very sound basis in fundamental physics. You can skip this sentence if you are unfamiliar with electrical units, but for those in the know, an eV is technically defined as the kinetic energy acquired by a single electron when it is accelerated through a potential difference of one Volt. As an everyday example, the electrons supplied by a 1.5-V battery each emerge with 1.5 eV of kinetic energy. A common 9-V battery consists of six 1.5-V batteries connected in series, so their electrons emerge with 9 eV of energy. On an atom-by-atom basis, chemical reactions involve energies of a few eV. For example, when dynamite is detonated, the energy released is equivalent to 9.9 eV per molecule.

Nuclear reactions are much more energetic than chemical ones, typically involving energies of *millions* of electron-volts (MeV). We will see many reactions involving MeVs in this book. If a nuclear reaction liberates 1 MeV per atom involved (nucleus, really) while a chemical reaction liberates 10 eV per atom involved, the ratio of the nuclear to chemical energy releases will be 100,000. This begins to give you a hint as to the compelling power of nuclear weapons. An "ordinary" bomb that contains 1,000 pounds of chemical explosive could be replaced with a nuclear bomb that utilizes only 1/100 of a pound of a nuclear explosive, presuming that the weapons detonate with equal efficiency. Thousands of *tons* of conventional explosive can be replaced with a few tens of *kilograms* of nuclear explosive. Nuclear fission weapons like those used at Hiroshima and Nagasaki involved reactions which liberated about 200 MeV per reaction, so a nuclear explosion in which even only a small amount of the "explosive" actually reacts (e.g., one kilogram) can be incredibly devastating.

It did not take physicists long to appreciate that natural radioactivity was accompanied with substantial energy releases. In 1903, Pierre Curie and a collaborator, A. Laborde, found that just one gram of radium released on the order of 100 physical calories of heat energy per hour. Rutherford and Soddy were also on the same track. In a May, 1903, paper titled "Radioactive Change," they wrote that (expressed in modern units) "the total energy of radiation during the disintegration of one gram of radium cannot be less than 10^8 calories and may be between 10^9 and 10^{10} calories. The union of hydrogen and oxygen liberates approximately 4×10^3 calories per gram of water produced, and this reaction sets free more energy for a given weight than any other chemical change known. The energy of radioactive change must therefore be at least 20,000 times, and may be a million times, as great as the energy of any molecular change." Another statistic Rutherford was fond of quoting was that a single gram of radium emitted enough energy during its life to raise a 500-t weight a mile high.

The moral of these numbers is that nuclear reactions liberate vastly more energy per reaction than any chemical reaction. As Rutherford and Soddy wrote: "All these considerations point to the conclusion that the energy latent in the atom must be enormous compared with that rendered free in ordinary chemical change." That enormity would have profound consequences.

2.1.4 Isotopes, Mass Spectroscopy, and the Mass Defect

In this section I give a brief history of the discovery of isotopes, and a description of the notations now used to designate them. This material is somewhat technical, but the concept of isotopy is an important one that will run throughout this book.

In modern terminology, an element's location in the periodic table is dictated by the number of protons in the nuclei of its atoms. This is known as the *atomic number*, and is designated by the letter Z. Atoms are usually electrically neutral, so the atomic number also specifies an atom's normal complement of electrons. Chemical reactions involve exchanges of so-called valence electrons, which are the outermost electrons of atoms. Quantum physics shows us that the number of electrons in an atom, and hence the number of protons in its nucleus, accounts for its chemical properties. The periodic table as it is published in chemistry texts is deliberately arranged so that elements with similar chemical properties (identical numbers of valence electrons) appear in the same column of the table.

The number of neutrons in a nucleus is designated by the letter N, and the total number of neutrons plus protons is designated by the letter A: $A = N + Z$. A is known as the *mass number*, and also as the *nucleon number*; the term *nucleon* means either a proton or a neutron. By specifying Z and A, we specify a given isotope. Be careful: A is also used to designate the *atomic weight* of an element (or isotope) in grams per mole. The atomic weight and nucleon number of an isotope are always close, but the difference between them is important. The nucleon number is always an integer, but the atomic weight will have decimals. For example, the nucleon number of uranium-235 is 235, but the atomic weight of that species is 235.0439 g/mol. The term *nuclide* is also sometimes encountered, and is completely synonymous with *isotope*.

The general form for isotope notation is

$$\ _{Z}^{A}X. \tag{2.9}$$

In this expression, X is the symbol for the element involved. The subscript is always the atomic number, and the superscript is always the mass number. For example, the oxygen that you are breathing while reading this passage consists of three stable isotopes: $^{16}_{8}O$, $^{17}_{8}O$, and $^{18}_{8}O$. All oxygen atoms have eight protons in their nuclei, but either eight, nine, or ten neutrons. These nuclides are also referred to as oxygen-16 (O-16), oxygen-17 (O-17), and oxygen-18 (O-18). By far the most common isotope of oxygen is the first one: 99.757 % of naturally-occurring oxygen is O-16, with only 0.038 and 0.205 % being O-17 and O-18, respectively. Three isotopes that will prove very important in the story of the Manhattan Project are uranium-235, uranium-238, and plutonium-239: $^{235}_{92}U$, $^{238}_{92}U$, and $^{239}_{94}Pu$.

The concepts of atomic number and isotopy developed over many years. The foundations of modern atomic theory can be traced back to 1803, when English chemist John Dalton put forth a hypothesis that all atoms of a given element are identical to each other and equal in weight. An important development in Dalton's time came about when chemical evidence indicated that the masses of atoms of various elements seemed to be very nearly equal to integer multiples of the mass of hydrogen atoms. This notion was formally hypothesized about 1815 by English physician and chemist William Prout, who postulated that all heavier elements are aggregates of hydrogen atoms. He called the hydrogen atom a "protyle," a forerunner of Ernest Rutherford's "proton." Parts of both Dalton's and Prout's hypotheses would be verified, but other aspects required modification. In particular, something looked suspicious about Prout's idea from the outset, as some elements had atomic weights that were *not* close to integer multiples of that of hydrogen. For example, chlorine atoms seemed to weigh 35.5 times as much as hydrogen atoms.

The concept of isotopy first arose from evidence gathered in studies of natural radioactive decay chains (Sect. 2.1.6). Substances that appeared in different decay chains through different modes of decay often seemed to have similar properties, but could not be separated from each other by chemical means. The term "isotope" was introduced in 1913 by Frederick Soddy, who had taken a position at the University of Glasgow. Soddy argued that the decay-chain evidence suggested that "the net positive charge of the nucleus is the number of the place which the element occupies in the periodic table". Basing his hypothesis on the then-current idea that the electrically neutral mass in nuclei was a combination of protons and electrons, Soddy went on to state that the "algebraic sum of the positive and negative charges in the nucleus, when the arithmetical sum is different, gives what I call "isotopes" or "isotopic elements," because they occupy the same place in the periodic table." The root "iso" comes from the Greek word "isos," meaning "equal," and the *p* in *tope* serves as a reminder that it is the number of protons which is the same in all isotopes of a given element. In the same paper, Soddy also developed an ingenious argument to show that the electrons emitted in beta-decay had to be coming from *within the nucleus*, not from the "orbital" electrons.

True understanding of the nature and consequences of isotopy came with the invention of *mass spectroscopy*, an instrumental technique for making extremely precise measurements of atomic masses. In his 1897 work, J. J. Thomson measured the ratio of the electrical charge carried by electrons to their mass by using electric and magnetic fields to deflect them and track their trajectories. In 1907, Thomson modified his apparatus to investigate the properties of positively-charged (ionized) atoms, and so developed the first "mass spectrometer." In this device, electric and magnetic fields were configured to force ionized atoms to travel along separate, unique parabolic-shaped trajectories which depended on the ions' masses. The separate trajectories could be recorded on a photographic film for later analysis.

In 1909, Thomson acquired an assistant, Francis Aston, a gifted instrument-maker (Fig. 2.6). Aston improved Thomson's instrument, and, in November, 1912, obtained evidence for the presence of two isotopes of neon, of mass numbers 20 and 22 (taking hydrogen to be of mass unity). The atomic weight of neon was known to be 20.2. Aston reasoned that this number could be explained if the two isotopes were present in a ratio of 9:1, as is now known to be the case. (There is a third isotope of neon, of mass 21, but it comprises only 0.3 % of natural neon.) Aston tried to separate the two neon isotopes using a technique known as *diffusion*. As described in Chap. 1, this refers to the passage of atoms through a porous membrane. Aston passed neon through clay tobacco pipes, and did achieve a small degree of enrichment.

Following a position involving aircraft research during World War I, Aston returned to Cambridge, and in 1919 he built his own mass spectrometer which incorporated some improvements over Thomson's design. In a series of papers published from late that year through the spring of 1920, he presented his first results obtained with the new instrument. These included a verification of the two

Fig. 2.6 Francis Aston (1877–1945) *Source* http://commons.wikimedia.org/wiki/File:Francis_William_Aston.jpg

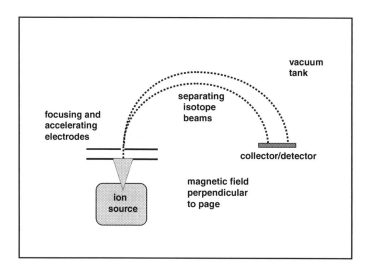

Fig. 2.7 Principle of mass spectroscopy. Positive ions are accelerated by an electric field and then directed into a magnetic field which emerges perpendicularly from the page. Ions of different mass will follow different circular trajectories, with those of greater mass having larger orbital radii

previously-detected neon isotopes, and a demonstration that chlorine comprised a mixture of isotopes of masses 35 and 37 in an abundance ratio of about 3:1. In later years (1927 and 1937), Aston developed improved instruments, his so-called second and third mass spectrometers.

The principle of Aston's mass spectrometer is sketched very simplistically in Fig. 2.7. Inside a vacuum chamber, the sample to be investigated is heated in a small oven. The heating will ionize the atoms, some of which will escape through a narrow slit. The ionized atoms are then accelerated by an electric field, and directed into a region of space where a magnetic field of strength B is present. The magnetic field is arranged to be perpendicular to the plane of travel of the positively-charged ions, that is, perpendicular to the plane of the page in Fig. 2.7; the electrical coils or magnet poles for creating the field are not sketched in the diagram. The magnetic field gives rise to an effect known as the Lorentz Force Law, which causes the ions to move in circular trajectories; an ion of mass m and net charge q that is moving with speed v will enter into a circular orbit of radius $r = mv/qB$.

If all ions are ionized to the same degree and have the same speed, heavier ones will be deflected somewhat less than lighter ones; that is, they will have larger-radius orbits. Only two different mass-streams are sketched in Fig. 2.7; there will be one stream for each mass-species present. The streams will be maximally separated after one-half of an orbit, where they can be collected on a film. Present-day models incorporate electronic detectors which can feed data to a computer for immediate analysis.

During his career, Aston discovered over 200 naturally-occurring isotopes, including uranium-238. Surprisingly, he does not have an element named after him, but he did receive the 1922 Nobel Prize for Chemistry. Aston's work showed that John Dalton's 1803 conjecture had been partially correct: atoms of the same element behave identically as far as their chemistry is concerned, but the presence of isotopy means that not all atoms of the same element have the same weight. Similarly, Aston found that Prout's conjecture that the masses of all atoms were integer multiples of that of hydrogen, if one substitutes "isotopes" for "atoms," was also *very nearly* true. But that *very nearly* proved to involve some very important physics.

What is meant by *very nearly* here? As an example, consider the common form of iron, Fe-56, nuclei of which contain 26 protons and 30 neutrons. Had Prout been correct, the mass of an iron-56 atom should be 56 "mass units," if one neglects the very tiny contribution of the electrons. (A technical aside: 56 electrons would weigh about 1.4 % of the mass of a proton. We are also assuming, for sake of simplicity, that protons and neutrons each weigh one "mass unit"; neutrons are about 0.1 % heavier than protons.) Mass spectroscopy can measure the masses of atoms to remarkable precision; the actual weight of an iron-56 atom is 55.934937 atomic mass units (see Sect. 2.5 for a definition of the atomic mass unit). The discrepancy of $55.934937 - 56 = -0.065063$ mass units, what Aston called the "mass defect," is significant, amounting to about 6.5 % of the mass of a proton. This mass defect effect proved to be systemic across the periodic table: *all stable atoms are less massive than one would predict on the basis of Prout's whole-number hypothesis.* Iron has a fairly large mass defect, but by no means the largest known (Fig. 2.8). The mass-defect is *not* an artifact of protons and neutrons having slightly different masses; if one laboriously adds up the masses of all of the constituents of atoms, the defects are still present. The unavoidable conclusion is that when protons and neutrons assemble themselves into nuclei, they give up some of their mass in doing so.

Physicists now quote mass defects in terms of equivalent energy in MeVs, thanks to $E = mc^2$. One mass unit is equivalent to 931.4 MeV, so the iron-56 mass defect amounts to just over 60 MeV. Because this is a mass *defect*, it is formally cited as a negative number, -60.6 MeV. The capital Greek letter delta (as in "Defect") is now used to designate such quantities: $\Delta = -60.6$ MeV.

Where does the mass *go* when nature assembles nuclei? Empirically, nuclei somehow have to hold themselves together against the immense mutual repulsive Coulomb forces of their constituent protons; some sort of nuclear "glue" must be present. To physicists, this "glue" is known synonymously as the "strong force" or as "binding energy," and is presumed to be the "missing" mass transformed into some sort of attractive energy. The greater the magnitude of the mass defect, the more stable will be the nucleus involved. Figure 2.8 shows a graph of the mass defects of 350 nuclides that are stable or have half-lives greater than 100 years, as a function of mass number A. The deep valley centered at $A \sim 120$ attests to the great stability of elements in the middle part of the periodic table; negative values of Δ connote intrinsic stability. The gap between $A \sim 210$ and 230 is due to the

Fig. 2.8 Mass defect in MeV versus mass number A for 350 nuclides with half-lives $\geq$100 years; $1 \leq A \leq 250$. The dashed arrows indicate a splitting of a heavy nucleus ($A \sim 240$) into two nuclides of $A \sim 120$. The energy released would be $\sim(40) - 2(-90) \sim 220$ MeV

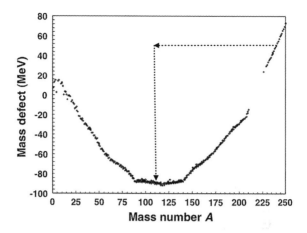

fact that there are no long-lived isotopes of elements between bismuth ($Z = 83$) and thorium ($Z = 90$). Isotopes with $A > 230$ could be said to have a "mass surplus." Consistent with the idea that *negative* Δ-values connote stability, all such *positive* Δ-valued nuclei eventually decay.

The two forgoing paragraphs actually muddle the concepts of mass defect and binding energy for sake of simplifying the description. Strictly, these are separate (but related) quantities. At a qualitative level, the details of the technical distinctions between them do not really add to the central concept that "lost mass" transforms to "binding energy." For sake of completeness, however, further details are discussed in Sect. 2.5, which can be considered optional.

Figure 2.8 can be used to estimate the energy released in hypothetical nuclear reactions. This is discussed in greater detail in Sect. 2.1.6, but the essence is straightforward: Add up the Δ-values of all of the input reactants (be careful with negative signs!), and then subtract from that result the sum of the Δ-values of the output products. The arrows in Fig. 2.8 show an example: a hypothetical splitting of a nucleus with $A \sim 240$ into two nuclei of $A \sim 120$. The input Δ-value is $\sim +40$ MeV. The sum of the output Δ-values is approximately $(-90$ MeV$) + (-90$ MeV$) \sim -180$ MeV. The difference between these is $\sim (+40) - (-180) \sim +220$ MeV, a large amount of energy even by nuclear standards. In late 1938 it was discovered that reactions like this are very real possibilities indeed.

There exist 266 apparently permanently stable, naturally occurring isotopes of the various elements, and about a hundred more "quasi-stable" ones with half-lives of a hundred years or greater. A compact way of representing all these nuclides is to plot each one as a point on a graph where the x-axis represents the number of neutrons, and the y-axis the number of protons. All isotopes of a given element will then lie on a horizontal line, since the number of protons in all nuclei of a given element is the same. This is shown in Fig. 2.9 for the 350 stable and quasi-stable nuclei of Fig. 2.8. Clearly, stable nuclei follow a very well-defined $Z(N)$ trend. Nature provides nuclei with neutrons to hold them together against the

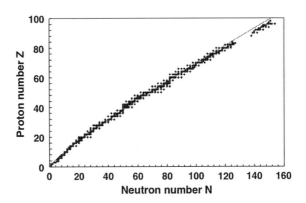

Fig. 2.9 Proton number Z versus neutron number N for 350 isotopes with half-lives ≥100 years, showing the narrow "band of stability" for nuclides. The trendline is described by the equation $Z \sim 1.264\, N^{0.87}$. As discussed in Sect. 2.1.6, "neutron-rich" nuclei would lie below the band of points and decay to stability along leftward-diagonally-upwards trajectories by β^- decay. Conversely, "neutron-poor" nuclei would lie above the band of points and decay along rightward-diagonally-downwards trajectories by β^+ decay; see Fig. 2.12

mutual repulsion of their protons, but she is economical in doing so. Mass represents energy ($E = mc^2$), and Nature is evidently unwilling to invest more mass-energy to stabilize nuclei than is strictly necessary.

Note also that the points in the graph curve off to the right; this indicates that the vast majority of nuclei, except for a very few at the bottom-left of the graph, contain more neutrons than protons; this effect is known as the *neutron excess*. We will revisit such plots in Sect. 2.1.6.

We now return for one section to pick up the story of early nuclear physics, after which will follow another tutorial on nuclear reactions.

2.1.5 Alpha Particles and the Nuclear Atom

In the spring of 1907, Rutherford returned to England to take a position at Manchester University. When he arrived there, he made a list of promising research projects, one of which was to pin down the precise nature of alpha particles. Based on experiments where the number of alphas emitted by a sample of radium had been counted and the charge carried by each had been determined, he had begun to suspect that they were ionized helium nuclei. However, he needed to trap a sample of alphas for confirming spectroscopic analysis. Working with student Thomas Royds, Rutherford accomplished this with one of his typically elegant experiments.

In the Rutherford-Royds experiment, a sample of radon gas was trapped in a very thin-walled glass tube, which was itself surrounded by a thicker-walled tube.

The space between the two tubes was evacuated, and the radon was allowed to decay for a week. The energetic radon alphas could easily penetrate through the 1/100-mm thick wall of the inner tube. During their flights they would pick up electrons, become neutralized, and then become trapped in the space between the tubes. The neutralized alphas were then drawn off for analysis, and clearly showed a helium spectrum. Rutherford and Royds published their finding in 1909. In the notation described in the preceding section, alpha particles are identical to helium-4 nuclei: $^{4}_{2}$He.

The discovery for which Rutherford is most famous is that atoms have nuclei; this also had its beginnings in 1909. One of the projects on Rutherford's to-do list was to investigate how alpha particles "scattered" from atoms when they (the alphas) were directed through a thin metal foil. At the time, the prevailing notion of the structure of an atom was of a cloud of positive electrical material within which were embedded negatively-charged electrons. Thomson had determined that electrons weighed about 1/1,800 as much as a hydrogen atom; since hydrogen was the lightest element, it seemed logical to presume that electrons were small in comparison to their host atoms. This picture has been likened to a pudding, with electrons playing the role of raisins inside the body of the pudding. Another line of atomic structure evidence came from the chemistry community. From the bulk densities of elements and their atomic weights, it could be estimated that individual atoms behaved as if they were a few Ångstroms in diameter ($1 \text{ Å} = 10^{-10}$ m; see Exercise 2.1). The few Ångstroms presumably represented the size of the overall cloud of positive material.

Rutherford had been experimenting with the passage of alpha-particles through metal foils since his earliest days of radioactivity research, and all of his experience indicated that the vast majority of alphas were deflected by only a very few degrees from straight-line paths as they barreled their way through a layer of foil. This observation was in line with theoretical expectations. Thomson had calculated that the combination of the size of a positively-charged atomic sphere and the kinetic energy of an incoming alpha (itself also presumably a few Ångstroms in size) would be such that the alpha would typically suffer only a small deflection from its initial trajectory. Deflections of a few degrees would be rare, and a deflection of 90° was expected to be so improbable as to never have any reasonable chance of being observed. In the Thomson atomic model, a collision between an alpha and an atom should not be imagined as like that between two billiard balls, but rather more like two diffuse clouds of positive electricity passing through each other. The alphas would presumably strike a number of electrons during the collision, but the effect of the electrons' attractive force on the alphas would be negligible due to the vast difference in their masses, a factor of nearly 8,000. Electrons played no part in Rutherford's work.

Rutherford was working with Hans Geiger (Fig. 2.10) of Geiger counter fame, who was looking for a project to occupy an undergraduate student, Ernest Marsden, another New Zealand native. Rutherford suggested that Geiger and Marsden check to see if they could observe any large-angle deflections of alphas

Fig. 2.10 Hans Geiger
(1882–1945) in 1928. *Source*
http://commons.
wikimedia.org/wiki/
File:Geiger,Hans_1928.jpg

when they passed through a thin gold foil, fully expecting a negative result. Gold was used because it could be pressed into a thin foil only about a thousand atoms thick. To Geiger and Marsden's surprise, a few alphas, about one in every 8,000, were *bounced backward* toward the direction from which they came. The number of such reflections was small, but was orders of magnitude more than what was expected on the basis of Thomson's model. Rutherford was later quoted as saying that the result was "almost as incredible as if you had fired a 15-in. shell at a piece of tissue paper and it came back and hit you." Geiger and Marsden published their anomalous result in July, 1909. The work of detecting the scattered alpha-particles was excruciating. A Geiger counter could have been used to *detect* the alphas, but they had to be *seen* to get detailed information on their direction of travel. This was done by having the scattered alphas strike a phosphorescent screen; a small flash of light (a "scintillation") would be emitted, and could be counted by an observer working in a darkened room. Geiger and Marsden counted thousands of such scintillations.

So unexpected was Geiger and Marsden's result that it took Rutherford the better part of 18 months to infer what it meant. The conclusion he came to was that the positive electrical material within atoms must be confined to much smaller volumes than had been thought to be the case. The alpha-particles (themselves also nuclei) had to be similarly minute; only in this way could the electrical force experienced by an incoming alpha be intense enough to achieve the necessary repulsion to turn it back if it should by chance strike a target nucleus head-on; the vast majority of alpha nuclei sailed through the foil, missing gold nuclei by wide margins. The compaction of the positive charge required to explain the scattering experiments was stunning: down to a size of about 1/100,000 of an Ångstrom. But, atoms as a whole still behaved *in bulk* as if they were a few Ångstroms in diameter. Both numbers were experimentally secure and had to be accommodated.

This, then, was the origin of our picture of atoms as miniature solar systems: very small, positively-charged "nuclei" surrounded by orbiting electrons at distances out to a few Ångstroms. This configuration is now known as the "Rutherford atom."

A sense of the scale of Rutherford's atom can be had by thinking of the lone proton that forms the nucleus of an ordinary hydrogen atom as scaled up to being two millimeters in diameter, about the size of an uncooked grain of rice. If this enlarged proton is placed at the center of a football field, the diameter of the lowest-energy electron orbit (that which comes closest to the nucleus) would reach to about the goal lines. In giving us nuclei and being credited with the discovery of the positively-charged protons that they contain, Rutherford bequeathed us atoms that are largely empty space.

The first public announcement of this new model of atomic structure seems to have been made on March 7, 1911, when Rutherford addressed the Manchester Literary and Philosophical Society; this date is often cited as the birthdate of the nuclear atom. The formal scientific publication came in July, and directly influenced Niels Bohr's famous atomic model which was published two years later. Rutherford's nucleus paper is a masterpiece of fusion of experimental evidence and theoretical reasoning. After showing that the Thomson model could not possibly generate the observed angular distribution of alpha scatterings, he demonstrated that the nuclear "point-mass" model gave predictions in accord with the data. Rutherford did not use the term "nucleus" in his paper; that nomenclature seems to have been introduced by Cambridge astronomer John Nicholson in a paper published in November, 1911. The term "proton" was not introduced until June, 1920, but was coined by Rutherford himself.

With the understanding that scattering events were the results of such nuclear collisions, Rutherford's analysis could be applied to other elements in the sense of using an observed scattering distribution to infer how many fundamental "protonic" charges the element possessed; this helped to place elements in their proper locations in the periodic table. Elements had theretofore been defined by their atomic weights (A), but it was the work of researchers such as Rutherford, Soddy, Geiger, and Marsden which showed that it is an element's atomic number (Z) that dictates its chemical identity.

The atomic weights of elements were still important, however, and very much the seat of a mystery. Together, chemical and scattering evidence indicated that the atomic weights of atoms seemed to be proportional to their number of protonic charges. Specifically, atoms of all elements weighed about twice as much or more as could be accounted for on the basis of their numbers of protons. For some time, this extra mass was thought to be due to additional protons in the nucleus which for some reason contained electrons within themselves, an electrically neutral combination. This would give net-charge nuclei consistent with the scattering experiments, while explaining measured atomic weights. By the mid-1920s, however, this proposal was becoming untenable: the Uncertainty Principle of quantum mechanics ruled against the possibility of containing electrons within so small a volume as a single proton, or even a whole nucleus. For many years before

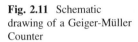

Fig. 2.11 Schematic drawing of a Geiger-Müller Counter

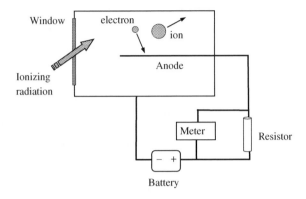

its discovery, Rutherford speculated that there existed a third fundamental constituent of atoms, the neutron. As described in Sect. 2.2, he would live to see his suspicion proven by one of his own students. That atoms are built of electrons orbiting nuclei comprised of protons and neutrons is due very much to Rutherford and his collaborators and students.

Having mentioned Hans Geiger, it is worthwhile to describe briefly the operation of his eponymous counter, as its use will turn up in other contexts in this book. The original version of the Geiger counter was invented by Geiger and Rutherford in 1908. In 1928, Geiger and a student, Walther Müller, made some improvements on the design, and these devices are now properly known as Geiger-Müller counters.

Geiger-Müller counters operate by detecting ionizing radiation, that is, particles that ionize atoms in a sample of gas through which they pass. Fundamentally, the counter consists of a metal case which is closed at one end and which has a thin plastic "window" at the other end (Fig. 2.11). Inside the tube is an inert gas, usually helium. The case and a metallic central anode are connected to a battery which makes (in the Figure) the tube negative and the anode positive. Energetic particles such as alpha or beta rays penetrate through the window, and ionize atoms of the inert gas. The liberated electrons will be attracted to the anode, while the ionized atoms are attracted to the case. The net effect is to create an electrical current, which is shunted by a resistor to pass through a meter. By incorporating a speaker in the electronics, the current can be converted to the familiar "clicks" one hears on news reports and in movies.

2.1.6 Reaction Notations, Q-Values, Alpha and Beta Decay, and Decay Chains

In this and the following section we again break with chronological order to fill in some background on notations for nuclear reactions, and the details of alpha and beta-decays and naturally-occurring decay schemes. We will pick up with more of Rutherford's work in Sect. 2.1.7. The present section is somewhat technical, but involves no computations; the concept of a Q-value will prove very important later on.

The notation for writing a nuclear reaction is very similar to used that for describing a chemical reaction. *Reactants* or *input nuclides* are written on the left side of a rightward-pointing arrow, and *products* or *output nuclides* are placed on the other side, like this:

$$reactants \rightarrow products. \tag{2.10}$$

Decades of experimental evidence indicate that there are two rules that are *always* obeyed in nuclear reactions:

(i) The total number of input nucleons must equal the total number of output nucleons. *The numbers of protons and neutrons may (and usually do) change, but their sum must be conserved.*

(ii) Total electric charge must be conserved. Protons count as one unit of positive charge. Beta decays involve nuclei which create within themselves and then eject either an electron or a positively-charged particle with the same mass as an electron, a so-called *positron*. The charges of these ejected particles must be taken into account in ensuring charge conservation (negative or positive one unit), but they are not considered to be nucleons and so are not counted when applying rule (i). Positrons are also known as beta-positive (β^+) particles, while ordinary electrons are also known as beta-negative particles (β^-).

As an example of a typical reaction, here is one that will be discussed in more detail in Sect. 2.1.7: alpha-bombardment of nitrogen to produce hydrogen and oxygen:

$$_2^4He + {}_7^{14}N \rightarrow {}_1^1H + {}_8^{17}O. \tag{2.11}$$

Verification that both rules are followed can be seen in that (i) $4 + 14 = 1 + 17$, and (ii) $2 + 7 = 1 + 8$. In this type of bombardment reaction, the notational convention is to write the lighter incoming reactant first on the left side, followed by the target nucleus. Note that a hydrogen nucleus, $_1^1H$, is simply a proton. A proton is sometimes written as just "p", but this book will always employ the more explicit $_1^1H$ notation. On the output side, the lighter product is usually written first. In a "four-body" reaction such as this, a more compact shorthand notation that puts the target nucleus first is sometimes employed:

$$_7^{14}N\left(_2^4He, {}_1^1H\right)_8^{17}O. \tag{2.12}$$

In this format, the convention can be summarized as:

$$target\ (projectile,\ light\ product)\ heavy\ product \tag{2.13}$$

In any reaction where the input and input reactants are different, experiments show that *mass is not conserved*. That is, the sum of the input masses will be different from the sum of the output masses. Mass can either be created or lost; what happens depends on the nuclides involved. The physical interpretation of this relates to Einstein's famous $E = mc^2$ equation. If mass is lost (sum of output masses < sum of input masses), the lost mass will appear as kinetic energy of the output products. If mass is gained (sum of output masses > sum of input masses), energy must be drawn from somewhere to create the mass gained, and the only source available is the kinetic energy of the "bombarding" input reactant. Nuclear physicists always express the mass gain or loss in units of energy equivalent, almost always in MeV. Such energy gains or losses are termed *Q-values*. If $Q > 0$, kinetic energy is created by consuming input-reactant mass, whereas if $Q < 0$, input-particle kinetic energy has been consumed to create additional output mass. The technical definition of Q is

$$Q = (\text{sum of input masses}) - (\text{sum of product masses}), \qquad (2.14)$$

quoted in units of equivalent energy (one atomic mass unit = 931.4 MeV). When applied to computing the energy consumed or liberated in a reaction, this definition gives the same result as the graphical method illustrated in Fig. 2.8.

The alpha-nitrogen reaction above has $Q = -1.19$ MeV. To approach the nucleus and initiate the reaction, the incoming alpha-particle must possess at least this much kinetic energy. In fact, the alpha must possess considerably more than 1.19 MeV of kinetic energy due to an effect that is not accounted for in computing the Q-value alone, the so-called *Coulomb barrier*; this is discussed in Sect. 2.1.8.

Returning for a moment to mass spectroscopy, the development of means to measure precise masses for isotopes was a crucial step forward in the progress of nuclear physics. With precise masses and knowledge of Einstein's $E = mc^2$ equivalence, the energy liberated or consumed in reactions could be predicted. Measurements of the kinetic energies of reaction products would then serve as checks on the mass values. Conversely, for a reaction where the mass or identity of some of the particles involved was not clear, measurements of the kinetic energies could be used to infer what was happening. On reflecting on these connections, you might wonder how Rutherford measured such kinetic energies; after all, tracking a nucleus is obviously not the same as using a radar gun to measure the speed of a car or a baseball. Experimenters had to rely on proxy measurements such as how far a particle traveled through air or a stack of thin metal foils before being brought to a stop. If precise mass defects are known from mass spectroscopy, the energy liberated (or consumed) in a reaction can be computed, and the numbers can be used to calibrate a range-versus-energy relationship. This combination of theory, experimental technique, and instrumental development is an excellent example of scientific cross-fertilization.

2.1.6.1 Alpha Decay

Ernest Rutherford decoded alpha-decay as a nucleus spontaneously transmuting itself to a more stable mass-energy configuration by ejecting a helium nucleus. In doing so, the original nucleus loses two protons and two neutrons, which means that it ends up two places down in atomic number on the periodic table and has four fewer nucleons in total. Alpha-emission is a common decay mechanism in heavy elements, and can be written in the arrowed notation as

$$_{Z}^{A}X \xrightarrow{\alpha} {}_{Z-2}^{A-4}Y + {}_{2}^{4}\text{He}. \tag{2.15}$$

Here, X designates the element corresponding to the original nucleus, and Y that of the "daughter product" nucleus. Sometimes the half-life is written below the arrow; for example, the alpha-decay of uranium-235 can be written as

$$_{92}^{235}\text{U} \xrightarrow[7.04 \times 10^8 \text{ year}]{\alpha} {}_{90}^{231}\text{Th} + {}_{2}^{4}\text{He}. \tag{2.16}$$

As always, electrical charge and nucleon number are conserved. In such decays, the total mass of the output products is always less than that of the input particles: Nature spontaneously seeks a lower mass-energy configuration ($Q > 0$). The energy release in alpha-decays is typically $Q \sim 5 - 10$ MeV, the majority of which appears as kinetic energy of the alpha-particle itself. Helium nuclei tend to be ubiquitous in nuclear reactions as they have a large mass defect and so are very stable.

As a tool to induce nuclear reactions, the Curies and Rutherford often utilized alpha particles emitted in radium decay:

$$_{88}^{226}\text{Ra} \xrightarrow[1,599 \text{ year}]{\alpha} {}_{86}^{222}\text{Rn} + {}_{2}^{4}\text{He}. \tag{2.17}$$

This decay has a Q-value of $+4.87$ MeV, which explains how the $_{7}^{14}\text{N}\left(_{2}^{4}\text{He}, _{1}^{1}\text{H}\right)_{8}^{17}\text{O}$ reaction described above can be made to happen.

2.1.6.2 Beta Decay

Two types of beta decay occur naturally. Look back at Fig. 2.9, which illustrates the narrow "band of stability" of long-lived isotopes. If an isotope should find itself with too many neutrons for the number of protons that it possesses (or, equivalently, too few protons for its number of neutrons), it will lie to the right of the band of points. Conversely, should it have too few neutrons for the number of protons that it possesses (or too many protons for its number of neutrons), it will lie to the left of the band of points.

Suppose that a nucleus is too neutron-rich for its number of protons. Purely empirically, it has been found that Nature deals with this by having a neutron spontaneously decay into a proton. But this, by itself, would represent a net creation of electric charge, and hence a violation of charge conservation. So, a negative electron is created in the bargain to render no net charge created. Nucleon number is conserved; remember that electrons do not count as nucleons. The electron is also known as a β^- particle, and the reaction can be symbolized as $n \rightarrow p + e^-$, or, equivalently, $n \rightarrow p + \beta^-$. The number of neutrons drops by one while the number of protons grows by one, so the number of nucleons is unchanged. The overall effect is

$$_Z^A X \xrightarrow{\beta^-} {}_{Z+1}^{A}Y + {}_{-1}^{0}e^-. \tag{2.18}$$

Note that a "nucleon-like" notation has been appended to the electron to help keep track of the charge and nucleon numbers. *The result of β^- decay is to move a nucleus up one place in the periodic table.* It was Henri Becquerel who showed, in 1900, that the negatively-charged beta-rays being observed in such decays were identical in their properties to Thomson's electrons.

If a nucleus is neutron-poor, a proton will spontaneously decay into a neutron. But this would represent a loss of one unit of charge, so Nature creates a *positron*—an anti-electron—to maintain the charge balance: $p \rightarrow n + e^+$, or $p \rightarrow n + \beta^+$. Here the overall effect is

$$_Z^A X \xrightarrow{\beta^+} {}_{Z-1}^{A}Y + {}_{1}^{0}e^+. \tag{2.19}$$

The result of β^+ decay is to move a nucleus down one place in the periodic table.

As shown in Fig. 2.12, decay mechanisms can be represented graphically in the (Z, N) grid format of Fig. 2.9. Also shown in Fig. 2.12 is the effect of *neutron capture*. In this process, a nucleus absorbs an incoming neutron to become a heavier isotope of itself; this will be important in later discussions of fission and the synthesis of plutonium.

Fig. 2.12 Decay and neutron-capture transmutation trajectories of an original nucleus of Z protons and N neutrons. This (Z, N) arrangement is as that in Fig. 2.9

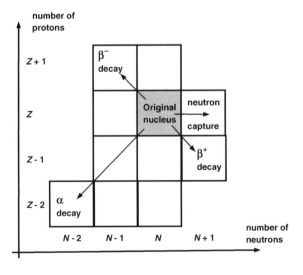

2.1.6.3 Natural Decay Schemes

The work of the Curies and Rutherford and their various collaborators culminated in the understanding that three lengthy decay sequences occur spontaneously in Nature. All three begin with an isotope of thorium or uranium, and terminate with three different isotopes of lead. These are illustrated in Fig. 2.13, which is of the same form as Fig. 2.12.

As an example of the use of decay-chain notation, consider again Rutherford's one-minute thorium "emanation" of Sect. 2.1.2. The observed one-minute half-life probably represented the decay of radon-220:

$$\overset{228}{\underset{90}{}}\text{Th} \xrightarrow[\text{1.9 year}]{\alpha} \overset{224}{\underset{88}{}}\text{Ra} \xrightarrow[\text{3.63 days}]{\alpha} \overset{220}{\underset{86}{}}\text{Rn} \xrightarrow[\text{55.6 s}]{\alpha} \overset{218}{\underset{84}{}}\text{Po} \xrightarrow[\text{3.1 min}]{\alpha} \overset{214}{\underset{82}{}}\text{Pb} \rightarrow \dots \quad (2.20)$$

The first step in this chain, thorium-228, is itself a decay product of uranium. The identification of Rn-220 as Rutherford's thorium emanation is strong, but not absolutely secure: many heavy-element half-lives are on the order of a minute; as described in a McGill University website, it is not clear exactly what isotopes Rutherford detected, as the concept of isotopy had not yet been established in 1900.

The work accomplished by the world's first generation of radiochemists was staggering. Uranium and thorium ores contain constantly varying amounts of a number of isotopes which are created by decays of heavier parent isotopes, and which themselves decay to lighter daughter products until they arrive at stable

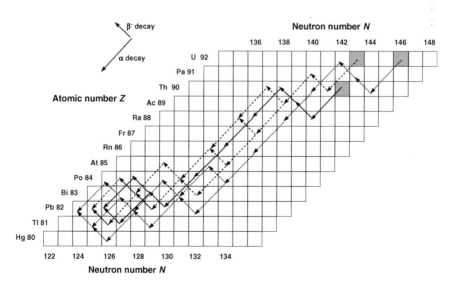

Fig. 2.13 Natural radioactive decay sequences. Nuclei of uranium-238, uranium-235, and thorium-232 all decay to isotopes of lead (Pb-206, 207, and 208, respectively) via sequences of alpha and beta decays. These sequences are respectively indicted by the chains of *dotted*, *dashed*, and *solid lines*

neutron/proton configurations. Only by isolating samples of individual elements and subjecting them to mass-spectroscopic analyses could individual isotopes be characterized.

2.1.7 Artificial Transmutation

Rutherford's last great discovery came in 1919. This was his realization that it was possible to set up experimental situations wherein atoms of a given element could be *transmuted* into those of another, when bombarded by nuclei of yet a third. The idea of elemental transmutation was not new; after all, this is precisely what happens in natural alpha and beta-decays. What was new was the realization that transmutations could be induced by human intervention.

The work that led to this discovery began around 1915, and was carried out by Ernest Marsden. As part of an experimental program involving measurements of reaction energies, Marsden bombarded hydrogen atoms with alpha-particles produced by the decay of samples of radon gas contained in a small glass vials. A hydrogen nucleus would receive a significant kick from a collision with an alpha-particle and be set into motion at high speed. These experiments were done by sealing the alpha source and hydrogen gas inside a small chamber, as sketched in Fig. 2.14. At one end of the chamber was a small scintillation screen which could be viewed through a microscope, as had been done in the alpha-scattering experiments. By placing thin metal foils just behind the screen, Marsden could determine the ranges, and hence the energies, of the struck protons. So far, there is nothing unusual here; these experiments were routine work that involved the use of known laws of conservation of energy and momentum to cross-check and interpret measurements.

Breakthroughs favor an attentive and experienced mind, and Marsden's was ready. His crucial observation was to notice that when the experimental chamber was evacuated, the radon source itself seemed to give rise to scintillations like

Fig. 2.14 Sketch of Rutherford's apparatus for the discovery of artificial transmutation

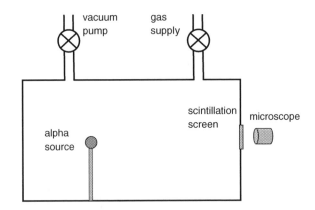

those from hydrogen, *even though there was no hydrogen in the chamber*. The implication seemed to be that hydrogen was arising in radioactive decay, an occurrence that had never before been observed. Marsden returned to New Zealand in 1915, and Rutherford, heavily occupied with research for the British Admiralty, could manage only occasional experiments until World War I came to an end in late 1918. In 1919, he turned to investigating Marsden's unexpected radium/hydrogen observation, and was rewarded with yet another pivotal discovery.

Look again at Fig. 2.14. Rutherford placed a source of alpha particles within a small brass chamber which could be evacuated and then filled with a gas with which he wished to experiment. As Rutherford reported in his June, 1919, discovery paper, he set out to investigate the phenomenon that "a metal source, coated with a deposit of radium-C [bismuth-214], always gives rise to a number of scintillations on a zinc sulphide screen far beyond the range of the α particles. The swift atoms causing these scintillations carry a positive charge and are deflected by a magnetic field, and have about the same range and energy as the swift H atoms produced by the passage of α particles through hydrogen. These 'natural' scintillations are believed to be due mainly to swift H atoms from the radioactive source, but it is difficult to decide whether they are expelled from the radioactive source itself or are due to the action of α particles on occluded hydrogen."

Rutherford proceeded by investigating various possibilities as to the origin of the hydrogen scintillations. No vacuum pump is ever perfect; some residual air would always remain in the chamber no matter how thoroughly it had been pumped down. While hydrogen is normally a very minute component of air (about half a part per million), more could be present if the air contained water vapor. Suspecting that the alpha particles might be striking residual hydrogen-bearing water molecules, Rutherford began by introducing dried oxygen and carbon dioxide into the chamber, observing, as he expected, that the number of scintillations decreased. Surprisingly, however, when he admitted *dry air* into the chamber, the number of hydrogen-like scintillations *increased*. This suggested that hydrogen was arising not from the radium-C itself, but from some interaction of the alpha particles with air.

The major constituents of air are nitrogen and oxygen; having eliminated oxygen, Rutherford inferred that nitrogen might be involved. On admitting pure nitrogen into the chamber, the number of scintillations increased yet again. As a final test that hydrogen was not somehow arising from the radioactive source itself, he found that on placing thin metal foils close to the radioactive source, the scintillations persisted, but their range was reduced in accordance with what would be expected if the alpha particles were traveling through the foils before striking nitrogen atoms; the scintillations were evidently arising from within the volume of the chamber. As Rutherford wrote, "it is difficult to avoid the conclusion that the long-range atoms arising from collision of α particles with nitrogen are … probably atoms of hydrogen …. If this be the case, we must conclude that the nitrogen atom is disintegrated under the intense forces developed in a close collision with a swift α particle, and that the hydrogen atom which is liberated formed a constituent part of the nitrogen nucleus".

In modern notation, the reaction is written as

$$_2^4\mathrm{He} + {}_7^{14}\mathrm{N} \rightarrow {}_1^1\mathrm{H} + {}_8^{17}\mathrm{O} + \gamma. \tag{2.21}$$

The "γ" here indicates that this reaction also releases a gamma-ray. The gamma-ray plays no role in the interpretation of Rutherford's experiment through conservation of charge and mass numbers, but is included here for sake of completeness; it will play a role in the discussion of the discovery of the neutron in Sect. 2.2.

Because atoms are mostly empty space, only about one alpha per hundred thousand induces such a reaction. Nuclear physicists speak of the *yield* of a reaction, which is the fraction of incident particles that cause a reaction. Yield values on the order of 10^{-5} in alpha-induced reactions are not uncommon.

Why does such a reaction not take place with oxygen, say,

$$_2^4\mathrm{He} + {}_8^{16}\mathrm{O} \rightarrow {}_1^1\mathrm{H} + {}_9^{19}\mathrm{F} + \gamma? \tag{2.22}$$

The Q-value of this reaction is -8.1 MeV. Radium-C alphas have energies of about 5.5 MeV, which is not enough to make the reaction happen. In the case of nitrogen, the Q-value is about -1.2 MeV, so the alphas are sufficiently energetic to make that reaction occur.

Rutherford and Marsden's discovery opened yet another experimental venue: Could alpha-particles induce transmutations in any other elements? What products could be created? What yields were involved? As discussed in the next section, however, there was a serious natural limitation to further experiments.

Later in 1919, Rutherford moved from Manchester University to Cambridge University to fill the position of Director of the Cavendish Laboratory, which had become vacant upon the retirement of J. J. Thomson. Rutherford would remain at Cambridge until his death in October, 1937, nurturing another generation of nuclear experimentalists. He died just 2 years before the discovery of nuclear fission, which would lead, in a few more years, to the development and use of nuclear weapons.

2.1.8 The Coulomb Barrier and Particle Accelerators

Consider again Rutherford's alpha-bombardment of nitrogen, the first artificial transmutation of an element (neglecting the gamma-ray):

$$_2^4\mathrm{He} + {}_7^{14}\mathrm{N} \rightarrow {}_1^1\mathrm{H} + {}_8^{17}\mathrm{O}. \tag{2.23}$$

Neglecting the fact that this reaction has a negative Q-value, a simple interpretation of this equation is that if you were to mix helium and nitrogen, say at room-temperature conditions, hydrogen and oxygen would result spontaneously. But even if Q were positive, this would not happen because of an effect that is not accounted for in merely writing down the reaction or in computing the Q-value: the so-called "Coulomb barrier" problem.

Electrical charges of the same sign repel each other. This effect is known as "the Coulomb force" after French physicist Charles-Augustin de Coulomb, who performed some of the first quantitative experiments with electrical forces in the late 1700s. Because of the Coulomb force, nitrogen nuclei will repel incoming alpha-particles; only if an alpha has sufficiently great kinetic energy will it be able to closely approach a nitrogen nucleus. Essentially, the two have to collide before stronger but shorter-range "nuclear forces" between nucleons that effect trans-mutations can come into play. The requisite amount of kinetic energy that the incoming nucleus must possess to achieve a collision is called the "Coulomb barrier."

For an alpha-particle striking a nitrogen nucleus, the barrier amounts to about 4.2 MeV, a fairly substantial amount of energy. An atom or molecule at room temperature will typically possess only a fraction of an eV of kinetic energy (about 0.025 eV on average), not nearly enough to initiate the reaction. Rutherford was able to induce the nitrogen transmutation because his radium-C alphas possessed over 5 MeV of kinetic energy.

The following material examines the physics of the Coulomb barrier. Readers who wish to skip this material should proceed to the paragraph following Eq. (2.31).

Figure 2.15 sketches two nuclei that are undergoing a collision. One of them, of atomic number Z_1, is presumed to be bombarding a fixed target nucleus of atomic number Z_2. Since the charge on each proton is the same magnitude as the electron charge "e," the charges within the nuclei will be $+eZ_1$ and $+eZ_2$.

According to Coulomb's law, if the centers of the two nuclei are distance d apart, the system will possess a potential energy PE given by

$$PE = \frac{(eZ_1)(eZ_2)}{4\pi\varepsilon_0 d} = \frac{e^2 Z_1 Z_2}{4\pi\varepsilon_0 d}, \qquad (2.24)$$

where ε_O is a physical constant, 8.8544×10^{-12} C^2/(J-m).

To effect a collision, the incoming nucleus needs to approach to a distance d which is equal to the sum of the radii of the two nuclei. To achieve such an approach, the incoming nucleus must start with an amount of kinetic energy which

Fig. 2.15 Colliding nuclei. The nucleus on the right is presumed to be fixed while the one on the left approaches

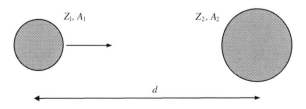

is at least equal to the potential energy of the system at the moment of contact in order that it will not have been brought to a halt beforehand by the Coulomb repulsion. Equation (2.24), when evaluated for the value of d corresponding to the two nuclei just touching, gives the Coulomb barrier. We thus need a general way to evaluate (2.24) for when two nuclei are just touching.

Empirically, scattering experiments show that the radii of nuclei can be expressed in terms of their mass numbers according as the expression

$$radius \sim a_0 A^{1/3}, \tag{2.25}$$

where $a_0 \sim 1.2 \times 10^{-15}$ m. Designating the mass numbers as A_1 and A_2, we have

$$Coulomb\ barrier \sim \left(\frac{e^2}{4\,\pi\,\varepsilon_0 a_0}\right) \frac{Z_1 Z_2}{\left(A_1^{1/3} + A_2^{1/3}\right)}. \tag{2.26}$$

The value of e is 1.6022×10^{-19} C; substituting this, ε_0, and $a_0 \sim 1.2 \times 10^{-15}$ m into (2.26) gives the bracketed factor as

$$\frac{e^2}{4\,\pi\,\varepsilon_0 a_0} = 1.9226 \times 10^{-13}\ \text{J.} \tag{2.27}$$

This can be expressed more conveniently in terms of MeV; 1 MeV = 1.6022×10^{-13} J:

$$\frac{e^2}{4\,\pi\,\varepsilon_0 a_0} = 1.2\ \text{MeV.} \tag{2.28}$$

We can then write (2.26) as

$$Coulomb\ barrier \sim \frac{1.2(Z_1 Z_2)}{\left(A_1^{1/3} + A_2^{1/3}\right)}\ \text{MeV.} \tag{2.29}$$

For an alpha-particle striking a nitrogen nucleus, this gives, as claimed above,

$$Coulomb\ barrier \sim \frac{1.2(2 \times 7)}{(4^{1/3} + 14^{1/3})} \sim \frac{16.8}{(1.587 + 2.410)} \sim 4.2\ \text{MeV.} \tag{2.30}$$

Now imagine trying to induce a reaction by having alpha-particles strike nuclei of uranium-235. The experiment would be hopeless if you are using an alpha whose kinetic energy is of the typical 5–10 MeV decay energy:

$$Coulomb\ barrier \sim \frac{1.2(2 \times 92)}{(4^{1/3} + 235^{1/3})} \sim \frac{220.8}{(1.587 + 6.171)} \sim 28.5\ \text{MeV.} \tag{2.31}$$

If one is using alphas created in natural decays, it is practical to carry out bombardment experiments with target elements only up to $Z \sim 20$. By the mid-1920s this was becoming a serious problem: researchers were literally running out of elements to experiment with. The curiosity-driven desire to bombard heavier

Fig. 2.16   Left Rolf Wideröe (1902–1996). Right Ernest Lawrence (1901–1958). Sources AIP Emilio Segre Visual Archives; http://commons.wikimedia.org/wiki/File:Ernest_Orlando_ Lawrence.jpg

elements thus generated a technological challenge: Was there any way that the alpha (or other) particles could be accelerated once they had been emitted by their parent nuclei? It was this challenge that gave birth to the first generation of particle accelerators.

The first practical particle acceleration scheme was published by Norwegian native Rolf Wideröe (Fig. 2.16) in a German electrical engineering journal in 1928. The essence of Wideröe's proposal is pictured in Fig. 2.17. Two hollow metal cylinders are placed end-to-end and connected to a source of variable-polarity voltage. This means that the cylinders can be made positively or nega-tively charged, and the charges can be switched as desired. A stream of protons (say) is directed into the leftmost cylinder, which is initially negatively charged. This will attract the protons, which will speed up as they pass through the cylinder.

Just as the bunch of protons emerges from the first cylinder, the voltage polarity is switched, making the left cylinder positive and the right one negative. The protons then get a push from the first cylinder while being pulled into the second one, which further accelerates them. By placing a number of such units back-to-back, substantial accelerations can be achieved; this is the principle of a *linear accelerator*. Obviously, many of the incoming particles will be lost by crashing into the side of a cylinder or because their speed does not match the frequency of the polarity shifts of the voltage supplies; only a small number will emerge from the last cylinder. But the point here is not necessarily efficiency; it is to generate *some* high-speed particles which could surmount the Coulomb barriers of heavy target nuclei. The longest linear accelerator in the world is now the Stanford Linear

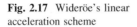

Fig. 2.17 Wideröe's linear
acceleration scheme

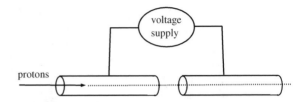

Accelerator in California, which can accelerate electrons to 50 *billion* electron-volts of kinetic energy over a distance of 3.2 km (2 miles).

Wideröe's work came to the attention of Ernest Orlando Lawrence (Fig. 2.16), an experimental physicist at the University of California at Berkeley. Lawrence and collaborator David Sloan built a Wideröe device, which by late 1930 they had used to accelerate mercury ions to kinetic energies of 90,000 eV. While experimenting with the Wideröe design, however, Lawrence had an inspiration that was to have profound consequences. He desired to achieve higher energies, but was daunted by the idea of building an accelerator that would be meters in length. How could the device be made more compact?

In Sect. 2.1.4 a description was given of how Francis Aston utilized the Lorentz force caused by a magnetic field to separate ions of different masses in his mass spectrometer. Lawrence's new device, which he called a *cyclotron*, also made use of this force law, but in a way that simultaneously incorporated Wideröe's alternating-voltage acceleration scheme.

Lawrence's cyclotron is sketched in Fig. 2.18, which is taken from his application for a patent on the device. Here the voltage supply is connected to two D-shaped metal tanks placed back-to back; they are known to cyclotron engineers as "Dees." The entire assembly must be placed within a surrounding vacuum tank to avoid deflective effects of collisions of the accelerated particles with air molecules.

The source of the ions (usually positive) is placed between the Dees. In the diagram, the ions are initially directed toward the upper Dee, which is set to carry a negative charge to attract them. If the voltage polarity is not changed and there is nothing to otherwise deflect the ions, they would crash into the edge of the Dee. But Lawrence knew from Aston's work that if the assembly were placed between the poles of a magnet (with the magnetic field again emerging from the page), the Lorentz force would try to make the ions move in circular paths. The net result of the combination of the ions' acceleration toward the charged Dee and the Lorentz force is that the ions move in outward-spiraling trajectories. If the magnetic field is strong, the spiral pattern will be "tight", and the ions will get nowhere near the edge of the Dee in their first orbit. As ions leave the upper Dee, the polarity is switched in order to attract them to the lower Dee. Switching and acceleration continues (for microseconds only) until the ions strike a target at the periphery of one of the Dees.

Lawrence and graduate student Nils Edlefsen first reported on the cyclotron concept at a meeting of the American Association for the Advancement for

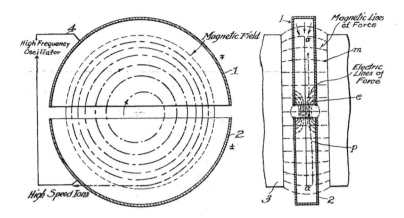

Fig. 2.18 Schematic illustration of Lawrence's cyclotron concept in top and side view, from his patent application. *Source* http://commons.wikimedia.org/wiki/File:Cyclotron_patent.png

Science held in September, 1930, but they had no results available at that time. By May, 1931, Lawrence had a 4.5-in. diameter device in operation (Fig. 2.19); he and student M. Stanley Livingston reported at an American Physical Society meeting that they were able to accelerate hydrogen molecule-ions (H_2^+) to energies of 80,000 eV using only a 2,000 V power supply. Later the same year, Lawrence achieved MeV energies with an eleven-inch cyclotron. By 1932 he had constructed a 27-in. device which achieved an energy of 3.6 MeV (Fig. 2.20), but had bigger plans yet. Lawrence was as adept at fundraising as he was at electrical engineering, and by 1937 had constructed a 37-in. model capable of accelerating deuterons (nuclei of "heavy hydrogen," ^2_1H) to energies of 8 MeV. By 1939 he had brought into operation a 60-in. model that required a 220-t magnet, and which could accelerate deuterons to 16 MeV. In 1942 he brought online his 184-in. diameter cyclotron, which is still operating and can accelerate various types of particles to energies exceeding 100 MeV. Along the way, Lawrence established the University of California Radiation Laboratory ("Rad Lab"), which is now the Lawrence Berkeley National Laboratory (LBNL).

Particle accelerators allowed experimenters to surmount the Coulomb barrier and so open up a broad range of energies and targets to experimentation. Lawrence's ingenuity earned him the 1939 Nobel Prize for Physics, and a variant of his cyclotron concept would play a significant role in the Manhattan Project. Today's giant accelerators at the Fermi National Accelerator Laboratory (Fermilab) and the European Organization for Nuclear Research (CERN) are the descendants of Widerøe's and Lawrence's pioneering efforts, and still use electric and magnetic fields to accelerate and direct particles.

Lawrence's development of the cyclotron occurred just as a pivotal discovery was unfolding in Europe: the existence of the neutron. This is the topic of the next section.

Fig. 2.19 Left Lawrence's original 4.5-in. cyclotron. Middle Lawrence at the controls of his later 184-in. cyclotron. Right Lawrence, Glenn Seaborg (1912–1999), and Robert Oppenheimer (1904–1967). *Sources* Lawrence Berkeley National Laboratory, courtesy AIP Emilio Segre Visual Archives

Fig. 2.20 M. Stanley Livingston and Ernest O. Lawrence at the Berkeley 27-in. cyclotron. *Source* Lawrence Berkeley National Laboratory, courtesy AIP Emilio Segre Visual Archives

2.2 Discovery of the Neutron

The discovery of the neutron in early 1932 by Ernest Rutherford's protégé, James Chadwick, was a critical turning point in the history of nuclear physics. Within two years, Enrico Fermi would generate artificially-induced radioactivity by neutron bombardment, and five years after that, Otto Hahn, Fritz Strassmann, and Lise Meitner would discover neutron-induced uranium fission. The latter would lead directly to the *Little Boy* uranium-fission bomb, while Fermi's work would lead to reactors to produce plutonium for the *Trinity* and *Fat Man* bombs.

The experiments which led to the discovery of the neutron were first reported in 1930 by Walther Bothe (Fig. 2.21) and his student, Herbert Becker, who were working in Germany. Their research involved studying the gamma radiation which is produced when light elements such as magnesium and aluminum are bombarded by energetic alpha-particles. In such reactions, the alpha particles often interact with a target nucleus to yield a proton and a gamma-ray, as Ernest Rutherford had found when he first achieved an artificially-induced nuclear transmutation:

$$^4_2\text{He} + ^{14}_7\text{N} \rightarrow ^1_1\text{H} + ^{17}_8\text{O} + \gamma. \tag{2.32}$$

The mystery began when Bothe and Becker found that boron, lithium, and particularly beryllium gave evidence of gamma emission under alpha bombardment, *but with no accompanying protons being emitted*. A key point here is that they were certain that some sort of energetic but electrically neutral "penetrating radiation" was being emitted; this radiation could penetrate foils of metal but could not be deflected by a magnetic field as charged particles would be. Gamma-rays were the only electrically neutral form of penetrating radiation known at the time, so it was natural for them to interpret their results as evidence of gamma-ray emission despite the anomalous lack of protons.

Bothe and Becker's beryllium result was picked up by the Paris-based husband-and-wife team of Frédéric Joliot and Irène Curie (the daughter of Pierre and Marie; Fig. 2.3), hereafter referred to as the Joliot-Curies. In January, 1932, they reported that the presumed gamma-ray "beryllium radiation" was capable of knocking protons out of a layer of paraffin wax that had been put in its path. The situation is shown schematically in Fig. 2.22, where the supposed gamma-rays are labeled as "mystery radiation."

At Cambridge, this interpretation struck Chadwick as untenable. He had searched for neutrons for many years with no success, and suspected that Bothe and Becker and the Joliot-Curies had stumbled upon them. He immediately set about to reproduce, re-analyze, and extend their work. In his recreation of the Joliot-Curies' work, Chadwick's experimental setup involved polonium (the alpha source)

Fig. 2.21 Left Walter Bothe (1891–1957); Right James Chadwick (1891–1974). *Sources* Original drawing by Norman Feather, courtesy AIP Emilio Segre Visual Archives; http://commons.wikimedia.org/wiki/File:Chadwick.jpg

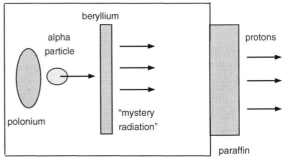

deposited on a silver disk 1 cm in diameter placed close to a disk of pure beryllium
2 cm in diameter, with both enclosed in a small vessel which could be evacuated.
In comparison to the gargantuan particle accelerators of today, these experiments
were literally table-top nuclear physics.

Let us first assume that Bothe and Becker and the Joliot-Curies were correct in
their interpretation that α-bombardment of beryllium creates gamma-rays. To
account for the lack of protons created in the bombardment, the Joliot-Curies
hypothesized that the reaction was

$$^{4}_{2}\text{He} + {}^{9}_{4}\text{Be} \rightarrow {}^{13}_{6}\text{C} + \gamma. \tag{2.33}$$

The Q-value of this reaction is 10.65 MeV. Polonium decay yields alpha particles with kinetic energies of about 5.3 MeV, so the emergent γ-ray can have at
most an energy of about 16 MeV. A more detailed analysis which accounts for the
energy and momentum transmitted to the carbon atom shows that the energy of the
gamma ray comes out to be about 14.6 MeV. The 14.6-MeV gamma-rays then
strike protons in the paraffin, setting them into motion. Upon reproducing the
experiment, Chadwick found that the struck protons would emerge with maximum
kinetic energies of about 5.7 MeV.

The problem, Chadwick realized, was that if a proton was to be accelerated to
this amount of energy by being struck by a gamma-ray, conservation of energy and
momentum demanded that the gamma-ray would have to possess about 54 MeV of
energy, nearly four times what it could have! This strikingly high energy demand
is a consequence of the fact that photons do not possess mass. Relativity theory
shows that massless particles do carry momentum, but much less than a "material"
particle of the same kinetic energy; only an extremely energetic gamma-ray can
kick a proton to a kinetic energy of several MeV. Analyzing a collision between a
photon and a material particle involves relativistic mass-energy and momentum
conservation; details can be found in Reed (2007). The results of such an analysis
show that if a target nucleus of rest-energy E_t (that is, mc^2 equivalent energy) is to
be accelerated to kinetic energy K_t by being struck head-on by a photon of energy
E_γ which then recoils backwards (this transfers maximum momentum to the struck
nucleus), then the energy of the photon must be

$$E_\gamma = \frac{1}{2}\left[K_t + \sqrt{2\,E_t\,K_t}\,\right]. \tag{2.34}$$

For a proton, $E_t \sim 938$ MeV; with $K_t \sim 5.7$ MeV, the value of E_γ works out to about 54 MeV, as claimed above. Remarkably, the Joliot-Curies had realized that this discrepancy was a weak point in their interpretation, but attributed it to the difficulty of accurately measuring the energy of their "gamma rays." Another clue that led Chadwick to suspect a material particle as opposed to a high-energy photon was that the "beryllium radiation" was more intense in the forward direction than in the backward direction; if the radiation was photonic, it should have been of equal intensity in all directions.

Before invoking a mechanism involving a (hypothetical) neutron, Chadwick devised a further test to investigate the remote possibility that 54-MeV gamma-rays could be being created in the α-Be collision. In addition to having the "beryllium radiation" strike protons, he also arranged for it to strike a sample of nitrogen gas. If struck by such a photon, a nucleus of nitrogen should acquire a kinetic energy of about 450 keV [A nitrogen nucleus has a rest energy of about 13,000 MeV; check the consistency of these numbers with Eq. (2.34)]. From prior experience, Chadwick knew that when an energetic particle travels through air it produces ions, with about 35 eV required to produce a single ionization, which yields one pair of ions. A 450 keV nitrogen nucleus should thus generate some (450 keV/35 eV) $\sim$ 13,000 ion pairs. Upon performing the experiment, however, he found that some 30,000–40,000 ion pairs would typically be produced, which implied kinetic energies of about 1.1–1.4 MeV for the recoiling nitrogen nuclei. Such numbers would in turn require the nitrogen nuclei to have been struck by gamma-rays of energy up to $\sim$90 MeV, a value completely inconsistent with the $\sim$54 MeV indicated by the proton experiment. Upon letting the supposed gamma-rays strike heavier and heavier target nuclei, Chadwick found that "if the recoil atoms are to be explained by collision with a quantum, we must assume a larger and larger energy for the quantum as the mass of the struck atom increases." The absurdity of this situation led him to write that, "It is evident that we must either relinquish the application of conservation of energy and momentum in these collisions or adopt another hypothesis about the nature of the radiation."

After refuting the Joliot-Curies' interpretation, Chadwick provided a more physically realistic one. This was that if the protons in the paraffin were in reality being struck by neutral *material* particles of mass equal or closely similar to that of a proton, then the kinetic energy of the striking particles need only be on the order of the kinetic energy that the protons acquired in the collision. As an everyday example, think of a head-on collision between two equal-mass billiard balls: the incoming one stops, and the struck one is set into motion with the speed that the incoming one had. This is the point at which the neutron makes it debut.

Chadwick hypothesized that instead of the Joliot-Curie reaction, the α-Be collision leads to the production of carbon and a neutron via the reaction

$$_2^4\text{He} + {}_4^9\text{Be} \rightarrow {}_6^{12}\text{C} + {}_0^1 n. \tag{2.35}$$

$_0^1 n$ denotes a neutron: it carries no electric charge but it does count as one nucleon. In this interpretation, a ^{12}C atom is produced as opposed to the Joliot-Curies' proposed ^{13}C. Since the "beryllium radiation" was known to be electrically neutral, Chadwick could not invoke a charged particle such as a proton or electron to explain the reaction. Hypothesizing that the neutron's mass was similar to that of a proton (he was thinking of neutrons as being electrically neutral combinations of single protons and single electrons), Chadwick was able to show that the kinetic energy of the ejected neutron would be about 10.9 MeV. A subsequent neutron/proton collision will be like a billiard-ball collision, so it is entirely plausible that a neutron which begins with about 11 MeV of kinetic energy would be sufficiently energetic to accelerate a proton to a kinetic energy of 5.7 MeV, even after the neutron battered its way out of the beryllium target and through the window of the vacuum vessel on its way to the paraffin. As a check on his hypothesis, Chadwick calculated that a neutron of kinetic energy 5.7 MeV striking a nitrogen nucleus should set the latter into motion with a kinetic energy of about 1.4 MeV, which was precisely what he had measured in the ion-pair experiment!

Further experiments with other target substances showed similarly consistent results. Chadwick estimated the mass of the neutron as between 1.005 and 1.008 atomic mass units; the modern figure is 1.00866. The accuracy he obtained with equipment which would now be regarded as primitive is nothing short of awe-inspiring. Chadwick reported his discovery in two papers. The first, titled "Possible Existence of a Neutron," was dated February 17, 1932, and was published in the February 27 edition of *Nature*. An extensive follow-up analysis dated May 10 was published in the June 1 edition of the *Proceedings of the Royal Society of London*. Chadwick was awarded the 1935 Nobel Prize in Physics for his discovery. While later experiments showed that the neutron is a fundamental particle in its own right (as opposed to being a proton/electron composite), that development does not affect the above analysis.

Why is the discovery of the neutron regarded as such a pivotal event in the history of nuclear physics? The reason is that neutrons do not experience any electrical forces, so they experience no Coulomb barrier. With neutrons, experimenters now had a way of producing particles that could be used to bombard nuclei without being repelled by them, no matter what the kinetic energy of the neutron or the atomic number of the target nucleus. It was not long before such experiments were taken up. Neutrons would prove to be the gateway to reactors and bombs, but, at the time, Chadwick anticipated neither development. In the February 29, 1932, edition of the *New York Times*, he is quoted as stating that, "I am afraid neutrons will not be of any use to any one."

About 18 months after Chadwick's dismissal of the value of neutrons, an idea did arise as to a possible application for them: As links in the progression of a nuclear chain reaction. This notion seems to have occurred inspirationally to a Hungarian-born engineer, physicist, and inventor named Leo Szilard (Fig. 2.23), a personal friend and sometimes collaborator of Albert Einstein.

Fig. 2.23 Leo Szilard
(1898–1964) *Source* http://
commons.wikimedia.org/
wiki/File:Leo_Szilard.jpg

Szilard was living in London in the fall of 1933, and happened to read a description of a meeting of the British Association for the Advancement of Science published in the September 12 edition of the London *Times*. In an article describing an address to the meeting by Rutherford on the prospects for reactions that might be induced by accelerated protons, the *Times* quoted Rutherford as stating that, "We might in these processes obtain very much more energy than the proton supplied, but on the average we could not expect to obtain energy in this way. It was a very poor and inefficient way of producing energy, and anyone who looked for a source of power in the transformation of the atoms was talking moonshine." Historian of science John Jenkin has pointed out that Rutherford's private thoughts on the matter may have been very different, however. Some years before World War II, Rutherford evidently advised a high government official that he had a hunch that nuclear energy might one day have a decisive effect on war.

Szilard reflected on Rutherford's remarks while later strolling the streets of London. From a 1963 interview:

> Pronouncements of experts to the effect that something cannot be done have always irritated me. That day as I was walking down Southampton Row and was stopped for a traffic light, I was pondering whether Lord Rutherford might not prove to be wrong. As the light changed to green and I crossed the street, it suddenly occurred to me that if we could find an element which is split by neutrons and which would emit two neutrons when it absorbed one neutron, such an element, if assembled in sufficiently large mass, could sustain a nuclear chain reaction, liberate energy on an industrial scale, and construct atomic bombs. The thought that this might be possible became an obsession with me. It led me to go into nuclear physics, a field in which I had not worked before, and the thought stayed with me.

It did not take Szilard long to get up to speed in his new area. Envisioning a chain reaction as a source of power and possibly as an explosive, he filed for patents on the idea in the spring and summer of 1934. His British patent number 630,726, "Improvements in or relating to the Transmutation of Chemical Elements," was issued on July 4, 1934 (curiously, the date of Marie Curie's death), and referred specifically to being able to produce an explosion given a sufficient mass of material. To keep the idea secret, Szilard assigned the patent to the British Admiralty in February, 1936. The patent was reassigned to him after the war, and finally published in 1949.

2.3 Artificially-Induced Radioactivity

Irène and Frédéric Joliot-Curie must have been deeply disappointed at their failure to detect the neutron in early 1932, but scored a success almost exactly two years later when they discovered that normally stable nuclei could be induced to become radioactive upon alpha-particle bombardment. In early 1934, the Joliot-Curies were performing some follow-up experiments involving bombarding thin foils of aluminum with alpha-particles emitted by decay of polonium, the same source of alphas used in the neutron-discovery reaction. To their surprise, their Geiger counter continued to register a signal after the source of the alpha particles was removed. The signal decayed with a half-life of about 3 min. Performing the experiment in a magnetic field led them to conclude that positrons were being emitted, that is, that β^+ decays were occurring.

They proposed a two-stage reaction to explain their observations. First was formation of phosphorous-30 by alpha-capture and neutron emission:

$$^{4}_{2}\text{He} + {}^{27}_{13}\text{Al} \rightarrow {}^{1}_{0}n + {}^{30}_{15}\text{P}. \tag{2.36}$$

The phosphorous-30 nucleus subsequently undergoes positron decay to silicon; the modern value for the half-life is 2.5 min (the emitted beta-particle is omitted here; it is the decay product that is important):

$$^{30}_{15}\text{P} \xrightarrow[\text{2.5 min}]{\beta^+} {}^{30}_{14}\text{Si}. \tag{2.37}$$

To be certain of their interpretation, the Joliot-Curies dissolved the bombarded aluminum in acid; the small amount of phosphorous created could be separated and chemically identified as such. That the radioactivity "carried with" the separated phosphorous and not the aluminum verified their suspicion. Bombardment of boron and magnesium showed similar effects. They first observed the effect on January 11, 1934, and reported it in the January 15 edition of the journal of the French Academy of Sciences; an English version appeared in the February 10 edition of *Nature*. The discovery of artificially-induced radioactivity opened up the whole field of synthesizing short-lived isotopes for medical treatments. Emilio Segrè, one of Enrico Fermi's students, described this development as one of the most important discoveries of the century.

Induced radioactivity had almost been discovered in California, where Ernest Lawrence's cyclotron operators often noticed that their detectors kept registering a signal after the cyclotron had been shut down following bombardment experiments. Thinking that the detectors were misbehaving, they arranged circuitry to shut them down simultaneously with the cyclotron. The history of nuclear physics, particularly events surrounding the discovery of fission, is replete with such missed chances.

2.4 Enrico Fermi and Neutron-Induced Radioactivity

Surprisingly, neither the Joliot-Curies nor James Chadwick particularly experimented with using neutrons as bombarding particles. Norman Feather, one of Chadwick's collaborators, did carry out some experiments with light elements, and found that neutrons would disintegrate nitrogen nuclei to produce an alpha-particle and a boron nucleus:

$$\ _0^1 n + \ _7^{14}N \rightarrow \ _2^4 He + \ _5^{11}B. \tag{2.38}$$

The same type of reaction also occurs with elements such as oxygen, fluorine, and neon, but apparently neither British nor French researchers carried out experiments with heavy-element targets.

The idea of systematically using neutrons as bombarding particles did occur to a physicist at the University of Rome, Enrico Fermi (Fig. 2.24). Fermi had established himself as a first-rate theoretical physicist at a young age, publishing his first paper while still a student. As a postdoctoral student with quantum mechanist Max Born, Fermi had prepared an important review article on relativity theory while in his early twenties, and a few years later made seminal contributions to statistical mechanics. At the young age of 26 he was appointed to a full professorship at the University of Rome, and in late 1933 he developed a quantum–mechanically-based theory of beta decay. He was to prove equally gifted as a nuclear experimentalist.

The reticence of Chadwick and the Joliot-Curies to carry out neutron-bombardment experiments may seem strange, but was understandable in view of the low yields expected. Chadwick estimated that he produced only about 30 neutrons for every million alpha-particles emitted by his sample of polonium. If the neutrons interacted with target nuclei with similarly low yields, virtually nothing could be expected to result. Otto Frisch, one of the co-interpreters of fission, later remarked that, "I remember that my reaction and probably that of many others was that Fermi's was a silly experiment because neutrons were so

Fig. 2.24 Left Enrico Fermi (1901–1954). Right The Fermi family (Laura, Giulio, Nella and Enrico) arrive in America, January, 1939. *Sources* University of Chicago, courtesy AIP Emilio Segre Visual Archives; AIP Emilio Segre Visual Archives, Wheeler Collection

much fewer than alpha particles." But this overlooked the fact that neutrons would not experience a Coulomb barrier.

Fermi desired to break into nuclear experimentation, and saw his opening in this under-exploited possibility. He began work in the spring of 1934 with a group of students and collaborators including Edoardo Amaldi, Franco Rasetti, chemist Oscar D'Agostino, and Emilio Segrè (Fig. 2.25), who would later write an very engaging biography of Fermi, titled *Enrico Fermi: Physicist.*

In the early 2000s a group of Italian historians, Giovanni Acocella, Francesco Guerra, Matteo Leone, and Nadia Robotti found Fermi's original laboratory notebooks (and some of his neutron sources!) from the spring of 1934, so there is now available a very detailed record of his work. Much of the material in this section is adapted from their analysis of Fermi's notes.

Fermi's first challenge was to secure a strong neutron source. In this sense he was fortunate in that his laboratory was located in the same building as the Physical Laboratory of the Institute of Public Health, which was charged with

Fig. 2.25 Some of Fermi's collaborators. Left to right Oscar D'Agostino (1901–1975), Emilio Segrè (1905–1989), Edoardo Amaldi (1908–1989), and Franco Rasetti (1901–2001). *Source* Agenzia Giornalistica Fotovedo, courtesy AIP Emilio Segre Visual Archives

controlling radioactive substances in Italy. The Laboratory held many radium sources that had been used for cancer treatments, and Fermi used them as a source of radon gas. When mixed with powdered beryllium, the radon gave rise to a copious supply of neutrons. Radon is produced in the decay of radium,

$$\underset{88}{226}\text{Ra} \xrightarrow[1,599 \text{ year}]{\alpha} \underset{86}{222}\text{Rn} + \underset{2}{4}\text{He}. \tag{2.39}$$

The radon daughter product has a very short half-life, which means a correspondingly great flux of alpha-particles from the decay

$$\underset{86}{222}\text{Rn} \xrightarrow[3.82 \text{ days}]{\alpha} \underset{84}{218}\text{Po} + \underset{2}{4}\text{He}. \tag{2.40}$$

After being harvested from the decaying radium, the radon gas was captured in inch-long glass vials which contained powdered beryllium. The radon-produced alphas in (2.40) then gave rise to neutrons via the same reaction that Bothe and Becker, the Joliot-Curies, and Chadwick had experimented with:

$$\underset{2}{4}\text{He} + \underset{4}{9}\text{Be} \rightarrow \underset{6}{12}\text{C} + \underset{0}{1}n. \tag{2.41}$$

This series of reactions yields neutrons with energies of up to about 10 MeV, more than energetic enough to escape through the thin walls of the glass vials and so bombard a sample of a target element. Fermi estimated that his sources yielded about 100,000 neutrons per second. Because the neutrons generated by his radon-beryllium sources tended to be emitted in all directions, Fermi usually formed samples of the target elements to be investigated into cylinders which could be placed around the sources in order to achieve maximum exposure. The cylinders were made large enough so that after being irradiated they could be slipped around a small handmade Geiger counter.

Fermi's goal was to see if he could induce artificial radioactivity with neutron bombardment. Possibly anxious to see if he could induce *heavy-element* radioactivity, his first target was the heavy element platinum (atomic number 78). Fifteen minutes of irradiation gave no discernible signal. Perhaps inspired by the Joliot-Curies' experience, he then turned to aluminum. Here he did succeed, and found a different half-life than they had. The reaction involved ejection of a proton from the bombarded aluminum, leaving behind magnesium,

$$\underset{0}{1}n + \underset{13}{27}\text{Al} \rightarrow \underset{1}{1}\text{H} + \underset{12}{27}\text{Mg}. \tag{2.42}$$

The magnesium beta-decays back to aluminum with a half-life of about 10 min:

$$\underset{12}{27}\text{Mg} \xrightarrow[9.5 \text{ min}]{\beta^-} \underset{13}{27}\text{Al}. \tag{2.43}$$

After aluminum, Fermi tried lead, but with negative results. His next attempt was with fluorine, irradiation of which produced a very short-lived heavier isotope of that element:

$$_0^1n + {}_9^{19}\text{F} \rightarrow {}_9^{20}\text{F} \xrightarrow[11\,\text{s}]{\beta^-} {}_{10}^{20}\text{Ne}. \tag{2.44}$$

Guerra and Robotti have pinpointed the date of Fermi's first success with aluminum as having occurred on March 20, 1934. Fermi announced his discovery five days later in the official journal of the Italian National Research Council, and an English-language report dated April 10 appeared in the May 19 edition of *Nature*. By late April, the Rome group had performed experiments on about 30 elements, 22 of which yielded positive results, including the four medium-weight elements antimony ($Z = 51$), iodine (53), barium (56), and lanthanum (57).

Fermi and his co-workers found that, as a rule, light elements exhibited three reaction *channels*: a proton or an alpha could be ejected, or the element might simply capture the neutron to become a heavier isotope of itself and then subsequently decay. In all three cases, the products would undergo β^- decay. Aluminum is typical in this regard:

$$_0^1n + {}_{13}^{27}\text{Al} \rightarrow \begin{cases} {}_1^1\text{H} + {}_{12}^{27}\text{Mg} & \xrightarrow[9.5\,\text{min}]{\beta^-} {}_{13}^{27}\text{Al} \\[2mm] {}_2^4\text{He} + {}_{11}^{24}\text{Na} & \xrightarrow[15\,\text{h}]{\beta^-} {}_{12}^{24}\text{Mg} \\[2mm] {}_{13}^{28}\text{Al} & \xrightarrow[2.25\,\text{min}]{\beta^-} {}_{14}^{28}\text{Si}. \end{cases} \tag{2.45}$$

With a *heavy*-element target, the result is typically the latter of the above channels. Gold is characteristic in this regard:

$$_0^1n + {}_{79}^{197}\text{Au} \rightarrow {}_{79}^{198}\text{Au} \xrightarrow[2.69\,\text{days}]{\beta^-} {}_{80}^{198}\text{Hg}. \tag{2.46}$$

By the early summer of 1934, Fermi had prepared improved sources, which he estimated were yielding about a million neutrons per second. Based on work with these new sources, he published a stunning result in the June 16, 1934, edition of *Nature*: that his group was producing *transuranic* elements, that is, ones with atomic numbers greater than that of uranium. Since uranium was the heaviest-known element, this meant that they believed that they were synthesizing new elements. If true, this would be a remarkable development.

Fermi's radical assertion was based on the fact that uranium could be activated to produce beta-decay upon neutron bombardment. The results were complex, however, with evidence for half-lives of 10 s, 40 s, 13 min, and at least two more half-lives of up to one day. Whether this was a chain of decays or some sort of parallel sequence was unknown. Whatever sequence was occurring, however, the initial step was presumably the formation of a heavy isotope of uranium, followed by a beta-decay as in the gold reaction above:

$$_0^1n + {}_{92}^{238}\text{U} \rightarrow {}_{92}^{239}\text{U} \xrightarrow{\beta^-} {}_{93}^{239}X, \tag{2.47}$$

where X denotes a new, transuranic element.

The 13-min decay was convenient to work with, and the Rome group managed to separate chemically its decay product from the bombarded uranium. Analysis showed that the decay product did not appear to be any of the elements between lead ($Z = 82$) and uranium. Since no natural or artificial transmutation had ever been observed to change the identity of a target element by more than one or two places in the periodic table, it would have seemed perfectly plausible to assume that a new element was being created.

To isolate the product of the 13-min activity, Fermi and his group began with manganese dioxide as a chemical carrier. The rationale for this was that if element 93 were actually being created, it was expected that it would fall in the same column of the periodic table as manganese ($Z = 25$), and so the two should have similar chemistry (see Fig. 3.2). The Romans' analysis came in for criticism, however, from a German scientist, Ida Noddack (Fig. 2.26). Noddack was a well-regarded chemist who in 1925 had participated in the discovery of rhenium; she would be nominated for a Nobel Prize on three occasions. In a paper published in September, 1934, Noddack criticized Fermi on the grounds that numerous elements were known to precipitate with manganese dioxide, and that he should have checked for the possibility that elements of lower atomic numbers than that of lead were being produced. In what would prove to be a prescient comment, Noddack remarked that, "When heavy nuclei are bombarded by neutrons, it is conceivable that the nucleus breaks up into several large fragments, which would of course be isotopes of known elements but would not be neighbors of the irradiated element." Noddack's "breaking up" is now known as "fission."

Noddack was ahead of her time in suggesting that heavy nuclei might fission. Ironically, Fermi was probably both inducing fissions *and* creating transuranic elements. Nuclei of the most common isotope of uranium, U-238 (>99 % of natural uranium), are fissile when bombarded by the very fast neutrons that Fermi was using, but when struck by slow neutrons tend to capture them and subsequently decay to neptunium and plutonium. These processes will be discussed at length in subsequent sections.

That Noddack's idea was not taken seriously has sometimes been construed as an example of blatant sexism. But the reasons were much more prosaic. She offered no supporting calculations of the energetics of such a proposed splitting, and years of experience with nuclear reactions had always yielded products that were near the bombarded elements in atomic number. Nobody had any reason to anticipate such a splitting. Otto Frisch thought Noddack's paper was "carping criticism." In any case, by the summer of 1934, Fermi's group had developed an improved rhenium-based chemical analysis of the 13-min uranium activation which appeared to strengthen the transuranic interpretation.

Fermi's next discovery would prove pivotal to the eventual development of plutonium-based nuclear weapons. In the fall of 1934, his group decided that they needed to more precisely quantify their assessments of activities induced in various elements; previously they had assigned only qualitative "strong-medium-

Fig. 2.26 Ida Noddack
(1896–1978) *Source* http://
commons.wikimedia.org/
wiki/File:Ida_Noddack-
Tacke.png

weak" designations. As a standard of activity, they settled on a 2.4-min half-life induced in silver:

$$\ _{0}^{1}n \ + \ _{47}^{107}\mathrm{Ag} \ \rightarrow \ _{47}^{108}\mathrm{Ag} \ \underset{2.39\,\mathrm{min}}{\overset{\beta^{-}}{\longrightarrow}} \ _{48}^{108}\mathrm{Cd}. \tag{2.48}$$

However, they soon ran into a problem: the activity induced in silver seemed to depend on where in the laboratory the sample was irradiated. In particular, silver irradiated on a wooden table became much more active then when irradiated on a marble-topped one. To try to discern what was happening, a series of calibration experiments was undertaken, some of which involved investigating the effects of "filtering" neutrons by interposing layers of lead between the neutron source and the target sample.

Fermi made the key breakthrough on October 22, 1934: "One day, as I came into the laboratory, it occurred to me that I should examine the effect of placing a piece of lead before the incident neutrons. Instead of my usual custom, I took great pains to have the piece of lead precisely machined. I was clearly dissatisfied with something; I tried every excuse to postpone putting the piece of lead in its place. When finally, with some reluctance, I was going to put it in its place, I said to myself: "No, I don't want this piece of lead here; what I want is a piece of paraffin." It was just like that with no advance warning, no conscious prior reasoning. I immediately took some odd piece of paraffin and placed it where the piece of lead was to have been."

To Fermi's surprise, the presence of the paraffin caused the level of induced radioactivity to increase. Further experimentation showed that the effect was characteristic of filtering materials which contained hydrogen; paraffin and water were most effective. Within a few hours of the discovery, Fermi developed a working hypothesis: That by being slowed by collisions with hydrogen nuclei, the neutrons would have more time in the vicinity of target nuclei to induce a reaction. Neutrons and protons have essentially identical masses, and, as with a billiard-ball collision, a head-on strike would essentially bring a neutron to a stop. Since atoms always have random motions due to being at a temperature that is above absolute zero, the incoming neutrons will never be brought to dead stops, but in practice only a few centimeters of paraffin or water are needed to bring them to an average speed characteristic of the temperature of the slowing medium. This process is now called "thermalization." Nuclear physicists define "thermal" neutrons as having kinetic energy equivalent to a temperature of 298 K, or 77 °F—not much warmer than the average daily temperature in Rome in October. The speed of a thermal neutron is about 2,200 m/s, and the corresponding kinetic energy is about 0.025 eV, much less than the ~ 10 MeV of Fermi's radon-beryllium neutrons. Thermal neutrons are also known as "slow" neutrons; those of MeV-scale kinetic energies are, for obvious reasons, termed "fast." The water or paraffin is now known as a "moderator"; graphite (crystallized carbon) also works well in this respect. Fermi's wooden lab bench, by virtue of its water content, was a more effective moderator than was his marble-topped one.

Be sure to understand what is meant by "fast" and "slow" neutrons. When uranium is bombarded by neutrons, what happens depends very critically on the kinetic energies of the neutrons. Fast and slow neutrons lie at the heart of why nuclear reactors and bombs function differently, and why a bomb requires "enriched" uranium to function. This is a complex topic with a number of interconnecting aspects; the following chapter is devoted to a detailed analysis of the ramifications of this fast-versus-slow issue.

Following Fermi's serendipitous discovery, his group began re-investigating all elements which they had previously subjected to *fast* (energetic) neutron bombardment. Extensive results were reported in a paper published in the spring of 1935. For some target elements, the effect was dramatic: activity in vanadium and silver were increased by factors of 40 and 30, respectively, over that achieved by unmoderated neutrons. Uranium also showed increased activation, by a factor of about 1.6.

Fermi's hypothesis that slower neutrons have a greater chance of inducing a reaction is now quantified in the concept of a reaction *cross-section*. This is a measure of the cross-sectional area that a target nucleus effectively presents to a bombarding particle that results in a given reaction. Because of a quantum–mechanical effect known as the de Broglie wavelength, a target nucleus will *appear* larger to a slower bombarding particle than to a faster one, sometimes by factors of hundreds. Each possible reaction channel for a target nucleus will have its own characteristic run of cross-section as a function of bombarding-particle energy.

Cross-sections are designated with the Greek letter sigma (σ), equivalent to the English letter "s," which serves as a reminder that they have units of surface area. The fundamental unit of cross-section is the "barn"; 1 bn $= 10^{-28}$ m^2. This miniscule number is characteristic of the geometric cross-sectional area of nuclei, which is given approximately in terms of the mass number by the empirical relationship

$$\textit{Geometric cross–section} \sim 0.0452\, A^{2/3} \text{ (barns)}. \tag{2.49}$$

As an example, Fig. 2.27 shows the "radiative capture cross-section" for aluminum-27 when bombarded by neutrons of energies from 10^{-11} to 10 MeV. (Al-27 is the one stable isotope of that element.) In this reaction, the aluminum absorbs the neutron, sheds some energy via a gamma-ray, and eventually decays to silicon via the last branch of the three-channel reaction in (2.45); both "capture" and "radiation" occur, hence the name of the cross-section. Both scales of the graph are logarithmic; this is done in order to accommodate a wide range of energies and cross-sections. Thermal neutrons have log (Energy) ~ -7.6 when the energy is measured in MeV. For an Al-27 nucleus, the geometric cross-section is about 0.407 barns, or log(area) $= -3.91$. For fast neutrons, reaction cross-sections are typically of the size of the geometrical cross-sectional area of the target nucleus.

The spikes in Fig. 2.27 are known as "resonance capture lines." Just as atomic orbital electrons can be excited to different energy levels, so can the protons and neutrons within nuclei; resonance energies correspond to the bombarding particles having just the right energies to excite nucleons to higher energy levels. As the number of nucleons grows, so does the complexity of the structure of the resonance spikes; for a more dramatic example, see the graph for the uranium-238 neutron capture cross-section in Fig. 3.11.

Fermi was awarded the 1938 Nobel Prize for Physics for his demonstration of the existence of new radioactive elements produced by neutron irradiation. Fermi's

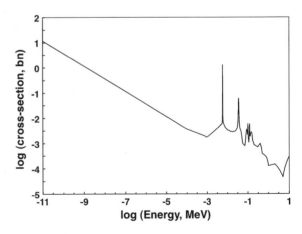

Fig. 2.27 Radiative capture cross-section for neutrons on aluminum-27

Fig. 2.28 Arthur Dempster (1886–1950). *Source* http://commons.wikimedia.org/wiki/File:Arthur_Jeffrey_Dempster_-_Portrait.jpg

wife and children were Jewish, and he and his family used the excuse of the trip to Stockholm to escape the rapidly deteriorating fascist political situation in Italy by subsequently emigrating to America, where he had arranged for a position at Columbia University. The American branch of the Fermi family was established on January 2, 1939.

Before proceeding to the story of the discovery of nuclear fission, a brief but important intervening discovery needs to be mentioned. This is that uranium possesses a second, much less abundant isotope than the U-238 that Fermi had assumed was the sole form of that element. In 1931, Francis Aston had run uranium hexafluoride through his mass spectrometer and concluded that only an isotope of mass number 238 was present. In the summer of 1935, Arthur Dempster of the University of Chicago (Fig. 2.28) discovered evidence for a lighter isotope of mass number 235. Dempster estimated U-235 to be present to an extent of less than one percent of the abundance of its sister isotope of mass 238. Within a few years that one percent would prove very important indeed.

2.5 Another Look at Mass Defect and Binding Energy (Optional)

In Sect. 2.1.4, the concepts of mass defect and binding energy were treated as interchangeable. They are, however, separate but related quantities. The strict definition of binding energy is described in this section, which can be considered optional.

Atomic masses are measured in terms of *mass units*. Abbreviated simply as *u*, the mass unit is defined as one-twelfth of the mass of a neutral carbon-12 atom. Chemists will know the mass unit as a "Dalton," and older readers will be more familiar with the term *atomic mass unit* (*amu*). The numerical value is

$$1\,u = 1.660539 \times 10^{-27}\,\text{kg}. \tag{2.50}$$

The masses of the proton, neutron, and electron in mass units are

$$m_p = 1.00727646677 \, u \tag{2.51}$$

$$m_n = 1.00866491597 \, u \tag{2.52}$$

$$m_e = 5.4857990943 \times 10^{-4} \, u \tag{2.53}$$

An important conversion factor here is that the energy equivalent of one mass unit is 931.494 MeV; this comes from $E = mc^2$. Give this value the symbol ε:

$$\varepsilon = 931.494 \, \text{MeV}. \tag{2.54}$$

A neutral atom comprises Z protons, Z electrons, and N neutrons. In the notation of (2.51)–(2.53), one would naively expect the mass of the assembled atom to be equal to $Z(m_p + m_e) + N(m_n)$ mass units. However, naturally-occurring "assembled" neutral atoms are always lighter than what this argument predicts. The *binding energy* E_B of an atom is defined as the difference between the naively-predicted mass and the "true" measured mass m_U of the assembled atom (in mass units), all expressed as an energy equivalent:

$$E_B = \left[Z(m_p + m_e) + N(m_n) - m_U \right] \varepsilon. \tag{2.55}$$

Substituting (2.51)–(2.54) into (2.55) gives the binding energy as

$$E_B = \left[938.783 \, Z + 939.565 \, N - 931.494 \, m_U \right] \text{MeV}. \tag{2.56}$$

As an example, for iron-56 ($m_U = 55.934937$):

$$E_B = 938.783(26) + 939.565(30) - 931.494(55.934937) = 492.2 \, \text{MeV}.$$

To correct a misleading statement in Sect. 2.1.4, it is this binding energy that holds nuclei together, not the mass-defect; a positive mass defect does not by itself connote instability. Stable atoms will have positive E_B values, but, conversely, a positive E_B value does not necessarily denote intrinsic stability. For example, uranium-235 has $E_B = 1{,}784$ MeV, but is unstable against alpha-decay, which is fundamentally a quantum-mechanical effect that cannot be understood on the basis of energy considerations alone.

For heavy elements, E_B values are large. To display them graphically it is more convenient to plot the binding emery *per nucleon*, E_B/A, versus the nucleon number A. This is shown in Fig. 2.29 for the same 350 nuclides as in Fig. 2.8. This plot immediately tells us that for $A > \sim 25$, each nucleon in a nucleus is "glued" into the structure of the nucleus to the extent of about 8 MeV/nucleon. This plot is known as "the curve of binding energy." For Fe-56, $E_B/A = 8.79$ MeV/nucleon.

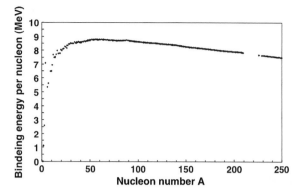

Fig. 2.29 Binding energy curve for 350 stable and quasi-stable nuclei

Exercises

2.1 Consider an element with atomic weight A grams per mole and density ρ grams per cubic centimeter. If atoms are imagined to be hard spheres of radius R packed edge-to edge, each atom will effectively occupy a cube of volume $8R^3$. Show that R can be expressed approximately as

$$R \sim 0.59(A/\rho)^{1/3}\text{Å}.$$

Apply this result to lithium; $(\rho, A) = (0.534 \text{ g/cm}^3, 7 \text{ g/mol})$, and uranium; $(\rho, A) = (18.95 \text{ g/cm}^3, 238 \text{ g/mol})$. [Ans: $R \sim 1.4$ Å in each case]

2.2 Show that a 100 food-calorie snack is equivalent to about 2.6×10^{18} MeV.

2.3 Take radium to have an atomic weight of 226 g/mol. The energy of each radium alpha-decay is 4.78 MeV. If all of the atoms in one gram of radium decay and all of that energy could be used to lift a mass m to a height $h = 1$ mile $= 1,609$ m, how much mass could be so lifted (hint: mgh)? You will find Rutherford's estimate of 500 t to be optimistic, but the answer, 129,000 kg ~ 143 t, is still impressive.

2.4 Consider one gram of freshly-isolated radium-226 ($t_{1/2} = 1,599$ year). If each alpha-decay liberates 4.78 MeV, how much energy will be emitted in the first year after the sample is isolated? [Ans: 884 kJ]

2.5 A serving of a sports drink contains 50 mg of potassium to help athletes restore their electrolyte levels. However, one naturally-occurring isotope of potassium, K-40, is a beta-decayer with a half-life of 1.25 billion years. This isotope is present to the level of 1.17 % in natural potassium. If the average atomic weight of potassium is 39.089 g/mol, what level of beta-activity will you consume with one serving—at least until you excrete it? [Ans: 158 decay/s]

2.6 Given the empirical relationship $R \sim a_O A^{1/3}$ between nucleon number and nuclear radius with $a_O = 1.2 \times 10^{-15}$ m, verify Eq. (2.49) for the geometric cross-section of a nucleus.

2.7 In a particle accelerator, it is desired to fire calcium atoms, (Z, A) = (20, 40), into a stationary uranium target, (Z, A) = (92, 238), in an effort to synthesize nuclei of high atomic number. What Coulomb barrier will have to be overcome? [Ans: 230 MeV]

Further Reading

Books, Journal Articles, and Reports

G. Acocella, F. Guerra, N. Robotti, Enrico Fermi's discovery of neutron-induced artificial radioactivity: The recovery of his first laboratory notebook. Phys. Perspect. **6**(1), 29–41 (2004)

E. Amaldi, O. D'Agostino, E. Fermi, B. Pontecorvo, F. Rasetti, E. Segrè, Artificial radioactivity produced by neutron bombardment—II. Proc. Roy. Soc. Lond. Ser. A **149**(868), 522–558 (1935)

F. Aston, Neon. Nature **104**(2613), 334 (1919)

F. Aston, A positive ray spectrograph. Phil. Mag. Ser. 6, 38(228), 707–714 (1919)

F. Aston, The constitution of atmospheric neon. Phil. Mag. Ser. 6, 39(232), 449–445 (1920)

F. Aston, The mass spectra of chemical elements. Phil. Mag. Ser. 6, 38(233), 611–625 (1920)

F. Aston, Constitution of thallium and uranium. Nature **128**(3234), 725 (1931)

L. Badash, Radioactivity before the Curies. Am. J. Phys. **33**(2), 128–135 (1965)

L. Badash, The discovery of thorium's radioactivity. J. Chem. Educ. **43**(4), 219–220 (1966)

L. Badash, The discovery of radioactivity. Phys. Today **49**(2), 21–26 (1996)

L. Badash, J.O. Hirschfelder, H.P. Broida (eds.), *Reminiscences of Los Alamos 1943–1945* (Reidel, Dordrecht, 1980)

H. Becquerel, Sur les radiations émises par phosphorescents. Comptes Rendus **122**, 420–421 (1896a)

H. Becquerel, Sur les radiationes invisibles émises par les corps phosphorescents. Comptes Rendus **122**, 501–503 (1896b)

H. Becquerel, Contribution à l'étude du rayonnement du radium. Comptes Rendus **130**, 120–126 (1900a)

H. Becquerel, Déviation du rayonnement du radium dans un champ électrique. Comptes Rendus **130**, 809–815 (1900b)

B. Boltwood, Note on a new radio-active element. Am. J. Sci. Ser. **4**(24), 370–372 (1907)

W. Bothe, H. Becker, Künstliche Erregung von Kern-γ-Strahlen. Z. Angew. Phys. **66**(5/6), 289–306 (1930)

A. Brown, *The Neutron and the Bomb: A Biography of Sir James Chadwick* (Oxford University Press, Oxford, 1997)

D.C. Cassidy, *A Short History of Physics in the American Century* (Harvard University Press, Cambridge, 2011)

J. Chadwick, Possible Existence of a Neutron. Nature **129**(3252), 312 (1932a)

J. Chadwick, The Existence of a Neutron. Proc. Roy. Soc. Lond. **A136**, 692–708 (1932b)

Curie, I., F. Joliot, F.: Émission de protons de grande vitesse par les substances hydrogénnés sous l'influence des rayons g très pénétrants. Comptes Rendus 194, 273-275 (1932)

Curie, I., F. Joliot, F.: Un nouveau type de radioactivite. Comptes Rendus 198, 254-256 (1934)

Curie, P., Curie, S.: Sur une substance nouvelle radio-active, contenue dans la pitchblende. Comptes Rendus 127, 175-178 (1898). Marie Curie's initial appears here as "S," as in "Sklodowski."

Curie, P., Curie, Mme. P., Bémont, G.: Sur une nouvelle substance fortement radio-active, contenue dans la pitchblende. Comptes Rendus 127, 215-1217 (1898)

P. Curie, A. Laborde, Sur la chaleur dégagée spontanément par les sels de radium. Comptes Rendus **136**, 673–675 (1903)

A.J. Dempster, Isotopic constitution of uranium. Nature **136**(3431), 180 (1935)

N. Feather, Collision of neutrons with nitrogen nuclei. Proc. Roy. Soc. Lond. **136A**(830), 709–727 (1932)

B.T. Feld, G.W. Szilard, K. Winsor, The collected works of Leo Szilard, vol. I (Scientific Papers MIT Press, London, 1972)

E. Fermi, Radioactivity induced by neutron bombardment. Nature **133**(3368), 757 (1934a)

E. Fermi, Possible production of elements of atomic number higher than 92. Nature **133**(3372), 898–899 (1934b)

E. Fermi, E. Amaldi, O. D'Agostino, F. Rasetti, E. Segrè, Artificial radioactivity produced by neutron bombardment. Proc. Roy. Soc. Lond. Ser. A **146**(857), 483–500 (1934)

O.R. Frisch, The discovery of fission. Phys. Today **20**(11), 43–52 (1967)

H. Geiger, E. Marsden, On a diffuse reflection of the α-particles. Proc. Roy. Soc. **A82**(557), 495–500 (1909)

F. Guerra, M. Leone, N. Robotti, Enrico Fermi's discovery of neutron-induced artificial radioactivity: Neutrons and neutron sources. Phys. Perspect. **8**(3), 255–281 (2006)

F. Guerra, N. Robotti, Enrico Fermi's discovery of neutron-induced artificial radioactivity: The influence of his theory of beta decay. Phys. Perspect. **11**(4), 379–404 (2009)

F. Guerra, M. Leone, N. Robotti, The discovery of artificial radioactivity. Phys. Perspect. **14**(1), 33–58 (2012)

J. Hughes, 1932: The annus mirabilis of nuclear physics. Phys. World **13**, 43–50 (2000)

J.G. Jenkin, Atomic energy is "moonshine": What did Rutherford really mean? Phys. Perspect. **13**(2), 128–145 (2011)

F. Joliot, I. Curie, Artificial production of a new kind of radio-element. Nature **133**(3354), 201–202 (1934)

Knolls Atomic Power Laboratory, Nuclides and Isotopes Chart of the Nuclides, 17th edn. http://www.nuclidechart.com/

H. Kragh, Rutherford, radioactivity, and the atomic nucleus. http://arxiv.org/abs/1202.0954

F. Kuhn, Jr, Chadwick calls neutron 'difficult catch'; his find hailed as aid in study of atom. New York Times, Feb 29, 1932, pp. 1, 8

W. Lanouette, B. Silard, *Genius in the Shadows: A Biography of Leo Szilard. The Man Behind the Bomb* (University of Chicago Press, Chicago, 1994)

E.O. Lawrence, N.E. Edlefsen, On the production of high speed protons. Science **72**(1867), 376 (1930)

E.O. Lawrence, D.H. Sloan, The production of high speed mercury ions without the use of high voltages. Phys. Rev. **37**, 231 (1931)

E.O. Lawrence, M.S. Livingston, A method for producing high speed hydrogen ions without the use of high voltages. Phys. Rev. **37**, 1707 (1931)

D. Neuenschwander, The discovery of the nucleus. Radiations 17(1), 13–22 (2011). http://www.spsnational.org/radiations/

D. Neuenschwander, The discovery of the nucleus, part 2: Rutherford scattering and its aftermath. The SPS Observer XLV(1), 11–16 (2011). http://www.spsobserver.org/

J.W. Nicholson, The spectrum of nebulium. Mon. Not. R. Astron. Soc. **72**, 49–64 (1911)

I. Noddack, Über das Element 93. Zeitschrift fur Angewandte Chemie 47(37), 653–655 (1934). An English translation prepared by H. G. Graetzer is available at http://www.chemteam.info/Chem-History/Noddack-1934.html

R. Peierls, *Bird of Passage: Recollections of a Physicist* (Princeton University Press, Princeton, 1985)

F. Perrin, Calcul relative aux conditions éventuelles de transmutation en chaine de l'uranium. Comptes Rendus **208**, 1394–1396 (1939)

D. Preston, *Before the Fallout: From Marie Curie to Hiroshima* (Berkley Books, New York, 2005)

B.C. Reed, Chadwick and the discovery of the neutron. Radiations, 13(1), 12–16 (Spring 2007)

B.C. Reed, *The Physics of The Manhattan Project* (Springer, Berlin, 2011)

R. Rhodes, *The Making of the Atomic Bomb* (Simon and Schuster, New York, 1986)

E. Rutherford, Uranium radiation and the electrical conduction produced by it. Phil. Mag. Ser. 5, xlvii, 109–163 (1899)

E. Rutherford, A radio-active substance emitted from thorium compounds. Phil. Mag. Ser. 5, 49(296), 1–14 (1900)

E. Rutherford, The cause and nature of radioactivity—part I. Phil. Mag. Ser. 6, 4(21), 370–396 (1902)

E. Rutherford, F. Soddy, Radioactive change. Phil. Mag. Ser. 6, v, 576–591 (1903)

E. Rutherford, H. Geiger, An electrical method of counting the number of α particles from radioactive substances. Proc. Roy. Soc. Lond. **81**(546), 141–161 (1908)

E. Rutherford, T. Royds, The nature of the α particle from radioactive substances. Phil. Mag. Ser. 6, xvii, 281–286 (1909)

E. Rutherford, The scattering of α and β particles by matter and the structure of the atom. Phil. Mag. Ser. 6, xxi, 669–688 (1911)

E. Rutherford, Collision of a particles with light atoms. IV. An anomalous effect in nitrogen. Phil. Mag. Ser. 6, 37, 581–587 (1919)

E. Segrè, *Enrico Fermi* (Physicist University of Chicago Press, Chicago, 1970)

F. Soddy, Intra-atomic charge. Nature **92**(2301), 399–400 (1913)

G. Squires, Francis Aston and the mass spectrograph. J. Chem. Soc. Dalton Trans. **23**, 3893–3899 (1998)

J.J. Thomson, Cathode rays. Phil. Mag. Ser. 5, 44(269), 293–316 (1897)

J.J. Thomson, On rays of positive electricity. Phil. Mag. Ser. 6, 13(77), 561–575 (1907)

S. Weinberg, *The Discovery of Subatomic Particles* (Cambridge University Press, Cambridge, 2003)

R. Wideröe, Uber ein neues Prinzip zur Herstellung hoher Spannungen. Arkiv fur Electrotechnik **21**(4), 387–406 (1928)

Websites and Web-based Documents

Lawrence Berkeley National Laboratory. http://www.lbl.gov/Science-Articles/Research-Review/Magazine/1981/81fchp1.html

McGill University website on Rutherford. http://www.physics.mcgill.ca/museum/emanations.htm

Chapter 3
The Discovery and Interpretation of Nuclear Fission

If the story of the four-year progression from Enrico Fermi's discovery of neutron-induced radioactivity to the discovery of fission were cast as a suspense novel, it would probably be considered too concocted to be credible. Apparently reasonable, mutually-supporting assumptions, experimental near-misses, and unprecedented interpretations of data all conspired to create a situation of immense confusion before the truth was revealed by chemical detective work. Historian of science Ruth Sime has described the discovery of fission as an example of the illogical progress of scientific discovery.

Fortunately, histories of physics are not bound by the same literary conventions as detective stories, so there is no harm in giving away the ultimate resolution of the case up front; having the outcome in mind will help in understanding the travails with which the discoverers of fission struggled. This Chapter opens, then, with a description of what happens in the fission process.

3.1 The Discovery of Fission

In nuclear fission, a uranium or similar heavy-element nucleus, when struck by a bombarding neutron, breaks up into two lighter nuclei, accompanied by the release of a great deal of energy and the essentially instantaneous release of two or three neutrons. Fission was first detected when barium and krypton showed up as a result of neutron bombardment of uranium:

$$\, _{0}^{1}n + \, _{92}^{235}U \rightarrow \, _{56}^{141}Ba + \, _{36}^{92}Kr + 3 \, (_{0}^{1}n). \tag{3.1}$$

An astute reader will have noticed that this reaction involves U-235, not the more common U-238 isotope. This is a crucial part of the story, which will take some time to explain in the balance of this section and the following ones.

This reaction is unlike any previously described in this book (or, in 1938, known) in a number ways. First, neither of the products are near uranium in the periodic table. Since alpha and beta decays or neutron bombardments had always led to changes in the atomic number of the bombarded element by at most one or

B. C. Reed, *The History and Science of the Manhattan Project*,
Undergraduate Lecture Notes in Physics, DOI: 10.1007/978-3-642-40297-5_3,
© Springer-Verlag Berlin Heidelberg 2014

two, radiochemists naturally concentrated on looking for such "nearby" products, never anticipating such a radical departure from accumulated experience. Second, the energy released in this reaction, about 170 MeV, is enormous even by the violent standards of nuclear reactions. Most of this energy appears as kinetic energy of the fission products. Third, the reaction liberates three "secondary neutrons," as Leo Szilard had anticipated.

A less obvious consequence of this reaction is that since fission products are typically very neutron-rich for their Z-values (recall that the neutron excess increases with atomic number), they will undergo a series of successive beta-decays until they achieve stability. It is the *products* of fission that are responsible for radioactive fallout persisting long after the destructive effects of the energy release have made themselves felt. Fission bombs are *not* simply scaled-up ordinary chemical bombs. (Spent reactor fuel rods also contain fission products, but, unless the rods are recycled, the products remain stuck in them.)

Why did Fermi and his collaborators fail to discover high-energy fission fragments in 1934? The culprit was the nature of their radon-beryllium neutron sources. In addition to being an alpha-emitter, radon is a fairly prolific gamma-ray emitter, and these gamma-rays caused unwanted background signals in Geiger counters when the latter were placed near the neutron sources. Consequently, the procedure the Rome group adopted was to irradiate target samples and then literally run them down a hallway to a detector in a room far from the neutron source. Since the goal of the experiments was to detect delayed effects (the induced half-lives were often on the order of minutes), this procedure would not affect their results. But any high-energy fission fragments that might have been detected would have been brought to rest by the time the sample arrived at the detector. What the Romans attributed to decays of transuranic elements were beta-decays from the fission fragments, although the transuranics were no doubt also being created, as will be explained later. Fermi and his group had concentrated on a 13-min decay product of uranium bombardment. A common fission product is barium, and a particular isotope of this element, Ba-131, has a half-life of 14.6 min; this may have been what they were detecting. Fermi never expected fission to happen and so never considered that his experimental arrangement might be biasing him against detecting it. Retrospect is always perfect.

Fermi's claim that transuranic elements could be created through neutron bombardment stimulated great interest within the nuclear research community. In addition to Frédéric and Irène Joliot-Curie, the other main leaders of that community were Otto Hahn and Lise Meitner at the Kaiser Wilhelm Institute for Chemistry in Berlin (Fig. 3.1). Hahn, a radiochemist, and Meitner, a physicist, had known each other and collaborated on-and-off for 30 years. In 1918 they had discovered the rare element protactinium ($Z = 91$), and by the 1930s had accumulated between them years of experience with the chemistry and physics of radioactive elements. Meitner became interested in Fermi's experiments, and in 1935 convinced Hahn to renew their collaboration in order to sort out exactly how uranium transmuted under slow neutron bombardment. To help with the work they brought on board chemist Fritz Strassmann.

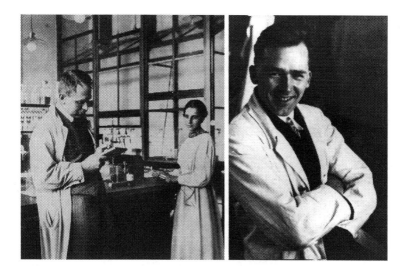

Fig. 3.1 *Left* Lise Meitner (1878–1968) and Otto Hahn (1879–1968) in their laboratory, 1913. *Right* Fritz Strassmann (1902–1980). *Sources* http://commons.wikimedia.org/wiki/File:Otto_ Hahn_und_Lise_Meitner.jpg; *AIP Emilio Segre Visual Archives, gift of Irmgard Strassmann*

To understand the Berlin group's assignments of identities for putative new elements, it is helpful to understand the nature of their chemical procedures. In the 1930s, heavy elements such as thorium, protactinium, and uranium were thought to be "transition" elements occupying the seventh row of the periodic table; this was before they were recognized as a separate group of "actinide elements" whose chemical properties are more similar to each other then they are to elements above them in the columns of the table in which they were presumed to reside. It is now understood that as one moves along the actinide row, additional electrons fill an inner electron shell, the $5f$ shell, as opposed to an outer shell. The outer-shell electron configuration remains the same as one moves along the row, which explains why these elements have such similar chemistry.

It was presumed that elements 93, 94, and so on would have chemical properties analogous to elements above them in the columns of the table in which they were expected to reside. Those elements are successively rhenium (presumably above $Z = 93$), osmium (above 94), iridium (95), platinum (96), and gold (97); see Fig. 3.2. The anticipated new elements were given the tentative names eka-rhenium (EkaRe), eka-osmium, and so forth; the root "eka" is from the Greek for "beyond." In line with this expectation, Hahn, Meitner, and Strassmann separated their induced radioactivities from uranium by precipitating them out of solutions with transition-metal compounds, and naturally assumed that the activities were due to the sought-after transuranic elements.

By 1937 the situation had become extremely muddled. The Berlin team had identified no less than 9 distinct half lives arising from uranium bombardment, many more than Fermi had detected. These were thought to involve a number of

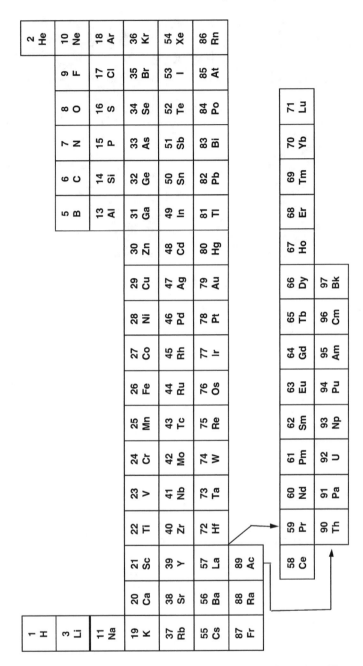

Fig. 3.2 A simplified periodic table of the elements, with their atomic numbers. Elements 58–71 are the so-called "rare earth" elements, which have chemical properties similar to each other, as do the "actinide" elements in the *bottom row*. Otherwise, elements in a given column ("families") have similar chemical properties. Elements up to number 112 have now been named, but play no role in the events described in this book

new transuranic elements, with atomic numbers up to 97. The activities were assigned to three possible reaction processes:

$$n +_{92} U \rightarrow (_{92}U + n) \xrightarrow[10\,s]{\beta^-}$$

$$_{93}\text{EkaRe} \xrightarrow[2.2\text{min}]{\beta^-} {}_{94}EkaOs \xrightarrow[59\,\text{min}]{\beta^-} {}_{95}EkaIr \xrightarrow[66\,h]{\beta^-} {}_{96}EkaPt \xrightarrow[2.5\,h]{\beta^-} {}_{97}EkaAu?$$

(3.2)

$$n +_{92} U \rightarrow (_{92}U + n) \xrightarrow[40\,s]{\beta^-} {}_{93}\text{EkaRe} \xrightarrow[16\,\text{min}]{\beta^-} {}_{94}EkaOs \xrightarrow[5.7\,h]{\beta^-} {}_{95}EkaIr?$$

(3.3)

$$n +_{92} U \rightarrow (_{92}U + n) \xrightarrow[23\,\text{min}]{\beta^-} {}_{93}\text{EkaRe}.$$

(3.4)

Chemically, these identifications seemed secure, but Meitner struggled to understand the corresponding physics. How could the neutrons be exciting three different energy levels in uranium? Such a situation had never been observed before. Also contrary to all previous experience were the three extended decay sequences, with the first two appearing to involve "inherited" excited energy levels.

To confound the situation further, in October, 1937, Irène Curie and Paul Savitch in Paris identified an approximately 3.5-h beta-decay half-life resulting from slow-neutron bombardment of uranium, an activity which the Berlin group had not found. Curie and Savitch suggested that the decay might be attributable to thorium, element number 90. If this were true, it would mean that thermal neutrons—slowed to the point of possessing less than a single electron-volt of kinetic energy—were somehow capable of prompting uranium nuclei to eject alpha-particles. The presumed reaction was

$$_{0}^{1}n + {}_{92}^{238}U \rightarrow {}_{2}^{4}He + {}_{90}^{235}Th \xrightarrow[3.5\,h]{\beta^-} {}_{91}^{235}Pa.$$

(3.5)

While such a reaction is energetically possible, the chance of an alpha-ejection could be computed from quantum mechanics, and was found to be extremely unlikely. (The half-life for natural U-238 alpha-decay is 4.5 billion years.) Thorium 235 is in fact a beta-decayer, but has a half-life of about 7.2 min; no known isotope of thorium has a half-life in the vicinity of 3.5 h. In Berlin, Meitner asked Strassmann to search for thorium. He did so, but in a way that overlooked the fission product that was actually present and being mistaken for thorium, of which he found no evidence. Ironically, in Rome in 1935, Edoardo Amaldi had tried looking for evidence of alpha-emitting reactions in bombarded uranium. But to do so he had to filter out the natural alpha-decay activity of uranium, which he did by wrapping his samples in thin aluminum foils on the rationale that any alphas arising from short half-life decays should be energetic enough to pass through the foils on their way to his detector. He detected no alphas, but the foils also blocked fission products.

October, 1937, was also notable for a more somber event. On the 19th of that month, Ernest Rutherford passed away following a fall at his home in Cambridge. With Rutherford's passing it could be said that the first great era of nuclear physics had come to a close. Element 104, Rutherfordium, is now named in his honor; its most stable known isotope, ^{267}Rf, has a spontaneous-fission half-life of about 2 h.

Further work by Curie and Savitch resulted in a paper published in September, 1938, wherein they argued that their 3.5-h beta-emitter seemed to have chemical properties similar to that of lanthanum, element 57. Lanthanum is in the same column of the periodic table as actinium, element 89, which is only three places away from uranium, so it seemed sensible to attribute the activity to actinium, or perhaps a new transuranic element; they would not have dared propose that they were actually detecting lanthanum. While one could propose producing actinium directly, say via a reaction such as

$$\frac{1}{0}n + \frac{238}{92}U \rightarrow \frac{7}{3}Li + \frac{232}{89}Ac, \tag{3.6}$$

the problem remains that if the probability of ejecting an alpha-particle is unlikely to begin with, that of ejecting a lithium nucleus will be even less. One might then posit modifying the original alpha-producing reaction to be followed by a positron decay to produce actinium, perhaps

$$\frac{1}{0}n + \frac{238}{92}U \rightarrow \frac{4}{2}He + \frac{235}{90}Th \xrightarrow[?]{\beta+} \frac{235}{89}Ac. \tag{3.7}$$

Here the problem is that the positron-decay part of this reaction is energetically unfavorable, having $Q = -3.95$ MeV. While Curie and Savitch's *chemistry* was indicating thorium or actinium, all of the known *physics* of their proposed decay schemes seemed improbable. Curie and Savitch may have been detecting La-141, which is now known to have a half-life of 3.9 h; another possibility is yttrium-92, which has a 3.54-h half-life; yttrium is also in the same column of the periodic table as lanthanum and actinium.

A few months before this confusion arose, Lise Meitner had been forced to flee Berlin. Born into a Jewish family in Austria, she had assumed that her Austrian citizenship protected her against German anti-Semitic laws. That protection ended with the German annexation of Austria in March, 1938. On July 13 of that year she fled to Holland with only 10 Marks in her purse and literally the clothes on her back. She eventually made her way to Sweden, where she was given a position at the Nobel Institute for Experimental Physics, but she was not too warmly received nor particularly well supported. While she continued to collaborate with Hahn and Strassmann by letter, her career was essentially destroyed, and her pension was stolen by the Nazi government.

In later years, Hahn would largely disavow Meitner's contributions, even to the point of suggesting that her considerations of physics impeded the discovery of fission. Fritz Strassmann set the record straight: "What does it matter that Lise Meitner did not take *direct* part in the discovery? ... [She] has been the intellectual leader of our team and therefore she was one of us, even if she was not actually present for the 'discovery of fission.'" Hahn was awarded (solely) the 1944 Nobel Prize for Chemistry; Meitner and Strassmann did not share in the recognition, although Element 109, Meitnerium, is now named in her honor.

In Berlin, Hahn and Strassmann continued to look for Curie and Savitch's 3.5-h activity. In a series of letters to Meitner through October and November, Hahn chronicled their progress. By October 25 he had become convinced of the existence of the 3.5-h activity, but suspected that a radium isotope (element 88) might be involved. By November 2 he was becoming convinced that two or three radium isotopes were being created. If so, that implied a two-stage double-alpha decay, say

$$\ _0^1 n + \ _{92}^{238} U \rightarrow \ _2^4 He + \ _{90}^{235} Th \rightarrow \ _2^4 He + \ _{88}^{231} Ra, \qquad (3.8)$$

an even more improbable sequence than the Curie-Savitch process. On November 8, Hahn and Strassmann sent a paper to *Naturwissenschaften* reporting half-lives for three radium and three actinium isotopes. By mid-December, Hahn and Strassmann had further refined their chemical techniques, and came to the startling conclusion that what they had thought were radium isotopes were actually *barium* isotopes (element 56). Since radium and barium are in the same column of the periodic table, it is not surprising that it took them some time to sort this out.

Hahn wrote to Meitner late in the evening of Monday, December 19, to explain the situation and to seek her opinion as to how neutron bombardment of uranium might yield a product of atomic weight only about half that of uranium. Hahn apparently did not yet realize that their results meant that uranium nuclei were bursting into two or more fragments. An indication of the extent to which Meitner was still considered part of the team is indicated by a second letter from Hahn on the 21st, which included the sentiment "How beautiful and exciting it would be just now if we could have worked together as before."

Hahn's first letter reached Meitner in Stockholm on Wednesday, December 21. Excited by such an unexpected turn of events, she replied at once: "Your radium results are very startling ... At the moment the assumption of such a thorough-going breakup seems very difficult to me, but in nuclear physics we have experienced so many surprises that one cannot unconditionally say: it is impossible." Hahn would receive her reply on December 23. Anxious not to be scooped, Hahn and Strassmann did not wait for Meitner's reply before submitting a paper to the journal *Naturwissenschaften* on December 22. The paper was written by Hahn, who hedged by referring to "alkaline earth" elements, although he did specifically mention the barium finding. The paper would be published on January 6, 1939.

On December 23, the same day that Hahn received Meitner's reply to his letter of the 19th, Meitner traveled from Stockholm to spend Christmas with some friends in the town of Kungälv, near Göteborg on the west coast of Sweden. Her nephew, Otto Frisch, a nuclear physicist—and another refugee from their native Austria—was then working at Niels Bohr's Institute for Theoretical Physics in Copenhagen. He traveled to Sweden to spend Christmas with his aunt, arriving also around the 23rd (Fig. 3.3).

Fig. 3.3 Otto Frisch
(1904–1979). *Source*
Photograph by Lotte Meitner-
Graf, London, courtesy AIP
Emilio Segre Visual Archives

At some time in the following few days—likely some time after Christmas—
Meitner and Frisch went for a walk in the snow, he on skis and she keeping up on
foot. He had hoped to interest her in an experiment he was planning, but she
instead drew him into a discussion of Hahn's letter of the 19th. Hahn's manuscript
had also been forwarded to her from her home in Stockholm. As Frisch relates in
his memoires, they sat down on a tree trunk, and began to calculate on scraps of
paper. Working from a theoretical model of nuclei that had been developed some
years previously by George Gamow and Niels Bohr, the so-called liquid-drop
model (see below), Meitner and Frisch knew that uranium nuclei with their many
protons are near the limit of intrinsic stability beyond which no additional number
of neutrons can inhibit them from spontaneously breaking up. Uranium nuclei are
somewhat like wobbly drops, liable to fragment in response to a modest provocation such as the impact of a neutron. Should a uranium nucleus break in two, the
resulting fragments would experience a mutually repulsive Coulomb force, and fly
away from each other at high speeds.

How much energy might be liberated in such a process? Meitner had the mass-
defect curve of Sect. 2.1.4 committed to memory, and quickly calculated that the
two fragments formed by the breakup of a uranium nucleus would total to a mass
less than that of a uranium nucleus by about one-fifth of the mass of a proton,
equivalent to about 200 MeV. This substantial amount of energy would appear to
the outside world in the form of the kinetic energy of the fission fragments. Thus
was the process of fission conceived in a snowy Swedish forest.

On average, the energy liberated in the fission of uranium nuclei is about
170 MeV. One gram of uranium contains some 2.5×10^{21} atoms, so fission of one
kilogram of atoms will liberate some 4.4×10^{26} MeV, or 7×10^{13} Joules.
Explosion of a thousand metric tons (10^6 kg; a so-called *kiloton*) of TNT liberates
some 4.2×10^{12} Joules, which means that one kilogram of uranium is equivalent
to about 17 kilotons of conventional explosive.

One can only wonder what it must have felt like to be one of the only two
people in the world who at that moment realized that a fundamentally new

physical process had been discovered. For Meitner especially there must have been a torrent of mixed emotions. On one hand was the realization that a rich new area of physics was opening up, while on the other was the revelation that phenomena which for several years she had attributed to transuranic elements were likely to have been the products of neutron-rich fission fragments decaying toward stability. Only the 23-min decay of reaction (3.4) would prove to be yielding a transuranic element. But it was those years of "transuranic" groundwork that had paved the way to the discovery of fission.

In another of the curious confluences of events that seem to characterize nuclear history, it was around this time (December 24, to be precise) when Enrico Fermi and his family set out for America from Southampton, England. Fermi would know nothing of these developments until he met Niels Bohr some three weeks later in New York. Since his Nobel Prize was made for the presumed discovery of elements 93 and 94, the discovery of fission would prompt Fermi to add a footnote to the published version of his Nobel lecture.

In the meantime, Otto Hahn had also begun thinking that what he had previously assumed to be transuranic elements might be lighter elements, and on the 27th phoned the editor of *Naturwissenschaften* to append a comment to this effect to his and Strassmann's paper. The next day he wrote Meitner again, pleading with her to consider whether the energetics of the proposed splitting made sense. The mail must have been speedy, for she replied on the 29th, asking whether the "actinium" products were correspondingly turning out to be lanthanum. Back in Stockholm for New Year's Eve, Meitner again wrote Hahn: "We have read and considered your paper very carefully; *perhaps* it is energetically possible for such a heavy nucleus to break up." On New Year's Day, Frisch returned to Copenhagen, promising to keep in telephone contact with his aunt as they drafted a paper based on the work they had begun on a tree trunk a few days earlier.

In his memoirs, Frisch relates that in all the excitement, he and Meitner missed an important point: the possibility of a chain reaction. A Danish colleague, Christian Møller, suggested to him that the fission fragments might contain enough energy to each eject a neutron or two, which might go on to cause other fissions. That the fragments would be neutron-rich in comparison to stable nuclides of the same atomic number made this possibility very real. Frisch's immediate response was that if such were the case, no deposits of uranium ore should exist as they would have blown themselves up long ago. But he then realized that this argument was too naïve: ores contained other elements which might absorb neutrons, and many neutrons might simply escape before causing another fission. Leo Szilard's chain-reaction vision of five years earlier had taken its first steps toward possible reality.

Meitner wrote Hahn again on January 3 to congratulate him and Strassmann, and to express her frustration at having to watch developments from afar: "I am now almost *certain* that the two of you really do have a splitting to Ba and I find that to be a truly beautiful result, for which I most heartily congratulate you and Strassmann … And believe me, even though I stand here with very empty hands, I am nevertheless happy for these wondrous findings."

In early January, 1939, the focus of fission research shifted briefly to Copenhagen, and then primarily to America. On the 3rd, Frisch caught up with Niels Bohr to apprise him of the situation. The conversation was brief; Bohr was preparing to spend a semester at the Institute for Advanced Study in Princeton, New Jersey. According to Frisch, Bohr's reaction was to wonder why he had not thought of fission before; Frisch would later depict Bohr as hitting himself on the forehead and exclaiming, "Oh what idiots we have all been. Oh but this is wonderful! This is just as it must be! Have you and Lise Meitner written a paper about it?" Bohr promised not to disclose the discovery until their paper had been prepared.

Bohr and Frisch conversed again on January 6 to review the calculation of the near-instability of uranium. Frisch also discussed the situation with theoretician George Placzek, who recalled that Irène Curie had told him in the fall of 1938 that she had found light elements from uranium bombardment, but did not trust herself to publish it. The next morning, Frisch met Bohr just before the latter's departure, and handed him a two-page draft of the paper he was coauthoring with Meitner.

Curiously, Frisch had not thought of setting up an experiment to detect the expected high-energy fission fragments until Placzek encouraged him to do so. (For that matter, neither had Hahn and Strassmann.) He did so on Friday, January 13, and immediately detected the fragments. Frisch is thus credited with being the first person to set up an experiment to deliberately demonstrate fission. He is also credited with coining the term "nuclear fission," after having asked an American biologist working in Copenhagen, William Arnold, what term was used for the process of cell division: "binary fission". Extending Hahn and Strassmann's work, Frisch also tested thorium, which proved to act like uranium in that it would fission under bombardment by *fast* (unmoderated) neutrons, but to act unlike uranium in that it did *not* do so when bombarded with *slow* (moderated) neutrons. Frisch consequently prepared two papers. The first was co-authored with Meitner and described their Christmastime insights, while the second described his own experiments. Both were sent to *Nature* on January 16; the joint paper was published on February 11, and the experimental one on February 18. The uranium/thorium asymmetry would prove to be a crucial observation a few weeks later, when Niels Bohr worked to understand the underlying physics of the process.

Bohr sailed to America, accompanied by his son Erik and collaborator Léon Rosenfeld of the University of Liège in Belgium. Bohr had a blackboard installed in his stateroom, and during the transatlantic voyage he and Rosenfeld, despite seasickness, began to develop a theoretical understanding of fission. They arrived in New York on the afternoon of Monday, January 16, where they were met by Enrico and Laura Fermi. Bohr had business in New York, and Rosenfeld left directly for Princeton. But Bohr had not told Rosenfeld of his promise to Frisch to keep the news of fission quiet until Meitner and Frisch's paper had been submitted, and Rosenfeld spilled the beans that evening at a meeting of the Princeton Physics Journal Club. When Bohr heard that word of fission was out, he hastily drafted his own note to *Nature* on January 20 to assert Meitner and Frisch's priority; it would be published on February 25.

Bohr's January 20 paper is worth attention. In it he gives a description of how the fission process could be envisioned. This was based on his "liquid drop" model, which involved conceiving of nuclei as acting like droplets of liquid that could be distorted when perturbed by some external cause, such as a bombarding particle. In ordinary (non-fission) reactions, Bohr speculated that the energy of the bombarding particle was distributed in the target nucleus among various modes of vibration in a manner resembling the thermal agitation of a liquid drop. If a large part of the energy should come to be concentrated on some particle at the surface of the nucleus, then that particle will be ejected, akin to the evaporation of a molecule from a drop. In a fission reaction, Bohr reasoned, the distribution of energy would have to result in a mode of vibration of the nucleus that involved a considerable deformation of the surface (Fig. 3.4). He deduced, purely qualitatively as yet, that in heavy nuclei the energy sufficient to distort the surface to the point where two lobes formed that would repel each other and so cause the nucleus to fission must be of the same order of magnitude as the energy necessary for the escape of a single particle from a lighter nucleus. The concept of a requisite deformation energy would soon find rigorous quantitative expression as the *fission barrier* described in Sect. 3.3.

The first demonstration of fission in America occurred at Columbia University. On Wednesday, January 25, Bohr, while on his way to attend a conference in Washington (about which more below), stopped at Columbia to find Fermi. Fermi was out, and Bohr instead encountered one of Fermi's graduate students, Herbert Anderson (Fig. 3.5). As Anderson told the story, Bohr came right over to him, grabbed him by the shoulder, and whispered in his ear "Young man, let me explain to you about something new and exciting in physics". Anderson, who was preparing a thesis on neutron scattering, instantly understood the significance of what Bohr related.

Bohr went on his way, and Anderson went to find Fermi, who had already heard the news through a contact at Princeton. Fermi had to leave for Washington as well, and so was not present that evening when Anderson set up an experiment to detect the high-energy fission fragments with an ionization chamber he had prepared for his thesis work. The ionization pulses were readily apparent, and the experiment was witnessed by Professor John Dunning (Fig. 3.6). Anderson states that Dunning telegraphed the news to Fermi in Washington, but it is not clear if he

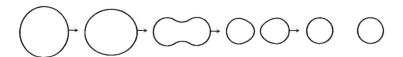

Fig. 3.4 Schematic representation of steps in the progression of "droplet fission." An initially spherical nucleus (*left*) is perturbed by some agency such as a bombarding neutron (not shown), and begins to distort. In the middle diagram, two lobes have formed, which will force each other apart due to electrostatic repulsion, leading to fission. Adapted with the kind permission of the author from R. M. Eisberg, *Fundamentals of Modern Physics* (New York: John Wiley and Sons, 1961), p. 616

Fig. 3.5 Enrico Fermi and his reactor group at the University of Chicago. This photograph was taken December 2, 1946, on the fourth anniversary of the operation of Fermi's CP-1 reactor (Chap. 5). *Back row (l–r)* Norman Hilberry, Samuel Allison, Thomas Brill, Robert Nobles, Warren Nyer, Marvin Wilkening. Middle row (l–r): Harold Agnew, William Sturm, Harold Lichtenberger, Leona Woods, Leo Szilard. *Front row (l–r)* Enrico Fermi, Walter Zinn, Albert Wattenberg, Herbert Anderson. *Source* http://commons.wikimedia.org/wiki/File:Chicago PileTeam.png

Fig. 3.6 *Left* John Dunning (*left*) and Eugene Booth inspect a cyclotron at Columbia University. *Right* George Gamow (1904–1968). *Sources* AIP Emilio Segre Visual Archives; AIP Emilio Segre Visual Archives, Physics Today Collection

actually did so. In Paris the next day, Frédéric Joliot read Hahn and Strassmann's paper, and also detected evidence of fission.

While word of the discovery had been spreading privately at Princeton and Columbia since Bohr and Rosenfeld's arrival on the 16th, the public coming-out of the news came on January 26. What had drawn Fermi and Bohr to Washington was the Fifth Washington Conference on Theoretical Physics. These conferences were co-hosted by George Washington University (GWU) and the Carnegie Institution of Washington (a prestigious private research institution), and were mainly organized by George Gamow (Fig. 3.6) and Edward Teller (Fig. 4.13), both of whom were then at GWU. The topic of the 1938 meeting was to be low-temperature physics, but that agenda quickly found itself derailed.

The conference, scheduled for January 26–28, opened at two o'clock on the afternoon of Thursday, the 26th. Gamow opened the proceedings by introducing Bohr, who related Hahn and Strassmann's discovery and Meitner and Frisch's interpretation. The news electrified the fifty-odd participants, some of whom departed promptly to perform their own experiments. Today, a plaque outside Room 209 of GWU's Hall of Government commemorates Bohr's historic announcement.

The next deliberate demonstration of fission in America seems to have occurred on Saturday morning, January 28, at Johns Hopkins University in Baltimore. Apparently tipped off by a colleague attending the conference, R. D. Fowler and R. W. Dodson tested both uranium and thorium, and verified Frisch's observation that slowing neutrons with paraffin increased the fission rate in uranium, but had no effect on that for thorium. That evening at the Carnegie Institution, Richard Roberts and colleagues Lawrence Hafstad and Robert Meyer demonstrated fission, with Bohr, Fermi, Rosenfeld, Teller, and others in attendance. In the public domain, the *New York Times* reported on the discovery in its Sunday, January 29, edition, noting that scientists at the Washington meeting thought that it might be twenty or twenty-five years before the phenomenon could be put to use. In Berkeley, Luis Alvarez, a member of Ernest Lawrence's Radiation Laboratory staff, read of the discovery in the *San Francisco Chronicle* and passed the word to his graduate student, Philip Abelson, who verified the finding on January 31. Abelson detected iodine as a decay product of tellurium, which was itself a direct fission product, and over subsequent weeks identified a number of other fission products. Independently, Alvarez also verified that slow neutrons were more effective in causing fission than fast ones. The Johns Hopkins, Carnegie, and Berkeley reports all appeared in the February 15, 1939, edition of the *Physical Review*. The Columbia group's first paper did not appear until the March 1 edition, but contained the first quantitative cross-section measurements for both fast and slow neutrons.

Fission can happen in a number of ways, but it is always accompanied by a tremendous release of energy. For example, the Hahn and Strassmann discovery reaction,

$$\,^{1}_{0}n + \,^{235}_{92}U \rightarrow \,^{141}_{56}Ba + \,^{92}_{36}Kr + 3\left(\,^{1}_{0}n\right), \qquad (3.9)$$

assuming that three secondary neutrons are emitted, releases just over 170 MeV of energy. The vast majority of this is carried off in the form of kinetic energy by the barium and krypton fission products, but the neutrons carry off on average about

2 MeV each, a number that will prove to be important. The neutron-rich fission products then decay by a series of beta decays,

$$\underset{56}{^{141}}Ba \xrightarrow[18.3\ min]{\beta^-} \underset{57}{^{141}}La \xrightarrow[3.9\ h]{} \beta^- \underset{58}{^{141}}Ce \xrightarrow[32.5\ days]{\beta^-} \underset{59}{^{141}}Pr, \qquad (3.10)$$

and

$$\underset{36}{^{91}}Kr \xrightarrow[8.6\ s]{\beta^-} \underset{37}{^{91}}Rb \xrightarrow[58\ s]{\beta^-} \underset{38}{^{91}}Sr \xrightarrow[9.5\ h]{\beta^-} \underset{39}{^{91}}Y \xrightarrow[58.5\ days]{\beta^-} \underset{40}{^{91}}Zr. \qquad (3.11)$$

As some 30 different elements are produced by uranium fission, it is no wonder that Hahn, Meitner, and Strassmann had observed a confusing plethora of lengthy decay chains. Their detection of the barium-krypton fission channel was presumably a result of their use of barium chemistry.

Intuitively, one might expect that if the channel that a particular reaction chooses to follow is some sort of random process, then the distribution of masses of fission fragments should be more-or-less symmetric around $A \sim 235/2 \sim 118$. But this is far from the case in reality, which shows that fission can happen in a great multiplicity of ways, leading to dozens of possible products. Curiously, an *equal* splitting of the uranium nucleus is quite rare, although not impossible. Figure 3.7 shows the distribution of fragment masses from thermal-neutron fission of U-235 as a function of mass number A. Fragment masses with $A \sim 90$ and 140 are clearly preferred; equal splitting is actually somewhat unlikely. As the energy of the bombarding neutron increases, the distribution becomes more centrally peaked, but there is as yet no known reason from fundamental physics as to why thermal-neutron yields are so asymmetric.

As Leo Szilard and doubtlessly many others realized, there would have to be on average *at least one* neutron liberated per fission if there was to be any hope of sustaining a neutron-moderated chain reaction. As Herbert Anderson later wrote, "Nothing known then guaranteed the emission of neutrons. Neutron emission had to be observed experimentally and measured quantitatively." Soon after the

Fig. 3.7 Logarithm of percent yield of fission-fragment mass as a function of mass number for thermal-neutron fission of uranium-235. Data from T. R. England and B. F. Rider, *Evaluation and Compilation of Fission Prioduct Yields: 1993*, Los Alamos National Laboratory report LA-UR-94-3106. Data available at <http://ie.lbl.gov/fission/235ut.txt>

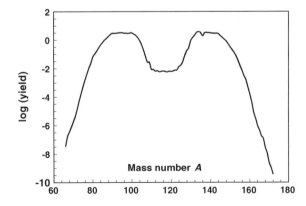

discovery of fission, a number of research teams began looking for evidence of such "secondary" neutrons, and proof of their presence was not long in coming. On March 16, two independent groups at Columbia submitted reports to *The Physical Review* reporting their discovery: Anderson, Fermi & Hanstein, and Szilard & Zinn. Both groups estimated about two neutrons emitted per each captured; their papers were published on April 15. Anderson, Fermi, and Hanstein were able to determine from their experiment that the total thermal-neutron absorption cross-section for natural uranium (fission, radiative capture, and other processes) is about 5 barns; this is in respectable agreement with the modern value of about 7.6 barns.[1] Szilard and Zinn configured their experiment to detect the emission of any *fast* neutrons as a consequence of fission, and indeed observed them. Szilard recalled later his reaction upon detecting the neutrons: "That night, there was very little doubt in my mind that the world was headed for grief." Confirming evidence for secondary neutrons soon came in from Europe: in Paris on April 7, Hans von Halban, Frédéric Joliot, and Lew Kowarski submitted a paper to *Nature* (published April 22) in which they reported 3.5 ± 0.7 neutrons liberated per fission. The modern value for the average number of secondary neutrons liberated by U-235 when it is fissioned by thermal neutrons is about 2.4.

   The physics of fission is a complex topic, and is the subject of the following three sections.

3.2   The Physics of Fission I: Nuclear Parity, Isotopes, and Fast and Slow Neutrons

The observation that the propensity of uranium to fission depended on the velocity of bombarding neutrons and that uranium and thorium differed in their responses to slow-neutron bombardment catalyzed several crucial revelations on the part of Niels Bohr in early February, 1939. Working with remarkable haste, Bohr prepared and sent off a paper to the *Physical Review* dated February 7, 1939; it was published in the February 15 edition alongside the reports of fission detected in various American laboratories. In this paper, Bohr developed arguments to show that it was likely the rare isotope uranium-235 that must be responsible for *slow-neutron* fission in that element, and to explain why thorium did not exhibit slow-neutron fission.

   Bohr's argument comprised two interlinked components. The first involved his liquid-drop model of the preceding section. In his new paper, Bohr linked this argument to some earlier experiments of Meitner, Hahn, and Strassmann wherein they examined the radiative-capture response of uranium to neutrons of varying speeds. This work had revealed a rich forest of very strong resonance capture lines

---

[1] The 7.6-barn figure is a weighted average of the (total—elastic scattering) cross sections for thermal neutrons on U-235 (683 bn) and U-238 (2.7 bn), using abundances 0.0072 and 0.9928.

for neutrons of energies of from a few to thousands of eV; look ahead to Fig. 3.11. Based on some arguments from an area of statistical mechanics known as dispersion theory, Meitner, Hahn, and Strassmann had concluded that these resonances were likely attributable to the abundant isotope, U-238. However, the resonance captures were not associated with any corresponding increase in the fission cross-section, which led Bohr to infer that nuclei of the 238 isotope must consequently be very stable, since the liquid-drop model indicated that they would be expected to fission upon taking in such energetic neutrons. If U-238 does not fission under intermediate-energy neutron bombardment, it would certainly not be expected to do so under slow-neutron bombardment. Thus, Bohr reasoned that U-235 could be the only candidate for slow-neutron fission.

Bohr's second argument helped to clarify what was happening with thorium by looking at the situation from the point of view of what is known as "nuclear parity." Nuclear physicists classify the "parity" of nuclei according to the evenness or oddness of the number of protons and neutrons that they contain, always expressed in the order *protons/neutrons*, or Z/N. In this scheme, uranium-235 is an even/odd nucleus ($Z = 92$, $N = 143$), whereas uranium-238 is classified as even/even ($Z = 92$, $N = 146$). The parity distribution of the 266 known stable nuclides is summarized in Table 3.1.

Clearly, Nature has a preference for even/even nuclides. Virtually identical numbers of stable nuclides are even/odd or odd/even, and only a handful are odd/odd. The latter are all light nuclei; specifically, heavy hydrogen, lithium-6, boron-10, and nitrogen-14. This distribution is now considered to be a manifestation of the nature of the forces exerted between pairs of nucleons. Speaking somewhat loosely, a nucleus wherein every nucleon has a "partner" will exist in a more stable mass-energy state (greater mass defect) than one where unpaired nucleons are present. This is described below via an analogy, but, for the moment, let us return to Bohr's argument.

Bohr pointed out that uranium consists of two isotopes, one even/odd $\left(^{235}_{92}U\right)$ and one even/even $\left(^{238}_{92}U\right)$, whereas thorium has only one stable isotope, $^{232}_{90}Th$, of even/even parity. If it is the even/odd isotope that is responsible for slow-neutron fission in uranium, then one might *not* expect to see slow-neutron fission in thorium, as it lacks an isotope of such parity. This was consistent with what Otto Frisch and others had observed. Purely qualitatively, another way of paraphrasing Bohr's analysis is to say that when an even/odd nucleus (such as U-235) absorbs a neutron, it will find itself in a more excited energy state—and hence more prone to fission—than would an even/even nucleus. Numerical details on this point will follow shortly.

Table 3.1 Distribution of stable-nuclide parities	Z/N	Number of stable isotopes
	Even/Even	159
	Even/Odd	53
	Odd/Even	50
	Odd/Odd	4

The second-last paragraph of Bohr's paper presented an important hypothesis concerning *fast*-neutron fission, a speculation which was likely largely overlooked at the time with all the attention being devoted to *slow* neutrons. It is worth examining this part of his argument in some detail.

Quantum–mechanical considerations indicated that as the energy of bombarding neutrons increases (that is, as they become faster), the fission cross-section should generally decrease (see, for example, Fig. 3.12 for the case of U-235). For very fast (MeV) neutrons, the cross-section should never exceed the geometric cross-section of the nucleus itself, which for uranium is about 1.7 barns (Eq. 2.49). Since U-238 did not fission under intermediate-energy neutron bombardment, it would certainly not be expected to do so when struck by *fast* ones because of the lower cross-section to be expected at higher energies. (This reasoning appears to contradict the argument above that U-238 should not suffer slow-neutron fission, as the quantum–mechanical conclusion would lead one to expect a greater chance of fission for lower-energy neutrons. As described in the following pages, however, there are a number of further levels of subtlety to this story yet to be revealed.) On the other hand, Bohr pointed out, U-235 might have a chance of sustaining fast-neutron fission in view of its apparently very large cross-section for slow neutrons, that is, there might be sufficient "remaining" cross-section for fast neutrons despite the expected decrease in cross-section with increasing neutron energy.

While Bohr left unstated the question of what might happen if U-235 could be separated from U-238 and bombarded with fast neutrons, the possible implications of this question had not gone unnoticed. Philip Morrison, a student of Robert Oppenheimer, recalled that "when fission was discovered, within perhaps a week there was on the blackboard in Robert Oppenheimer's office a drawing ... of a bomb."

The levels of argument that Bohr wove into a two-page paper are dizzying. In his own words, it all reduced to "allowing us to account both for the observed yield of the process concerned for thermal neutrons and for the absence of any appreciable effect for neutrons of somewhat higher velocities. For fast neutrons ... because of the scarcity of the isotope concerned [U-235] the fission yields will be much smaller than those obtained from neutron impacts on the abundant isotope [U-238]." The details of Bohr's analysis would be revised as further experimental data accumulated, but by the spring of 1939, general outlines of understanding of the response of different uranium isotopes to neutron bombardment and the prospects for a chain reaction were beginning to become clear, at least in principle.

The parity distribution in Table 3.1 deserves further comment, and at least a pseudo-explanation. If we presume that these numbers reflect some underlying fundamental physics regarding stability of nuclei, a simple interpretation of the large number of even/even isotopes is that nuclei achieve their greatest stability when all of their nucleons can find a partner with which to "pair-bond." Given the large number of even/even isotopes, this pairing could be explained with two possible schemes: (i) neutrons are happy to pair with protons, with excess neutrons then being equally content to pair with others of their own kind (the neutron excess

$N - Z$ is ≥ 0 for all isotopes except ordinary hydrogen and helium-3), or (ii) nucleons prefer to pair with other nucleons of their own kind. In comparing (i) and (ii), note that (i) refers to *neutrons*, and (ii) to *nucleons*.

To decide between these two possibilities, look to the small number of odd/odd isotopes. In an odd/odd isotope, the number of excess neutrons $N-Z$ will always be *even*. If case (i) is in play, high stability would be achieved, as no neutron or proton would be left unpaired. But we would then have to explain why nature discriminates so strongly against odd/odd isotopes. On the other hand, if case (ii) is in play, then the small number of odd/odd isotopes is readily explained by the fact they would always contain one unpaired nucleon of each type, apparently the least-stable overall configuration. The intermediate even/odd and odd/even cases fit perfectly into possibility (ii), in that there is only one unpaired nucleon in either case. The conclusion must be that nucleons prefer to pair with others of their own kind, with much less (if not no) attraction at all to the other kind.

With the pair-bonding preferences of nuclei somewhat understood, we can now get to the promised analogy regarding the effect of odd and even numbers of nucleons on nuclear stability, and with it a quantification of the binding energy released upon neutron absorption.

Imagine neutrons as guests at a party, which is the nucleus itself. (Protons play no role here, as fission is a *neutron*-initiated phenomenon.) As with a human party, it is desirable for every guest to have a partner with whom to converse. Additional guests are welcome so long as the total number does not grow so large as to attract a visit from the beta-decay stability police. To attract new neutrons to the party, those already present can be thought of as each willing to give up a small amount of their mass to make room for newcomers. Particularly preferable would be a new guest whose addition would make the total number of guests even, so that everyone then has a partner. For a large party (a heavy nucleus), measured nuclear masses indicate that the already-present guests are collectively willing to sacrifice an amount of mass equivalent to about 6.5 MeV of energy to achieve an even number of guests. That liberated energy appears in the form of excitation energy of the nucleus. The party becomes louder, and some of the guests might fission out the door (taking protons with them) to form sub-parties. On the other hand, a newcomer whose addition would make the total number of guests odd is also welcome (neutrons never repel other particles), but less so in that those already present are somewhat less willing to make room for an odd-one-out. In this case they are willing to sacrifice only about 5 MeV mass equivalent, and the nucleus becomes less roiled than if it had sustained an odd-to-even neutron-number transition.

This scheme predicts that a nucleus which transforms from an odd/odd (or even/odd) to an odd/even (or even/even) configuration by absorbing a neutron will liberate more energy than one that goes from an odd/even (or even/even) to an odd/odd (or even/odd) configuration. The difference is about 1.5 MeV. This is precisely what happens when a uranium-235 nucleus (even/odd) takes in a neutron, versus what happens when a uranium-238 nucleus (even/even) does so. As described in more detail in the next section, the extra 1.5 MeV is enough to cause a neutron-bombarded U-235 nucleus to fission, whereas a U-238 nucleus simply absorbs the neutron and subsequently beta-decays.

The changes of parity upon neutron absorption and the corresponding energy releases can be summarized as

$$neutron + \left\{ \begin{array}{c} even/even \\ odd/even \end{array} \right\} \rightarrow \left\{ \begin{array}{c} even/odd \\ odd/odd \end{array} \right\} + 5 \text{ MeV} \qquad (3.12)$$

and

$$neutron + \left\{ \begin{array}{c} even/odd \\ odd/odd \end{array} \right\} \rightarrow \left\{ \begin{array}{c} even/even \\ odd/even \end{array} \right\} + 6.5 \text{ MeV}. \qquad (3.13)$$

The energy values are approximate; exact numbers depend on the isotopes involved. An example of the first type of reaction is $_0^1 n + _{92}^{238} U \rightarrow _{92}^{239} U$, and an example of the second type is $_0^1 n + _{92}^{235} U \rightarrow _{92}^{236} U$. For practical purposes, there are no cases of odd/odd → odd/even transitions to have to consider as there are no appreciably long-lived heavy odd/odd isotopes.

This scheme predicts that uranium-239 would appear to be a favorable candidate as a fissile material, as it would transmute from being even/odd to being even/even upon absorbing a neutron. This is true, but since U-239 has a beta-decay half life of only 23.5 min, it is not practical for consideration as a weapons material. The anticipated decay product of U-239, Np-239, would *not* be a favorable candidate, as its neutron-absorption transmutation would be odd/even to odd/odd. But, if Np-239 beta-decays to Pu-239, the latter would undergo an even/odd to even/even transmutation under neutron absorption, exactly as does U-235. As is described in Sect. 3.8, precisely this line of reasoning occurred to Louis Turner of Princeton University in early 1940, who realized that if Pu-239 should prove to be a reasonably stable decay product of neutron bombardment of U-238, it could open an alternate route to obtaining bomb-quality, fast-neutron-fissile material.

Analogies can never take the place of rigorous physical reasoning or experiments, but they can be helpful in organizing empirical data and serving as a basis on which to make qualitative predictions. Even for nuclear physicists, these parity arguments are still largely in the realm of empirical knowledge. At present, particle physics can only just predict the masses of individual fundamental particles from theories of the underlying physics of nuclear forces, let alone the mass of an entire nucleus.

3.3 The Physics of Fission II: The Fission Barrier and Chain Reactions

Niels Bohr's insight of February, 1939, that it was likely the rare isotope of uranium of mass number 235 that was responsible for slow-neutron fission was but the first step in an extensive chain of experimental and theoretical investigations into the fission process that unfolded over the following year. Verification of Bohr's hypothesis would come about a year later, as described in Sect. 3.6. The emphasis in the present section is on exploring how understanding of the roles played by different isotopes under fast and slow neutron bombardment developed in view of the parity argument presented in the preceding section.

Upon his arrival in America, Bohr began collaborating with John Wheeler, a young Assistant Professor at Princeton University (Fig. 3.8). Bohr and Wheeler had known each other since 1934, when Wheeler began a postdoctoral year at Bohr's Institute for Theoretical Physics in Copenhagen. In the September 1, 1939, edition of *The Physical Review*, they published an extensive analysis of the energetics of fission. For the purposes of this discussion, the results of this historic work can be summarized in three statements. These are:

(i) There exists a natural limit $Z^2/A \sim 48$ beyond which nuclei are unstable against disintegration by spontaneous fission. This arbitrary-looking expression arises from a combination of parameters used to fit a theoretical curve to the mass-defect data of Fig. 2.8.

(ii) In order to induce a nucleus with a Z^2/A value less than this limit to fission, it must be supplied with a necessary "activation energy," a quantity also known as the "fission barrier." This is the case for both isotopes of uranium.

(iii) The Z/N parity of an isotope plays a significant role in determining whether or not an isotope is slow-neutron fissile.

Points (ii) and (iii) are the key ones for understanding why uranium-235 makes an excellent bomb material while uranium-238 does not. Point (i) does address an

Fig. 3.8 John Wheeler
(1911–2008). *Source* AIP
Emilio Segre Visual Archives

interesting empirical question, however: Why does nature stock the periodic table with only about 100 elements? Nuclei have $A \sim 2Z$, so $Z^2/A \sim 48$ corresponds to a limiting Z of approximately 96, about the right value.

First consider point (ii). Bohr and Wheeler's analysis indicated that *any* otherwise stable nucleus can be induced to fission under neutron bombardment. However, any specific isotope possesses a characteristic *fission barrier*. This means that a certain minimum energy has to be supplied to deform the nucleus sufficiently to induce the process sketched in Fig. 3.4 to proceed. This activation energy can be supplied by a combination of two factors: (i) in the form of kinetic energy carried in by the bombarding neutron, and (ii) from binding energy liberated when the target nucleus absorbs the bombarding particle and becomes a different nuclide with its own characteristic mass. Both factors play roles in understanding uranium fission.

Figure 3.9 shows theoretically-computed fission barrier energies in MeV as a function of nucleon mass number A. Barrier energies vary from a maximum of about 55 MeV for isotopes with $A \sim 90$ down to a few MeV for the heaviest elements such as uranium and plutonium. For elements heavier than plutonium ($A \sim 240$), half-lives for α and β decays and spontaneous fission tend to be so short as to make them impractical candidates for weapons materials despite their low fission barriers.

In the discussion which follows, the situation for uranium is examined in detail.

In 1936, Bohr had developed a conceptual model of nuclei that is now known as the "compound nucleus" model. Based on this model, Bohr and Wheeler posited that fission is not an instantaneous process, but rather that the incoming neutron and target nucleus first combine to form an intermediate compound nucleus. Two cases are relevant for uranium:

$$\ {}_0^1 n + {}_{92}^{235}U \rightarrow {}_{92}^{236}U, \tag{3.14}$$

Fig. 3.9 Fission barrier versus mass number

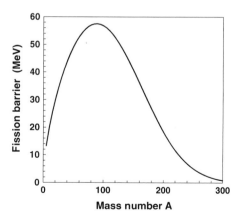

and

$$\,_0^1 n + \,_{92}^{238}U \rightarrow \,_{92}^{239}U. \qquad (3.15)$$

In accordance with their even/odd → even/even and even/even → even/odd parity changes, the Q-values of these reactions are respectively 6.55 and 4.81 MeV. If the bombarding neutrons are "slow," that is, if they bring essentially no kinetic energy into the reactions, then the nucleus of ^{236}U formed in reaction (3.14) will find itself in an excited state with an internal energy of about 6.6 MeV, while the ^{239}U nucleus formed in reaction (3.15) will have a like energy of about 4.8 MeV. In comparison, the fission barriers for ^{236}U and ^{239}U are respectively about 5.7 and 6.4 MeV. *It is the differences between the Q-values and the barrier energies that are crucial here.* In the case of ^{236}U, the Q-value *exceeds* the fission barrier by nearly 0.9 MeV. *Any bombarding neutron, no matter how little kinetic energy it has, can induce fission in* ^{235}U. On the other hand, the Q-value of reaction (3.15) falls some 1.6 MeV short of the fission barrier, which means that to fission ^{238}U by neutron bombardment requires supplying neutrons of at least this amount of energy. ^{235}U is known as a "fissile" nuclide, while ^{238}U is termed "fissionable."

In Fig. 3.10, Q–$E_{Barrier}$ is plotted as a function of target-nuclide mass number A for uranium and plutonium isotopes. The upper line is for plutonium isotopes, and the lower for uranium isotopes. Here the effect of the different parity changes appears quite strikingly as high Q–$E_{Barrier}$ values for targets whose mass numbers are odd; both uranium and plutonium possess even numbers of protons.

It appears that both ^{232}U and ^{233}U would make good candidates for weapons materials. ^{232}U is untenable, however, as it has a 70-year alpha-decay half-life. For practical purposes, ^{233}U is not convenient as it does not occur naturally, and has to be created by neutron bombardment of thorium in a reactor that is already producing plutonium. Aside from its fission-barrier issue, ^{234}U has such a low natural abundance as to be of negligible consequence (~ 0.006 %), and ^{236}U does not occur naturally at all. ^{237}U is close to having Q–$E_{Barrier} \geq 0$, but has only a 6.75-day half-life against beta-decay. In the case of plutonium, isotopes of mass numbers 236, 237, 238 and 241 have such short half-lives against various decay processes as to render them too unstable for use in a weapon even if one went to

Fig. 3.10 Energy release minus fission barrier for fission of target isotopes of uranium (*lower line*) and plutonium (*upper line*)

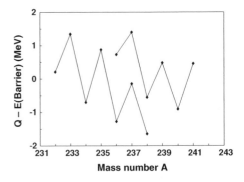

the trouble of synthesizing them in the first place: 2.87-day alpha-decay, 45-day electron capture, 88-day alpha-decay, and 14-day beta-decay, respectively. ^{240}Pu turns out to have such a high spontaneous fission rate that its very presence in a bomb core actually presents a danger of causing an uncontrollable premature detonation (Chap. 7). Plutonium-239 is the only isotope of that element that is suitable as a weapons material.

Taken together, Bohr and Wheeler's points (i) and (ii) provided the first real understanding as to why only a very few isotopes at the heavy end of the periodic table are subject to fission by slow neutrons: yet heavier ones are too near the Z^2/A limit to remain stable for very long against spontaneous breakup, while for lighter ones the fission barrier is too great to be overcome by the binding energy released upon neutron absorption. There is a very narrow "window of fissility" at the high-Z end of the periodic table.

The issue of the unsuitability of uranium-238 as a *weapons* material is actually somewhat more subtle than the above argument lets on. The average kinetic energy of secondary neutrons liberated in the fission of uranium nuclei is about 2 MeV, and about half of them have energies greater than the ~ 1.6 MeV excitation energy of the $n + {}^{238}U \rightarrow {}^{239}U$ reaction. In view of this, it would appear that ^{238}U might make a viable weapons material. Why does it not? The problem depends on what happens when fast neutrons strike ^{238}U nuclei.

When a neutron strikes and is scattered by a target nucleus (that is, if the neutron is "deflected" and goes on its way, as opposed to being absorbed or causing a fission), the collision may happen in one of two ways: elastically or inelastically. In elastic scattering the kinetic energy of the incoming neutron is unaffected. If the collision is inelastic, the neutron loses kinetic energy. The "lost" energy goes into leaving the struck nucleus in an excited energy state, analogous to a chemical reaction that leaves an electron in a higher-energy orbit.

Averaged across the range of energies of fission-produced neutrons, the effective inelastic-scattering cross-section for neutrons against ^{238}U is about 2.6 barns. In comparison, the equivalent effective fission cross-section for neutrons against ^{238}U is about 0.31 barns. The ratio of these cross-sections, $2.6/0.31 \sim 8.4$, indicates that a fast neutron striking a ^{238}U nucleus is about eight times as likely to be inelastically scattered as it is to induce a fission. Experimentally, neutrons of energy 2.5 MeV that inelastically scatter from ^{238}U have their kinetic energy reduced to a most probable value of about 0.275 MeV as a result of a *single* scattering. As a result, the vast majority of neutrons striking ^{238}U nuclei are promptly slowed to energies below the 1.6-MeV fission threshold. To make the situation worse, ^{238}U has an appreciable radiative-capture cross-section for neutrons of energy less than about 1 MeV; for energies below about 0.01 MeV, the capture cross-section is characterized by a dense forest of resonances with cross-sections of up to thousands of barns. These trends are illustrated in Figs. 3.11, 3.12, 3.13; the curves in Fig. 3.13 terminate at about 0.03 MeV at the low-energy end. At the time of Bohr and Wheeler's work, physicists were aware of the presence of these intermediate-energy capture resonances, but did not have the

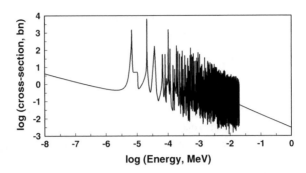

Fig. 3.11 Total neutron-capture cross-section for uranium-238 as a function of neutron energy in MeV, from 10^{-8} MeV to 1 MeV. Both scales are logarithmic. Many of the resonance capture spikes are so finely spaced that they cannot be resolved here. Data from Korean Atomic Energy Research Institute file pendfb7/U238:102. Only about 1 % of the available data is plotted here

benefit of the detailed data we now enjoy access to. It is this inelastic scattering effect that resolves the apparent contradiction involving neutron speeds in the preceding section.

In short, the non-utility of ^{238}U as a weapons material is due *not* to a lack of fission cross-section for fast neutrons, but rather to a parasitic combination of inelastic scattering and a fission threshold below which that isotope has an appreciable capture cross-section for slowed neutrons. The presence of even small amounts of ^{238}U in a *fast-neutron* environment will consequently suppress any chain reaction. ^{235}U and ^{239}Pu possess inelastic scattering cross-sections as well, but they differ from ^{238}U in that they have no fission threshold. Slow neutrons will fission ^{235}U and ^{239}Pu, and fission strongly dominates over capture for them. All of these isotopes also elastically scatter neutrons, but this is of no concern here as that process does not degrade the neutrons' kinetic energies.

To put further understanding to this fast-fission poisoning effect of ^{238}U, consider the following example. Suppose that 2-MeV fission-generated secondary neutrons lost only half their energy due to inelastic scattering. At 1 MeV, the fission cross-section of ^{235}U is about 1.22 bn, while the capture cross-section of ^{238}U is about 0.13 bn. In a sample of natural uranium, the ^{238}U to ^{235}U abundance ratio is 140:1, so capture will dominate fission by a factor of about [0.13(140)/ 1.22] = 15. *The net result is that only ^{235}U can sustain a fast-neutron chain reaction, and it is for this reason that the lighter isotope must be laboriously isolated from its more populous sister isotope if one aspires to build a "fast-fission" uranium bomb.* Further numerical details regarding chain reactions will be discussed in the next section.

Despite its non-fissility, ^{238}U did play a crucial role in the Manhattan Project. The ^{239}U nucleus formed in reaction (3.15) above sheds its excess energy in a series of two beta-decays, ultimately giving rise to plutonium-239:

$$^{239}_{92}U \xrightarrow[23.5\,\text{min}]{\beta^-} {}^{239}_{93}Np \xrightarrow[2.36\,days]{\beta^-} {}^{239}_{94}Pu. \tag{3.16}$$

Fig. 3.12 Fission cross-section for uranium-235; scales are as in Fig. 3.11. Data from Korean Atomic Energy Research Institute file pendfb7/U235:19. For thermal neutrons (log $E = -7.6$), the cross-section is 585 barns. Only about 5 % of the available data is plotted here

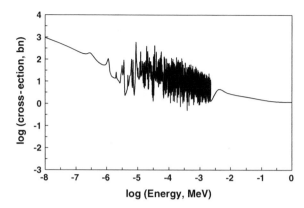

Fig. 3.13 ^{239}Pu, ^{235}U, and ^{238}U fission cross-sections and ^{238}U capture and inelastic-scattering cross-sections as functions of bombarding neutron energy. The curves terminate at about 0.03 MeV on the left end to avoid the complex array of resonance spikes evident in Figs. 3.11 and 3.12

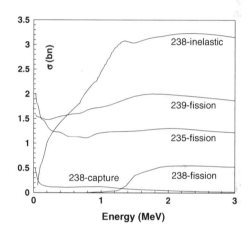

Like ^{235}U, ^{239}Pu is fissile under slow-neutron bombardment. The reaction

$$\, ^1_0 n + \, ^{239}_{94} Pu \rightarrow \, ^{240}_{94} Pu \qquad (3.17)$$

has a Q-value of 6.53 MeV, but the fission barrier of ^{240}Pu is only about 6.1 MeV. Slow neutron bombardment of Pu-239 can lead to two outcomes: fission (cross-Section 750 bn), or neutron absorption (cross-Section 270 bn), which produces semi-stable Pu-240. This latter isotope has an α-decay half-life of only 6560 years.

3.4   The Physics of Fission III: Summary

The preceding two sections covered a great deal of material that involves a number of interconnecting issues. This section offers a brief summary comparison of the possibilities for reactors and bombs.

First, consider trying to establish a chain-reaction using *slow* neutrons. The neutrons emitted in fissions will be *fast*, but are subject to the U-238 inelastic-scattering and capture problem. In order to have any hope of keeping the reaction going, the neutrons have to be moderated in order to (i) avoid being captured by U-238 nuclei, while (ii) taking advantage of the enormous fission cross-section of U-235 nuclei for thermal neutrons (585 barns; see below). However, a bomb based on such a scheme would weigh tons and be impractical to deliver to a target in any way; essentially, it would be a reactor. More important, the neutrons would be so slow that the reaction would grow at a rate not much faster than an ordinary chemical reaction. The result would be that the device would heat itself, melt, and disperse, which would allow neutrons to escape and cause the reaction to shut down. A slow-neutron bomb would create an expensive fizzle, not a "bang".

To create a reaction violent enough to warrant making a bomb requires using *fast* neutrons. In this case, the only isotope that *might* be able to sustain a fast-neutron chain reaction is U-235, but this would require separating the two isotopes of uranium in kilogram quantities. Even if the separation could be achieved, there was no guarantee in 1939 that some unanticipated effect might not arise that could render a bomb unworkable. It is not surprising that Niels Bohr thought that a weapon based on uranium fission would be impractical or impossible.

Quantifying some of the fast-versus-slow issues can help in developing a deeper appreciation of them. Some of these quantifications are straightforward, and are discussed in the following paragraphs.

One quantification involves comparing the possibilities for fast and slow-neutron chain-reactions by making use of available modern values for the cross-sections. Table 3.2 shows values for fission and radiative capture cross-sections (in barns) for U-235 and U-238 for *fast* neutrons. The last column of the Table shows overall effective cross–sections, computed by taking into account the natural fractional abundances of the isotopes: 0.0072 for U-235, and 0.9928 for U-238. For example, for fast neutrons, the overall fission cross-section is calculated as $1.235(0.0072) + 0.308(0.9928) = 0.315$. In this case, capture overwhelms fission, so there is no possibility of maintaining a chain reaction. *A fast-neutron reaction cannot be maintained with ordinary-abundance uranium.* To maintain a fast-neutron reaction, the abundance ratio must be changed to increase the fraction of U-235 in order that fission can compete against capture. The break-even point is a fractional U-235 abundance of about 0.66, an extremely difficult level of enrichment to achieve (bomb-grade uranium is defined as 90 % pure ^{235}U). Even then, fission would be only just as probable as capture.

Table 3.2 Fast-neutron cross-sections (barns)

Process	U-235	U-238	Overall cross-section weighted by abundance
Fission	1.235	0.308	0.315
Capture	0	2.661	2.642

Table 3.3 Slow-neutron cross-sections (barns)

Process	U-235	U-238	Overall cross-section weighted by abundance
Fission	584.4	0	4.208
Capture	98.8	2.717	3.409

The situation for slow (thermal) neutrons is summarized in Table 3.3. For slow neutrons, fission is the dominant process (although not overwhelmingly so), due to the enormous fission cross-section for U-235. This is what makes possible the idea of a chain-reaction using natural-abundance uranium as fuel, as some types of commercial reactors do. The hard part is to slow down the high-energy neutrons emitted in fissions of U-235 nuclei without having them be captured or otherwise lost while they are being slowed.

A second quantification relates to the issue of neutron speed. This is analyzed more fully in Sect. 3.7, but the key point is that the energy liberated by a nuclear bomb proves to be proportional to the inverse-square of the time required for a neutron to travel from where it is born in a fission to where it is likely to cause a subsequent fission. For a slow-neutron bomb, the energy yield works out to be only about 10^{-8} of that which would be liberated by a fast-neutron bomb. For a 20-kiloton TNT equivalent fast-neutron bomb, this implies that a slow-neutron bomb would release less energy than one pound of TNT. There is simply no point in making a slow-neutron bomb.

In summary, Bohr and Wheeler's work provided a solid theoretical foundation for understanding the fission process and its possibilities. But in 1939 a huge gulf lay between theoretical understanding and any possible practical application of the phenomenon. That gulf could only be filled with further experimental data on cross-sections and secondary neutron numbers, and consideration of large-scale techniques for isotope separation. Nobody could yet speak definitively regarding the prospects for a chain reaction or a bomb. But that did not mean that the possibilities could not be considered hypothetically.

3.5 Criticality Considered

Following the discovery that uranium fission did give rise to secondary neutrons, a number of physicists began to consider the conditions necessary for achieving a chain reaction, at least in theory.

Even if one has a fission to begin a putative chain reaction, the secondary neutrons that are liberated are by no means guaranteed to strike other nuclei. Some of the neutrons will inevitably reach the surface of the sample of uranium and escape, particularly if the sample is small. As the size of the sample increases, the probability that a given neutron will escape decreases, and while the probability never goes strictly to zero (unless the sample becomes infinitely large), it will eventually become low enough that a neutron is more likely to cause a subsequent fission than to escape. The key concept becomes that of a *critical mass*: the minimum mass of uranium that has to be assembled in one place in order to have a self-sustaining reaction which *in principle* continues until all of the uranium has fissioned, or (more likely) the material heats itself up and disperses.

Technically, *criticality* is said to obtain if the density of neutrons (that is, the number of neutrons per cubic meter) within the sample is increasing with time. Whether or not this condition is fulfilled depends on the density of the fissile material, its cross-sections for fission and scattering, and the number of neutrons emitted per fission. To analyze the evolution of the number of neutrons in a reactor or bomb core requires the use of time-dependent diffusion theory, which is covered in a number of technically-oriented texts. Diffusion theory goes back to classical thermodynamics, and was a well-established branch of theoretical physics by 1939.

First out of the gate was French physicist Francis Perrin (Fig. 3.14), who published a brief paper in the May 1, 1939, edition of the journal *Comptes Rendus*. Perrin applied diffusion theory to an assemblage of natural-abundance uranium in its oxide form (U_3O_8), assuming fast (unmoderated) neutrons. With rough estimates for some of the relevant parameters, he arrived at a critical mass of 40 metric tons (40,000 kg). Perrin also analyzed how this figure could be reduced by surrounding the material with a tamper. The purpose of a tamper is to reflect escaped neutrons back into the fissile material, giving them another chance at causing fissions; the net effect is to lower the critical mass. In the case of a bomb, the tamper also briefly retards the expansion of the exploding core, allowing criticality to hold for a few tenths of a microsecond longer than if no tamper were present; this yields a more efficient explosion. Perrin's 40-ton figure has no real relevance for a bomb, where pure U-235 is used. However, he did clearly establish the relevant diffusion physics, and introduced the notion of tampering.

Not far behind Perrin was German physicist Siegfried Flügge of the Kaiser Wilhelm Institute for Chemistry in Berlin, who published a much lengthier analysis in the June 9 edition of *Naturwissenschaften*. Also considering U_3O_8,

Fig. 3.14 Francis Perrin (1901–1992) in 1951. *Source* AIP Emilio Segre Visual Archives, Physics Today Collection

Flügge deduced the astounding figure that if all of the uranium in one cubic meter of U_3O_8 were to fission, the energy released could raise a cubic kilometer of water to a height of 27 km. Flügge assumed that both uranium isotopes fissioned; if fission of only U-235 is assumed, the correct height is much less, but still impressive: about 370 m. Flügge did not estimate a critical mass, but gave a figure for the critical radius of greater than 50 cm, again based on estimated parameters.

For English-language readers, the most accessible of the early criticality papers is one published by Rudolf Peierls (Fig. 3.15) of the University of Birmingham in October, 1939. Peierls was an outstanding theoretical physicist who had been born in Germany and emigrated to England in 1933. Like Otto Frisch, Peierls was Jewish; both men would come to be concerned about fission research being done in Germany.

How Peierls came to be in Birmingham is worthy of a brief digression, which also serves to introduce another important person in this history. In the fall of 1937, Marcus Oliphant, an Australian native who had been one of Rutherford's many students, was appointed head of the physics department at Birmingham. One of Oliphant's first faculty recruits was Peierls, to whom he offered a permanent Professorship. Peierls leapt at the opportunity, especially as it carried a salary over twice what he had been receiving in a non-permanent position at Cambridge. Peierls took up his position in the fall of 1937, and became a naturalized British citizen in February, 1940.

In mid-1939, with the threat of war clearly looming in Europe, Oliphant made another valuable acquisition: Otto Frisch, who was then still in Copenhagen. Not bothering with formalities, Oliphant simply invited Frisch over for a summer vacation, and found him work as an auxiliary lecturer.

Oliphant's strategic disregard for proper channels manifested itself in other productive ways. Working on radar research for the British Admiralty, he found Peierls' knowledge of electromagnetism an invaluable resource. But as an enemy

Fig. 3.15 *Left* Genia (1908–1986) and Rudolf (1907–1995) Peierls in New York 1943. *Right* Marcus Oliphant (1901–2000). *Sources* Photograph by Francis Simon, courtesy AIP Emilio Segre Visual Archives, Francis Simon Collection; AIP Emilio Segre Visual Archives, Physics Today Collection

alien, Peierls could have no official contact with classified work. Oliphant circumvented the problem by the artifice of posing questions to Peierls in the guise of their being purely academic exercises. Both men were well aware of the fiction, but it worked. Oliphant would later play a seminal role in prodding American physicists to accelerate their country's fission-bomb efforts.

In his memoirs, Peierls described how he read Perrin's paper and realized that he could refine the calculation. Given the potential military applications, he had some doubts about publishing his analysis openly, and claims that he consulted Frisch on the advisability of doing so. Confident that Bohr had shown that an atomic bomb was not a realistic proposition, Frisch saw no reason not to publish. Frisch makes no reference to such a conversation in his own memoirs, but a few months later the two would find themselves in a very different circumstance.

Peierls developed explicit formulae for estimating the critical mass in two extreme cases. These were when the number of neutrons generated per fission was very close to one, or much greater than one. In the region of practical interest, where the number is about 2.5, the two expressions turn out to not differ greatly in their predictions, and a sensible estimate can be obtained by averaging the two results. When applied in this way with modern parameter values to U-235, the predicted critical radius comes within 5 % of what later, more sophisticated, Los Alamos diffusion theory predicted. Curiously, Peierls did not bother to substitute any numbers into his expressions; his paper was entirely analytic. One cannot help but wonder if he would have published had he been in possession of even approximately accurate values for the relevant cross-sections for U-235 by itself.

3.6 Bohr Verified

Niels Bohr's February, 1939, suggestion that it was the rare 235 isotope of uranium that was responsible for slow-neutron fission begged for experimental test. The only sure way to test the hypothesis would be to isolate pure, separated samples of U-235 and U-238, and subject them both to neutron bombardment. The

Fig. 3.16 Alfred Nier
(1911–1994) *Source*
University of Minnesota,
courtesy AIP Emilio Segre
Visual Archives

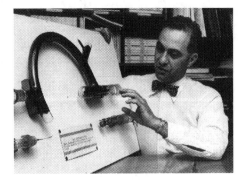

only practical method of isotope separation known at the time was mass spectroscopy. The task of preparing the samples came into the hands of a superb mass spectroscopist, Alfred Nier of the University of Minnesota (Fig. 3.16).

Nier had come to the attention of uranium physicists with a paper he had published in the January 15, 1939, edition of *Physical Review* in which he reported the discovery of a third isotope of uranium, U-234. Despite the fact that this isotope is present to the extent of only about one atom per every 18,000 atoms of U-238 in a sample of natural uranium, Nier's mass spectrometer was sensitive enough to achieve the detection. Nier met Enrico Fermi at an American Physical Society meeting held in Washington in April, 1939, at which time Fermi encouraged him to try to separate small samples of uranium isotopes in order to put Bohr's theory to experimental test. Busy with teaching and other projects, Nier did not take up the challenge until prodded again by Fermi in October of the same year.

In order to achieve sufficient isotopic separation, Nier had to build a new mass spectrometer, which he completed in February, 1940. His first successful separation runs were carried out on February 28 and 29. Nier glued the minute samples to a letter which he posted by airmail special delivery to Columbia University, where they were promptly subjected to slow-neutron bombardment.

The samples Nier isolated were truly miniscule. He did two separation runs, of 10 and 11 hour durations, which he predicted yielded 0.17 and 0.29 micrograms of U-238, respectively, assuming that all of the ions stuck to the collector. The amounts of U-235 would have been 1/140 as much, or about 1.2 and 2.1 *nano*grams. To collect a full kilogram at a rate of 2.1 nanograms per 11 hours of operation would require some 600 million years of continuous operation, a testament to Niels Bohr's opinion of the impracticality of a U-235 bomb.

At Columbia, the U-235 samples clearly showed evidence for slow-neutron fission, while the U-238 samples showed none at all. Despite the minute sample sizes, the Columbia team was able to estimate the slow-neutron fission cross-section for U-235 as 400–500 barns; the modern value is 585. These results were reported in a paper published in the March 15, 1940, edition of the *Physical Review*, which listed Nier, Eugene Booth, John Dunning, and Aristide von Grosse as authors. Their paper closed with the observation that "These experiments emphasize the importance of uranium isotope separation on a larger scale for the investigation of chain reaction possibilities in uranium." The concept of isotopy, barely 25 years old, was about to assume enormous importance. Unfortunately, Nier's samples were too small to test for fast-neutron fission.

A follow-up paper published a month later reported further results based on operating the mass spectrometer with increased ion currents. This time the U-238 samples comprised 3.1 and 4.4 micrograms, over 10 times as much as had been obtained in the earlier runs. This was enough to allow testing U-238 for fission by both slow and fast neutrons, and that isotope was verified to fission *only* under fast-neutron bombardment. The paper did not report the energy of the fast neutrons; it must have been greater than the ~ 1.6-MeV threshold for U-238 fission discussed in Sect. 3.3. Slow-neutron fissility of U-235 was again verified, but the sample of

U-235 was too small to test for fast-neutron fission. Nier later wrote that had his budget been a few hundred dollars richer, he could have afforded better vacuum pumps, which would have facilitated obtaining a sample of U-235 large enough for the fast-neutron test.

The Nier/Columbia work received some remarkable public exposure. In the May 5, 1940, edition of the *New York Times*, science reporter William Laurence—who would later witness the *Trinity* test and the Nagasaki bombing—was overly optimistic in stating that the prospect of nuclear power was perhaps just a few months to a year distant, but otherwise gave a fairly clear description of the Columbia work, the Bohr/Wheeler nuclear parity argument, the role of slow neutrons in sustaining a chain reaction, and the fact that one pound of U-235 would be equivalent to some 15 kilotons of conventional explosive. According to Laurence, "reliable sources" indicated that the Nazi government in Germany had ordered its greatest scientists to concentrate their energies on the uranium issue. In the June edition of *Harpers Magazine*, John O'Neill reported on the work, explaining the effects of fast and slow neutrons, the neutron-absorbing effects of U-238, how a chain-reaction could work, and the difficulty of isotope separation. While some of O'Neill's speculations were over-the-top (nuclear-powered automobiles would put gasoline stations out of business), he did raise the possibility of explosives: "But ... if we use too pure a sample of Uranium 235 the process may take place at such a rapid rate that all the energy ... may be given off ... before control processes can become operative. If this condition were brought about ... we should then have not an atomic power source but an atomic energy explosive."

The Nier/Columbia work verified Bohr's hypothesis, although it did leave open the question of the fast-neutron fissility of U-235. But even as Nier and his collaborators were undertaking their work, Otto Frisch and Rudolf Peierls were considering that very question.

3.7 The Frisch-Peierls Memorandum and the MAUD Committee

It is rare for a scientific manuscript to have an impact on world affairs, but such was the importance of what has come to be called the "Frisch-Peierls Memorandum" of March, 1940. This document directly initiated British investigations which resulted in a consensus that nuclear weapons were not only feasible, but within close enough grasp to likely affect the outcome of the war. The British efforts would have significant impact on American opinions beginning about the summer of 1941, and would strongly influence a report delivered to President Roosevelt later that year. Our focus here is with the memorandum's background and technical content.

At Birmingham, Otto Frisch, like Rudolf Peierls, was barred from war research, and had plenty of free time to pursue his own interests. Aware of Bohr's prediction regarding slow-neutron fission being caused by U-235, he began to contemplate how the theory might be put to a test. Months before Alfred Nier and his collaborators performed their experiments, Frisch concluded that one approach would be to prepare a sample of uranium in which the proportion of U-235 had been artificially increased, that is, to prepare a sample of *enriched* uranium. If Bohr was correct, then the enriched sample should show an increased rate of fission under slow-neutron bombardment when compared to an unenriched sample.

Frisch began to research isotope enrichment methods, and soon zeroed in on the thermal-diffusion method described briefly in Chap. 1. This was also known as the Clusius-Dickel method, after the two German scientists, Klaus Clusius and Gerhard Dickel, who had then only recently (1938) developed and successfully applied it to enriching neon and chlorine isotopes. Frisch went so far as to have the Birmingham physics department's glassblower prepare a diffusion tube, although the experiment did not succeed. But his attention soon became drawn in a much more compelling direction.

Unexpectedly, Frisch received an invitation from the Royal Society for Chemistry to write a review article on radioactivity and subatomic phenomena for the 1940 edition of their Annual Report on the Progress of Chemistry. In his memoirs, Frisch relates that the winter of 1940 was unusually cold and snowy in Birmingham, and that he prepared the report while wrapped in a winter coat, sitting before a fire in a room which did not get warmer than 42 °F during the day and fell to below freezing at night. The image of Frisch with his coat and typewriter has a certain charm, but the preparation of the review must have been a more extended effort, as it includes references to papers published as late as December, 1940.

Ironically, Frisch opened the introduction to his paper with the statement that "The year 1940 has produced no spectacular progress in nuclear physics. The "boom" in papers about nuclear fission ... has almost faded out." Much of the report is concerned with the decay products of various bombardment reactions, with only a brief mention given to verification of Bohr's speculation that U-235 was responsible for slow-neutron fission. The possibility of a chain reaction is raised in one lone sentence, only to be dismissed. Later, Frisch wrote that when he prepared the report, he truly believed that an atomic bomb was impossible. But writing the report evidently caused his thoughts to turn back to his enrichment experiment, and at some point he began to wonder if, in the event that he could produce enough pure or highly enriched U-235, would it be possible to make a truly explosive chain reaction based on *fast* neutrons as opposed to *slow* ones? Making a rough estimate of the fission cross-section of U-235 and using the critical-size formula that had been published by Peierls the previous October, Frisch estimated, to his surprise, a critical mass on the order of a pound.

Frisch's memoirs give the impression that he worked out the critical mass first, and then discussed the result with Peierls. On considering the expected efficiency of a single Clusius-Dickel tube, they estimated that a cascade of 100,000 such tubes might be sufficient to produce enough U-235 for a bomb in a matter of weeks. As Frisch wrote: "At that point we stared at each other and realized that an atomic bomb might after all be possible."

In his own memoirs, Peierls states that Frisch approached him in February or March of 1940 with the question: "Suppose someone gave you a quantity of pure 235 isotope of uranium—what would happen?" In Peierls' telling, they then worked out the critical mass together, arriving at the figure of about a pound. They then went on to estimate, with what Peierls described as a "back of the proverbial envelope" calculation, how much energy the reaction might liberate until the uranium dispersed itself. The result was equivalent to thousands of tons of ordinary explosive. Peierls related that, in a classic understatement, they said to themselves: "Even if this plant costs as much as a battleship, it would be worth having." Alarmed at the idea that German scientists might be thinking along the same lines, Frisch and Peierls felt it their duty to inform the British government of the possibility of atomic weapons, but in a way that would keep their work secret in case German researchers *hadn't* yet thought of it. They decided to prepare a memorandum on the matter, which Peierls typed up himself rather than entrust to a secretary. They kept only one carbon copy.

Frisch and Peierls actually prepared two memoranda. The first, titled "Memorandum on the Properties of a Radioactive Super-bomb," was a relatively brief qualitative description intended for government officials. The second, titled "On the construction of a "super bomb" based on a nuclear chain reaction in uranium," ran to seven pages and was more technically detailed. Both are reproduced in Robert Serber's *Los Alamos Primer*, although many reprintings of the technical memorandum contain transcription errors when compared to a copy of the original held by the Bodleian Library of Oxford University.

The two memoranda still make for fascinating reading. The non-technical one lays out in a few pages all of the key factors concerning how such a bomb might operate, as well as the associated strategic implications. After describing how there exists a critical mass and how such a device could be triggered by rapidly bringing together two otherwise perfectly safe sub-critical pieces of uranium, Frisch and Peierls describe some of the military implications: "As a weapon, such a bomb would be practically irresistible. There is no material or structure that could be expected to resist the force of the explosion." The ethics of nuclear warfare are touched upon: "the bomb could probably not be used without killing large numbers of civilians, and this may make it unsuitable as a weapon for use by this country." On civil defense and deterrence strategy: "no shelters are available that would be effective and could be used on a large scale. The most effective reply would be a counter-threat with a similar bomb. Therefore it seems to us important to start production as soon and as rapidly as possible, even if it is not intended to use the bomb as a means of attack." Without realizing it, Frisch and Peierls were drafting a script for the later Cold War.

The estimate of about a pound for the critical mass appears in the technical memorandum. As discussed by Bernstein (2011), this serious underestimate was caused by an overestimate of the effective fast-neutron fission cross-section for U-235: Frisch and Peierls assumed 10 barns, as opposed to the true value of about 1.24. The critical mass is approximately proportional to the inverse-square of the cross-section, so an error of a factor of eight in the cross-section has a significant effect. The true critical mass is more on the order of 100 pounds, although this can be reduced by use of a tamper (as Perrin had anticipated), a refinement Frisch and Peierls did not explore.

Frisch and Peierls' critical mass estimate was erroneous, but their underlying physics was entirely sound. The technical memorandum contained one mathematical formula, which was presented without derivation: an expression for the energy E liberated by a bomb whose core is of mass M and radius R_{core}. This appeared as

$$E \sim 0.2M \left(\frac{R_{core}^2}{t^2} \right) \left(\sqrt{\frac{R_{core}}{R_{crit}}} - 1 \right). \tag{3.18}$$

In this expression, R_{crit} is the critical radius for the fissile material involved, and t is the average time that a neutron travels between being emitted in a fission and causing another fission. For fast neutrons in uranium, this is about 10 ns (see Table 7.1). If values are entered in the formula in meter-kilogram-second units, the result will be in Joules, and can be converted to kilotons (kt) TNT equivalent through the conversion factor 1 kt $\sim 4.2 \times 10^{12}$ Joules. What is remarkable about this expression is that it can be shown to be exactly equivalent to what is predicted by modeling an exploding bomb core with neutron-diffusion theory, as Robert Serber did in his 1943 *Los Alamos Primer*. Peierls must have worked out the diffusion theory "in the background" on the back of his proverbial envelope. This expression also exemplifies what generations of physics professors have told their students: work out your problem analytically first, and then substitute numerical values at the end of the derivation. That way, if a numerical value changes, it is easy to recompute the result. Frisch and Peierls were surely aware that the numbers they adopted were at best approximations which would have to be refined through further experiments.

The Frisch-Peierls energy formula shows very directly why the energy that would be liberated in a *slow*-neutron bomb would not be worth the effort of making such a device, as alluded to in Sect. 3.4. The critical radius R_{crit} depends purely on nuclear parameters such as the fission cross-section and the density of uranium; it is not affected by the speed of the neutrons. For a bomb core of a given size, the only factor in the expression that is affected by the neutron speed v is the time t, which is inversely proportional to v. (A core which contains a moderator to slow neutrons will be bigger than one that does not, but the point here is a quick order-of-magnitude estimate.) A neutron's speed is proportional to the square root of its kinetic energy K. With all other factors held constant, the ratio of energies liberated in slow-neutron and fast-neutron reactions will then compare as

$$\frac{E_{slow}}{E_{fast}} = \left(\frac{t_{fast}}{t_{slow}} \right)^2 = \left(\frac{v_{slow}}{v_{fast}} \right)^2 = \frac{K_{slow}}{K_{fast}}. \tag{3.19}$$

Taking $K_{slow} \sim 0.025$ eV and $K_{fast} \sim 2$ MeV gives $E_{slow}/E_{fast} \sim 10^{-8}$, as claimed in Sect. 3.4.

Towards the end of their technical memorandum, Frisch and Peierls emphasized a crucial qualitative difference between fission bombs and ordinary explosives. This is that in addition to the immense destructive effect of the explosion itself, the blast will distribute highly radioactive fission products over a wide area, plus material from the bomb casing that is rendered radioactive by neutron capture. Frisch and Peierls estimated that since a bomb would generate radioactivity equivalent to hundreds of *tons* of radium, it would be dangerous for anybody to enter the devastated area for several days following the explosion.

At the time they prepared their memoranda, Peierls had only recently been naturalized, and Frisch was still an enemy alien; they were not sure how to get their ideas to appropriate officials. They took their documents to Oliphant, who forwarded them to Sir Henry Tizard, Chairman of the Committee on the Scientific Survey of Air Warfare. British historian Ronald Clark found the non-technical memorandum among Tizard's papers some years later, and deduced that the documents reached him on March 19, 1940. In another confluence of fission-history events, this was just four days after the publication date of the Nier-Columbia verification that it was indeed U-235 that was responsible for slow-neutron fission.

As it happened, Tizard had already been in contact regarding fission with George P. Thomson, a professor of physics at Imperial College, London (and son of J. J. Thomson of electron-discovery fame). In April, 1939, when Hans von Halban and his collaborators had published their discovery of approximately three neutrons emitted per fissioning uranium nucleus, Thomson had begun to consider the possibility of achieving a chain reaction if a sufficient mass of uranium could be brought together. Tizard was initially very skeptical of the idea that any practical form of bomb could be made with uranium, but had to take the possibility seriously.

James Chadwick initially also very much doubted the idea of a uranium bomb, but began to reconsider with publication of the Bohr-Wheeler theory in September, 1939. In early December he wrote to Professor Edward Appleton, the Secretary of the Department of Industrial and Applied Research, to express his concern. Appleton referred Chadwick to Thomson, who replied that he had done some work on slow-neutron fission, but that it did not look promising. By February, 1940—about the time Frisch and Peierls were reconsidering the matter—Thomson had almost come to the conclusion that atomic energy was not worth pursuing as a war effort; his group at Imperial College had tried to a achieve a chain reaction, but had been unsuccessful. Despite this, Chadwick began to ready his cyclotron at Liverpool to make measurements of the fission cross-section of uranium for *fast* neutron bombardment.

It was against this background that that the Frisch-Peierls memorandum reached Tizard, who prevailed upon Thomson to convene a committee to investigate the matter. Thomson served as chair; the members included, among others, Chadwick and Oliphant. Frisch and Peierls, being refugees, were barred from serving on the committee, and so were initially excluded from learning what happened in response to their memoranda. They appealed to Thompson, and it was agreed that they could serve as consultants. In March, 1941, the work of the committee was split into two groups, a Policy Committee and a Technical Committee; Frisch and Peierls were allowed to serve on the latter.

Thomson's group came to be called the MAUD Committee. This unusual name had a curious provenance. In April, 1940, Germany occupied Denmark. As this was happening, Niels Bohr sent a telegram to Otto Frisch through Lise Meitner, the six concluding words of which were "Tell Cockcroft and Maud Ray Kent." Cockcroft was John Cockcroft of Cambridge University, but the identity of "Maud Ray Kent" was a mystery. One theory was that by changing the "y" to an "i", "Maud Ray Kent" became an anagram for "radium taken." Somebody suggested MAUD as a cover name for the committee, and the appellation stuck. Officially, it had periods between the letters (M.A.U.D.), but I will use the simplified form. The mystery was not resolved until after Bohr escaped from Denmark to Sweden in late 1943, and then made his way to England. Maud Ray lived in Kent, and had at one time served as a governess for Bohr's children.

The MAUD committee held its first meeting on April 10, 1940, in the committee room of the venerable Royal Society in London. Within weeks, the Battle of Britain would be in full engagement. By the summer of 1940, research under MAUD auspices was underway at the universities of Liverpool (cross-section measurements), Birmingham (uranium chemistry), Cambridge and Oxford (separation methods), and at Imperial Chemical Industries. Peierls spent the summer studying isotope separation methods, and reported in September that the most promising approach looked to be gaseous diffusion through a mesh of fine holes; experiments along this line were being conducted by another refugee scientist, Franz Simon, at Oxford. By December, Simon's group was far enough along to estimate parameters for an actual production plant. For an output of 1 kg of U-235 per day, some 70,000 square meters (17 acres) of diffusion membrane would be required; the plant would cover some 40 acres, and consume power at a rate of about 60 megawatts. The estimate of the plant area would prove strikingly accurate: the K-25 diffusion plant in Tennessee would cover about 46 acres. Estimates of the cost of plant construction and the necessary number of operators proved far too low, but the important thing was that, in Britain at least, thoughts on atomic bombs were moving toward practical engineering considerations.

At the same time as enrichment techniques were being considered, James Chadwick's cross-section measurements were tending toward confirming Frisch and Peierls' theoretical analysis. Chadwick's initial skepticism began to turn to a gnawing worry. From a 1969 interview: "I remember the spring of 1941 ... I realized then that a nuclear bomb was not only possible—it was inevitable. ... I had many sleepless nights. ... And I had then to start taking sleeping pills. It was the only remedy. I've never stopped since then. It's been 28 years, and I don't think I've missed a single night in all those 28 years."

By March, 1941, Rudolf Peierls was convinced that a bomb was distinctly possible, writing that "there is no doubt that the whole scheme is feasible ... and that the critical size for a U sphere is manageable." On April 9, he reported his conclusion to a meeting of the MAUD committee. In early summer, the committee began to prepare its final report to Tizard.

The MAUD report would have a significant impact in America. So as not to get chronologies too far out of alignment, however, a description of the details of the report is deferred to Sect. 4.4. For now, our story goes back across the Atlantic to pick up on some important contemporaneous events occurring in America.

3.8 Predicting and Producing Plutonium

At about the time that Otto Frisch and Rudolf Peierls were re-evaluating the possibility of uranium bombs, an idea for extracting atomic energy in an indirect way from the apparently inert U-238 isotope was also being developed. The idea occurred to Princeton University physicist Louis Turner (Fig. 3.17), who had published a magisterial review article on nuclear fission in early 1940. Turner wrote up his speculation in a brief paper dated May 29, 1940, which he submitted to the *Physical Review*. In accordance with wartime censorship guidelines, he voluntarily withheld publication until after the war; it eventually appeared in April, 1946.

To understand Turner's idea, look back at how isotope parities change upon neutron absorption (Sect. 3.2):

$$neutron + \left\{ \begin{array}{c} even/even \\ odd/even \end{array} \right\} \rightarrow \left\{ \begin{array}{c} even/odd \\ odd/odd \end{array} \right\} + 5\ \mathrm{MeV} \tag{3.20}$$

and

$$neutron + \left\{ \begin{array}{c} even/odd \\ odd/odd \end{array} \right\} \rightarrow \left\{ \begin{array}{c} even/even \\ odd/even \end{array} \right\} + 6.5\ \mathrm{MeV}. \tag{3.21}$$

Fig. 3.17 Louis Turner (1898–1977). *Source* Argonne National Laboratory, courtesy AIP Emilio Segre Visual Archives, Physics Today Collection

When a U-238 nucleus (*even/even*) takes in a neutron, it becomes U-239 (*even/odd*). Turner's insight was based on the understanding that neutron-rich nuclei tend to suffer β^- decays and transmute to elements of greater atomic number, as Fermi thought he had achieved with uranium bombardment in late 1934. As mentioned in Sect. 3.2, Turner realized that U-239 might then undergo one or two such decays, creating new transuranic elements:

$$n + {}^{238}_{92}U \rightarrow {}^{239}_{92}U \xrightarrow[?]{\beta^-} {}^{239}_{93}X \xrightarrow[?]{\beta^-} {}^{239}_{94}Y \xrightarrow[?]{\beta^-} ?? \tag{3.22}$$

Here, X and Y represent the new elements. The succession of products, U-239, X-239, and Y-239, are respectively of parity *even/odd*, *odd/even*, and *even/odd*. The first and last of these are precisely the parities that tend to release greater amounts of binding energy when they themselves absorb neutrons. Turner speculated that these products might consequently be thermal-neutron fissile, and particularly drew attention to the possible *even/odd* Y-239 decay product, as it would be of the same parity as U-235 (U-239, by being neutron rich, would likely decay quite promptly). If neutron bombardment of U-238 did generate such a product and it proved stable, it could be separated from the bombarded uranium by ordinary chemical means, and hence provide a path for extracting nuclear energy from U-238. Leo Szilard would remark in a 1946 address that "With this remark of Turner, a whole landscape of the future of atomic energy arose before our eyes in the Spring of 1940 and from then on the struggle with ideas ceased and the struggle with the inertia of Man began." With timing that Szilard would have appreciated, this very landscape was already being opened on the other side of the country.

One of the first confirmations of uranium fission had been at Ernest Lawrence's Radiation Laboratory at Berkeley, and work on elucidating the nature of that process continued there. In the March 1, 1939, edition of the *Physical Review*, Edwin McMillan (Fig. 3.18) reported on an experiment where a thin foil of uranium was placed against a stack of aluminum foils, and then exposed to neutrons from a cyclotron. Fission products ejected from the uranium were collected in the aluminum foils, from which they could be chemically extracted and their decay schemes studied.

McMillan observed that following the neutron bombardment, the uranium itself (not the fission products) appeared to be exhibiting two beta-decays, with half lives of approximately 25 min and 2 days. He attributed the 25-min decay to an isotope of uranium formed by neutron capture, a suggestion that had initially been made by Meitner, Hahn, and Strassmann in 1937.

In June, Emilio Segrè confirmed that the 25-min decayer (by then refined to 23 min) was indeed a uranium isotope (U-239), and also determined that since the 2-day decayer could be chemically separated from uranium, it must be a different element. Segrè's suspicion was that the product of the 23-min decay was a long-lived isotope of element 93, while the 2-day source was an isotope of some rare-earth element, presumably a fission product. If the 23-min decay product was truly

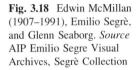

Fig. 3.18 Edwin McMillan (1907–1991), Emilio Segrè, and Glenn Seaborg. *Source* AIP Emilio Segre Visual Archives, Segrè Collection

an isotope of element 93, it would mean that a transuranic element had finally been synthesized. A loose end in Segrè's interpretation, however, was that if the 2-day source was a fission product, it behaved anomalously in that it remained stuck in the bombarded uranium as opposed to being ejected; this was something to be followed up.

The next installment in this story is a brief paper prepared a year later by McMillan and Philip Abelson. Their paper was dated May 27, 1940, just two days before Louis Turner's speculation on the possible fissility of uranium decay products. However, unlike Turner's paper, McMillan and Abelson's report was published promptly, in the June 15 edition of the *Physical Review*. They reported two key observations. These were that the McMillan/Segrè 2-day source (refined to 2.3 days) did *not* in fact behave like a rare-earth element, and that there was a clear relationship between the decay of the 23-min substance and the growth of the 2.3-day decayer: the latter was evidently a decay product of the former. The reaction and decay scheme appeared to be exactly as Turner had speculated:

$$\ _{0}^{1}n + \ _{92}^{238}U \rightarrow \ _{92}^{239}U \xrightarrow[23.5\ \mathrm{min}]{\beta^-} \ _{93}^{239}X \xrightarrow[2.36\ \mathrm{days}]{\beta^-} \ _{94}^{239}Y \qquad (3.23)$$

The names *neptunium* and *plutonium* were not yet assigned to X and Y.

Given the potential of Y-239 as a source of atomic energy, it seems surprising that McMillan and Abelson published their result. They may have been unaware of Turner's speculation when they prepared their paper, and the possibility of Y-239 as a weapons material might simply not have occurred to them, although this seems hard to imagine. James Chadwick was so upset with the publication, however, that he placed an official protest through the British Embassy.

McMillan and Abelson's work came to the attention of Glenn Seaborg, who in the summer of 1939 had been appointed as an Instructor of Chemistry at Berkeley after completing his Ph.D. at the same institution. Seaborg resolved to search for the product of the 2.3-day beta-decay of element 93, which was presumed to be an isotope of element 94. McMillan had detected indications of a long-lived alpha-activity building up in a sample of purified element 93; Seaborg suspected that the alphas might be a decay signature of element 94. Fortunately, Seaborg was a meticulous diary-keeper, and bequeathed history a nearly day-by-day record of his life and work.

Seaborg teamed up with fellow faculty member Joseph Kennedy and graduate student Arthur Wahl, who would study element 93 for his doctoral thesis. With access to Lawrence's 60-in. cyclotron, the group was able to create their samples of element 93 by bombarding uranium targets, which were usually in the form of uranium-nitrate-hydride, UNH. The bombardment method depended on what other experiments were underway with the cyclotron. Two methods were used, and both played roles in the discovery of plutonium. In one, which was first used on August 30, 1940, a beryllium target was bombarded with deuterons to create neutrons via the reaction $^{2}_{1}H + ^{9}_{4}Be \rightarrow ^{1}_{0}n + ^{10}_{5}B$. The neutrons would then be thermalized with paraffin and allowed to strike the target, presumably giving rise to elements 93 and 94 via the above sequence. The ultimate goal was to detect the presence of element 94 via its own alpha-decay:

$$^{1}_{0}n + ^{238}_{92}U \rightarrow ^{239}_{92}U \xrightarrow[23.5\,\text{min}]{\beta^{-}} ^{239}_{93}Np \xrightarrow[2.36\,\text{days}]{\beta^{-}} ^{239}_{94}Pu \xrightarrow[24,100\,\text{y}]{\alpha} ^{235}_{92}U \qquad (3.24)$$

By October, Kennedy had developed a counter capable of detecting alpha particles in the presence of background beta decays, and by late November Wahl had perfected a technique for isolating very pure samples of element 93. They were ready to begin their search for element 94.

The second bombardment method, first used around December 14, 1940, involved direct exposure of the uranium to accelerated deuterons. Various reaction channels are possible in this case, but a representative one is

$$^{2}_{1}H + ^{238}_{92}U \rightarrow 2\left(^{1}_{0}n\right) + ^{238}_{93}Np \xrightarrow[2.103\,\text{days}]{\beta^{-}} ^{238}_{94}Pu \xrightarrow[87.7\,\text{y}]{\alpha} ^{234}_{92}U \qquad (3.25)$$

While this method generates plutonium as well, it gives rise to the short-lived Pu-238 isotope, not the even/odd Pu-239 isotope generated by direct neutron bombardment of uranium. This deuteron-bombardment reaction was historically important, however, as it was Pu-238 that Seaborg and his group first isolated; this process is considered to be the discovery reaction for plutonium. Evidence for the 2.1-day decay of neptunium-238 was detected just before Christmas, 1940. By January 5, 1941, Wahl had proven that the alpha-emitting material was definitely not element 93, and that it had chemical properties similar to rare earth elements such as thorium and actinium. By the end of January, 1941, the group felt sufficiently confident of their results to prepare a brief paper announcing the discovery of element 94, based on the fact that the 88-year alpha decayer was chemically separable from both uranium and element 93. Dated January 28, 1941, the paper was withheld from publication until April, 1946, but established priority for the discovery. Our main interest here, however, is with the creation of Pu-239 via reaction (3.24).

Seaborg's goal was to produce sufficient plutonium to test its slow-neutron fissility. On January 31, 1941, he began a "practice run" neutron bombardment of over 500 grams of UNH. Within a few days a sample of element 93 of initial radioactivity 480 microcuries had been extracted, an amount equivalent to about 2 nanograms. On February 23, bombardment of a 1.2-kg sample of UNH was commenced, and proceeded intermittently until March 3. This was to be the sample from which slow-neutron fissility of the new element would be tested.

On March 6, the sample of element 93 extracted from this second bombardment gave a beta-decay count estimated at 76 millicuries, which corresponds to a mass of about 0.3 micrograms. The sample was allowed to sit for three weeks, by which time, with its half-life of only 2.3 days, essentially all of the 93 would have beta-decayed to element 94. The first test of element 94's slow-neutron fissility was carried out on March 28 using paraffin-thermalized neutrons generated in Lawrence's 37-in. cyclotron. The result was that the new element did indeed seem to be slow-neutron fissile, with a cross-section estimated to be about one-fifth that of uranium-235. The sample geometry was poor, however (it was too thick), and since it was covered in a drop of glue, the true cross-section would likely be greater.

By May 12, Wahl had succeeded in further purifying and thinning the minute sample of element 94, and a second slow-neutron experiment was begun on the 17th. This time the results gave a cross-Section 1.7 times greater than that for U-235, in fair accord with the modern-day value of about 1.3. They were also able to estimate the alpha-decay half life at roughly 30,000 years. The slow-neutron fissility was reported in a paper dated May 29, 1941, which also had to wait until 1946 for publication. On May 19, Seaborg related the result to Ernest Lawrence, who promptly phoned Arthur Compton at the University of Chicago with the news. Compton had just finished preparing a report on behalf of the National Academy of Sciences concerning possible military applications of atomic fission (Sect. 4.3). If element 94 bred from U-238 was indeed so fissile, Seaborg and his team had just increased the amount of potential bomb material by a factor of over 100.

Plutonium is one of the most unusual elements known. As described by former Los Alamos National Laboratory Director Siegfried Hecker, it seems an element at odds with itself. With little provocation, its density can change by as much as 25 percent; it can be brittle or malleable; expands when solidifying from a liquid; tarnishes within minutes; reacts vigorously with oxygen, hydrogen, and water; its own alpha-decay causes self-irradiation damage that can fundamentally change its crystalline properties; and its corrosion products can spontaneously combust in air. As was discovered at Los Alamos in the spring of 1944, plutonium is further unusual in that it exhibits five different "allotropic forms" between room temperature and its melting point: that is, it exhibits different crystal structures as a function of temperature (Fig. 3.19; six such forms are now known). The allotropes all have different densities and mechanical properties, which can affect alloying properties and corresponding critical masses. To top it off, plutonium is, as Glenn Seaborg described it, "fiendishly toxic, even in small amounts."

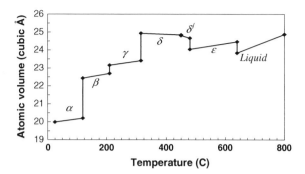

Fig. 3.19 Effective volume of individual plutonium atoms (cubic Ångstroms) versus temperature (C). Allotropic phases are identified by Greek letters. Vertical line segments correspond to phase transitions. Note that density *increases* with temperature in the δ and δ-primed phases; that is, Pu contracts within those phases. Data from F. W. Schonfeld and R. E. Tate, "The Thermal Expansion Behavior of Unalloyed Plutonium," Los Alamos report LA-13034-MS (September 1996)

These unusual properties of plutonium were entirely unknown in the spring of 1941, but in time they and others would cast into serious doubt the idea of using plutonium in a weapon. These complications, however, should not detract from appreciation of the incredible adroitness of Seaborg and his collaborators with microchemical experimentation and their intimate understanding of the radio-chemistry of heavy elements.

Due to the efforts of investigators like Nier and Seaborg, it was appreciated by the spring of 1941 that two routes to fissile material for nuclear weapons were possible: isolating uranium-235, and breeding plutonium by neutron bombardment of uranium-238. But only micrograms of either U-235 or Pu-239 had been isolated; to secure the kilograms that would be necessary to make a bomb would require an industrial-scale effort. In London and Washington, the organization of such efforts was coming under increasing official scrutiny. These considerations are the subject of Chap. 4.

Exercises

3.1 Assume uranium oxide, U_3O_8, to be composed entirely of U-238 and O-16. What will be its atomic weight? The modern value for the density of U_3O_8 is 8,380 kg/m^3. If every atom of uranium in a cubic meter of U_3O_8 fissions with a release of 170 MeV of energy, how high could one cubic kilometer of water be raised? How does your result compare with Siegfried Flügge's estimate of 27 km? [Ans: 842 gr/mol, and about 50 km. The discrepancy is due to the fact that Flügge took the density to be about 4,200 kg/m^3]

3.2 In investigating the energetics of fission, an important factor is the electrostatic self-potential-energy of the nucleus. From electromagnetic theory, the self-potential U_{self} of a sphere of radius R throughout which a total electrical charge Q is uniformly distributed is given by

$$U_{self} = \frac{3\,Q^2}{20\,\pi\,\varepsilon_o R}.$$

For a nucleus of Z protons, $Q = Ze$. Empirically, the radii of nuclei depends on their nucleon number as $R \sim a_o A^{1/3}$, where $a_o \sim 1.2 \times 10^{-15}$ m. Hence the self-potential can be written as

$$U_{self} = \frac{3\,e^2}{20\,\pi\,\varepsilon_o a_o}\left(\frac{Z^2}{A^{1/3}}\right).$$

Show that the group of physical and numerical constants here reduces to a value of 0.72 MeV. This quantity is usually abbreviated as a_C, the "Coulomb energy constant" for nuclei.

3.3 Refer to the previous problem. As sketched below, suppose that a spherical nucleus of atomic number Z and nucleon number A fissions into two identical spherical nuclei each of atomic number $Z/2$ and nucleon number $A/2$. Nuclei are essentially incompressible, so the radius of each product nucleus must be $2^{-1/3}$ times that of the original nucleus in order to conserve volume.

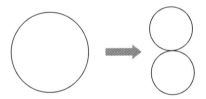

Show that the self-potential energy of the fissioned system when the product nuclei are just touching is given by

$$U_{fission} = \frac{17}{12(2^{2/3})}\left(a_C\frac{Z^2}{A^{1/3}}\right)$$

HINT: Do not forget the potential energy contributed by now having two charges $(Ze/2)$ a distance $2(2^{-1/3})R_{original}$ apart; recall the Coulomb potential energy $Q_1 Q_2 / 4\pi\varepsilon_o d$ for two charges separated by distance d. Apply your results to the case of $Z = 92$ and $A = 235$ to show that the potential energy of the fissioned system is about 100 MeV *less* than the original system. The "lost" 100 MeV must appear in the form of kinetic energy of the product nuclei.

3.4 How would you classify the parity of protactinium, $^{231}_{91}Pa$? Based on Eqs. (3.12) and (3.13), would you expect this isotope to behave like U-235 or

U-238 under neutron bombardment? Experimentally, protactinium fissions only under fast-neutron bombardment [Grosse, Booth, and Dunning; *Phys. Rev.* **56**, 382 (1939)].

Further Reading

Books, Journal Articles, and Reports

P. Abelson, Cleavage of the uranium nucleus. Phys. Rev. **55**, 418 (1939)

H.L. Anderson, E.T. Booth, J.R. Dunning, E. Fermi, G.N. Glasoe, F.G. Slack, The fission of uranium. Phys. Rev. **55**, 511–512 (1939a)

H.L. Anderson, E. Fermi, H.B. Hanstein, Production of neutrons in uranium bombarded by neutrons. Phys. Rev. **55**, 797–798 (1939b)

H.L. Anderson, The legacy of Fermi and Szilard. Bull. Atom. Sci. **30**(7), 56–62 (1974)

R.D. Baker, S.S. Hecker, D.R. Harbur, *Plutonium: a Wartime Nightmare but a Metallurgist's Dream*. Los Alamos Science, Winter/Spring 1983, 142–151

J. Bernstein, *Plutonium: a history of the world's most dangerous element* (Joseph Henry Press, Washington, 2007)

J. Bernstein, A memorandum that changed the world. Am. J. Phys. **79**(5), 440–446 (2011)

N. Bohr, Neutron capture and nuclear constitution. Nature **137**(3461), 344–348 (1936)

N. Bohr, Disintegration of heavy nuclei. Nature **143**(3617), 330 (1939a)

N. Bohr, Resonance in uranium and thorium disintegrations and the phenomenon of nuclear fission. Phys. Rev. **55**, 418–419 (1939b)

N. Bohr, J.A. Wheeler, The mechanism of nuclear fission. Phys. Rev. **56**, 426–450 (1939)

A. Brown, *The Neutron and the Bomb: A Biography of Sir James Chadwick* (Oxford University Press, Oxford, 1997)

D.C. Cassidy, *A Short History of Physics in the American Century* (Harvard University Press, Cambridge, 2011)

R.W. Clark, *Tizard* (The M.I.T. Press, Cambridge, 1965)

K. Clusius, G. Dickel, Neus Verfahren zur Gasentmischung und Isotopentrennung. Naturwissenschaften **26**, 546 (1938)

K. Clusius, G. Dickel, Zur trennung der chlorisotope. Naturwissenschaften **27**, 148–149 (1939)

R.H. Condit, Plutonium: an introduction. Lawrence Livermore national laboratory report UCRL-JC-115357 (1993). <www.osti.gov/bridge/product.biblio.jsp?osti_id=10133699>

E. Crawford, R.L. Sime, M. Walker, A Nobel Tale of Postwar Injustice. Phys. Today **50**(9), 26–32 (1997)

I. Curie, P. Savitch, Sur les radioéléments formes dans l'uranium irradié per les neutrons. J. Phys. Radium **8**(10), 385–387 (1937)

I. Curie, P. Savitch, Sur les radioéléments formes dans l'uranium irradié per les neutrons. II. J. Phys. Radium **9**(9), 355–359 (1938)

N.I. Fetisov, Spectra of neutrons inelastically scattered on U-238. At. Energ. **3**, 995–998 (1957)

S. Flügge, Kann der Energieinhalt der Atomkerne technisch nutzbar gemacht werden? Naturwissenschaften **27**, 402–410 (1939)

K.W. Ford, John Wheeler's work on particles, nuclei, and weapons. Phys. Today **62**(4), 29–33 (2009)

R.D. Fowler, R.W. Dodson, Intensely ionizing particles produced by neutron bombardment of uranium and thorium. Phys. Rev. **55**, 417–418 (1939)

O.R. Frisch, Physical evidence for the division of heavy nuclei under neutron bombardment. Nature **143**(3616), 276 (1939)

O.R. Frisch, Radioactivity and subatomic phenomena. Annu. Rep. Prog. Chem. **37**, 7–22 (1940)

O.R. Frisch, The Discovery of Fission. Phys. Today **20**(11), 43–52 (1967)

O.R. Frisch, *What little I remember* (Cambridge University Press, Cambridge, 1979)

M. Gowing, *Britain and atomic energy 1939–1945* (St. Martin's Press, London, 1964)

H.G. Graetzer, Discovery of nuclear fission. Am. J. Phys. **32**(1), 9–15 (1964)

G.K. Green, L.W. Alvarez, Heavily ionizing particles from uranium. Phys. Rev. **55**, 417 (1939)

O. Hahn, F. Strassmann, Über die Enstehung von Radiumisotopen aus Uran beim Bestrahlen mit schnellen und verlangsamten Neutronen. Naturwissenschaften **26**, 755–756 (1938)

O. Hahn, F. Strassmann, Concerning the existence of alkaline earth metals resulting from neutron irradiation of uranium. Naturwissenschaften **27**, 11–15 (1939) [Title translated]

P. Halpern, Washington: a DC circuit tour. Phys. Perspect. **12**(4), 443–466 (2010)

D. Hawkins, Manhattan District History. Project Y: the Los Alamos project. Volume I: Inception until August 1945. Los Alamos, NM: Los Alamos Scientific Laboratory (1947; Los Alamos publication LAMS-2532). Available online at <http://library.lanl.gov/cgi-bin/getfile?LAMS-2532.htm>

S.S. Hecker, Plutonium and its alloys: from atoms to microstructure. Los Alamos Sci. **26**, 290–335 (2000)

L. Hoddeson, P.W. Henriksen, R.A. Meade, C. Westfall, *Critical Assembly: A Technical History of Los Alamos During the Oppenheimer Years* (Cambridge University Press, Cambridge, 1993)

R.L. Kathren, J.B. Gough, G.T. Benefiel, The plutonium story: the journals of Professor Glenn T. Seaborg 1939–1946. Battelle Press, Columbus, Ohio (1994). An abbreviated version prepared by Seaborg is available at <http://www.escholarship.org/uc/item/3hc273cb?display=all>

J.W. Kennedy, G.T. Seaborg, E. Segrè, A.C. Wahl, Properties of 94(239). Phys. Rev. **70**(7–8), 555–556 (1946)

Knolls Atomic Power Laboratory, Nuclides and Isotopes: Chart of the Nuclides, 17th edn; <http://www.nuclidechart.com/>

W.L. Laurence, Vast power source in atomic energy opened by science. The New York Times, May 5, 1940, pp. 1 and 51

E. McMillan, Radioactive recoils from uranium activated by neutrons. Phys. Rev. **56**, 510 (1939)

E. McMillan, P.H. Abelson, Radioactive element 93. Phys. Rev. **57**, 1185–1186 (1940)

L. Meitner, O.R. Frisch, Disintegration of uranium by neutrons: a new type of nuclear reaction. Nature **143**(3615), 239–240 (1939)

New York Times: Atom Explosion Frees 200,000,000 Volts; New Physics Phenomenon Credited to Hahn. 29 Jan 1939, p. 2

A.O. Nier, The isotopic constitution of uranium and the half-lives of the uranium isotopes. I. Phys. Rev **55**, 150–153 (1939)

A.O. Nier, Some reminiscences of mass spectroscopy and the manhattan project. J. Chem. Educ. **66**(5), 385–388 (1989)

A.O. Nier, E.T. Booth, J.R. Dunning, A.V. Grosse, Nuclear fission of separated uranium isotopes. Phys. Rev. **57**, 546 (1940a)

A.O. Nier, E.T. Booth, J.R. Dunning, A.V. Grosse, Further experiments on fission of separated uranium isotopes. Phys. Rev. **57**, 748 (1940b)

J.J. O'Neill, Enter atomic power. Harpers Magazine **181**, 1–10 (1940)

R. Peierls, Critical conditions in neutron multiplication. Proc. Camb. Philos. Soc **35**, 610–615 (1939)

R. Peierls, *Bird of passage: recollections of a physicist* (Princeton University Press, Princeton, 1985)

F. Perrin, Calcul relative aux conditions éventuelles de transmutation en chaine de l'uranium. Comptes Rendus **208**, 1394–1396 (1939)

B.C. Reed, Simple derivation of the Bohr-Wheeler spontaneous fission limit. Am. J. Phys. **71**(3), 258–260 (2003)

B.C. Reed, Rudolf Peierls' 1939 analysis of critical conditions in neutron multiplication. Phys. & Soc. **37**(4), 10–11 (2008)

B.C. Reed, The Bohr-Wheeler spontaneous fission limit: an undergraduate-level derivation. Eur. J. Phys. **30**, 763–770 (2009)

B.C. Reed, A desktop-computer simulation for exploring the fission barrier. Nat. Sci. **3**(4), 323–327 (2011a)

B.C. Reed, *The Physics of the Manhattan Project* (Springer, Berlin, 2011b)

R. Rhodes, *The Making of the Atomic Bomb* (Simon and Schuster, New York, 1986)

R.B. Roberts, R.C. Meyer, L.R. Hafstad, Droplet formation of uranium and thorium nuclei. Phys. Rev. **55**, 416–417 (1939)

F.W. Schonfeld, R.E. Tate, The thermal expansion behavior of unalloyed plutonium. Los Alamos report LA-13034-MS, Sept 1996

G.T. Seaborg, E.M. McMillan, J.W. Kennedy, A.C. Wahl, Radioactive element 94 from deuterons on uranium. Phys. Rev. **69**(7–8), 366–367 (1946a)

G.T. Seaborg, A.C. Wahl, J.W. Kennedy, Radioactive element 94 from deuterons on uranium. Phys. Rev. **69**(7–8), 367 (1946b)

E. Segrè, An unsuccessful search for transuranic elements. Phys. Rev. **55**, 1104–1105 (1939)

E. Segrè, *Enrico Fermi, Physicist* (University of Chicago Press, Chicago, 1970)

R. Serber, *The Los Alamos Primer: the First Lectures on How To Build an Atomic Bomb* (University of California Press, Berkeley, 1992)

R.L. Sime, Lise Meitner and the Discovery of Fission. J. Chem. Educ. **66**(5), 373–376 (1989)

R.L. Sime, *Lise Meitner: A Life in Physics* (University of California Press, Berkeley, 1996)

R.L. Sime, The Search for Transuranium Elements and the Discovery of Nuclera Fission. Phys. Perspect. **2**(1), 48–62 (2000)

R. Stuewer, Bringing the news of fisison to America. Phys. Today **38**(10), 49–56 (1985)

R. Stuewer, The origin of the liquid-drop model and the interpretation of nuclear fission. Perspect. Sci. **2**, 76–129 (1994)

R. Stuewer, Gamow, alpha decay, and the liquid-drop model of the nucleus. Astron. Soc. Pac. Conf. Ser. **129**, 30–43 (1997)

L. Szilard, W.H. Zinn, Instantaneous emission of fast neutrons in the interaction of slow neutrons with uranium. Phys. Rev. **55**, 799–800 (1939)

L.A. Turner, Nuclear fission. Rev. Mod. Phys. **12**(1), 1–29 (1940)

L.A. Turner, Atomic energy from U^{238}. Phys. Rev. **69**(7–8), 366 (1946)

H. von Halban, F. Joliot, L. Kowarski, Number of neutrons liberated in the nuclear fission of uranium. Nature **143**(3625), 680 (1939)

S. Weinberg, *The Discovery of Subatomic Particles* (Cambridge University Press, Cambridge, 2003)

Websites and Web-based Documents

Lawrence Berkeley National Laboratory: <http://www.lbl.gov/Science-Articles/Research-Review/Magazine/1981/81fchp1.html>

Chapter 4
Organizing the Manhattan Project, 1939–1943

Effective organization and administration were vital to the success of the Manhattan Project. Between late 1939 and the end of the war, government funding of the Project would grow by a factor of over 300,000 from an initial investment of $6,000 to nearly $2 billion. Without aggressive, competent, and committed leaders of great personal integrity drawn from the ranks of civilian scientists and engineers, industrial executives, military officers, and government officials to oversee such an undertaking, the possibilities for inefficiency, lack of results, mismanagement, and outright waste would have been rife. It is a testament to the quality of these people that the record reveals both spectacular success and not even minor examples of such malfeasance. Without these individuals the Project could never have been mounted and carried out as effectively as it was.

Examining the history of the administration of the Project is valuable not only for getting a sense of how its leaders kept many threads of activity on track and coordinated, but also for dispelling the popular myth that America paid little attention to possible military applications of nuclear fission until after the Japanese attack on Pearl Harbor in December, 1941. The reality was much different. The ominous possibilities for nuclear energy were recognized soon after the discovery of fission, and research to explore the relevant properties of uranium began in 1939. This background research may not hold the drama of starting up a reactor or detonating a bomb, but it was vital for determining if these things could be done.

This chapter relates the administrative history of the Project from the fall of 1939 to early 1943, when the Army's Manhattan Engineer District (MED) began to oversee the construction and operation of vast facilities for enriching uranium, synthesizing plutonium, and establishing the parameters of bomb physics and design. This history is related largely in chronological order, with occasional diversions for coherence.

As described in Chap. 3, the understanding that uranium fissioned under slow-neutron bombardment was beginning to become established by mid-1939, and the notion that there appeared to be two possible methods of liberating nuclear energy on a large scale was gaining currency by the spring of 1941. One method would be to isolate a sample of the fissile U-235 isotope from a supply of uranium ore, and use it to create an explosive fast-neutron reaction. The other would be to construct

B. C. Reed, *The History and Science of the Manhattan Project*,
Undergraduate Lecture Notes in Physics, DOI: 10.1007/978-3-642-40297-5_4,
© Springer-Verlag Berlin Heidelberg 2014

some sort of slow-neutron reactor to breed plutonium, which could also be used to make a bomb. The purpose of this chapter is to explore how experimental nuclear physics was transformed into a project to produce a practical nuclear weapon. Our story for this chapter opens in early 1939, with physicists' first attempts to alert government officials to the potentialities of fission.

4.1 Fall 1939: Szilard, Einstein, the President, and the Uranium Committee

The first formal contact between nuclear scientists and government representatives occurred on March 18, 1939, when, at a meeting set up by Columbia University Dean of Science George Pegram, Enrico Fermi met with naval officers in Washington to explain the possibilities of using chain reactions as power sources or in bombs. One of the officers present was Admiral Stanford Hooper, technical assistant to the Chief of Naval Operations; also present was Ross Gunn, a civilian physicist working for the Naval Research Laboratory who would become involved with the liquid thermal diffusion project for uranium enrichment. The group decided to donate $1,500 to Columbia to help advance Fermi's research.

By 1939, Leo Szilard was living in New York, where, although independently wealthy, he maintained a part-time appointment at Columbia University. Szilard was much more alarmed than Fermi at the possibility of nuclear fission being turned into a powerful weapon, and felt that responsible government officials needed to be alerted to the issue. He discussed the matter with fellow émigré Eugene Wigner (Fig. 4.1), a brilliant theoretical physicist and chemical engineer who had been on the faculty of Princeton University since 1930. In 1936, Wigner had predicted that scientists would figure out how to release nuclear energy; he would later make significant contributions to reactor engineering.

Fig. 4.1 Eugene Wigner (1902–1995), at the time of his receiving the Nobel Prize (1963). *Source* http://commons.wikimedia.org/wiki/File:Wigner.jpg. *Right* In this 1946 photo, Albert Einstein and Leo Szilard re-enact the preparation of a letter to President Roosevelt. *Source* Courtesy Atomic Heritage Foundation, http://www.atomicheritage.org/mediawiki/index.php/File:Einstein_Szilard.jpg

Both Szilard and Wigner had grown up in Hungary, and had witnessed the rise of totalitarianism in their native country and in Germany. On the rationale that a possible strategy would be to deny Germany access to uranium ore, they decided to warn the government of Belgium of the issue. Some of the world's richest uranium ores were in the Congo, which was then a colony of Belgium. But how could two Hungarian scientists living in America deliver such a warning? On recalling that Albert Einstein was a personal friend of Belgium's queen mother, they decided to enlist his help. On July 16, 1939, six years to the day before the *Trinity* test, Szilard and Wigner drove to Einstein's summer home on Long Island. Szilard explained the possibility of an explosive chain reaction, which apparently came as a revelation to Einstein.

Wigner suggested that a letter written by refugees on a security issue to a foreign government might not be appropriate, so they decided that Einstein—the only one with a name famous enough to be recognized—would prepare a letter to the Belgian ambassador, with a covering letter to the State Department. Einstein drafted a letter in German, which Wigner translated, had typed up, and sent to Szilard. A few days later, however, Szilard came into contact with Alexander Sachs, an economist with the Lehman Brothers financial firm. Sachs had also trained as a biologist, and was a personal friend of and advisor to President Roosevelt. Sachs suggested that a better approach would be a letter directly to the President, and he offered to deliver one personally.

Sachs is little-known outside Manhattan Project scholarship circles, but one of the most valuable sources of information on the early history of the Project is a "Documentary Historical Report" that he prepared in August, 1945. This 27-page report covers the period from the Szilard/Einstein letter to when the project was placed under the oversight of the National Defense Research Committee in June, 1940 (Sect. 4.2). Sachs wrote in a peculiarly florid manner, but was an exceptionally perceptive and foresightful observer of the rapidly-evolving world situation of the time.

Szilard, this time accompanied by Edward Teller, visited Einstein again on July 30 to revise their original work. Einstein dictated another letter, which addressed not only the issue of Congolese uranium ores, but also the possibility of a significantly destructive new type of bomb.

The text of Einstein's letter follows:

Albert Einstein
Old Grove Rd.
Nassau Point
Peconic, Long Island
August 2nd 1939

F. D. Roosevelt
President of the United States
White House
Washington, D.C.

Sir:

Some recent work by E. Fermi and L. Szilard, which has been communicated to me in manuscript, leads me to expect that the element uranium may be turned into a new and important source of energy in the immediate future. Certain aspects of the situation which has arisen seem to call for watchfulness and, if necessary, quick action on the part of the Administration. I believe therefore that it is my duty to bring to your attention the following facts and recommendations:

In the course of the last four months it has been made probable—through the work of Joliot in France as well as Fermi and Szilard in America—that it may become possible to set up a nuclear chain reaction in a large mass of uranium, by which vast amounts of power and large quantities of new radium-like elements would be generated. Now it appears almost certain that this could be achieved in the immediate future.

This new phenomenon would also lead to the construction of bombs, and it is conceivable—though much less certain—that extremely powerful bombs of a new type may thus be constructed. A single bomb of this type, carried by boat and exploded in a port, might very well destroy the whole port together with some of the surrounding territory. However, such bombs might very well prove to be too heavy for transportation by air.

The United States has only very poor ores of uranium in moderate quantities. There is some good ore in Canada and the former Czechoslovakia, while the most important source of uranium is Belgian Congo.

In view of the situation you may think it desirable to have more permanent contact maintained between the Administration and the group of physicists working on chain reactions in America. One possible way of achieving this might be for you to entrust with this task a person who has your confidence and who could perhaps serve in an inofficial capacity. His task might comprise the following:

(a) to approach Government Departments, keep them informed of the further development, and put forward recommendations for Government action, giving particular attention to the problem of securing a supply of uranium ore for the United States;

(b) to speed up the experimental work, which is at present being carried on within the limits of the budgets of University laboratories, by providing funds, if such funds be required, through his contacts with private persons who are willing to make contributions for this cause, and perhaps also by obtaining the co-operation of industrial laboratories which have the necessary equipment.

I understand that Germany has actually stopped the sale of uranium from the Czechoslovakian mines which she has taken over. That she should have taken such early action might perhaps be understood on the ground that the son of the German Under-Secretary of State, von Weizsäcker, is attached to the Kaiser-Wilhelm-

Institute in Berlin where some of the American work on uranium is now being repeated.

Yours very truly,
Albert Einstein

Sachs secured a meeting with the President for October 11, 1939. In a summarizing cover letter of his own, he explained how the discovery that uranium could be split by neutrons could lead to the creation of a new source of energy, the possibility of creating "tons" of radium for use in medical treatments, and the "eventual probability of bombs of hitherto unenvisaged potency and scope." He suggested that with the danger of a German invasion of Belgium, it was urgent that arrangements be made with the mining firm of Union Minière du Haut-Katanga, whose head office was in Brussels, to make available supplies of uranium to the United States. He also urged acceleration of experimental work in America. Since such work could no longer be carried out within the limited budgets of university physics departments, he proposed that "public-spirited executives in our leading chemical and electrical companies could be persuaded to make available certain amounts of uranium oxide and quantities of graphite, and to bear the considerable expense of the newer phases of the experimentation." Sachs also suggested that Roosevelt designate an individual or committee to serve as a liaison between the scientists and the government (Fig. 4.2).

After hearing Sachs out, the President allegedly remarked, "Alex, what you are after is to see that the Nazis don't blow us up." Roosevelt ordered his Secretary, General Edwin M. Watson, to act as the White House's liaison on the issue, and to

Fig. 4.2 President Roosevelt signs the declaration of war against Japan, December 8, 1941 *Source* http://commons.wikimedia.org/wiki/File:Franklin_Roosevelt_signing_declaration_of_ war_against_Japan.jpg

work with the Director of the National Bureau of Standards, Lyman J. Briggs, to put together an advisory committee.

Sachs met with Briggs the next day, and hey assembled an Advisory Committee on Uranium, which came to be known simply as the Uranium Committee. The initial members were Briggs himself as Chair, plus Colonel Keith Adamson of the Army and Commander Gilbert C. Hoover of the Navy; Adamson and Hoover were ordnance experts whom Sachs had briefed just prior to meeting with the President. The name, membership, organizational structure, and responsibilities assigned to this committee would change many times over the course of the war (Figs. 4.3 and 4.4). In surveying the administrative history of the Project, the various incarnations of the Uranium Committee serve as helpful focal points. The names and acronyms of various Manhattan committees can be difficult to keep straight; for quick refresher summaries, see the Glossary.

The committee held its first meeting at the Bureau of Standards on October 21; Einstein did not attend. At Sachs' initiative, Enrico Fermi, Leo Szilard, Edward Teller, and Eugene Wigner were invited; also present were physicists Fred Mohler of the Bureau of Standards and Richard Roberts of the Carnegie Institution.

Despite the skepticism of the military officers present as to the possibility of revolutionary new weapons or sources of power, Briggs argued that the world situation and American interests must be taken into account in what he called "the equation of probabilities." The War and Navy Departments contributed $6,000 for the purchase of four tons of graphite, paraffin, cadmium, and other supplies in order that Fermi could carry out neutron absorption experiments at Columbia. The committee also appointed a Science Advisory Sub-Committee, whose members were Harold Urey (Chair; Columbia University), Gregory Breit (University of Wisconsin), George Pegram (Columbia), Merle Tuve (Carnegie Institution), Jesse

Fig. 4.3 Some of the Manhattan Project's administrators, at the Bohemian Grove meeting of September, 1942 (Sect. 4.9). *Left* to *Right* Major Thomas Crenshaw, Robert Oppenheimer, Harold Urey, Ernest Lawrence, James Conant, Lyman Briggs, Eger Murphree, Arthur Compton, Robert Thornton (University of California), Col. Kenneth Nichols. *Source* Lawrence Berkeley National Laboratory, courtesy AIP Emilio Segre Visual Archives

Fig. 4.4   April, 1940. *Left* to *right* Ernest Lawrence, Arthur Compton, Vannevar Bush, James Conant, Karl Compton, Alfred Loomis. *Source* http://commons.wikimedia.org/wiki/File:LawrenceCompton BushConantComptonLoomis.jpg

Beams (University of Virginia), and Ross Gunn. Many of these men would play prominent roles in the Manhattan Project. Urey was recognized as a world leader in techniques of isotope separation; in May 1940 he would be granted a contract to investigate application of thermal diffusion, chemical separation, and centrifugation to enriching uranium. Breit was an outstanding theoretical physicist, and Beams was conducting research on high-speed centrifuges.

Leo Szilard and Enrico Fermi spent considerable time over the summer of 1939 to considering how a chain-reacting mass of uranium and graphite might be configured. Szilard, again well ahead of his time, followed up with a memorandum to Briggs on October 26, urging the purchase of 100 metric tons of graphite and 20 metric tons of uranium oxide in order to get experiments underway as soon as possible. This was not done at the time, and, as the Manhattan Project progressed, Szilard was to experience no end of frustration with what he saw as bureaucratic inertia and official foot-dragging.

Briggs' committee reported to President Roosevelt on November 1 with a brief two-page letter. After opening with a rather technical summary of the process of fission, the letter related that a chain reaction was a possibility which could eventually prove to be a power source for submarines, and noted that if a nuclear reaction should be explosive, "it would provide a possible source of bombs with a destructiveness vastly greater than anything now known." The letter recommended that four tons of graphite be procured for experiments, which, if successful, would lead to a requirement for 50 t of uranium oxide; no mention was made of the $6,000 allocated to Columbia. They also recommended that the main committee be enlarged by the addition of Karl Compton, President of the Massachusetts Institute of Technology (and brother of physics Nobel Laureate Arthur Compton), Sachs, Einstein, and Pegram. Also added to the group at some point before the summer of 1940 was Admiral Harold G. Bowen, Director of the Naval Research Laboratory.

Watson acknowledged Briggs's report on November 17, indicating that the President would keep it on file for reference. Not until February 8, 1940, did

Watson follow up, asking Sachs and Briggs if there was anything new to report. Briggs replied on February 20 to indicate that the $6,000 that had been authorized the preceding October had been transferred to Columbia, and that he was waiting to be informed of results of the work.

Through the fall and winter of 1939/1940, scientists had been far from idle, however. Sachs' Historical Report lists several areas of experimental and theoretical research that were ongoing at the time: slow neutron reactions; fast neutron reactions; uranium isotope studies; isotope separation by diffusion, centrifugation, and other means; and production of uranium metal. Groups were active at Columbia, Princeton, the Carnegie Institution, Harvard, Yale, MIT, the University of Virginia, and George Washington University. Sachs did not mention the work on creating and isolating plutonium that was also underway at Berkeley; he may not have been aware of it.

In response to Watson's February 8 request for an update, Sachs responded on the 15th that he felt that the tone of the November 1 report had been too academic, and that possible practical applications should have been emphasized first. He promised Watson another letter from Einstein within a month. Einstein's letter, dated March 7, indicated that work on fission was being accelerated in Germany, and that Szilard had prepared a manuscript on how to set up a chain reaction. Sachs transmitted the letter to Roosevelt on March 15, about the time that the Frisch-Peierls memorandum began its journey up the chain of command in England (Sect. 3.7).

Watson replied to Sachs on March 27 to the effect that the Briggs Committee was awaiting a report on work being carried out at Columbia. Sachs had occasion to meet with Roosevelt in early April, and reiterated the importance of having Belgian ores shipped to the United States, as well as the urgency of having government or foundation funds allocated in such a way as to promote long-term research planning. Roosevelt and Watson both sent letters to Sachs on April 5, asking that another meeting be organized. Sachs encouraged Einstein to attend; he demurred, but did write Briggs on April 25 to express his conviction that the scale and speed of uranium work should be increased, and seconded a proposal by Sachs that a "Board of Trustees" be formed to solicit funds to support the work.

The pace of activity began to pick up in the spring of 1940. The Uranium Committee held its second meeting at the Bureau of Standards on Saturday, April 27, by which time Alfred Nier and his collaborators had verified that it was indeed U-235 that was responsible for slow-neutron fission. Briggs reported to Watson on May 9 that the committee was not prepared to recommend a large-scale experiment to attempt a chain reaction until the results of experiments on the neutron-absorption properties of graphite being conducted at Columbia were in, which was expected to be within a week or two. In the meantime, Fermi and Szilard were beginning to conceive of a reactor wherein a three-dimensional lattice of blocks of uranium would be distributed within a moderator.

On May 10, the same day that Germany invaded Belgium and Winston Churchill became Prime Minister of Great Britain, Sachs drafted a memorandum to himself which recorded that the next stages of the work would be to carry out a survey of nuclear constants (e.g., absorption and fission cross-sections and the like) to narrow down limits of experimental error, and then to undertake a "large-scale" experiment to demonstrate whether or not a chain reaction could be set up and maintained. The cost of these steps was estimated at $30,000 to $50,000, and $250,000 to $500,000, respectively. Still dissatisfied that work was being impeded by organizational difficulties, Sachs wrote to FDR the next day to again raise the idea of a non-profit corporation to raise funds to support research.

Sachs learned from Pegram that Fermi and Szilard had found the neutron absorption cross-section of carbon to be encouragingly small, and on May 13 wrote to Briggs with this news and a plea that the project needed to be accelerated while being kept secret. (A small absorption cross-section would mean less possibility of losing the neutrons necessary to maintain a chain reaction). Two days later, Sachs wrote to Watson to apprise him of the situation, and to suggest that the President establish a "Scientific Council of National Defense" which would be invested with authority to develop defense-related technical projects. Sachs followed up with another letter to Watson on May 23, wherein he reiterated the need to secure the Union Minière ore, and again proposed that the Uranium Committee be supplemented by a non-profit organization. Sachs raised the uranium issue yet again with the President at a White House conference on defense economics held in late May. Word must have got back to Briggs, as on June 5 he authorized Sachs to approach Union Minière to gather information on ore stocks, costs, and anticipated mine extraction rates.

In addition to drawing the attention of government officials to the prospects for nuclear energy, émigré European physicists were also instrumental in alerting the American scientific community to the need to censor publications concerning developments that could become of military importance. At a meeting of the Division of Physical Sciences of the National Research Council in April, 1940, Gregory Breit (Fig. 4.5) suggested the formation of a committee to control publication in all American scientific journals, a concept completely at odds with the historic practice of open scientific publication and debate. Various subcommittees were set up to deal with publications in a number of fields. The first one, chaired by Breit, was devoted to considering uranium fission. Well before any formal military involvement in nuclear fission, scientists had begun to police their own publication practices.

At this point, Alexander Sachs leaves our story. But one last inclusion in his Historical Report is worth mentioning. This is a five-page aide-memoir to himself prepared on April 20, 1940, under the convoluted title "Import of War Developments for Application to National Defense of Uranium Atomic Disintegration." This document opens with the observation that superior technology had enabled Nazi forces to overrun a number of European countries, and that other countries which had not brought their defenses up to the same level of technological quality could expect the same fate. He then remarked that uranium research may prove as important to national defense as the most advanced chemical and electrical research then being undertaken. Anticipating that a chain reaction would be successfully demonstrated and that war between America and Japan was likely, Sachs

Fig. 4.5 *Left* Gregory Breit (1899–1981) at the 1939 meeting of the American Physical Society. *Right* Vannevar Bush (1890–1974). *Sources* Photo by Esther Mintz, courtesy AIP Emilio Segre Visual Archives, Esther Mintz Collection; Harris and Ewing, News Service, Massachusetts Institute of Technology, *Courtesy* AIP Emilio Segre Visual Archives

argued that nuclear-propelled American naval vessels, particularly aircraft carriers armed with nuclear-bomb carrying aircraft, could easily extend their range to Japan without the need for refueling. This remarkable analysis was written over 19 months before Pearl Harbor, and some 31 months before Enrico Fermi's first demonstration of a chain reaction.

4.2 June 1940: The National Defense Research Committee; Reorganization I

In June, 1940, Lyman Briggs' Uranium Committee underwent a significant change of venue within governmental administration, as well as a change in membership. On June 27 of that year, President Roosevelt established the National Defense Research Committee (NDRC), which was charged with supporting and coordinating research conducted by civilian scientists which might have military applications. The NDRC was the brainchild of Vannevar Bush (Fig. 4.5), whom Roosevelt appointed to be its Director. A veteran of many years of government science administration, Bush had earned a Ph.D. jointly from the Massachusetts Institute of Technology (MIT) and Harvard University in 1917. During World War I he had worked with the National Research Council on the application of science to warfare, including development of submarines. After the war, Bush joined the department of Electrical Engineering at MIT, where he served as a faculty member. In 1932, he moved up to be Dean of Engineering, at which post he remained until 1938. While at MIT he developed, among other things, an early analog computer known as the differential analyzer. In 1939, Bush became President of the Carnegie Institution of Washington, as well as Chairman of the National Advisory Committee for Aeronautics (NACA). These positions enabled him to direct research toward military applications, and gave him a conduit for providing scientific advice to government officials.

During World War I, Bush had observed firsthand the lack of cooperation between civilian scientists and the military, and was determined that such inefficiency not repeat itself in the war which was engulfing Europe and would likely eventually involve America. In 1939, he began thinking of a federal-level agency to coordinate research, an idea he discussed with fellow NACA member James B. Conant, a distinguished chemist and President of Harvard University. Bush also ran the concept past his MIT colleague Karl Compton, as well as Frank Jewett, President of the National Academy of Sciences. Bush secured a meeting with President Roosevelt for June 12, 1940, and soon had his agency, which entered into official existence fifteen days later. Conant, Compton, and Jewett were made members of the new Committee, along with Richard Tolman, Dean of the graduate school at the California Institute of Technology. Compton was assigned responsibility for work in the area of radar, Conant for chemistry and explosives, Jewett for armor and ordnance, and Tolman for patents and inventions. Funded by and reporting directly to the President, the NDRC was remarkably free of Congressional and bureaucratic interference. In addition to its involvement in the Manhattan Project, the NDRC and its successor agency, the Office of Scientific Research and Development (OSRD; Sect. 4.4), were involved with the development of technologies such as radar, sonar, proximity fuses, and the Norden bomb sight.

On June 15, Briggs received a letter from President Roosevelt informing him that the Uranium Committee was being absorbed into the NDRC. On July 1, Briggs summarized the work of his Committee to that time in a letter to Bush. Fermi's measurements of neutron absorption in carbon looked promising as far as eventually obtaining a chain reaction was concerned, and the Science Advisory Subcommittee felt that there was justification to pursue work in two directions: (1) methods of separating U-235, and (2) further measurements towards determining the feasibility of a chain reaction in natural uranium. For item (1), an allotment of $100,000 had been made by the Army and Navy to investigate centrifugal and thermal diffusion methods; this work was being administered by the NRL. For item (2), Briggs recommended that the NDRC provide $140,000. An NDRC meeting held the next day included a resolution that the Committee on Uranium be constituted as a special committee of the NDRC, with membership of Briggs (chair), Beams, Breit, Gunn, Pegram, Sachs, Tuve, and Urey. Einstein, Bowen, Adamson, and Hoover had been dropped from the October 1939 incarnation of the group, but the minutes indicate that Bowen would continue to follow the activities of the committee. Bowen was apparently present at the meeting, however, as the minutes also record that he related that the Navy was coordinating a series of projects on isotope separation to the tune of $102,300. In addition, an Executive Committee of the committee on uranium was formed, comprising Briggs, Gunn, Pegram, Tuve, and Urey. It was also voted to "approve in principle" the $140,000 program proposed by Briggs, "and to direct the Chairman (Bush) to place the project in definitive form for later consideration."

An interesting document in NDRC records is a twelve-page memorandum dated August 14, 1940, apparently written by Briggs. This was evidently intended as a sort of history of the project to that time. It opens with a summary of what had been learned since the discovery of fission; how U-235 and U-238 differed in their response to neutron bombardment; how a controlled chain reaction might be achieved; the Einstein/Szilard/Sachs letter to President Roosevelt; the original $6,000 provided by the Army and Navy; the absorption of the uranium committee into the NDRC; Briggs's July 1 letter with its funding recommendation of $140,000; and that a special advisory group (Briggs, Urey, Tuve, Wigner, Breit, Fermi, Szilard and Pegram) which had met on June 13 recommended that funds be sought to support further measurements of nuclear constants and experiments with uranium and carbon. The memo proposed that the NDRC contract with Pegram to conduct research on the uranium–carbon chain-reaction problem. No salary was to be provided for Fermi and Pegram (who were employed by Columbia), but salaries of $4,000 and $2,400 per year were suggested for Szilard and Herbert Anderson, respectively.

With the NDRC in the picture, the pace of work within the United States' uranium project began to pick up. Between the fall of 1940 and the time of the Japanese attack on Pearl Harbor, the NDRC/OSRD let contracts totaling about $300,000 for fission and isotope-separation research to various universities (California, Chicago, Columbia, Cornell, Iowa State, Johns Hopkins, Minnesota, Princeton, Virginia), industrial concerns (Standard Oil Development Company), government agencies (National Bureau of Standards), and private research organizations (Carnegie Institution).

When the NDRC was established in the summer of 1940, the British MAUD committee was just beginning its work in response to the Frisch-Peierls memorandum; Edwin McMillan and Philip Abelson had just isolated a minute sample of element 93; and Louis Turner was speculating that neutron bombardment of U-238 might lead to a fissile form of element 94. In Britain, Rudolf Peierls reported to a meeting of the MAUD Technical Committee (Sect. 3.7) on April 9 that a fast-neutron fission bomb was feasible. A copy of the minutes of that meeting have been redacted from National Archives records of OSRD files, but in a typescript draft of an unpublished history of the bomb project prepared in May, 1943, Conant relates that at the MAUD meeting, James Chadwick stated that "The separation plant is the only large-scale project at present requiring consideration since the primary task of the Committee was to provide a military weapon." The coincidence of these various events is striking.

In the spring of 1941, Vannevar Bush began to receive complaints about the pace of the uranium committee's work. On March 17, Karl Compton wrote to Bush, referring to a presentation just two weeks earlier by Briggs on what Compton called the "#92 project". While it looked as if the project was moving ahead, there appeared to be a number of disquieting aspects: the English were "apparently farther ahead than we are," there was reason to believe that the Germans were very active in this area, and "very few of our own nuclear physicists are being put to work on the project and even those who are working on it are decidedly restive". Compton was concerned that the committee practically never met, that its conduct was extremely slow, that the work was being conducted in such secrecy that it was preventing people from knowing what was going on closely in related areas, and that Briggs was "slow, conservative, methodical and accustomed to operate at peace-time government bureau tempo". Harold Urey, a member of the project's Executive Committee, was just as disturbed, and baffled as to what could be done to improve the situation. Eugene Wigner, who had been working on the theory of chain reactions, described dealing with the Briggs Committee as like "swimming in syrup."

Compton proposed to let in on the project a group of the ablest theoretical physicists, and raised the question of whether the NDRC should take a more vigorous role as opposed to acting as a passive administrator. He further related that he and Ernest Lawrence had spoken that morning, and that Lawrence was under pressure from colleagues to see what could be done to speed up the work. Compton suggested that Bush appoint Lawrence as his deputy to explore and report on the situation, or, alternatively, arrange to assign Briggs a deputy to work full time on the project. Bush responded to Compton on the 21st to indicate that that he had met with Lawrence, and that he had called Briggs with the suggestion that Lawrence serve as a temporary consultant; the latter two were to meet that day.

Bush felt that he needed some independent advice on the uranium issue. On April 19, he asked Frank Jewett to appoint a committee under NAS auspices to review possible military aspects of fission. This would be the first of three such committees whose reports were to have far-reaching consequences.

4.3 May 1941: The First NAS Report

Jewett's committee was chaired by Arthur Compton, who was then Dean of Science at the University of Chicago. The other members were William D. Coolidge, who earlier in his career had made significant improvements to X-ray tubes and who had just retired as director of research at General Electric Research Laboratories; Ernest Lawrence; MIT theoretical physicist John Slater; Harvard physics theoretician and future Nobel Laureate (1977) John Van Vleck; and retired Bell Telephone Laboratories Chief Engineer Bancroft Gherardi. Due to illness, Gherardi never participated in any of the committee's activities; he passed away in August, 1941.

Compton's group met with Briggs, Breit, Gunn, Pegram, Tuve, and Urey in Washington on April 30, held a second meeting in Cambridge, Massachusetts, on May 5, and submitted their report to Jewett on May 17. Their seven-page document addressed the question of whether uranium research merited greater funds, facilities, and pressure in the light of then-current knowledge and the probability of applications in connection with national defense. The primary recommendation was that a strongly intensified effort should be spent on the problem during the following six months. While the committee felt that it would seem unlikely that nuclear fission could become of military importance within less than two years, they did comment that a chain reaction could become a determining factor in warfare if it could be produced and controlled.

The Compton report listed three possible military applications of uranium fission: (a) production of violently radioactive materials to be used as missiles "destructive to life in virtue of their ionizing radiation," (b) as a power source for submarines and other ships, and (c) violently explosive bombs. Discussion of the latter possibility concentrated mistakenly on *slow*-neutron fission of U-235, but it was predicted that the time required to separate an adequate amount of uranium would be from three to five years. It was pointed out, however, that element 94 could potentially be produced in abundance in a chain reaction. The day after the report was submitted, Emilio Segrè and Glenn Seaborg succeeded in isolating a sample of plutonium large enough that its fission cross-section for slow neutrons could be measured (Sect. 3.8).

While acknowledging that separation of a sufficiently large quantity of U-235 could become "a most important aspect of the problem," the bulk of the Compton committee's report was devoted to considering what resources would be needed for achieving a chain reaction in the ensuing months. The most urgent requirements for the following six months were considered to be full support for an intermediate-scale uranium/graphite experiment, a pilot plant for producing heavy water, support for investigating the properties of beryllium as a moderating agent, and maintaining work on isotope separation. The total cost was estimated at $350,000. If graphite proved to be a useable moderator, the cost of a full-scale experiment to produce a chain reaction was estimated to be as much as $1 million. If progress with the beryllium and heavy-water projects looked favorable at the end of the six-month period, further support should be extended for a subsequent stage of the beryllium experiment and a full-scale heavy water plant, at respective estimated costs of $130,000 and $800,000. The projected costs of America's wartime nuclear energy program were already reaching into million-dollar territory. In response to concerns with the pace of work, Compton's group praised the efforts of Briggs' committee, but suggested that a subcommittee be formed to plan and carry through the research programs, to confer on developments as they occurred, to see that information was made available to those who needed it, and to report as appropriate to the main committee.

Concerns with the report began to surface almost immediately. On May 28, Jewett solicited input from Robert Millikan, expressing concern that fundamental practical aspects of securing a chain reaction may have been minimized by physicists who were enthusiastic for going ahead. Could a chain reaction be used in practice, beyond highly special circumstances? What of limitations of physical space? What was known of the supply of materials, in particular the availability of uranium? Recognizing that even if the answers to these questions should be discouraging, Jewett opined that it might be wise to push experimentation on a large scale if for no other reason than to disprove over-enthusiastic claims: "At the same time it would be foolish to proceed solely on the basis of one-sided enthusiasm and a trust that in an eight-handed poker game the Lord will always enable us to draw the right two cards to complete a royal flush." Millikan responded on the 31st with the opinion that there seemed to be little if any hope in realizing a chain reaction in ordinary (natural) uranium, and that, if this proved so, it would be necessary to concentrate U-235, which would be a long and tedious process. While Millikan preferred that the energies of available personnel be concentrated on problems which would have a good chance of getting into practical use within two or three years at most, he did close with a suggestion that attempting a chain reaction with natural uranium would not be an expensive matter.

Jewett also solicited the opinion of Oliver Buckley, President of the Bell Telephone Laboratories. In his June 4 response, Buckley primarily emphasized the value of a chain reaction for naval propulsion, but added that if U-235 could be concentrated in quantity, it would have "enormous potentialities." But he also thought that the enrichment would be so difficult that there was no confidence of an early solution.

Jewett summarized his concerns in a letter to Vannevar Bush on June 6. While having a "lurking fear" that the Academy report might have been over-enthusiastic and not well balanced, he concluded that they should nevertheless go ahead on an enlarged program, with the proviso that major initial approvals and appropriations should be concentrated on the more fundamental aspects of establishing the possibilities of chain reaction. Any final approval for other phases of the matter, thought Jewett, should be reserved for a later time.

Bush's June 7 response to Jewett is worth quoting at some length. He related that Millikan was evidently unaware that "The British have apparently definitely established the possibility of a chain reaction with 238, which entirely changes the complexion of the whole affair." (About the British contribution, see Sect. 4.4.) Bush then proceeded to give Lyman Briggs some uncommon praise: "Briggs has been in a very difficult situation on this matter. I know of no project anywhere where there has been so much need for a balanced, reasoned approach which would, on the one hand, not neglect the possibilities of potential importance but unlikely to develop, and which, on the other hand, would not run wild as the result of unbridled speculation. I think Briggs has done exceedingly well to keep his balance, and to approach the matter on a basis which would seem to me to have good sense. Moreover, I think that Briggs is a grand person to have in the matter, and I have backed him up to the best of my ability, and I intend to do so in the future." On the other hand, Bush related concern with Ernest Lawrence, who was playing the role of a loose cannon: "I finally had to have a very frank talk with him in which I told him flatly that I was running the show, that we had established a procedure for handling it, that he could either conform to that as a member of the NDRC and put in his kicks through the internal mechanism, or he could be utterly on the outside and act as an individual in any way that he saw fit. He got into line and I arranged for him to have with Briggs a series of excellent conferences." Bush praised the Academy report, and also added that "[Briggs]... agrees to the enlargement of his section, the adding of a vice-chairman, the adding of a technical side, and in general the gearing up of the affair so as to handle the program to better advantage"; Briggs had been asked for his input on the personnel issue before a meeting scheduled for June 12. Bush also thought that there should be "at least one good sound engineer" on the enlarged Uranium Committee. The last paragraph of Bush's letter revealed some growing frustration: "As I have said many times, I wish that the physicist who fished uranium in the first place had waited a few years before he sprung this particular thing upon an unstable world. However, we have the matter in our laps and we have to do the best we can."

Briggs responded on June 11 with an estimate of Uranium Committee expenditures for fiscal year 1942, which would start on July 1. These covered a broad range of activities: a uranium–carbon experiment at Columbia; a uranium–carbon-beryllium experiment in Chicago; heavy water catalysis at Columbia; an experimental heavy water production set-up to be built by Standard Oil in Baton Rouge, Louisiana; work on centrifuges at Columbia and the University of Virginia; research on diffusion at Columbia; mass spectroscopy under Alfred Nier at the University of Minnesota; and miscellaneous administrative and experimental work at the Bureau of Standards. All of this would run to $583,000 for the first six months of the fiscal year. Costs for the balance of the year would depend on the outcomes of various experiments, but perhaps $1 million would be needed for a full-scale chain-reaction experiment and a heavy-water plant. The most immediate need was for $241,000 to acquire materials.

Despite Bush's knowledge of the British opinion that fission bombs were virtually certainly feasible, the NDRC voted the next day only to allocate the $241,000 for materials, and to increase the amount authorized to the University of Minnesota for preparation of 5 µg of U-235 by $500 (the amount of the original authorization does not appear in the minutes). The irony that Briggs is often accused of foot-dragging speaks for itself. Briggs submitted a revised proposal on July 8 which brought his request down to $357,000, mostly by decreasing requests for the Chicago and Columbia pile experiments. Approval of the revised request was voted at a meeting held on July 18. While funding was still a matter of fits and starts, the fortunes of the American uranium program were beginning to shift for the better in the early summer of 1941. The participants could not have been unaware of an increasingly perilous world situation: on June 22, Germany invaded Russia, adding a dramatic new dimension to the war.

4.4 July 1941: The Second NAS Report, MAUD, the OSRD, and Reorganization II

At the June 12 NDRC meeting discussed above, it was also voted to request to have the NAS once again review the proposed program, but this time by a committee which included individuals qualified to consider engineering aspects of the situation. Bush put the request to Frank Jewett the next day, and the Academy Committee went back to work, this time under the chairmanship of Coolidge (Compton was traveling at the time). To provide the relevant engineering perspective, the committee was augmented by Oliver Buckley (Bell Labs) and Lawrence Chubb, Director of the Westinghouse Electric Research Laboratories. The group acted quickly, meeting in Washington on July 1 with Briggs, Gregory Breit, and Sam Allison of the University of Chicago, and then with Pegram and Fermi at Columbia on July 2. They submitted their four-page report to Jewett on July 11, who passed it on to Bush on the 15th.

The report did not particularly address engineering aspects as Bush had requested, but rather related some crucial developments in nuclear physics. The last page of the report is an appendix written by Ernest Lawrence, who described how, since the May 17 report, experiments at Berkeley had verified that element 94 was formed via slow-neutron capture in U-238, and that the new and yet-unnamed transuranic element underwent slow-neutron fission. This opened up the prospect of what Lawrence called a "super bomb" if enough 94 could be produced. Given this development, the committee considered whether the prospect of military applications was such as to justify allocation of defense monies toward support of an intensified drive on producing atomic fission, concluding that "We are convinced that such support is not only sound but urgently demanded." The committee also gave Bush ammunition for reorganizing the project: "The efficient and expeditious conduct of this larger scale attack requires also ... a different pattern of organization from that of the work under the present Uranium Committee ... The project should be under a director able to devote his entire time to it." Costs were projected at over $1 million for salaries and materials for the first year, and the committee also suggested that an isolated laboratory be established at which to locate the relevant work.

Support for the committee's opinion was received from Enrico Fermi, who composed an eight-page report titled "Some Remarks on the Production of Energy by a Chain Reaction in Uranium." Dated June 29, 1941, Fermi describes a possible reactor design with lumps of natural-composition uranium metal or oxide distributed in a lattice-like array throughout a moderator, with carbon (graphite) mentioned specifically for the latter. This is precisely the arrangement he would use in his CP-1 reactor some 18 months later (Chap. 5). Fermi's admittedly uncertain figure for the amount of energy produced per gram of U-235 fissioned was about 80 billion joules, which is in quite good agreement with the 17 kilotons per kilogram calculated in Sect. 3.1 (~ 71 billion Joules per gram); he also estimated that a pile generating one megawatt of thermal energy would produce about one gram of element 94 per day. This proposal, however, was for an *uncooled* reactor; by using active cooling with fluid or gas piped through appropriate channels, the power level could be raised to tens of megawatts, which would increase plutonium production correspondingly.

One intriguing possibility pointed out by Fermi was cooling by liquid bismuth, which would have the advantage of breeding radioactive polonium through neutron bombardment via the reaction

$$_{0}^{1}n + {}_{83}^{209}Bi \rightarrow {}_{83}^{210}Bi \xrightarrow[5.0\ days]{\beta^{-}} {}_{84}^{210}Po.$$

In exactly this way, slugs of bismuth would be introduced into the pilot-scale X-10 reactor at Oak Ridge, Tennessee, and into the production reactors at Hanford, Washington, to breed polonium for use in neutron-generating "triggers" for the Hiroshima and Nagasaki bombs. Fermi also addressed the need for shielding, which could be accomplished with a surrounding barrier of water several feet thick. In a July 21 memo to Conant, Bush praised Fermi's report as the first time he had seen anything that approximated engineering data, and that it looked "to be good stuff."

While engineering issues were being considered, Bush was rearranging the administration of the NDRC. The NDRC could undertake to issue contracts for research, but lacked the authority to underwrite engineering development. To address this, he conceived of a higher-level umbrella organization, the Office of Scientific Research and Development (OSRD). The NDRC would continue, but as a sub-component of OSRD; Bush would Direct the OSRD, while Conant would take on Chairmanship of the NDRC and with it responsibility for the uranium project. The OSRD was established by Executive Order 8807, signed by President Roosevelt on June 28, 1941.

Beyond the National Academy reports and the growing restlessness of individual scientists, the single most important stimulus to the American fission project in the summer of 1941 came from outside the country's borders. A foreign bombshell was about to land in the lap of the newly-formed NDRC: the British MAUD report.

The spectacular success of the Manhattan Project under U. S. Army leadership and the fact that the bulk of its facilities were located on American soil have tended to cast the Project as an almost exclusively American affair. But such a view trivializes very important British contributions to the Project. Even General Groves, who has been quoted as characterizing the British contribution as "helpful but not vital," tempered his assessment with the observation that "I cannot escape the feeling that without active and continuing British interest there probably would have been no atomic bomb to drop on Hiroshima. The British realized from the start what the implications of the work would be. They realized that they must be in a position to capitalize upon it if they were to survive … and they must also have realized that by themselves they were unable to do the job. They saw in the United States a means of accomplishing their purpose."

American authorities were not unaware of progress in Britain; exchanges between the two countries on scientific matters were well-established before America entered the war. In late August, 1940, a mission headed by Henry Tizard left for a two-month visit to America, where they demonstrated progress that had been made in Britain with equipment relating to radar and proximity fuses. One of the results of this visit was the establishment in Washington of a formal organization to facilitate information exchange, the British Commonwealth Scientific Office. In the spring of 1941, Charles G. Darwin—a grandson of *the* Charles

Darwin—was appointed as its Director. Reciprocally, in February, 1941, Conant traveled to London to set up an office of the NDRC. Harvard physicist Kenneth Bainbridge, who would direct the *Trinity* test, attended the April 9 meeting of the MAUD committee at which Rudolf Peierls reported that a fast-neutron bomb was feasible. On July 1, Caltech physicist Charles Lauritsen attended another meeting, at which the main conclusions of the committee's final report were discussed. Lauritsen briefed Bush in Washington on July 10, just a few days before Bush received a draft copy of the report which had been transmitted to the NDRC London office. This was just before the second NAS report landed on Bush's desk. There were actually two MAUD reports, both authorized by George Thomson on July 15. The first, which is the one of interest here, was titled "Use of Uranium for a Bomb"; the second was "Use of Uranium as a Source of Power." Both are reproduced in Margaret Gowing's book on the British atomic energy program, and are still well worth reading.

The first part of the MAUD bomb report summarizes the situation in non-technical terms in a few pages. It opens with a description of why a critical mass exists for a fissile isotope, how a bomb could be triggered by bringing together two subcritical masses, the probable effects of the explosion (estimated as equivalent to 1,800 t of TNT for 25 pounds of U-235), and a discussion of materials and costs. A lengthy technical appendix describes how a fast-neutron chain reaction cannot not be sustained in U-238 due to the presence of inelastic scattering and absorption, how the efficiency of a bomb could be estimated, factors that affect the determination of critical mass, estimates of damage, and the characteristics of a diffusion plant. In mid-1941, the clarity of many British scientists' understanding of the basic elements of a fast-fission weapon was far ahead of that of most of their American counterparts. George Thomson personally handed Bush and Conant copies of the MAUD report on October 3, but under terms which did not permit its disclosure to the NAS Committee. Despite that injunction, Thomson had met with both the Uranium Committee and the NAS Committee to apprise them of the situation. The MAUD bomb report would have a significant, if officially unacknowledged, impact on the preparation of a third National Academy report in late 1941.

On July 30, Conant received from Briggs a letter describing how the Uranium Committee was being reorganized. Briggs would remain as Chair; George Pegram had agreed to serve as Vice Chair. The other members were to be Gregory Breit, Harold Urey, Samuel Allison, Henry Smyth of Princeton University (see the Preface), and Edward Condon of Westinghouse Electric. Briggs also expanded the breadth of the committee by adding four consultant subcommittees. These were to deal with the areas of Separation (i.e., enrichment), Power Production, Heavy Water, and Theoretical Aspects, and were respectively chaired by Urey, Pegram, Urey, and Fermi. The Separation group included Philip Abelson and Ross Gunn; the latter was also a member of the Power Production group. Merle Tuve, Alexander Sachs, and Albert Einstein had disappeared from the July 1940 makeup of the committee. Jesse Beams was also dropped from the main committee, although he would continue as a member of the Separation Group. Henceforth, the Uranium Committee would be known as Section S-1 of the OSRD.

Replying to Briggs the same day, Conant indicated that it would be necessary to communicate to Beams, Gunn, and Tuve that their services would not be needed in the newly organized section, and asked if Briggs would prefer to write them himself, or have Conant or Bush do it? Briggs opted for the latter. As a result, in a letter to Ross Gunn on August 14, Bush explained Gunn's removal from the main committee. The problem, as Bush described it, was that the Uranium Committee had been formed before the NDRC had taken it over: "At the time the NDRC started its work and formed Sections the policy was adopted of not having Army or Navy personnel directly appointed to membership on these Sections, but rather to provide the desired contacts by the system of liaison officers, and this has worked out well. The situation in the uranium committee was hence a bit of an anomaly, but we did not disturb it as it seemed to be working well." But with the recent reorganization, it was time to bring the Uranium Committee into line. Bush suggested that in place of serving on the committee, Gunn be nominated as the individual who would serve as the direct contact between the committee and the Navy. Gunn replied on the 18th, formally tendering his resignation from the committee, and indicating that there should be no objection on the part of the Navy to his serving as liaison. Ironically, at just this time Philip Abelson was beginning to test experimental liquid-diffusion columns at the Naval Research Laboratory. As is related in Chap. 5, there would be much more to come regarding the relationship between the Navy and the Manhattan Project.

British physicists continued to pressure their American counterparts to push ahead with a bomb project. On September 11, 1941, W. D. Coolidge wrote to Frank Jewett to describe a visit Marcus Oliphant had made to General Electric in Schenectady. Coolidge was astonished to learn that only 10 kg of U-235 would be needed for a fast-fission reaction equivalent to 1,000 t of high explosive, and remarked that this information, so far as he knew, was not available in the United States until after the second National Academy report of July 11. Jewett replied that he had already received the same information "indirectly," and that while he thought that the matter was fully understood by the S-1 Committee, he would send a copy of Coolidge's letter to Bush as a precaution. Conant felt that Oliphant's talking to Coolidge might have been a breach of secrecy, but many American scientists have credited Oliphant for spurring the S-1 program forward.

Oliphant also visited Berkeley and met with Ernest Lawrence, who was so impressed with British progress that he began thinking of how he might turn his 37-inch cyclotron into a large-scale mass spectrometer for separating isotopes. In September, Lawrence related Oliphant's story to Conant and Compton during a visit to Chicago, apparently not realizing that they already knew of it. Lawrence stressed the importance of element 94 to making a bomb, and again expressed his dissatisfaction at the slow pace of work in the United States. In his memoirs, Compton relates how he met with Conant and Lawrence in the living room of his house. After Lawrence had given his description of the prospect for fission bombs, Conant asked him: "Ernest, you say you are convinced of the importance of these fission bombs. Are you ready to devote the next several years of your life to getting them made?" After a brief hesitation, Lawrence's answer was "If you tell me this is my job, I'll do it."

4.5   October–November 1941: The Top Policy Group and the Third National Academy Report

October 9, 1941, was a pivotal day in the history of the American atomic bomb program. That morning, Vannevar Bush met with, among others, President Roosevelt and Vice-President Henry Wallace to inform them of developments. Bush summarized the meeting in a memo sent to James Conant later the same day. The most significant matter was that the President had made it clear that considerations of policy were to be restricted to a group comprising himself, the Vice-President, Secretary of War Henry Stimson, Army Chief of Staff General George C. Marshall (Fig. 4.6), and Bush and Conant. This group would come to be known as the Top Policy Group. From this point forward, American scientists would have to funnel their thoughts concerning policy issues on the fission weapons that they would create through Bush and Conant. That the President had charged a group to consider nuclear weapons policies indicated that the highest levels of leadership of the United States were beginning to understand the implications of a successful full-scale commitment to the uranium project.

During the meeting, Bush described British conclusions regarding critical mass, the size of necessary isotope-separation plants, costs, time schedules, and raw materials. The meeting endorsed interchange with the British on technical issues, and also considered post-war control of nuclear materials. Another significant matter was that Bush advocated that a broader program ought to be handled independently of the then-present organization, a notion with which the President

Fig. 4.6 General George C. Marshall (1880–1959) and Secretary of War Henry Stimson (1867–1950), ca. 1942 *Source* http://commons.wikimedia.org/wiki/File:George_marshall%26henry_stimson.jpg

agreed. Roosevelt instructed Bush to not proceed with any definite steps on the expanded plan until receiving further instructions, but Bush essentially emerged from the meeting with the authority to determine if a bomb could be made, and at what cost.

The same day, Bush also requested a third National Academy Report. This time he gave the committee very clear directions as to what he was after. He wrote to Arthur Compton, referring to having received a "communication from Britain" which dealt with the technical aspects of "the matter under consideration by the committee." The British report was available only to himself and Conant, but this being the case would have the advantage that the Academy committee's work would provide an independent check on things. While Bush acknowledged that the way in which the committee wished to conduct its study and prepare its report was a matter for itself to decide, he offered some topics for their consideration: critical mass; the mutual velocity of approach of the subcritical masses during bomb core assembly; efficiency; premature explosions; and isotope separation methods. Bush copied his instructions to Briggs, again adding that he was not able to pass on the British report. Despite Briggs' position as chair of the S-1 Committee, lines of authority were shifting toward Bush and Conant.

Compton's group went back to work, meeting with Fermi, Urey, Wigner, Seaborg, and others. On October 21, they held a meeting at the General Electric laboratories in Schenectady, which Robert Oppenheimer attended. They soon produced a draft report, which prompted a letter from Frank Jewett to Bush on November 3. Jewett was concerned that the draft mentioned a cost figure as great as $100 million, but, at the same time, noted that practically every element of the proposed research and development program possessed very fundamental uncertainties. Jewett felt that a stronger case would have to be in hand for when the time came to seek appropriations, but also argued that there was a case for expenditure of considerable money to resolve the uncertainties in order to develop a basis of proven technical information.

In a reply the next day marked "Personal," Bush argued in support of the program, pointing out that Conant's opinion had swung around entirely after initial skepticism. It was now crucial, he felt, "to bring to bear some good sound engineering brains on design." In accordance with what he had related to the President, Bush was also formulating further administrative reorganizations which would constitute a new group to handle development and pilot-plant experimentation, while leaving Briggs in charge of only a section devoted to physical measurements. He had in mind Ernest Lawrence to direct the new group, but was hesitant given the need for secrecy and the fact that Lawrence, in defiance of President Roosevelt's dictum, was stirring things up by talking about policy issues. Bush suggested that Jewett destroy the letter, but a copy does appear in OSRD files.

For its third report, the committee was expanded to include MIT chemical engineer Warren K. Lewis, Harvard explosives expert George Kistiakowsky (Fig. 4.7) and future (1966) Nobel chemistry Laureate Robert Mulliken of the University of Chicago. The full report can be found in OSRD records, and, like its MAUD counterpart, is still worth reading.

Fig. 4.7 *Left* George
Kistiakowsky (1900–1982);
Right Warren K. Lewis
(1882–1975). *Sources* AIP
Emilio Segre Visual
Archives; http://
en.citizendium.org/wiki/
Warren_K._Lewis

The Committee transmitted its report, dated November 6, to Jewett on November 17. In a brief cover letter, Compton reported that the committee was "unanimously of the opinion that the prosecution of this program is a matter of urgent importance." At sixty pages, the report comprised a number of interlinked sections, which can be can be summarized under four groups. First came a six-page cover letter to Jewett which summarized the conditions needed for a fission bomb, the expected effects of such bombs, estimates of how long it might take to produce them, and the costs involved. Second is a 20-page appendix, evidently written by Compton, which forms the technical core of the report. Calculations here dealt with critical radius, the effect of a surrounding tamper, and efficiency of the anticipated explosion. This appendix can be considered to be the parent document of Robert Serber's *Los Alamos Primer* (Sect. 7.2), and can still be recommended to a reader who seeks a description of the basic physics of fission weapons. (For an undergraduate-level analysis, see Reed 2007, 2009). Third is an appendix prepared by George Kistiakowsky which describes the probable destructive action of fission bombs. Lastly appears an 18-page report prepared by Mulliken which discusses the feasibility of various isotope separation methods.

The summary letter to Jewett gets right to the essence on its first page: "A fission bomb of superlatively destructive power will result from bringing quickly together a sufficient mass of element U235. This seems to be as sure as any untried prediction based upon theory and experiment can be." The critical mass of U-235 was estimated as hardly less than 2 kg nor greater than 100 kg, and the expected efficiency at between 1 and 5 %. It was difficult to assess the destructive capabilities of a fission weapon because the theory for describing high-pressure shock waves was not then well-advanced, but the committee conservatively estimated an equivalence of about 30 t of TNT per kg of U-235; this would prove to be a serious underestimate.

The committee took an interesting approach to analyzing the amount of U-235 deemed necessary to defeat Germany. Based on an estimate that some 500,000 t of TNT would be required to devastate military and industrial objectives in that country, they projected that some 1–10 t of U-235 would be required to do the same job with fission bombs, an analysis which clearly overlooked the psychological effect that even a few bombs might have. As for obtaining fissile material, centrifugal and diffusion separation methods were approaching the stage of practical tests. The committee estimated that if all possible effort were spent on the program, fission bombs might be available in significant quantity within three or four years; the bombing of Hiroshima would occur three years and nine months to the day from the date of the report. As to finances, the committee estimated a rough cost of $80 to $130 million, not including the cost of electrical power for operating the enrichment plants. Ultimately, the electromagnetic separation method alone would consume nearly four times this amount of funding.

Compton's letter closed with a series of recommendations. The immediate needs were to build and test trial units of centrifugal and diffusion separators, to secure samples of separated U-235 and U-238 for physical-constant measurements, and to begin work on the engineering aspects of enrichment plants. Finally, a suggestion that no doubt pleased Vannevar Bush (and, one suspects, may have been planted by him) was that it may be necessary to reorganize the entire program. In a separate letter to Bush on the same day, Compton offered some private advice on reorganization: assign key men responsibility for solving certain problems, and give them adequate funds to "get the answers in their own way." Urey, Lawrence, Beams, and Allison were suggested as appropriate "key men."

The difference between the third Academy report and its two predecessors is stunning. In his May, 1943, history, James Conant remarks (paraphrased): "A historian of science two generations hence who might come across the three National Academy reports might well be bewildered by the change. In July 1941 the Committee was speaking of the need for a successful demonstration of a controlled chain reaction. On November 6 the Committee concludes that the availability of bombs may determine military superiority." Conant attributed this shift in emphasis to a general feeling that war was felt to be much nearer and more inevitable in November than in May, and that advocates of a head-on attack on the uranium issue had become more vocal and determined. In a comment that presaged the postwar perception that the Manhattan Project was essentially an exclusively American affair, Conant wrote, somewhat disingenuously, that "It must be remembered that the British report ... had not been seen by any member of the National Academy Committee even by November." This is strictly true, but was a selective truth.

4.6 November 1941: Bush, FDR, Reorganization III, and the Planning Board

Vannevar Bush wasted no time in using the third Academy report to bolster what he had reported to President Roosevelt on October 9. On Thursday, November 27, he transmitted the report to the President and the Top Policy Group; Bush and the President evidently did not meet face-to-face that day. (Ironically, that date was about the time that a Japanese task force set sail on its mission to attack Pearl Harbor.) While advising the President that the cost of and time to produce bombs would be greater than the MAUD report had suggested—this was attributed to the Academy Committee having included some "hard headed engineers" in addition to physicists—Bush felt that the matter called for serious attention. He offered that he would again wait to be instructed by the President before taking any steps which made a commitment to any specific program, but that in the meantime he was forming an engineering group to study plans for possible production, as well as accelerating relevant research.

By presenting the MAUD and Academy reports as independent but mutually supportive, Bush played them brilliantly as political cards. A handwritten note from Roosevelt accompanying return of the report to Bush on January 19, 1942, has been taken by some historians to be essentially the initiating Presidential "OK" for the American atomic bomb program (Fig. 4.8).[1]

Bush and Conant proceeded with their reorganization. OSRD records contain a two-page handwritten memo from Conant to Bush; the date is uncertain, but a note in Conant's hand reads "must be in November 1941". The note is difficult to read in places, but one passage is fairly clear and harbingers further sidelining Lyman Briggs: "Hence take Briggs section out of NDRC and make it [a] research division of [a] new setup, at this point replace Briggs with a full-time man. ... Set up a development committee of chemical engineers with advisory group of top men". A rough draft organizational chart shows Bush at the top, with separate parallel research and development committees under him; Conant does not appear in the chart (Fig. 4.9). The growing momentum of the project is clear from Conant's

---

[1] Hewlett and Anderson attribute the note to January 19, 1942. The note can be found in M1392(1), 0945. However, another date suggests itself. The immediately preceding image on the DVD version of the microfilm records supplied to this author is a copy of a letter from Bush to FDR dated June 17, 1942, wherein Bush enclosed a June 13 report on the subject of "Atomic Fission Bombs"; see Sect. 4.9. The June 17 letter is also marked "V.B. OK FDR". Might the "January 19" note have been penned on June 19? The month on the note of Fig. 4.8 is indistinct. If it is June 19, a Presidential response within two days may seem speedy, but was by no means unprecedented. For example, as discussed in the text, on March 9, 1942, Bush sent FDR an extensive update on the status of the project; the record contains a note signed by Roosevelt on March 11, acknowledging return of the document to Bush (Fig. 4.11). But if Roosevelt annotated the June 17 letter, why would he have felt compelled to send a separate note two days later? The copy of Compton's November, 1941, report in the OSRD records bears no Presidential annotation, which could suggest the need for a separate acknowledgement. Mere proximity of FDR's note to the June 17 letter on the DVD supplied to this author is no guarantee of temporal closeness: my experience is that the documents in these records are often very mixed-up chronologically. If the note does refer to returning the third Academy report, this would represent a lapse of some seven weeks between the meeting and the return. But in the hectic days following the attack on Pearl Harbor this may not have been unreasonable; Roosevelt may have been further delayed because he was hosting Winston Churchill for the First Washington Conference, which ran from December 22, 1941, to January 14, 1942. It appears that convincing arguments can be mounted for either date. For this writer, coming across this minor confusion reminded him of advice he received many decades ago from an eighth-grade teacher: "When you write something, date it." To which I would add: And do it clearly and completely.

Fig. 4.8 Franklin Roosevelt
to Vannevar Bush, January
(?) 19, 1942

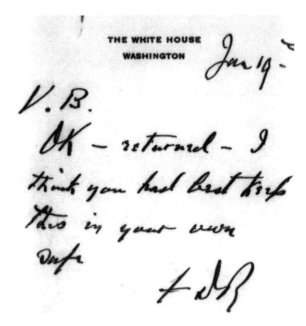

comments that "I believe we should from now on waste no more time," and "I do think you ought to move fast from now on." Lawrence was again suggested to oversee the research committee.

Bush had already been laying such plans. On November 26, he offered the position of Director of a Planning Board to Eger V. Murphree, a distinguished chemical engineer and Vice-President of Research and Development for the Standard Oil Development Company (SODC; Fig. 4.3). The Board would be charged with the responsibility of presenting Bush with recommendations for production and contracts for engineering studies. Murphree accepted the appointment, subject to his being free to select a group of consultants to serve as an advisory committee. His appointment was formalized in a letter from Bush on the

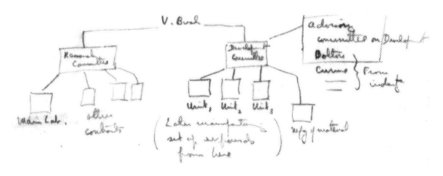

Fig. 4.9 Conant's draft organizational chart, ca. November, 1941

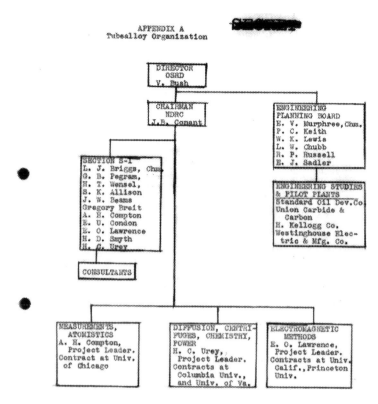

APPENDIX A
Tubealloy Organization

DIRECTOR
OSRD
V. Bush

CHAIRMAN
NDRC
J.B. Conant

ENGINEERING
PLANNING BOARD
E. V. Murphree, Chm.
P. C. Keith
W. K. Lewis
L. W. Chubb
R. P. Russell
E. J. Sadler

SECTION S-1
L. J. Briggs, Chm.
G. B. Pegram,
H. T. Wensel,
S. K. Allison
J. W. Beams
Gregory Breit
A. H. Compton
E. U. Condon
E. O. Lawrence
H. D. Smyth
H. C. Urey

ENGINEERING STUDIES
& PILOT PLANTS
Standard Oil Dev.Co.
Union Carbide &
Carbon
M. Kellogg Co.
Westinghouse Elec-
tric & Mfg. Co.

CONSULTANTS

MEASUREMENTS,
ATOMISTICS
A. H. Compton,
Project Leader.
Contract at Univ.
of Chicago

DIFFUSION, CENTRI-
FUGES, CHEMISTRY,
POWER
H. C. Urey,
Project Leader.
Contracts at
Columbia Univ.,
and Univ. of Va.

ELECTROMAGNETIC
METHODS
E. O. Lawrence,
Project Leader.
Contracts at Univ.
Calif.,Princeton
Univ.

Fig. 4.10 Manhattan Project organizational chart, ca. March, 1942

29th, which made clear that while the Board was free to consult with Briggs and the Academy Committee, Murphree was to report directly to Bush, with Bush and Conant sharing overall responsibility for the program. The new organizational structure, which effectively orphaned the S-1 Section, was laid out by Bush at an NDRC meeting on November 28, and is shown in Fig. 4.10, which is copied from a March, 1942, report to President Roosevelt. The term "tubealloy" was the British code name for U-235.

The reorganization was not the only step in Briggs' marginalization. On December 1, Bush received a report from Harold Urey, and a letter from Henry Smyth. Both urged speedier action. Urey had just returned from a visit to London, and reported that Chadwick believed there to be a 90 % chance of a practical fission bomb. Urey felt that "nothing else in the entire war effort should be placed ahead of it. If the Germans get this bomb the war will be over in a few weeks." Smyth was briefer but blunter, asking whether Briggs was in charge of the whole uranium work and free to call on the Uranium Section for advice or to ignore it at his discretion, or were recommendations to the NDRC supposed to represent the informed majority judgment of the members of the S-1 Section, merely transmitted by Briggs as chairman of the Section? Smyth understood from Briggs that the situation was the latter, but the infrequency of meetings and their highly informal character left him feeling that in practice the situation was more nearly the former.

Bush passed Smyth's letter to Briggs the next day, along with suggestions as to how to present the reorganization, essentially a *fiat accompli*, at a Section meeting scheduled for December 6. Physics research would continue, but S-1 members had to understand that policy issues were not their affair. Briggs should announce plans that he was to discuss with Compton regarding splitting the research into parts and putting key individuals in charge of each. Bush would ask eminent chemical engineers (Murphree, Keith and Lewis in Fig. 4.10) to advise him directly in regard to engineering points. Bush wanted a clear-cut division between scientific and engineering studies, but also interrelation between the two.

As Planning Board chair, Eger Murphree got to work promptly, meeting with Harold Urey on December 2 to review isotope enrichment methods. Based on analyses carried out at Columbia and experiments conducted by Jesse Beams at Virginia, Murphree estimated that some 20,000 centrifuges would be needed to produce one kilogram 235 per day, and advocated setting up a pilot plant with 10 such machines in series. John Dunning at Columbia was working on developing diffusion membranes by an etching process; for this method, Murphree estimated that a full-scale plant would require some 4,000 stages. The centrifuge method would eventually be abandoned, but the diffusion plant would be built (Sect. 5.4).

4.7 December 1941–January 1942: The Pile Program Rescued and Centralized

The one major aspect of the project left unaddressed by the third Academy Committee report was the possibility of developing reactors to synthesize plutonium. This work was salvaged by the personal intervention of Arthur Compton at the December 6 meeting of the S-1 Section held in Washington—the day before Pearl Harbor. Curiously, OSRD records contain no minutes from that meeting, at least that this author has been able to turn up. However, many of the details were related in a letter from Bush to Murphree on December 10, and a firsthand account was published by Compton in his postwar memoir *Atomic Quest*.

As shown in Fig. 4.10, the first main outcome of the meeting was the creation of three development programs, with each being lead by a Program Chief. Harold Urey would be responsible for research on separating uranium isotopes by diffusion and centrifugation, as well as work on heavy water. Lawrence, who now gained an official position and assigned duties, was charged with investigating electromagnetic methods of isotope separation. As Bush described Compton's role, it was to be concerned with fundamental atomic physics; in particular, measurements of material physical constants. Compton put it more immodestly as "My job was to be the design of the bomb itself." At this point, the formal meeting adjourned with the understanding that there would be another gathering in two weeks to shape plans more firmly.

After the meeting, Bush, Conant and Compton went to lunch at the Cosmos Club on LaFayette Square, where, as Compton relates it, he argued that further thought should be given to the production of plutonium as an alternative to enriching uranium. Conant expressed concern as to the time that would be needed to perfect the chemical extraction of plutonium from bombarded uranium, which would be complicated by intense radioactivity. Compton claims to have responded with "[Glenn] Seaborg tells me that within six months from the time plutonium is formed he can have it available for use in the bomb." Conant's reply to this was "Glenn Seaborg is a very competent young chemist, but he isn't that good." Just how capable Seaborg would prove to be can be judged from his subsequent record. After discovering and chemically characterizing plutonium, he developed separation techniques that could be scaled up from microgram to kilogram quantities; was involved in the discovery and analysis of several transuranic elements; earned a Nobel Prize (1951); served as Chairman of the Atomic Energy Commission from 1961 to 1971; and have an element named after him while he was still living. Compton relates that as a result of this conversation, he was given authority to see what could be done towards producing plutonium via a chain-reaction. Smyth refers to Compton's authorization as "an afterthought.".

Compton's advocacy of the pile program was well-founded. At Columbia, Fermi and Szilard had been continuing their experiments to determine how neutrons slowed down and diffused through graphite, with particular attention to measuring the net number of neutrons produced per each consumed in a fission. This is known as the reproduction factor, and is designated with the letter k. A k value of unity or greater is necessary to maintain a chain reaction. These experiments involved a graphite column (a "pile," as the arrangement came to be known) of dimensions 3 by 3 by 8 ft, with a source of neutrons placed inside. Strategically placed detectors mapped the number of neutrons present and their energy distribution. These investigations revealed that the high-speed neutrons emitted in fissions were practically all reduced to thermal velocities after traveling through about 40 cm of graphite.

Around July, 1941, the first "lattice" experiment was set up at Columbia. This comprised a graphite cube about eight feet on each side, with some seven tons of uranium oxide enclosed in iron cans distributed throughout. Larger such structures followed in the fall of that year, by which time Fermi was able to report a k value of 0.87. By May, 1942, a k-value of 0.98 would be achieved. At Chicago, Samuel Allison began similar experiments with the goal of investigating the possibility of enveloping a pile with a layer of beryllium to reflect neutrons back into the pile; beryllium was known to have an extremely small neutron-absorption cross-section. This method was ultimately not pursued due to the difficulty of obtaining sufficient beryllium, but Allison's experiments provided valuable checks on those being conducted at Columbia.

America's entry into the war galvanized the uranium project. On December 10, 1941, Murphree wrote again to Bush regarding progress with centrifugation, and to reiterate his argument for developing a pilot plant of 10 to 25 machines at an estimated cost of $75,000–$150,000. Bush responded on the 13th, urging Murphree to go ahead. The same day, Bush sent letters to Compton, Lawrence, and Urey, formally outlining the new organizational structure and their individual responsibilities. On December 16, nine days after Pearl Harbor, Bush met with Vice-President Wallace and Secretary of War Stimson to discuss the third Academy report, and to apprise them of the ongoing reorganization of the project. This meeting was another key step in the project's advance toward full-out status. In summarizing the meeting in a memo to Conant, Bush indicated that the group felt that work should proceed as fast as possible on fundamental physics, engineering planning, and pilot plants. The cost estimate had escalated to four to five million dollars. Perhaps the most important point of the discussion that day, however, was that Bush made clear that he felt that the Army should take over when full-scale construction was started, and to that end felt it would be appropriate to have an officer become familiar with the project. This seems to be the first time that this suggestion, which would have such profound consequences, was made at such a high level. Another advantage of bringing in the Army would be that the necessary budget could be hidden among that organization's enormous wartime appropriations under innocuous-sounding items such as "Expediting Production" or "Engineer Service—Army." While some key Congressmen and Senators were briefed on the Manhattan Project from time to time, the vast majority of legislators had no idea of its existence.

Arthur Compton was a religious man, and he went about his new responsibilities with an almost missionary zeal. In a December 20 "Urgent" letter to Bush, Conant, and Briggs, he laid out an ambitious plan for work at Columbia, Princeton, Chicago, and Berkeley. There were to be three major components: (1) theoretical problems regarding nuclear explosions, (2) production of a chain reaction, and (3) determination of physical constants relevant to components (1) and (2). His goals were to obtain a chain reaction by October 1, 1942; to have a pilot plant for the production of plutonium in operation by October 1, 1943; and to be producing useable quantities of plutonium by December 31, 1944. These estimates would prove reasonably durable. Enrico Fermi's CP-1 graphite reactor would first go critical on December 2, 1942; the X-10 pilot-scale reactor at Oak Ridge would achieve criticality on November 4, 1943; and two reactors at Hanford, Washington, would go critical with full fuel loads in December, 1944. Compton's projected budget for the first six months of 1942 came to $1.2 million, nearly half of which was for purchase of uranium alloy, graphite, and beryllium. He also proposed formally bringing Robert Oppenheimer into the theoretical work; Oppenheimer had prepared part of the criticality analysis in the third Academy report. Ernest Lawrence had been busy as well. Also on December 20, James Conant endorsed a recommendation by Lawrence to enter into a contract with the University of California for $305,000 for the coming six months. Lawrence's objective was to determine, within those six months, whether or not an electromagnetic method would be in the running as a uranium separation technique. His group had just prepared their first substantial sample of U-235: some 50 μg mixed with about 200 μg of U-238.

Also left unaddressed in the late-1941 reorganization of the uranium project was the question of centralizing any or all parts of the effort. Compton had ideas on this as well. From January 3–5, 1942, he held a planning session in Chicago with representatives from Berkeley, Columbia, and Princeton, and then another at Columbia on January 18. On the 22nd, he wrote to Conant to indicate that he and Lawrence proposed to concentrate "both programs"—presumably separation and pile studies—at Berkeley, with the approximate date of the transfer being February 10. The advantage to this would be the availability of the cyclotron and magnets at Berkeley, although there was some concern with the issue of the west coast being a target for Japanese bombs. However, Lawrence apparently underwent a change of heart, and wired Conant and/or Bush (the record is not clear whom) on the afternoon of January 24: "The latest of Compton's several successive proposals namely maintain status quo excepting moving Princeton to Chicago is in my judgment acceptable only as temporary arrangement. I sympathize with his difficulty in decision as there are numerous conflicting factors however we do need above all vigorous action." Later that evening came another wire from Lawrence: "Just learned Compton's decision to move Columbia [and] Princeton to Chicago which is much better than moving Princeton only." With this decision, all pile and physical-constants work would move to Chicago, but electromagnetic research would remain in Berkeley.

On February 20, Conant summarized progress on various fronts in a special report to Bush on the status of the S-1 Section. Contracts totaling over one million dollars had been authorized to 12 different institutions, mostly for periods of six months. A full re-evaluation of the whole program should be planned for July, 1942, but, as things then stood, four methods for acquiring fissile material were still in play: the electromagnetic, centrifugal, and diffusion methods for enriching uranium, and synthesizing plutonium via reactors. In just a few short weeks, Lawrence's electromagnetic method had shot to the top of the list of enrichment options. Use of his 184-in. magnet at Berkeley as a mass spectrometer was expected to be ready by July 1, and was expected to yield 0.1 to 10 g of U-235 per day by September. With a series of giant spectrographs, Lawrence was looking to produce perhaps a kilogram of U-235 per day by the summer of 1943. The centrifuge pilot plant was expected to be in operation by August 1, 1942. If this could be put into large-scale operation, it could produce one kilogram of U-235 every 10 days by July, 1943. The properties of plutonium were completely unknown, but a few micrograms for research purposes were expected from Lawrence's cyclotrons by June, 1942. Expenditures were expected to amount to about $3 million by August 1, but the costs of large-scale construction were, as Conant put it, "anybody's guess." If all methods continued to be pursued and a decision on which to retain was postponed to January, 1943, some $10 million would be called for.

Beginning in early 1942, Ernest Lawrence began making steady advances with electromagnetic separation. In early January, he produced 18 micrograms of material enriched to 25 % U-235; in February, three 75 µg samples enriched to 30 % were available. Lawrence chronicled his progress in a series of letters to Bush and Conant through the spring. On March 7, an ion source designed to give a beam of current 10 mA was delivering somewhat more than that to the collecting anode, and Lawrence felt ready to proceed to design and construct a 100 mA source and to begin planning for a 1 A setup based on ten such units. Six days later he reported that the 10 mA source was yielding 25 mA, that he was proceeding with the design of the 100 mA unit, and that he was considering design and construction of a "multiple" mass spectrograph using the 184 in. cyclotron magnet. This concept would involve a dozen separate ion sources, each rated for 0.1 to 0.5 A. With a one-amp beam corresponding to one gram of U-235 per day, quantity production would then be underway. Lawrence hoped to have four sources in operation by July, with the entire plant in operation by autumn at a cost of perhaps a half-million dollars. In time, Lawrence's cyclotrons, re-purposed as calutrons, would succeed in producing U-235 on a large scale, but a very bumpy road yet lay ahead.

4.8 Spring 1942: Time is Very Much of the Essence, and Trouble in Chicago

A reader who seeks a more definitive Presidential directive to proceed with an all-out project to develop nuclear weapons than that implied by the formation of the Top Policy Group in October, 1941, can look to a report on the status of the project which Vannevar Bush sent to Roosevelt, Stimson, Marshall, and Wallace on March 9, 1942. This report was an expanded version of Conant's February 20 report, and gives a detailed picture of the project at that time.

In a cover letter, Bush indicated that work was under way at full speed. The amount of fissile material necessary for a bomb appeared to be less than previously thought, the anticipated effects were predicted to be more powerful, and the possibility of actual production seemed more certain. While Bush felt that America may be engaged in a race with Germany toward realization of such weapons, he had no indication of the status of any German program; ironically, physicist Werner Heisenberg had reported on possible uses of atomic energy to a meeting of the Reich Research Council in Berlin only 11 days earlier. Bush advised that the program was rapidly approaching the pilot plant stage; by the summer of 1942, the most promising methods could be selected, and plant construction could be started. He urged that at that time, the whole matter should be turned over to the War Department. The amount of necessary "active material" (fissile material) was estimated to be 5 to 10 pounds, to which would be added a heavy casing. The effect of a single bomb was now estimated as equivalent to 2000 t of TNT. A twenty-unit centrifuge pilot plant was under construction, and it was estimated that a corresponding full-scale plant could be completed by December, 1943. A pilot-scale diffusion plant was being constructed by the British, and the electromagnetic method, "a relatively recent development," might offer a shortcut in both time and plant requirements in offering the possibility of practicable quantities of material by the summer of 1943. The report mentions power production (reactors) only briefly, as such developments were expected to be some years off; no mention was

Fig. 4.11 President
Roosevelt to Vannevar Bush,
March 11, 1942

THE WHITE HOUSE
WASHINGTON

March 11, 1942.

MEMORANDUM FOR DR. VANNEVAR BUSH:

I am greatly interested in your
report of March ninth and I am returning it
herewith for your confidential file. I
think the whole thing should be *pushed*
not only in regard to development, but also
with due regard to time. This is very much
of the essence. I have no objection to turning over future progress to the War Department on condition that you yourself are
certain that the War Department has made
all adequate provision for absolute secrecy.

F.D.R.

made of plutonium. Bush also summarized the organization of the project, pointing
out that the three Project Leaders under the Planning Board were all Nobel Laureates. To help maintain security, the project was subdivided; full information was
not being given to every worker. Despite this, Bush felt that the whole enterprise
was more vulnerable to espionage than was desirable, an additional reason for the
work to be placed under Army control as soon as actual production was embarked
upon. Roosevelt's March 11 response speaks for itself (Fig. 4.11).

Bush could not have asked for a clearer go-ahead. The project was now on a
fast track, with cost being no object. Within days, the wheels of Army bureaucracy
began to turn. On March 14, Harvey Bundy, a Special Assistant to Secretary of
War Stimson, wrote to Bush to confirm that General Marshall had authorized
Brigadier General Wilhelm Styer as the Army's contact for S-1. Styer was Chief of
Staff to Lieutenant General Brehon Somervell, who commanded the Army's
Services of Supply. Further details on the various levels of Army bureaucracy are
described in Sect. 4.9.

April 1, 1942 saw another extensive report from Conant to Bush, the result of a conference he and Briggs had that day with Compton regarding bomb theory and the plutonium program. Compton's schedule for achieving a chain reaction had been moved forward to November 1, and the expected date for production of experimental quantities of plutonium from a 5 MW pilot plant had advanced from early 1944 to May 15, 1943, a projection that would prove too optimistic. The projected date for quantity production of plutonium, 1 kilogram/day, had been pushed back to July, 1945, presuming the development of 100 MW plants; this prediction would be beaten. To support all of this, Conant requested an additional $422,000 in funding to July 1. Conant also raised the issue of sites for production plants, suggesting that one be chosen within the next month. It should be "located in a wilderness," and in such a locality that reactors or factories for any of the enrichment methods could be built. The location would need to be secure from espionage, take into account the safety of the workers and the surrounding population, have considerable electrical power available, and would require an adequate supply of cooling water if reactors were to be constructed. This would all require construction not only of the plants, but of living quarters, machines shops, laboratories, and other support facilities.

Conant called a meeting of Murphree, Briggs, and the Program Chiefs for Saturday, May 23; General Styer was also invited. On the 14th, Conant wrote to Bush to express concern that significant issues were approaching decision points, and that, unless the ultimate decision was a green light on everybody's hopes and ambitions, there would be some "disgruntled and disheartened" people who might "take the case to the court of public opinion, or at least the top physicists of the country." Conant was apparently unsure of his authority in the matter, however, for after suggesting that the group could act as a committee and send Bush a recommendation, perhaps with majority and minority reports, he wondered if Bush wanted Conant himself to act as a member of such a committee, or should he simply forward a report with a recommendation? Since the activities of the Planning Board were outside of his jurisdiction, what role it should play? He then outlined the scale of pending decisions. Fissile material preparation by centrifugation, diffusion, and electromagnetism were still in the running on about equal footings, as were reactors with or without heavy water. All would be entering pilot development within next six months, and production plants should be under design and construction even before the pilot plants were finished. What Conant called this "Napoleonic approach" could run to $500 million, the first mention of a cost in the multiple hundreds-of-millions range. Despite this potential cost, Conant felt that an all-out program might be justified: it seemed fairly certain that all methods would yield a weapon, which meant that the probability of the Germans developing such devices was also high. In reflection of President Roosevelt's dictum that time was of the essence, Conant observed that if they discarded some of the fissile-material production methods at that point, they may be unconsciously betting on the "slower horse"; a delay of even only three months could be fatal if within such time Germany could employ a dozen such bombs against England. Finally, he offered some thoughts as to the Army's eventual role, suggesting that while that organization might may be willing to take on production or even pilot plants, he felt that it ought not take over research. As to administration: "I do not believe Briggs should be brought back into the picture with any more authority. I am quite sure that Beams, Lawrence, and Murphree should go full time into the Army, probably as officers". He closed with a sense of exasperation: "If the whole matter were out of our hands, it would be a relief, but I am inclined to think a good deal would be lost and eventually it might come back again!" Bush replied on May 21, indicating that he would prefer that Briggs, Murphree, and the Program Chiefs constitute themselves as a committee to send a report through Conant. The committee should give summaries on issues such as an outline of the program for each method for next six months and the ensuing year, judgments on how many programs should be continued, and suggestions as to what parts of the program should be eliminated if there were limitations on people, money, and material.

Even as his influence within the project was diminishing, Lyman Briggs still had to deal with his share of headaches. On the day of the May 23 meeting, he received a letter from Gregory Breit in which he announced that he was leaving Compton's project and his position as coordinator of fast-neutron research, to which he had been appointed only in February. Breit was furious over what he saw as lax security at Chicago, and charged that the whole range of activity of S-1 had been discussed at meetings held at Columbia and Princeton. Without naming names, Breit alleged that there were some individuals at Chicago who were strongly opposed to secrecy, and that, while he had communicated his concerns to Compton, he feared that the latter's course of action was likely to be influenced by considerations regarding satisfying personal desires and ambitions. In anticipating that the bomb would exceed ordinary weapons by orders of magnitude in offensive power, Breit felt that it would be necessary to have adequate security not only during the war, but also for decades afterwards, and urged government control of the whole matter. He was particularly upset regarding research on bomb design, and urged centralizing that work in two or three locations isolated from the reactor project. He felt further that the University of Chicago should have no part in the bomb-design work, and that such work should be placed directly under the control of one of the armed services. In a sense, Breit jumped the gun: many of his ideas would come to be reality within a year. His resignation did, however, open the door for Robert Oppenheimer to head the fast-neutron work.

On Monday, May 25, Conant reported to Bush on the May 23 meeting, which had run from 9:30 a.m. to 4:45 p.m. The bottom line was that if the urgency of securing fissile material justified an all-out program, then the group recommended an extensive program to run from mid-1942 to mid-1943 in which every method would receive support: over $38 million for a 0.1 kg/day centrifuge plant to be ready by January, 1944; over $2 million for pilot-plant and engineering work on a 1 kg/day diffusion plant; $17 million for a 0.1 kg/day electromagnetic separation plant to be ready by September, 1943; $15 million for reactors to produce 0.1 kg/ day of element 94; over $3 million for a half-ton per month heavy water plant, and miscellaneous research valued at just over $2 million. Throwing in $5 million for contingencies brought the total to $85 million. If cuts were necessary, the reactor and electromagnetic programs could be decreased by \$10 million and \$2.5 million, respectively; delaying centrifuge construction to 1943 would save \$18 million, but cost six months in lost time. The group could not choose between cut options. It was predicted that with the full program, bombs could be expected to be ready by July, 1944.

In the meantime, Bush had not heard the last of concerns with the way things were being managed in Chicago. A May 26 letter from Leo Szilard expressed growing frustration with what he saw as the slowness of work on reactor development. In Szilard's opinion, graphite and uranium oxide of the required purity had been available for a pile in 1940 (in reality, this was not true), and the division of authority between Compton and Murphree was such that neither group could function properly. In a scathing indictment of the way the whole project had been managed since initial recommendations had been made in 1939, Szilard asked Bush if he "might not think that the war would be over by now, if these recommendations had been acted upon," and alluded to everybody's worst fears: "Nobody can tell now whether we shall be ready before German bombs wipe out American cities." Another letter dated the next day by Chicago physicist Edward Creutz echoed many of the same concerns. Conant spoke with Compton, who promised to calm Creutz down, and Bush wrote to Creutz and Szilard on June 1 to let them know that plans were being made to reorganize and expand the whole effort.

As a July 1 funding cutoff-date loomed and momentum gathered within the Army toward formal establishment of the Manhattan Engineer District (MED), large-scale decisions began to get made. On June 10, Bush conferred with Generals Marshall and Styer; they decided to proceed with the electromagnetic method and the "boiling project" (reactors), as they would cause the least disruption to critical materials. Styer was to study the impact of the proposed centrifuge and diffusion programs on other essential programs. In a summary memo to Conant the next day, Bush indicated that it was understood that Styer would inform other officers in the Services of Supply, and would plan to take over all production aspects of the project on July 1. The Planning Board would also be turned over to Styer at that time, who would be at liberty to modify the membership as he saw fit. On June 13, a few days prior to the formal establishment of the MED, Bush and Conant sent a 6-page status report to Wallace, Stimson, and Marshall; their approval signatures appear on the last page on the copy in OSRD records. From them the document would go to the President for final approval. The first paragraph got to the point that matters had proceeded to a point where a large-scale decisions were called for. The estimated explosive yield of fission bombs had been raised to several thousand tons of TNT, and it was estimated that with a suitably ambitious program, a small supply of such bombs could be ready by mid-1944, plus-or-minus a few months. To avoid the danger of concentrating on any one method, Bush echoed the full slate of recommendations of Conant's May 25 report. He also suggested that it was time to arrange for a committee to consider the military uses of the material produced. The uranium project was about to enter a significant new phase of its life.

4.9   June–September 1942: The S-1 Executive Committee, the Manhattan Engineer District, and the Bohemian Grove Meeting

Bush presented the June 13 report to President Roosevelt on Wednesday, June 17, 1942. His one-page cover letter, the archived copy of which bears an iconic "V.B. OK FDR", states that it was contemplated that all financing for the project would be handled by the Army's Chief of Engineers through the War Department. On the same day, General Styer telegraphed orders to Colonel James C. Marshall of the Syracuse (New York) Engineer District to report to Washington to take command of what was being called, for the time being, the DSM Project: Development of Substitute Materials.

In many histories of the Manhattan Project, Colonel Marshall suffers much the same fate as Lyman Briggs in being thrust into relative obscurity against the presence of more forceful personalities. Only three months after his appointment to the project, Marshall was replaced as its commander by a much more aggressive officer, Leslie Richard Groves (Fig. 4.12). Under Groves, Marshall retained the title of District Engineer until July, 1943, at which time Groves eased him out of that position in favor of Marshall's own deputy, Colonel Kenneth D. Nichols. This terminated Marshall's association with the Manhattan District, but, as described in what follows, he was by no means inactive during his tenure with the project.

The wartime organization of the Army was immensely complex. In March, 1942, a reorganization of the Army command structure saw the designation of three overall commands: Army Ground Forces (AGF), Army Air Forces (AAF), and Army Services of Supply, which later became the Army Service Forces (ASF). The latter is the one of interest here, and was under the command of Lieutenant General Brehon Somervell. A Lieutenant General carried three stars; beneath that rank came Major General (2 stars), and Brigadier General (1 star). Above Lieutenant General was General (4 stars, such as George C. Marshall, no relation to

Fig. 4.12 *Left* to *right* General Groves (1896–1970) and Robert Oppenheimer; a formal portrait of Groves; Kenneth D. Nichols (1907–2000). *Sources* http://commons.wikimedia.org/wiki/File:Groves_Oppenheimer.jpg http://commons.wikimedia.org/wiki/File:Leslie_Groves.jpg http://commons.wikimedia.org/wiki/File:Kenneth_D._Nichols.jpg

James Marshall). The Army Corps of Engineers (CE) was one of the operating divisions of the ASF. General Styer was Somervell's Chief of Staff, and the Chief of Engineers was Lieutenant General Eugene Reybold, who was appointed to that position on October 1, 1941. Within the Corps of Engineers lay the Construction Division, which was headed by Major General Thomas Robins. On March 3, 1942, Leslie Groves, then a Colonel, was appointed Deputy Chief of Construction.

Groves graduated fourth in his West Point class of November 1918, and also trained at the Army Engineer School, the Command and General Staff School, and the Army War College. His career in the Corps of Engineers was marked by steady advancement, and by 1942 his workload was enormous. Under Robins' supervision, he was responsible for overseeing all Army construction within the United States, as well as at off-shore bases. Camps, airfields, huge ordnance and chemical manufacturing plants, depots, ports, and even internment camps for Japanese-Americans all came under his purview. At the time the Army became involved in the Manhattan Project, the Corps of Engineers was engaging almost one million people under contracts consuming some $600 million per month; Manhattan was a drop in the bucket in comparison. This experience gave Groves intimate knowledge of how the War Department and Washington bureaucracies functioned, and of what contractors could be depended upon to competently undertake the design, construction, and operation of large plants and housing projects. Over the course of the war, the Corps of Engineers would place more than $12 billion worth of construction within the United States, including over 3,000 command installations and nearly 300 major industrial projects. In the spring of 1942, one of Groves' projects was the construction of the Pentagon, which was completed within sixteen months of ground being broken. The background of Groves' career is, however, getting somewhat ahead of the story. One of the most valuable sources of information on initial Army involvement in the Manhattan Project is a diary kept by Colonel Marshall, *Chronology of District X,* which runs from June 18, 1942, to January, 1943, with a few sporadic entries thereafter. Much of what is related in the following paragraphs is based on this diary.

Marshall's assignment was unusual. Normally, the Chief of Engineers oversaw projects through an "Engineer District." An individual designated as District Engineer reported to a Division Engineer, who headed one of eleven geographical divisions of the United States. But Marshall's new District had no geographical restrictions; in effect, he was to have all of the authority of a Division Engineer. While the terms Manhattan Project and Manhattan Engineer District are often used interchangeably (as in this book), it should be borne in mind that they are by no means the same. Marshall initially located his headquarters in New York City. When Groves was assigned to be Commanding General, he became senior to Marshall, and set up his headquarters in Washington. The District office itself remained in New York until Marshall's departure in 1943, at which time Colonel Nichols moved it to Oak Ridge. The term Manhattan Project never was an official one, and only came into general use after the war.

Styer briefed Marshall on his new assignment on the afternoon of June 18. On returning to the Office of the Chief of Engineers, Marshall informed a number of other officers, including Groves, of his new command. While Groves claims in his memoirs that he was "familiar" with the Project in its initial stages as a part of his overall responsibilities but knew little of its details, Marshall's diary makes it clear that he was in fact a very active participant from the outset. Groves often advised Marshall as to contractors and procedures, and was involved in suggesting the "Manhattan District" name. Marshall and Styer met with Vannevar Bush the next morning, at which time they saw Conant's May 25 report and Bush's June 17 letter to FDR. A meeting with Bush, Conant, and the Program Chiefs was set up for June 25. Groves was asked to undertake a survey of sites around the country that would have suitable power available to run the anticipated uranium enrichment plants and plutonium-producing reactors. Marshall also decided that day that he wished to have Nichols, his deputy at Syracuse, accompany him to his new District. Nichols held a Ph.D. in civil engineering, and would become one of the driving forces of the Manhattan Project.

While the Army was coming up to speed on the Project, Vannevar Bush moved to effect yet another rearrangement of the S-1 organization. Following a meeting with Marshall on June 19, he wrote to Briggs and Conant to apprise them of the appointment of an "S-1 Executive Committee" within the OSRD, which would replace the somewhat largish S-1 Section committee of Fig. 4.10. Conant would chair the group; the other members would be Briggs, Lawrence, Urey, Compton, and Murphree. Allison, Beams, Breit, Condon, and Smyth would continue to serve as consultants. The Planning Board would remain in existence, but would report in an advisory capacity to the Chief of Engineers. To set the stage for the planned June 25 meeting, responsibility for the various projects discussed in the May 25 report were divided between the new S-1 Executive Committee and the War Department. The Committee was to recommend contracts for centrifuge and diffusion pilot plants, research and development, a 5 g/day plant for the electromagnetic method, the heavy water project, and miscellaneous research. The War Department was to take on the 100 g/day centrifuge production plant, engineering and construction of a 1 kg/day diffusion plant, a 100 g/day electromagnetic plant, and a pilot-scale reactor to produce 100 g of plutonium per day.

Work on bomb physics also progressed during the summer of 1942. On May 19, Robert Oppenheimer wrote to Ernest Lawrence with the optimistic prediction that with a total of two or three experienced men and perhaps an equal number of younger ones, it should be possible to solve the theoretical problems of building a fast-fission bomb. Beginning in the second week of July, Oppenheimer gathered a group of theoretical physicists at Berkeley to consider the detailed physics of bomb design. The participants included some of the most outstanding physicists of the time, including Hans Bethe, Edward Teller, and Robert Serber, all of whom would work at Los Alamos (Fig. 4.13).

The discussions at Berkeley covered the entire spectrum of design issues, and even the possibility of fusion bombs. A particularly important issue was the danger that impurities in plutonium could cause a low-efficiency explosion. The problem was not the presence of impurities *per se*, but an indirect effect that harked back to the discovery of the neutron. Reactor-produced plutonium is a prolific alpha-

Fig. 4.13 *Left* Hans Bethe's (1906–2005) Los Alamos identity badge photo. Middle: Edward Teller (1908–2003) in 1958, as Director of the Lawrence Livermore National Laboratory. *Right* Robert Serber's (1909–1997) Los Alamos identity badge photo. Sources: http://commons.wikimedia.org/wiki/File:Hans_Bethe_ID_badge.png   http://commons.wikimedia.org/wiki/File:Edward_Teller_(1958)-LLNL-restored.jpg   http://commons.wikimedia.org/wiki/File:Robert_Serber_ID_badge.png

emitter, and as Bothe and Becker, the Joliot-Curies, and Chadwick had found, alpha particles striking nuclei of light elements tend to create neutrons, so-called (α, n) reactions. If chemical processing of plutonium left behind light-element impurities, this effect could give rise to a premature detonation. This issue would almost prove the undoing of the plutonium project (Sect. 7.7).

The June 25 Army/S-1 meeting was held at the OSRD, and saw a number of crucial decisions made. All of the major players were present: Marshall, Nichols, Styer, Bush, Conant, Lawrence, Compton, Urey, Murphree, and Briggs. Styer felt that the manufacturing plants should be set up somewhere between the Allegheny and the Rocky mountains to protect them from enemy coastal bombardment. Up to 150,000 kW of power would need to be available toward the end of 1943 to operate all of the electromagnetic, centrifuge, pile, and diffusion plants. The necessary site size for all of this was estimated to be some 200 square miles, preferably in the shape of a 10 by 20-mile rectangle. The construction and engineering firm of Stone and Webster (S&W) of Boston was suggested for site development, housing construction, engineering and construction of the centrifuge plant, and, if they were agreeable to the idea, to start work on a plant for Lawrence's electromagnetic method. Stone and Webster was already involved with the diffusion project through Eger Murphree, and Groves had contracted with them on a number of Army construction projects; it was apparently he who suggested the firm to Marshall. It was also decided to enter into a contract with University of Chicago to operate a pilot-scale reactor to be built by S&W in the Argonne Forest Preserve outside Chicago. That reactor, the X-10 pile, would ultimately be built in Tennessee (Sect. 5.2). Other suggested contractors were E. B. Badger and Sons, also of Boston, for a heavy-water plant to be set up in Trail, British Columbia, and, for the diffusion plant, the M. W. Kellogg Company of New Jersey, a firm with extensive experience in design and construction of petroleum refineries and chemical facilities.

A chronic issue in the plutonium project was that many of the staff at Compton's Metallurgical Laboratory felt that they themselves should direct the design, engineering, construction, and operation of the plants. From years of consulting experience, Compton knew that large industrial concerns typically divided responsibility for research, development, and production among separate departments. Compton described the reaction of many of his staff to bringing in a large industrial concern as "near rebellion." Utterly unaware of the scale and complexity of facilities that would be required, some of the scientists felt that they could supervise plant construction if Groves provided them with but fifty to one hundred engineers and draftsmen. Compton supported the decision to assign architect-engineer-manager responsibilities to Stone and Webster, a move that deeply angered many of his colleagues.

On June 27, Nichols met with John R. Lotz, the President of Stone and Webster, who was enthusiastic that his firm was being considered for construction of the plants, and also expressed interest in contracts to operate them. A formal meeting with Lotz and S&W engineers and managers was held in Groves' office on June 29 to hammer out details of the contract; Groves was also to see to approval of land purchases in Chicago and Tennessee (see below). On the same day, Marshall decided to establish his District Headquarters on the 18th floor of the Corps of Engineers North Atlantic Division building at 270 Broadway in New York City; Stone and Webster conveniently had offices in the same building.

By August, Compton was urging Marshall to select an operating contractor for the various plants. Since it was anticipated that the operator of the Argonne plutonium-extraction facility would also operate the works for the production plants, that organization should observe construction of the plant at Argonne. Compton suggested the DuPont corporation, SODC, or Union Carbide and Carbon as operator. Marshall, however, was reluctant to bring in any more firms for security reasons, and proposed instead that S&W add operations to their responsibilities, with the provision that they could secure technical assistance from other organizations.

Groves' survey indentified eastern Tennessee as a likely site for production plants (Fig. 4.14). The area was supplied with ample power by the Tennessee Valley Authority (TVA), which, by May, 1942, boasted an installed capacity of 1.3 million kilowatts (kW). This would grow to 2.5 million kW by the end of the war, at which time the TVA was supplying some 8 % of the nation's electricity. Over July 1–3, Marshall and Nichols visited the Knoxville area to inspect a site of roughly 17 by 7 miles near the town of Clinton. The area was promising not only for its good power and railroad accessibility, but also for possessing several parallel northeast-to-southwest valleys separated by 200–300 foot ridges. The ridges could be used to segregate different production areas, which would provide protection in case of a catastrophe at any one of them. The Clinch river provided

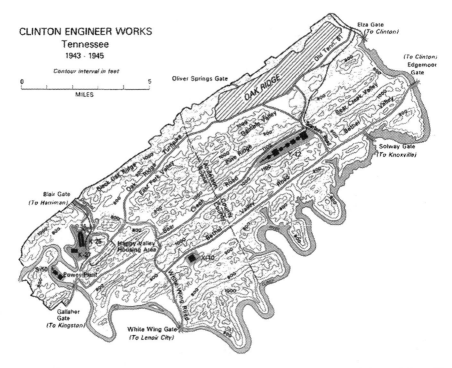

Fig. 4.14 The Clinton Engineer Works site, east-central Tennessee. Knoxville is located about 15 miles east of Solway gate. The locations of the Y-12, K-25, X-10, and S-50 facilities are marked. Tennessee highway 61 became the Oak Ridge Turnpike. *Source* http://commons.wikimedia.org/wiki/File:Oak_ridge_large.gif

natural boundaries on three sides of the area, and Tennessee State Highway 61 defined the north side.

One story surrounding the selection of the Clinton site—perhaps fictional—is that President Roosevelt met with Tennessee Senator Kenneth McKellar, an influential member of the Appropriations Committee, to ask him to devise means of hiding Project funding that would eventually amount to $2 billion. Roosevelt allegedly asked "Senator McKellar, can you hide two billion dollars for this supersecret national defense project?" McKellar's response is said to have been "Well, Mr. President, of course I can. And where in Tennessee do you want me to hide it?"

During the early days of the Manhattan District, Marshall and Nichols were on the road almost constantly. A selection entries from Marshall's diary from July through September gives a sense of what must have been a brutal pace. Immediately after the Knoxville trip, Nichols visited Chicago on July 6–7, where, along with Enrico Fermi and S&W representatives, he inspected Compton's Metallurgical Laboratory and the proposed 1,000-acre Argonne Forest site located about 10 miles west of the University. On the 9th, Marshall and Nichols were back in

Washington for another meeting with S & W representatives to discuss uranium supplies, priorities, the Trail plant, the Tennessee site, liaison with the British, use of silver as a substitute for high-priority copper in the magnets for Lawrence's cyclotrons, and funding issues. Marshall's diary for that day also recorded that Groves was perturbed with what he viewed as indefinite dates for when various parts of the project would get underway.

On July 13, Nichols and Groves prepared a memorandum requesting that the Corps' Real Estate Division secure a lease on the Argonne site. July 14 saw another conference with S&W to deal with contract legalities, purchasing procedures, and road relocations in Tennessee. It was estimated that the site would require housing for 5,000 people, a number which would prove to be a drastic underestimate. Stone and Webster wanted the site obtained by August 10 so that an administration area and 200 associated housing units could be built in October. On July 20, Marshall was in San Francisco to meet with local S&W engineers and Ernest Lawrence. Marshall and the engineers were concerned with a general lack of organization of the work at Berkeley, and encouraged the group there to begin construction of a pilot plant and design of a full-scale plant. Despite these misgivings, Marshall was of the opinion that Lawrence's electromagnetic approach was ahead of the other three fissile-material methods, and that it should be exploited to the fullest extent without delay. The next day, Nichols traveled to Boston to meet with Badger and Sons regarding timing and priorities for the Trail plant, a thorny issue by virtue of its own need for considerable quantities of copper. The following day he was back in Washington to confer with Groves regarding uranium ore and approval for the formal organization of the new District, which was to be named as soon as the site in Tennessee was chosen. Marshall favored "Knoxville District" as that would be their postal address, but Groves preferred something less revealing.

On July 29, the Real Estate Division got back to Nichols with an estimate of the cost of the Tennessee site. This would involve not only the direct cost of the land, but also relocation of cemeteries and utilities, road closures, and compensation for crop values. The total was estimated at $4.25 million for 83,000 acres, of which 3,000 were owned by TVA. Some 400 families would have to be relocated. The OSRD approved the acquisition the next day, despite the reluctance of some scientists to move to a hot climate. On July 31, however, Marshall told General Robins that he was unwilling to proceed with acquisition of the site or to begin any construction until Compton's pile process had proven itself. On August 6, Nichols was back in Boston to confer with S&W on supplier contracts with the Metal Hydrides Company, Mallinckrodt Chemical, the Consolidated Mining and Smelting Company of Canada, and DuPont. On the 11th, Marshall conferred again with Groves regarding drafting a General Order forming the new District. Groves still objected to "DSM"; they decided that "Manhattan" was the best place name they could use, and so the "Manhattan Engineer District" was born. The name began to appear in Marshall's diary the next day. On August 13, the day the District came into official existence, Marshall traveled to St. Louis to visit the Mallinckrodt Chemical Works to discuss a contract for purification of 300 t of uranium oxide. Apparently back in Washington on the 14th, he dealt with a contract for the Tennessee Eastman Corporation as an operating contractor for the Clinton site. On the 18th, Nichols was in California, where he learned that Lawrence was willing to proceed immediately with work on the design for the full-scale plant to be sited in Tennessee. While Marshall continued to hold off on acquiring the Tennessee site through August, much other groundwork was accomplished. On the 24th, he and Nichols conferred with Eger Murphree to discuss the idea of contracting with the SODC to operate the reactor pilot plant. The next day, Nichols visited Westinghouse in Pittsburgh to consider centrifuge design; he witnessed a meter-long centrifuge in operation, at least until its motor burned out at 25,000 rpm. Back in Washington for the 26th, Nichols and Marshall conferred again with Conant, Murphree, Urey, Compton, Lawrence, Briggs, and S&W representatives to review all production methods under consideration. A target date of August 1, 1943, was set for the electromagnetic pilot plant to be operation.

The reader has by now no doubt got the gist; dozens of other such events could be related. In subsequent months, the pace would only increase as the cost and complexity of Manhattan District activities grew. Any notion that the District was inactive until Groves assumed command is a serious misconception.

Of all of the conferences held between the S-1 Executive Committee and the Army, one of the most important occurred over September 13–14, 1942, at Bohemian Grove, an exclusive campground located within the Muir Woods National Monument just outside San Francisco. As Compton wrote, decisions made at that meeting were destined to shape the entire future development of the Project (Fig. 4.3).

The committee's first recommendation was to complete construction of the Argonne Forest site, and to locate Fermi's first critical pile there. A second pile was to be built there later for purposes of producing some plutonium, with the understanding that chemical processing plants to handle the separation of plutonium would be erected in Tennessee. Second, it was recommended that the Army and Stone and Webster enter into a subcontract with a chemical company to develop the separation facilities. Dow Chemical, Monsanto Chemical, and the Tennessee Eastman Corporation were suggested, but those facilities would ultimately be designed, constructed, and operated by DuPont. Third came a recommendation for the Army to enter into a commitment, estimated to cost $30 million, to build a 100 g/day U-235 electromagnetic separation plant of 100–400 vacuum tanks in Tennessee, although the committee reserved the right to recommend canceling orders for material at any time up to and including January 1, 1943. A sub-recommendation to this was that the OSRD sponsor construction of a pilot electromagnetic plant comprising five vacuum tanks, also to be located in Tennessee. Finally, it was voted to recommend to the Army that construction of the heavy-water plant in British Columbia should be completed by May 1, 1943. The diffusion and centrifuge methods were not considered at this meeting, at least as far as the minutes reflect the discussion.

Within a week of the Bohemian Grove meeting, Leslie Groves would be placed in command of the Manhattan District. What had been a demanding pace was about to become frantic.

4.10 September 17, 1942: Groves Takes Command

Despite the historical importance of Groves' appointment to take on overall command of the Manhattan District, the record of events surrounding that development is rather murky. Several different versions have been published.

The decision to place Groves in command was apparently made on September 16 by Somervell and Styer. When Groves later asked Styer about the circumstances, the latter's reply was that General (George C.) Marshall wanted Styer to take on the job, but Somervell objected to the prospect of losing Styer. Somervell discussed the matter with Marshall, who instructed him to come up with someone suitable, and Somervell and Styer decided that Groves would be appropriate. Styer may not have wanted to take on the job in any event, as apparently both he and Somervell were skeptical of the idea of a weapon based on atomic energy.

In his memoirs, Groves claims that he learned of his new assignment on the next morning, Thursday, September 17, 1942, when Somervell caught up with him just after Groves had finished testifying before a congressional committee on a military housing bill. Groves claims that he had been offered an overseas assignment, and was disappointed when Somervell told him he could not leave Washington because "The Secretary of War has selected you for a very important assignment, and the President has approved the selection." When Groves realized what Somervell had in mind, he claims that his response was "Oh, that thing." On meeting with Styer later that morning, Groves was also informed that he was to be promoted to Brigadier General. His response to this was to ask that he not be placed in official charge until the promotion had gone through, believing that this would put him in a stronger position to deal with the academic scientists involved in the project: it would be better if he were thought of as a General instead of as a Colonel. The promotion became official on September 23. Colonel Marshall was on the west coast on September 17, and the diary entry for that day was made by Nichols, who refers to himself and Groves visiting Styer to learn of the new arrangement. Marshall returned on the 19th; subsequent entries make no comment regarding his new, subordinate position, although he continued as District Engineer.

Groves later offered some comments on Marshall: "He was just too nice a person and was lacking in brashness and self-confidence necessary to fight and win his way in Washington against the opposition which such an enormous project would naturally encounter. He would present his case well but would accept adverse decisions from his seniors in government." In another version of the story, Hewlett and Anderson also allude to Marshall's nature, pointing to his indecision in delaying selection of the Tennessee site, and being happy to leave paper-pushing and priority haggling to Nichols in Washington. But they too skirt Groves's early involvement, having Somervell "casually" mentioning to Vannevar Bush that "he knew a Colonel Groves" who would be just the man to take over the project.

The text of Sommervell's one-page directive placing Groves in charge of the project read as:

September 17, 1942

Memorandum for the chief of engineers
SUBJECT: Release of Colonel L. R. Groves, C.E., for Special Assignment

1. It is directed that Colonel L. R. Groves be relieved from his present assignment in the Office of Engineers for special duty in connection with the DSM Project. You should, therefore, make the necessary arrangements in the Construction Division of your office so that Colonel Groves may be released for full time duty on this special work. He will report to the Commanding General, Services of Supply, for necessary instructions, but will operate in close conjunction with the Construction Division of your office and other facilities of the Corps of Engineers.
2. Colonel Groves' duty will be to take complete charge of the entire DSM project as outlined to Colonel Groves this morning by General Styer.
(a) He will take steps immediately to arrange for the necessary priorities.
(b) Arrange for a working committee on the application of the product.
(c) Arrange for the immediate procurement of the site of the TVA and the transfer of activities to that area.
(d) Initiate the preparation of bills of materials needed for construction and their earmarking for use when required.
(e) Draw up plans for the organization, construction, operation and security of the project, and after approval, take the necessary steps to put it into effect.

Brehon Sommervell
Lieutenant General
Commanding

Groves ran a remarkably tight headquarters. He and a staff of just a couple dozen administered the Manhattan Project from a small suite of offices on the fifth floor of the New War Building at the intersection of Twenty-First street and Virginia Avenue NW in Washington. The building is now part of the Department of State; because of renovations over the intervening years, the original offices no longer exist.

It is not uncommon to read Groves described as arrogant, arbitrary, insensitive, overbearing, and high-handed. More appropriate labels might be mission-focused, supremely competent, and able to get things done. His ability to juggle the multiple responsibilities of the Manhattan Project was remarkable. Colonel John Lansdale, Groves' head of security for the Project, offered this assessment of his superior: "General Groves was a man of extraordinary ability and capacity to get things done. Unfortunately, it took more contact with him than most people had to overcome a bad first impression. He was in fact the only person I have known who was every bit as good as he thought he was. He had intelligence, he had good judgment of people, he had extraordinary perceptiveness and an intuitive instinct for the right answer. In addition to this, he had a sort of catalytic effect on people. Most of us working with him performed better than our intrinsic abilities indicated."

The relationship between Groves and Kenneth Nichols was apparently somewhat strained, despite its productivity. After the war, Nichols offered this assessment:

> First, General Groves is the biggest S.O.B. I have ever worked for. He is most demanding. He is most critical. He is always a driver, never a praiser. He is abrasive and sarcastic. He disregards all normal organizational channels. He is extremely intelligent. He has the guts to make timely, difficult decisions. He is the most egotistical man I know. He knows he is right and so sticks by his decision. He abounds with energy and expects everyone to work as hard or even harder than he does … if I had to do my part of the atomic bomb project over again and had the privilege of picking my boss I would pick General Groves.

Groves' first meeting with Vannevar Bush was not auspicious. Styer had not had time to inform Bush of Groves' appointment, and Bush was reluctant to answer questions. After the meeting, Bush sent a note to Stimson's assistant Harvey Bundy, expressing doubt that Groves had sufficient tact for the job. The note closed with: "I fear we are in the soup." Another meeting between Groves and Bush two days later went much more smoothly; Groves later claimed that they became fast friends.

Groves got to work promptly in his new command. On September 18, his first full day in charge, he dispatched Nichols to New York to confer with Edgar Sengier of Union Minière to reach an agreement to purchase that firm's 1,200 ton stock of uranium-rich ore being held in storage in the United States. Nichols also made arrangements to ship to and store in the United States ores then being held in the Belgian Congo, and to assign those ores a prior right of purchase for the United States. Nichols and Marshall then visited the offices of Stone and Webster, where a $66 million estimate for engineering development for the four alternate production methods, construction of electromagnetic and reactor pilot plants, materials procurement, and town site development was hammered out. On the 19th, Groves issued a directive to purchase the Tennessee site. He also, in one step, resolved the issue of priority assigned to the project that had been a holdup for months. With a letter in hand addressed to himself which granted the project AAA priority—the highest possible—Groves appeared at the office of Donald Nelson, head of the War Production Board. Nelson initially refused to sign, but reversed himself when Groves said he would have to recommend to the President that the project be abandoned because the WPB was unwilling to cooperate. On the 21st, Groves and Marshall met again with Bush, where they learned that the Navy had been left out of the project at the explicit direction of the President. Despite that injunction, Groves and Nichols visited the NRL later the same day to see a 14-stage liquid thermal diffusion facility that was under construction. Ross Gunn was desirous of coordinating Navy efforts with the Army, but Groves' impression of the Navy effort was that it seemed to lack urgency.

One of the directives in Somervell's memo of September 17 was that Groves should arrange for a "working committee on the application of the product." This occurred on September 23, the day Groves was formally promoted to Brigadier General. At a meeting with Stimson, General Marshall, Conant, Bush, Styer, and Somervell, it was decided to appoint a Military Policy Committee (MPC), comprising Bush (as Chair; with Conant as his alternate), Styer, and Rear Admiral William Purnell of the Navy. The charge of the MPC was to determine general policies for the entire Project. Formally, Groves was to sit with the committee and act as an Executive Officer to carry out policies that it determined, but in practice the committee usually ended up reacting to what he had already done. Groves cut short his attendance at the meeting to undertake a tour of the Tennessee site. After returning to Washington, he and Nichols met with Stone and Webster officials on the 26th, at which time it was decided to approach DuPont to develop and operate the plutonium-extraction plants.

Founded in 1802, the E. I. du Pont de Nemours and Company (referred to here simply as DuPont) was considered to be "the colossus of American explosives and propellant production." The firm had vast experience with designing, constructing, and operating a wide variety of chemical processing facilities; by the end of the war, DuPont had built 65 % of total United States Ordnance Department powder production facilities. After some arm-twisting, DuPont accepted, on October 3, a contract to design and build the plutonium separation plants. Groves, impressed by the security advantages of DuPont's practice of building its own plants, soon began envisioning a much bigger role for the company in the Manhattan Project.

On October 2, Arthur Compton presented Groves with a proposal for development of four reactors. There were to be (1) the first experimental pile, to be in operation at the Argonne site by December 1; (2) a 10-MW water-cooled pilot reactor in Tennessee, to be in operation by March 15, 1943, for the purpose of generating small amounts of plutonium for testing development of the separation process; (3) a 100-MW liquid-cooled unit in Tennessee to be in operation by June 15, 1943; and (4) a helium-cooled 100-MW plant, also to be located in Tennessee, to be in operation by September 1, 1943. The plan for two 100-MW plants may seem redundant, but Compton wanted to insure adequate production given the uncertainties and problems that would inevitably arise. It was anticipated that the liquid-cooled plant could be constructed more quickly, but the form of cooling was not specified; both ordinary water and heavy water were still in the running. Groves assured Compton that a decision on an operating contractor would be made in about three weeks.

Three days later, Groves paid his first of many visits to Compton's Metallurgical Laboratory at the University of Chicago. Despite the high opinion he gained of the scientific competence of the Chicago group, he was horrified to learn that they blandly considered their estimate of the amount of fissile material needed for a bomb to be correct within a factor of ten, an enormous margin of uncertainty for an engineer. Groves related that he felt like a caterer who was being asked to prepare for a dinner for which anywhere between ten and a thousands guests might show up. They also discussed pile-cooling methods, eventually settling on (gaseous) helium, but that decision would be changed within three months.

Groves soon began pressuring DuPont to take a leading role in the plutonium program. He decided to relieve Stone and Webster of any responsibility for that program, having come to the opinion that every aspect of it, from design through operation of both piles and the separation facilities, should be overseen by a single firm. On October 31, Groves and Conant met with two DuPont Vice Presidents, Willis Harrington and Charles Stine. Groves pressed them to take on the pile program, stating that he felt that DuPont could handle all aspects of the project better than any other company in the country. Harrington and Stine were skeptical; chemistry, not nuclear physics, was DuPont's forte. On November 10, Groves, Nichols, and Compton visited DuPont's headquarters in Wilmington, Delaware, to put the case directly to the company's President, Walter Carpenter. Groves played to Carpenter's patriotic sympathies, emphasizing the importance that President Roosevelt, Secretary Stimson, and General Marshall attached to the plutonium work. In response to Carpenter's concern that the background knowledge necessary to design and build piles was not yet sufficient, Groves emphasized that the paramount importance of the project to the war effort required proceeding directly. Carpenter concluded that the company could not refuse, but the issue would have to be put before the firm's Executive Committee.

Groves returned to Wilmington on November 27 to meet with the Executive Committee, before which he reiterated the arguments made to Carpenter. The Committee concluded that the pile project would probably be feasible, but insisted as conditions of the company's involvement that a full review of the project be undertaken, and that the government be willing to indemnify DuPont against any losses or future liability claims due to the unusual hazards that would be involved. The issue of indemnification was serious. Concerned that liability claims for radiation-induced illnesses could begin cropping up twenty or thirty years in the future, DuPont insisted that a trust fund be set up to cover such claims; it would be funded to the extent of $20 million. Groves agreed, and a contract was signed on December 21. Since the company had no desire to produce plutonium after the war, it insisted that any patents revert to the Government, waived all profits, and accepted only payment for expenses plus a fixed fee of $1.00. The contract gave DuPont the option of leaving the project nine months after the end of the war, and allowed the company to continue to apply corporate pay scales to employees delegated to the project. Due to legalities regarding the duration of the contract, DuPont eventually netted only 68 cents of the dollar. Thirty-two members of the Pasco, Washington, Kiwanis Club subsequently each donated one cent to DuPont to make up the shortfall.

In late 1942, DuPont established a separate corporate division to organize its plutonium activities, the so-called TNX Division. Described as a "task force within a matrix organization," TNX would have two subdivisions: a Technical Division to carry out design, and a Manufacturing Division to advise DuPont's Engineering Division on construction of facilities and their operation.

Before Groves could get a review committee together, another problem arose on two fronts almost simultaneously. On November 3, Glenn Seaborg reported to Robert Oppenheimer his concern that, as described in Sect. 4.9, even very minor light-element impurities in plutonium could lead to an uncontrolled predetonation via (α, n) reactions. Seaborg's estimate that plutonium purity would have to be controlled to one part in 10^{11} could well put the entire pile project at risk. On November 14, Wallace Akers, technical chief of the British Directorate of Tube Alloys, informed James Conant that British scientists were concerned about exactly the same issue. Groves asked Lawrence, Compton, Oppenheimer, and Edwin McMillan to investigate the situation. They reported back on the 18th that the problem would perhaps not be quite as drastic as Seaborg feared, but DuPont engineers remained very concerned with the desired plutonium purity, which would have to be better than 99 %. Ultimately, the purity issue would come to be eclipsed by another problem: spontaneous fissioning of pile-produced plutonium. The latter proved an incredibly vexing issue, solution of which would demand remarkably ingenious bomb engineering. Seaborg anticipated this possibility as well, long before any pile-produced plutonium was available.

On November 18, Groves appointed a five-person review committee, heavily populated with DuPont representatives. The group was headed by Warren Lewis, the MIT chemical engineer who had been involved with the third National Academy report of a year earlier. The other members were Crawford Greenewalt (Fig. 4.15), a DuPont chemical engineer and former student of Lewis; Thomas Gary, a manager in the company's Engineering Department design division; and Roger Williams, an expert on plant operations in the company's Ammonia Department. The fifth member was to be Eger Murphree, but he had to withdraw on account of illness. Williams would be assigned overall responsibility for the TNX Division, and Greenewalt would be assigned to head the Technical Division. Greenewalt's job would come to involve almost continuous commuting between Wilmington and Chicago; Groves and Compton considered that he did a superb job. Greenewalt would go on serve as DuPont's President from 1948 to 1962.

The committee assembled in New York on Sunday evening, November 22, and began their work the next day with a visit to Columbia to review Harold Urey's gaseous diffusion research. On Thanksgiving day, November 26, they arrived in Chicago, where Compton presented them with a 150-page document titled "Report on the Feasibility of the 49 Project." This massive report explored all aspects of the proposed pile process: uranium-graphite designs utilizing helium, liquid bismuth, or water-cooling; a uranium-heavy water system where the heavy water would serve as both coolant and moderator; problems of extracting plutonium from irradiated uranium; health and safety issues; radioactive by-products; and proposed time and cost schedules. That evening, the committee left Chicago to see

Fig. 4.15 Crawford
Greenewalt (1902–1993) in
the late 1970s. *Source* AIP
Emilio Segre Visual
Archives, John Irwin Slide
Collection

Ernest Lawrence at Berkeley, where they witnessed calutrons in operation. On
their way back east, they stopped again in Chicago on December 2, where
Greenewalt, serving as the group's representative, witnessed the first criticality of
Enrico Fermi's CP-1 pile (Chap. 5).

The Lewis committee finished its report on Friday, December 4; it was submitted to Groves on the following Monday. Their main conclusions were somewhat surprising: Although the electromagnetic method was probably the most
immediately feasible approach for producing U-235, they felt that the diffusion
process probably had the best chance for ultimately producing it at the desired rate
of 25 kg per month, whereas the pile process (plutonium) might provide "the
possibility of earliest achievement of the desired result." The committee offered
five main recommendations: (1) Proceed immediately with the design and construction of a 4,600-stage diffusion plant to produce one kilogram of U-235 per day
(anticipated cost $150 million); (2) Expedite design and construction of a pilot-
scale pile and full-scale helium-cooled piles to produce 600 g of Pu-239 per day
($100 million); (3) Expedite development work on the electromagnetic method;
(4) Install a small electromagnetic plant to produce 100 g of U-235 for experimental purposes ($10 million); and (5) Construct a heavy water plant capable of
distilling two tons of that material per month ($15 million). In total, $315 million
should be available early in 1943, in addition to $85 million that was already
available from funds under the control of the Chief of Engineers. The explosive
power of a fission bomb was estimated at 12.5 kilotons TNT equivalent, a figure
which would prove close to *Little Boy's* yield at Hiroshima. As to predicted
availability, it was estimated that there was a small chance of production prior to
June 1, 1944, a "somewhat better" chance beginning before January 1945, and a
"good chance" during the first half of 1945. There was still fear that Germany may
be six months or a year ahead of America.

The Military Policy Committee met on December 10 to review the committee's report. This meeting would prove as crucial to the development of the Project as had the Bohemian Grove conference of three months earlier. The MPC endorsed all of the review committee's major recommendations, deciding to proceed with the kilogram-per-day diffusion plant and a 500-tank electromagnetic plant to obtain some early production of U-235, even though it would be in small quantities (0.1 kg/day). The Committee proposed that no intermediate-size piles be constructed, favoring instead that a full-scale pile program be undertaken directly at a site other than where the uranium plants would be located. With its proximity to Knoxville, the Clinton site appealed to neither Groves nor DuPont from a safety perspective; to site the piles there would require acquiring some 75,000 acres of land beyond what had already been taken. The X-10 pilot-scale reactor and separation facilities, which DuPont accepted a contract to design and construct on January 4, 1943, would, however, remain in Tennessee; DuPont referred to these facilities a "semi-works." The same day, Groves contracted with Westinghouse Electric, General Electric, and the Allis-Chalmers Manufacturing Company to produce components for the electromagnetic plant.

With removal of the production piles from Tennessee, Arthur Compton attempted to argue that the pilot-scale pile should be built at the Argonne site near Chicago so that personnel from the University would not have to relocate. This idea was vetoed by DuPont engineers, who feared that the scientists would interfere by insisting on endless design changes. Undeterred, Compton came back with yet another proposal: that his group should be allowed to build its own pile at the Argonne site in order to create enough plutonium for research purposes. Groves traveled to Chicago on January 11 to press for the Tennessee location. Compton reluctantly acquiesced, but the Chicago group did receive an unanticipated and initially unwanted consolation prize. Left unspecified in DuPont's January 4 contract was the question of who would operate the pilot plant. Despite their commitment to build and operate the main production plants, DuPont officials were reluctant to also agree to operate the pilot plant, preferring, as was corporate norm, to assign that task to research staff. DuPont proposed that the University operate the pilot plant, a suggestion which shocked Compton: universities do not normally operate industrial plants. In March, 1943, the University agreed to take on the operating responsibility, essentially doubling the size of its campus. The University remained as the operating contractor until July 1, 1945, when the task was taken over by the Monsanto Chemical Company.

4.11 December 1942: A Report to the President

On December 16, Bush carried the MPC decisions to President Roosevelt in a 29-page report on the project. The report was under Bush's name in his role as chairman of the MPC, but it had also gone through the Top Policy Group. This report is remarkable not only for its summary of the situation at the time, but particularly for its analysis of pending issues regarding international information interchange and postwar possibilities. In OSRD microfilm records, both the report itself and Bush's cover letter bear "OK FDR" scrawls.

Bush opened with a summary of work on the project to date: A site was being acquired near Knoxville to locate plants for the electromagnetic and diffusion methods. Another site was being procured in New Mexico for a secret bomb-design laboratory. Centrifuge work was being limited to research only as that method looked less promising then it had a few months earlier. A ten-stage pilot diffusion plant was under construction, with completion scheduled for the middle of 1943, and a 4,600-stage full-scale plant was being planned. An experimental pile had been constructed and operated (Chap. 5). Because of possible hazards with full-scale piles, the MPC considered it essential that the President authorize the War Department to enter into contracts where United States would assume all risks.

The latter part of the report was devoted to speculations on the possibilities of atomic power. These anticipated many issues still relevant today: "There remains, however, little doubt that man has available a new and exceedingly potent source of energy ... It is decidedly unfortunate, however ... that the operation of such a power-plant pile inevitably involves the incidental production of a material, which is, to a high degree of probability, a super explosive ... Certainly if, in the future nations are to construct and use power plants utilizing atomic power, and especially if a super-explosive is a possible by-product, the United States must be one of those nations."

The report also addressed implications in the areas of control of atomic energy and post-war international relations of what it called "a turning point in the technical history of civilization." Issues included the status of heavy water, uranium ores, accrual of patents to the Government, and relations with the British and Canadians. As to the latter, Bush reported that there had been complete scientific interchange between the British and American groups, but the subject was now entering a new phase with the involvement of the Army in developing production plants. Since the line between research and development was, in Bush's word, "nebulous," he felt that the situation demanded a "new and clear" (i.e., Presidential-level) directive on future U.S.—British relations in the atomic field. Since the British had no intention of engaging in production of U-235 or Pu-239, Bush felt that passing any American knowledge to the British in those areas would be of no use to them during the war. British research in the area of heavy water had been transferred to a group in Montreal operating under the auspices of the National Research Council of Canada, but Bush felt that not having that group available to the American program would "not hamper the effort at all fatally." The British were well-along in diffusion research, but here again Bush felt that a complete cessation of interchange in that field might somewhat hinder, but not seriously "embarrass" the United States' effort.

Having mustered his arguments, Bush summarized with a statement that hinted at postwar American isolationism: "it appears (a) that there would be no unduly serious hindrance to the whole project if all further interchange between the United

States and Britain in this matter were to cease, and (b) there would be no unfairness to the British in this procedure." Bush closed his discussion of the interchange issue by offering suggestions as to possible policy approaches. In a time-honored bureaucratic tactic, he presented three possibilities, with the politically unpalatable extremes presented first and second in order to pave the way for the third, which he evidently preferred. These were (a) cease all interchange; (b) have complete interchange in both research and development; or (c) restrict interchange only to information that the British could use directly. Within option (c), there would be no interchange on the purely American electromagnetic program, unrestricted interchange on the design and construction of the diffusion plant, research-only interchange (no plant design information) on the manufacture of plutonium and heavy water, and no interchange on the bomb-design laboratory that would be located at Los Alamos. Some historians have suggested that Bush viewed British interests as oriented more toward advantage in postwar commercial development of nuclear power than any wartime application, and so saw no justification for having an American-funded effort aid such development.

Bush's assessment of the implications of nuclear energy was sobering: "The whole development of atomic power, if it arrives as a new complication in an already complicated civilization, as now appears to be very probable as an event certainly of the next decade, may be an exceedingly difficult matter with which to deal wisely as between nations. On the other hand, it may be capable of maintaining the peace of the world."

President Roosevelt approved the recommendations on December 28, including the choice of interchange option (c), although the issue of relations with the British would prove to be far from settled. With the President's approval, work that had begun three years earlier with a commitment of $6,000 was approaching a cost anticipated to be in the range of $400 million.

By the spring of 1943, Groves and the Military Policy Committee were firmly in charge of the Manhattan Project. All OSRD research and development contracts were transferred to the Manhattan District as of May 1, 1943. Contracts had been let for the design, construction, and operation of enrichment facilities and plutonium-producing reactors, and the intricacies of bomb physics were being explored at Los Alamos. The Planning Board and the S-1 Executive Committee essentially disappear from the history at this point, although Groves did retain James Conant and Richard Tolman as personal scientific advisors.

A sense of how the times had changed was captured by Conant in his May, 1943, draft history of the project: "For eighteen months this highly secret war effort has moved at a giddy pace. New results, new ideas, new decisions and new organization have kept all concerned in a state of healthy turmoil. The time for "freezing design" and construction arrived a few weeks past; now, we must await the slower task of plant construction and large-scale experimentation. The new results when they arrive will henceforth be no laboratory affair, their import may well be world shattering. But as in the animal world, so in industry: the period of gestation is commensurate with the magnitude to be achieved."

Further Reading

Books, Journal Articles, and Reports

J.-J. Ahern, We had the hose turned on us! Ross Gunn and The Naval Research Laboratory's Early Research into Nuclear Propulsion, 1939–1946. International Journal of Naval History 2(1) (April 2003); http://www.ijnhonline.org/volume2_number1_Apr03/article_ahern_gunn_apr03.htm

S. Cannon, The Hanford Site Historic District—Manhattan Project 1943–1946, Cold War Era 1947–1990. Pacific Northwest national Laboratory (2002). DOE/RL-97-1047 http://www.osti.gov/bridge/product.biblio.jsp?osti_id=807939

R.P. Carlisle, J.M. Zenzen, Supplying the Nuclear Arsenal: American Production Reactors, 1942–1992 (Johns Hopkins University Press, Baltimore, 1996)

D.C. Cassidy, A Short History of Physics in the American Century (Harvard University Press, Cambridge, 2011)

J. Cirincione, Bomb Scare: The History and Future of Nuclear Weapons (Columbia University Press, New York, 2007)

R.W. Clark, The Birth of the Bomb: The Untold Story of Britain's Part in the Weapon that Changed the World (Phoenix House, London, 1961)

R.W. Clark, Tizard (The MIT Press, Cambridge, 1965)

K.P. Cohen, S.K. Runcorn, H.E. Suess, H.G. Thode, Harold Clayton Urey, 29(April), pp. 1893–5, January 1981. Biographical Mem. Fellows Roy. Soc. 29, 622–659 (1983)

A. L. Compere, W. L. Griffith, The U. S. Calutron Program for Uranium Enrichment: History, Technology, Operations, and Production. Oak Ridge National Laboratory report ORNL-5928 (1991)

A.H. Compton, Atomic Quest (Oxford University Press, New York, 1956)

B.T. Feld, G. W. Szilard, K. Winsor, The Collected Works of Leo Szilard. Volume I—Scientific Papers. (MIT Press, London, 1972)

L. Fine, J.A. Remington, United States Army in World War II: The Technical Services—The Corps of Engineers: Construction in the United States. Center of Military History, United States Army, Washington (1989)

S. Goldberg, Inventing a climate of opinion: Vannevar Bush and the decision to build the bomb. Isis 83(3), 429–452 (1992)

M. Gowing, Britain and Atomic Energy 1939–1945 (St. Martin's Press, London, 1964)

L.R. Groves, Now It Can be Told: The Story of the Manhattan Project (Da Capo Press, New York, 1983)

R.G. Hewlett, O.E. Anderson, Jr., A History of the United States Atomic Energy Commission, Vol. 1: The New World, 1939/1946. (Pennsylvania State University Press, University Park, PA 1962)

L. Hoddeson, P.W. Henriksen, R.A. Meade, C. Westfall, Critical Assembly: A Technical History of Los Alamos during the Oppenheimer Years (Cambridge University Press, Cambridge, 1993)

W. Isaacson, Einstein: His Life and Universe (Simon and Schuster, New York, 2007)

V.C. Jones, United States Army in World War II: Special Studies—Manhattan: The Army and the Atomic Bomb. Center of Military History (United States Army, Washington, 1985)

C.C. Kelly (ed.), Remembering the Manhattan Project: Perspectives on the Making of the Atomic Bomb and its Legacy (World Scientific, Hackensack, 2004)

C.C. Kelly (ed.), The Manhattan Project: The Birth of the Atomic Bomb in the Words of Its Creators, Eyewitnesses, and Historians (Black Dog & Leventhal Press, New York, 2007)

C.C. Kelly, R.S. Norris, *A Guide to the Manhattan Project in Manhattan* (Atomic Heritage Foundation, Washington, 2012)

D. Kiernan, *The Girls of Atomic City: The Untold Story of the Women Who Helped Win World War II* (Touchstone/Simon and Schuster, New York, 2013)

W. Lanouette, B. Silard, *Genius in the Shadows: A Biography of Leo Szilard. The Man Behind the Bomb* (University of Chicago Press, Chicago, 1994)

S. Lee, 'In no sense vital and actually not even important'? Reality and Perception of Britain's Contribution to the Development of Nuclear Weapons. Contemp. Br. Hist. **20**(2), 159–185 (2006)

J.C. Marshall, Chronology of District "X" 17 June 1942–28 October 1942

K.D. Nichols, *The Road to Trinity* (Morrow, New York, 1987)

R.S. Norris, *Racing for the Bomb: General Leslie R. Groves, the Manhattan Project's Indispensable Man* (Steerforth Press, South Royalton, VT, 2002)

B.C. Reed, Arthur Compton's 1941 Report on explosive fission of U-235: A look at the physics. Am. J. Phys. **75**(12), 1065–1072 (2007)

B.C. Reed, A brief primer on tamped fission-bomb cores. Am. J. Phys. **77**(8), 730–733 (2009)

B.C. Reed, Liquid thermal diffusion during the manhattan project. Phys. Perspect. **13**(2), 161–188 (2011)

R. Rhodes, *The Making of the Atomic Bomb* (Simon and Schuster, New York, 1986)

A. Sachs, Early History Atomic Project in Relation to President Roosevelt, 1939–1940. Unpublished manuscript, August 8–9, (1945)

E. Segrè, *Enrico Fermi, Physicist* (University of Chicago Press, Chicago, 1970)

R. Serber, *The Los Alamos Primer: The First Lectures on How To Build An Atomic Bomb* (University of California Press, Berkeley, 1992)

H.D. Smyth, *Atomic Energy for Military Purposes: The Official Report on the Development of the Atomic Bomb under the Auspices of the United States Government, 1940-1945* (Princeton University Press, Princeton, 1945)

F.M. Szasz, *British Scientists and the Manhattan Project: The Los Alamos Years* (Palgrave McMillan, London, 1992)

H. Thayer, *Management of the Hanford Engineer Works in World War II* (American Society of Civil Engineers, New York, 1996)

S.R. Weart, Scientists with a secret. Phys. Today **29**(2), 23–30 (1976)

A.M. Weinberg, Eugene Wigner, Nuclear Engineer. Phys. Today **55**(10), 42–46 (2002)

G.P. Zachary, *Endless Frontier: Vannevar Bush, Engineer of the American Century* (MIT Press, Cambridge, 1999)

Websites and Web-based Documents

Alexander Sachs letter to Roosevelt: A copy of Sachs' cover letter can be found by searching on "Alexander Sachs" at the site of the FDR library, http://www.fdrlibrary.marist.edu/. There were actually three letters from Einstein to Roosevelt between August, 1939 and April, 1940, plus another in March, 1945. Texts of all four letters can be found at http://hypertextbook.com/eworld/einstein.shtml

OSRD: A copy of the Executive Order establishing the OSRD can be found at http://www.presidency.ucsb.edu/ws/index.php?pid=16137#axzz1QbKXQHjp

Chapter 5
Oak Ridge, CP-1, and the Clinton Engineer Works

Of every dollar spent on the Manhattan Project, just over 60 cents went into the Clinton Engineer Works (CEW), the 90-square-mile tract in eastern Tennessee chosen in the fall of 1942 as the location for uranium enrichment plants and a pilot-scale nuclear reactor (Fig. 4.14). The responsibility borne by the CEW's commander, District Engineer Colonel Kenneth Nichols—who nominally administered production sites for the *entire* Project—was extraordinary. In 1942, the idea of using plutonium for a fission bomb was a tantalizing but wholly speculative prospect; only uranium looked certain as a bomb material. If plutonium proved unworkable (which it almost did), the success or failure of the Project would be decided at the Clinton site.

The roughly 140 pounds of enriched uranium that went into the Hiroshima *Little Boy* bomb would fit very comfortably inside a soccer ball. It might seem that if all one desired to do was to produce one or a few bombs, a small factory should be adequate. But the nature of the processes involved in enrichment are such that it simply does not work that way. To fulfill the task in any reasonable length of time means building factories that by their nature can turn out material for dozens or hundreds of bombs once they are in operation; one goes "all-in" or not at all. To produce that 140 pounds, the scale of CEW operations would grow to be enormous. The number of construction workers alone would peak at 45,000 in the spring of 1944, and by May, 1945, the entire CEW would employ just over 80,000 personnel.

This chapter describes the uranium enrichment complexes and the pilot-scale reactor constructed at the Clinton site. As described in Chap. 1, the enrichment facilities comprised two large complexes plus a third smaller one. The larger ones, code-named Y-12 and K-25, enriched uranium by electromagnetic and gaseous-diffusion processes, respectively. Work on these two facilities began in early 1943; together, they accounted for half the cost of the Manhattan Project. The third facility, known as S-50, utilized liquid thermal diffusion, and was not begun until mid-1944. All three contributed to enriching uranium for the Hiroshima bomb, although the lion's share of the burden was borne by Y-12 and K-25. The pilot-scale reactor, code-named X-10, produced about two-thirds of a pound of plutonium for research at Los Alamos.

B. C. Reed, *The History and Science of the Manhattan Project*,
Undergraduate Lecture Notes in Physics, DOI: 10.1007/978-3-642-40297-5_5,
© Springer-Verlag Berlin Heidelberg 2014

In 1942, rural eastern Tennessee was still very undeveloped. The population was sparse, most roads were unimproved, and services were minimal. If a secret production complex employing thousands of people was to be constructed, those people, many of them highly-educated and used to the amenities of big-city and university living, would need a place to call home. This chapter opens with a description of the town that was purpose-built to house the workers of the CEW: Oak Ridge.

5.1 Oak Ridge: The Secret City

Perhaps no other statistics speak more forcefully to the scale of the Clinton project than the growth of the town established to house its workers. In 1942, the city of Oak Ridge, Tennessee, did not exist. By mid-1945, it would be the fifth-largest city in the state, boasting a population of about 75,000. Located in the northeast corner of the Clinton reservation, it appeared on no maps at the time. Once the Clinton site was closed to public access as of April 1, 1943, Oak Ridge literally became a secret city, and was known to its residents by that term.

When the Stone and Webster company took on responsibility for constructing the electromagnetic enrichment plants along with site development and townsite design and construction in the fall of 1942, they envisioned a village to house some 5,000 inhabitants. By October 26, 1942, when S&W submitted its first plan for the area, the estimated population had grown to 13,000. With constant design changes and expansions of the electromagnetic plant (Sect. 5.3), S&W soon found its resources stretched. General Groves decided in late November to relieve S&W of town-design functions, although the firm would retain responsibility for overseeing construction, utility operations, and road maintenance. Design of housing units was contracted to the architectural firm of Skidmore, Owings, and Merrill of Chicago.

Oak Ridge grew up in three phases of development. The first, known as "East Town" from its location just southwest of the Elza Gate entrance to the reservation, was completed in early 1944 and contained over 3,000 family-type housing units, dormitories, 1,000 trailers, an administration building, stores, recreation areas, schools, churches, theatres, laundries, a cafeteria, and a hospital which would be the birth site of 2,910 babies in the first three years of its existence. Oak Ridge would acquire nearly 100 miles of paved streets; a further 200 miles would be laid to service the production sites.

By the fall of 1943, the population estimate had grown to 42,000, and phase two was begun about two miles west of East Town. By the summer of 1944, this had added some 4,800 family units, a number of barracks, and fifty dormitories which could house some 7,500 single residents. By early 1945, estimates were again revised upward to an ultimate population of 66,000. The third phase of development, built to both the east and west of the original site, saw the addition of 1,300 family units, 20 dormitories, hundreds of trailers, and associated services. By the time housing construction was finished in 1945, over 7,000 family houses, apartments containing over 9,000 dwelling units, 89 dormitories, 2,000 five-man "hutments," and seven trailer camps with a total capacity of about 4,000 occupants had been put up. To take care of the day-to-day needs of residents, two sewage-treatment plants, 130 miles of sewer mains, a steam plant, ten elementary schools, two junior-high schools, two senior-high schools, five nursery schools, nine shopping areas, and a number of temporary stores were erected. Trees cut down during building operations were turned into 163 miles of boardwalks. The name of the town came from the local name of the site, Black Oak Ridge, and was adopted in mid-1943 on the rationale that its rural-sounding connotation would help minimize outside curiosity. For residents, life was a bargain: rents were minimal and services heavily subsidized; household electricity use went unmetered. The cost of constructing Oak Ridge ran to just over $100 million, not including the building of construction camps which temporarily housed a further 14,000 inhabitants near the various enrichment plants.

To speed construction and minimize costs, Skidmore, Owings, and Merrill restricted plans to nine different types of pre-fabricated houses and three different apartment designs, all wood-frame structures. Many units incorporated interior and exterior panels of "cemesto," a sturdy, fireproof building material made of fiber board with pressed asbestos-cement panels bonded to both sides. One history of the area records that at one point, housing units fully equipped with appliances and furniture were being turned over from the contractors to the Government at a rate of one every thirty minutes. Intended to be only semi-permanent, many of the original cemesto homes still stand, now prized for their historic value and location close to the commercial center of town (Fig. 5.1).

Oak Ridge had not only to be constructed, but also managed and operated. For this, Groves approached the Turner Construction Company of New York, which he had used on other projects. Turner established a wholly-owned subsidiary, the Roane-Anderson Company, named after the two counties which the Clinton reservation straddled. On a cost-plus-fee basis, Roane-Anderson managed provision of services such as utilities, police and fire departments, medical personnel, trash collection, school maintenance, cemeteries, cafeterias, warehouses, deliveries of

Fig. 5.1 Typical cemesto homes and a trailer-housing area at Oak Ridge. *Source* http://www.y12.doe.gov/about/history/getimages

coal and ice, and granted concessions to private operators for grocery stores, drug stores, department stores, barber shops, and garages. The company also operated an extensive bus system, a railroad, and a motor pool. To take some 60,000 riders per day to and from the productions sites required 840 buses, making the system for a time the ninth-largest bus network in the United States. By early 1945, Roane-Anderson had over 10,000 employees on its payroll, although the number declined thereafter as services began to be contracted to other organizations.

During their wartime tenures, however, Oak Ridge and the Clinton Engineer Works had but two key functions: to build and operate the District's uranium plants and the X-10 reactor, and, from August, 1943, onwards, to serve as the site the administrative headquarters of the Manhattan Engineer District. In the following sections, we turn to descriptions of the enrichment plants and X-10 reactor. The first facility to go into operation (November, 1943) was X-10. However, X-10 descended directly from Enrico Fermi's experimental CP-1 reactor, which was located in Chicago. We begin, then, with the history of that program.

5.2 CP-1 and X-10: The Pile Program

It was described in Chap. 4 how Arthur Compton decided to centralize nuclear-pile research at the University of Chicago's Metallurgical Laboratory in early 1942. The first goal on the path to large-scale plutonium production would be to show that a self-sustaining chain reaction could be created and controlled. To this end, Enrico Fermi began moving his Columbia pile-research group to Chicago in early 1942 to join forces with Samuel Allison's group. Fermi himself commuted between New York and Chicago through the winter and early spring of 1942, moving permanently in April.

A serious issue was supply of critical materials. A chain-reacting pile would require several tons of uranium and hundreds of tons of graphite, both as pure as possible and with the uranium preferably in the form of pure metal as opposed to an oxide. While the purity requirements for graphite were stringent, there was at least an established graphite industry, and the Speer Carbon Company and the National Carbon Company were able to produce the necessary material. On the other hand, commercial use of uranium at the time was a relatively small-scale enterprise: the element was used only as a coloring agent in glasses and ceramics, as a source of radium, and in some specialty lamps produced by the Westinghouse Company. Westinghouse produced its uranium via a photochemical process which involved exposing large vats of solutions to sunlight, but this was far too slow for large-scale production. The Metal Hydrides Company of Beverly, Massachusetts developed a process for isolating uranium metal, but it emerged in a powdered form which caught fire when exposed to air. After considerable work, the problem of reducing uranium salts to a readily-handled metallic form metal was solved by Frank Spedding (Iowa State College) and Clement Rodden (National Bureau of

Standards), who devised a chemical process by which pure uranium metal could be produced by the ton. Large-scale production was contracted to the Mallinckrodt Chemical Company of Saint Louis.

As uranium and graphite began to become available, Fermi and his group built a succession of so-called sub-critical "exponential" piles. Between September 15 and November 15, 1942, sixteen piles were constructed at Chicago to help inform the decision of optimal lattice size and to test various batches of graphite and uranium. These piles used radium/beryllium neutron sources; in a pile large enough to sustain a chain-reaction, spontaneous and cosmic-ray- induced fissions would be sufficient to ensure self-start-up. By October, 1942, enough material was on hand to begin planning for a critical pile. As described in Chap. 4, the original intent had been to build the first chain-reacting pile at the Argonne Forest site outside Chicago. A building was to be ready by October 20, but labor disruptions threatened postponement. In early November, Fermi approached Compton with the idea of performing the experiment at the University itself.

Building an experimental nuclear reactor in the heart of a metropolitan area may sound like lunacy. But Fermi had done his calculations carefully, and was confident that the reaction could be safely controlled. A significant factor in this regard is that when fissile nuclei absorb neutrons, not all fissions occur instantaneously. A small fraction, about 1 %, are delayed by up to several seconds. If the reactor is operating just at criticality (reproduction factor $k = 1$), this delay allows enough time for adjustments to be made before the reaction runs out of control. Fermi also planned for a number of redundant safety systems that would allow for deliberate over-control of the pile. Compton, fearing that his superiors at the University would veto such a plan, decided to authorize it on his own responsibility. He described the plan at a meeting of the S-1 Executive Committee held on November 14, 1942, and wrote in his memoirs that James Conant's face went white. Given the delays at Argonne, however, it was decided to proceed. The site chosen was a 30 by 60-foot squash court under the west stands of the University's Stagg Athletic Field. According to some sources, mistranslations in Soviet reports had the reactor being located in a "pumpkin patch".

Critical Pile number one (CP-1; also known as Chicago Pile 1) was built in the shape of a somewhat flattened ellipsoid with an equatorial radius of 388 cm and a polar radius of 309 cm (Fig. 5.2). The original design called for a spherical shape, but the quality of materials, particularly the availability of pure uranium metal, permitted getting away with a somewhat smaller structure than was originally envisioned. Layers of solid graphite bricks alternated with ones within which slugs of uranium were embedded, with the slugs configured to form a cubical lattice of side length 21 cm as the pile was built up (Fig. 5.3). This length was the average displacement over which neutrons would become thermalized after successive strikes against carbon nuclei; there would be no use in making the lattice size any larger. The bottom layer of graphite lay directly on the floor of the squash court, with the assembly supported by a wooden framework. Herbert Anderson scoured Chicago lumberyards for what he called an "awesome number" of four-by-six-inch timbers. In case it would prove necessary to enclose and evacuate the pile to

Fig. 5.2 *Side-view* sketch of
the shape of CP-1 and its
equivalent ellipsoid. The
dimensions are from *side-
to-side* and *bottom-to-top* of
the ellipsoid. Adapted from
Fermi (1952)

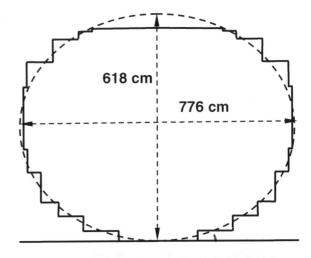

618 cm

776 cm

Fig. 5.3 Walter Zinn, *left*,
stands atop the partially
reconstructed CP-1/CP-2
reactor. *Photo Courtesy of
Argonne National
Laboratory;* http://
www.flickr.com/photos/
argonne/5963919079/

improve the reproduction constant, Anderson also arranged, with the Goodyear
Rubber Company, for a cubical rubber balloon 25 feet on a side. In practice, the
balloon enclosure was not used.

Construction began on November 16, with physicists and hired laborers
working twenty-four hour days in two twelve-hour shifts under the supervision of
Anderson (night shift) and Walter Zinn (day). Two special crews machined
graphite and pressed uranium oxide powder into solid slugs using a purpose-
designed die and a hydraulic press. Albert Wattenberg, who had joined Fermi's

group while a student at Columbia, recalled that between mid-October and early December, 90 hours work weeks were not uncommon, with crews often smoking on the job as a way to skip meals and save time. Two layers per shift was the normal rate of construction. Graphite was received from manufactures in the form of bricks of square 4.25-in. cross-section and lengths varying from 17 to 50 in. With planers and woodworking tools, the bricks were cut to 16.5-in. lengths and milled to smooth 4.125-in. cross-sections; surfaces were held to tolerances of 0.005 in., and lengths to 0.02 in. About 14 t of bricks could be processed per 8 hours of work. In all, CP-1 incorporated 385.5 t of graphite—some 40,000 bricks, at an average of about 20 pounds each.

The uranium was in the form of pure uranium metal (just over 6 t) and uranium oxide (about 40 t); the slugs of pure metal were placed in the center of the pile. Holes of diameter 3.25 in. were drilled into bricks on 21 cm centers to receive the slugs, some of which were cylindrical and some pseudo-spherical. A total of 19,480 slugs were pressed, with about 1,200 being produced every 24 hours. Fully one-quarter of the bricks needed to have holes drilled in them. Between 60 and 100 holes could be drilled per hour, but the drill bits would become dull after doing only about 60 holes; some 30 bits had to be resharpened every day.

The pile was arranged with ten horizontal slots into which cadmium-sheathed wooden rods could be inserted. With a thermal neutron capture cross-section of over 20,000 barns, cadmium-113 is a voracious neutron absorber; the rate of reactivity could be controlled by inserting and withdrawing rods as necessary. When construction was underway, all rods would be fully inserted and locked in place. Once per day, however, they would be temporarily removed and the neutron activity level measured, from which Fermi would compute design adjustments. As each layer was completed, Fermi computed an effective pile radius. A plot of the square of the effective radius (a measure of the surface area of the pile, through which neutrons could escape) divided by the number of neutron counts per minute (an indirect measure of the volume of the pile) versus the number of layers was essentially a descending straight line, as shown in Fig. 5.4. As the neutron flux became closer and closer to exponentially diverging, the surface-to-flux ratio would decline. By extrapolating the line to zero, Fermi could predict the layer at which criticality would occur.

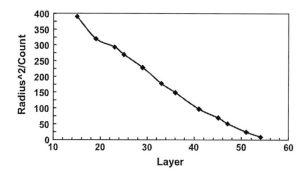

Fig. 5.4 Radius^2/count rate vs. number of completed layers. Data from Fermi (1952)

While any one cadmium rod was sufficient to bring the reaction below criticality at any time, multiple slots were provided so that several could be inserted. In addition, two safety rods (known as "zip" rods) and one automatic control rod were also incorporated into the design. During normal operation, all but one of the cadmium rods would be withdrawn from the pile. If neutron detectors signaled too great a level of activity, the vertically-arranged zip rods would be automatically released, accelerated by 100-pound weights. The automatic control rod could be operated manually, but was also normally under the control of a circuit which would drive it into the pile if the level of reactivity increased above a desired level, but which would withdraw it if the intensity fell below the desired level.

By late November, it was clear that the pile would become critical on the completion of its 56th layer. Fermi decided to add a 57th layer, which would be laid during the night of December 1–2. He instructed Anderson not to start the reaction that night in order that Laboratory staff and a representative of the visiting Lewis Committee could be present for the historic event on the next day.

The witnesses assembled at 8:30 on the morning of December 2 on a balcony at the north end of the squash court, about 10 feet above the floor. Including Fermi, 49 people were present to witness the dawn of the nuclear age. On the floor below, George Weil would handle the final cadmium rod, which was at layer 21. Atop the pile stood a three-man crew ready to dump buckets of liquid cadmium solution onto the pile as a last-resort emergency shutdown procedure. Fermi had calculated in advance the expected neutron intensity for each position of Weil's rod, and had a pocket slide rule on hand with which he checked readings against his predictions throughout the day. At 9:45, Fermi ordered the electrically-driven safety rods removed. The neutron count grew and steadied out. One of the safety rods was tied off to the balcony railing, with physicist Norman Hilberry standing by with an axe in order to cut the rope in case the automatic shutdown system should fail. According to some sources, the phrase "to scram" a reactor—execute an emergency shutdown—is an acronym for "safety control rod axe man."

Shortly after 10:00, Fermi ordered the automatic safety rod withdrawn. This was done, and again the neutron count grew and leveled off. At 10:37, he instructed Weil to pull the last rod out to 13 feet; the count again leveled within a few minutes. Fermi ordered another foot withdrawn, and then, at 11:00, another six inches. Additional withdrawals at 11:15 and 11:25 were not enough to achieve criticality, as Fermi had anticipated. Proceeding with caution, Fermi ordered the automatic control rod reinserted as a check; the intensity dropped accordingly. At about 11:35 the automatic rod was withdrawn and the cadmium rod adjusted outwards. Suddenly, a loud crash occurred. The threshold safety intensity had been set too low, and one of the zip rods had deployed itself. Fermi decreed that a lunch break was in order, and directed that all control rods be reinserted.

The group reassembled at 2:00 p.m. To check that the neutron flux returned to its pre-lunch reading, Fermi ordered all rods withdrawn except for Weil's. Satisfied with the neutron count, he then directed that one of the zip rods be inserted; the neutron count obediently declined. He then ordered Weil to withdraw the cadmium rod by one foot. On directing the zip rod to be removed, Fermi said to

Arthur Compton: "This is going to do it. Now it will become self-sustaining. The trace will climb and continue to climb; it will not level off."

Herbert Anderson recorded the time as 3:36 p.m. In his words:

> At first you could hear the sound of the neutron counter …. Then the clicks came more and more rapidly, and after a while they began to merge into a roar; the counter couldn't follow any more. That was the moment to switch the chart recorder [to a less-sensitive setting]. But when the switch was made, everyone watched in the sudden silence the mounting deflection of the recorder's pen. It was an awesome silence. Everyone recognized the significance of that switch; we were in the high-intensity regime. … Again and again, the scale of the recorder had to be changed to accommodate the neutron intensity which was increasing more and more rapidly. Suddenly Fermi raised his hand. "The pile has gone critical," he announced. No one present had any doubt about it.

Fermi allowed the pile to operate for 28 min before calling for a zip rod to be inserted. He estimated that at that point, the pile was operating at a power of about one-half of a Watt. Crawford Greenewalt recorded in his diary that "Fermi was as cool as a cucumber." Because of security regulations, no photographs of the completed pile were ever taken; Fig. 5.5 shows an artist's rendering of the startup, and Fig. 5.3 shows Walter Zinn standing atop the pile as it was being reconstructed in 1943 as pile CP-2 at the Argonne site.

A strip-chart recording of the neutron flux clearly shows the exponential growth characteristic of a self-sustaining reaction; this recording has been called "The Birth Certificate of the Nuclear Age" (Fig. 5.6).

Eugene Wigner presented Fermi with a bottle of Chianti, and a paper-cup toast was raised. Many of those present signed the wicker wrapping of the bottle. Arthur Compton excused himself to phone James Conant in Washington with the news. As Compton related their conversation:

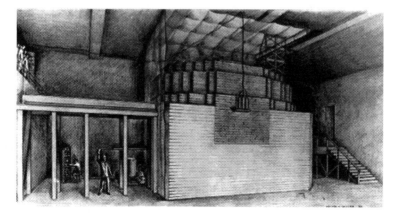

Fig. 5.5 Artist's conception of the startup of CP-1. *Source* http://commons.wikimedia.org/wiki/File:Chicagopile.gif

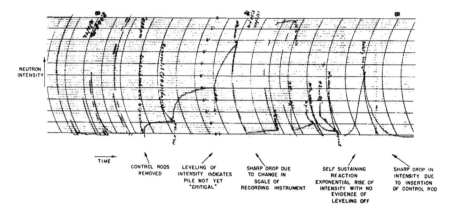

NEUTRON
INTENSITY

TIME

CONTROL RODS
REMOVED

LEVELING OF
INTENSITY INDICATES
PILE NOT YET
"CRITICAL"

SHARP DROP DUE
TO CHANGE IN
SCALE OF
RECORDING INSTRUMENT

SELF SUSTAINING
REACTION
EXPONENTIAL RISE OF
INTENSITY WITH NO
EVIDENCE OF
LEVELING OFF

SHARP DROP IN
INTENSITY DUE
TO INSERTION
OF CONTROL ROD

Fig. 5.6 Galvanometer tracing of CP-1 neutron intensity. *Source Courtesy of Argonne National Laboratory;* http://www.flickr.com/photos/argonne/7550395714/

"Jim, you'll be interested to know that the Italian navigator has just landed in the new world. The earth was not as large as he had estimated, and he arrived in the new world sooner than he had expected."

"Is that so? Were the natives friendly?" asked Conant.

"Everyone landed safe and happy," reported Compton.

On the twentieth anniversary of the achievement, Eugene Wigner offered a reflection:

Nothing very spectacular had happened. Nothing had moved and the pile itself had given no sound. Nevertheless, when the rods were pushed back and the clicking died down, we suddenly experienced a let-down feeling, for all of us understood the language of the counters. Even though we had anticipated the success of the experiment, its accomplishment had a deep impact on us. For some time we had known that we were about to unlock a giant; still, we could not escape an eerie feeling when we knew we had actually done it. We felt as, I presume, everyone feels who has done something that he knows will have very far-reaching consequences which he cannot foresee.

In the same essay, Wigner commented on the importance of the experiment:

Do we then exaggerate the importance of Fermi's famous experiment? I may have thought so some time in the past, but do not believe it now. The experiment *was* the culmination of the efforts to prove the chain reaction. The elimination of the last doubts in the information on which our further work had to depend had a decisive influence on our effectiveness in tackling the second problem of the Chicago project: the design and realization of a large-scale reactor to produce the nuclear explosive plutonium. This objective could now be pursued with all the energy and imagination which the project could muster.

Fermi computed the reproduction factor to be $k = 1.0006$. Because CP-1 was always operated at very low power, the level of radioactivity created was harmless. At the time, radiation doses were commonly measured in rems ("roentgen equivalent in man"). A lethal single-shot radiation dose for a human being is about 500 rems; the typical background dose for the entire population is about 0.2 rems per year. During CP-1 operation, the exposure level near the pile was about 0.05 rems per minute; at the sidewalk outside the building it was about one-thousandth as much. The effects of various radiation dose levels are discussed more extensively in Chap. 7.

Fermi's prediction that the pile could be safely controlled proved correct. When the pile was in steady-state operation under the control of a single cadmium rod, some 4 hours were required for the reactivity to rise by a factor of two if the rod was pulled out by one centimeter. Even if all rods were removed, the neutron intensity within the pile would have had a characteristic exponential rise time of about 2.6 minutes, which is not very short. Control could be so finely maintained that it was occasionally necessary to adjust a rod by a centimeter or two in response to the pile's reaction to changing atmospheric pressure. The temperature sensitivity of the reproduction factor, an important engineering consideration, could be measured by simply opening a window to allow outside air to cool the pile. Fermi described controlling the pile as being as easy as the minute steering adjustments one makes while driving a car on a straight road.

Because CP-1 was uncooled and unshielded, it was operated most of the time at half-Watt power, although it did operate briefly at 200 W on December 12. But before engineers could extrapolate to production-scale reactors, much more research on control and shielding systems was necessary. Consequently, after about three months, CP-1 was disassembled and moved to the Argonne site, where it was reassembled as CP-2 (Fig. 5.3). CP-2 was essentially cubical in shape, about 30 feet square in footprint by 25 feet high, and incorporated some 52 t of uranium and 472 t of graphite. CP-2 was also uncooled, but was shielded on all sides by a concrete wall five feet thick, and on its top by a six-inch layer of lead and 50 in. of wood. This shielding permitted operation at a power of a few kilowatts. The rebuilt pile first went critical in May, 1943, and was used for studies of neutron capture cross-sections, shielding, instrumentation, and as a training facility for later production operations. Also built at the Argonne site was the world's first heavy-water cooled and moderated reactor, CP-3, which began operation in May, 1944. This reactor consisted of an upright aluminum tank six feet in diameter, which was filled with about 6.5 t of heavy water. The tank was surrounded by a graphite "reflector," which was further surrounded by a lead shield and then a "biological shield" of concrete. The top of the structure was pierced with holes for experimental ports and control and fuel rods, and was shielded with removable bricks of alternating layers of iron and masonite. CP-3 reached its full operating power of 300 kW in July, 1944.

With a self-sustaining reaction having been demonstrated, attention could be turned to the construction of the X-10 pilot-scale pile. X-10 would have multiple missions: to produce plutonium to test chemical separation procedures and supply Los Alamos with fissile material for research, to train operating personnel for the eventual production-scale reactors, to serve as a platform for instrument development and cross-section research, and to conduct radiation-damage and biological radiation-effects studies. The history of how DuPont came to be the contractor for X-10 was described in Chap. 4; here, our concern is with the design and operation of the reactor itself. Evolution of design considerations for the production-scale reactors located at Hanford is discussed in Chap. 6.

Extrapolating from the operation of CP-1, Fermi estimated that a pure-uranium/graphite system could develop a reproduction constant of $k \sim 1.07$, a value great enough to keep open the possibilities of water-cooling production piles and air-cooling the anticipated pilot-scale unit. Crawford Greenewalt had been thinking of helium-cooling for the pilot unit, but air-cooling would be much simpler from an engineering perspective. By January, 1943, X-10's basic specifications had been developed: a 1,000 kW air-cooled, graphite-moderated pile of cubical shape. The anticipated power level was crucial. Plutonium production in a reactor is directly proportional to its operating power. A reactor fueled with natural uranium produces about 0.76 g of plutonium per day per megawatt (MW) of power produced, a mere one-third of the mass of a dime. If X-10 were to run for a full year at its 1,000 kW (=1 MW) rating, it would theoretically produce about 275 g of plutonium, assuming perfect chemical separation efficiency. It ultimately achieved better than this.

Formally, the X-10 reactor was under the administration of the University of Chicago's on-site Clinton Laboratories, which was located in a 112-acre site in the Bethel Valley of the Oak Ridge reservation (Fig. 4.14). X-10's core comprised a 73-layer graphite cube, 24 feet square on its base by 24 feet, 4 in. high. Right-angled notches were cut into the sides of the graphite bricks, which, when laid side-by-side, formed 1,248 horizontal diamond-shaped front-to-rear channels into which cylindrical aluminum-jacketed uranium slugs could be fed from the front face of the pile (Figs. 5.7, 5.8). X-10's 700 t of graphite was in the form of bricks of cross-section four inches, with lengths varying from eight to 50 in. The fuel channels were built on eight-inch centers; the pure uranium slugs were 1.1 in. in diameter and 4.1 in. long. A full fuel load would be about 120 t, but it was anticipated that the pile would go critical with about half that amount. The core was surrounded by a seven-foot thick shield made of a type of concrete that retained about 10 % of its weight of water upon setting; this helped to stop escaping neutrons. With the addition of layers of pitch to prevent the shielding from losing water along with special precast concrete blocks on the front face to align fuel channels, the full outside dimensions of the pile came to some 47 feet long by 38 feet wide by 35 feet high. A unique aspect of the design was a 20 by 24-in. "graphite thermal column" section of the core which could be lifted out to facilitate what were called "lattice dimension experiments."

After some period of operation, fuel slugs would be discharged from the back of the pile as new slugs were pushed in. Discharged slugs would fall through a chute into a pit containing 20 feet of water, where their intense short-lived radioactivity would be allowed to die off for a few weeks before they were transported to the chemical separation plant. To fuel the pile, workers rode in an elevator which spanned its front face (Fig. 5.8). While the X-10 reactor was not a model for the larger production reactors built at Hanford, some of its design features would find their way into those piles, particularly the procedures for fueling and handling discharged slugs. X-10 also served as a training ground for later Hanford

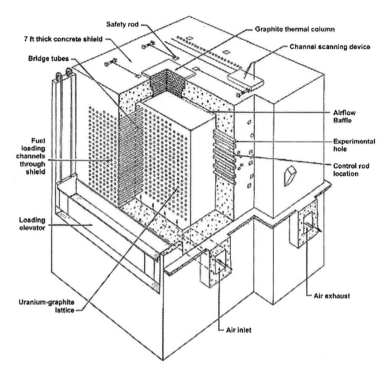

Fig. 5.7 Schematic drawing of X-10 pile. Not all horizontal and vertical channels are shown. *Courtesy of Oak Ridge National Laboratory, U.S. Dept. of Energy.* See http://info.ornl.gov/sites/publications/files/Pub20808.pdf

operators; a total of 183 DuPont employees would train at Clinton before moving to Washington state.

As with CP-1, the control system for X-10 was deliberately over-designed. Three sets of control rods were incorporated: regulating rods, shim rods, and safety rods. The latter were four eight-foot-long boron-steel rods suspended above the pile. They could be operated manually, but were held in place with electric brakes; in the event of a power failure they would passively fall into the pile. As an emergency backup system, hoppers above the pile could release small boron-steel balls into two other vertical channels. During normal operation, the pile would be controlled by two horizontal boron-steel regulating rods which entered from its right side. Four horizontal shim rods provided a means to compensate for variations too large to be handled by the regulating rods. The shim rods could effect a complete shutdown by themselves if necessary; they were connected to a weight-driven system which could drive them into the pile within five seconds in case of a power outage. Other channels served as test holes into which neutron monitors, irradiation samples, and small animals could be inserted. Fuel and experimental channels were equipped with plugs which were removed only when the power output was low enough to prevent a dangerous amount of radiation from escaping.

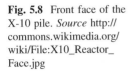

Fig. 5.8 Front face of the X-10 pile. *Source* http://commons.wikimedia.org/wiki/File:X10_Reactor_Face.jpg

The limiting factor in X-10's operation was the capacity of its forced-air cooling system. This initially consisted of two fans each capable of moving 30,000 cubic feet per minute (cfm), plus a stand-by steam-driven 5,000 cfm unit which would come on-line in the event of a power failure. Before being exhausted from a 200-foot stack, the heated air would be filtered and sprayed with water to remove any radioactivity that it might have picked up.

DuPont began excavation for the pile building on April 27, 1943. Concrete pouring began in June, and the Aluminum Company of America began "canning" uranium slugs. Graphite stacking began on September 1. By late October, construction and final mechanical testing was complete. Loading of fuel into the central portion of the pile began on the afternoon of November 3, with Enrico Fermi inserting the first slug. X-10 went critical at 5:07 on the morning of November 4, with only about 30 t of uranium inserted. After a week, the fuel load was increased to 36 t, and the power level reached 500 kW. Before November was out, five tons of fuel containing some 500 mg of plutonium had been discharged and sent off for chemical processing. In December, empty channels were blocked off with graphite plugs to force the airflow to be concentrated around the installed fuel; this permitted higher-temperature operation and raising the power level to about 800 kW. By February, 1944, the pile was producing irradiated uranium at a rate of about one-third of a ton per day; the efficiency of chemical separation of plutonium from uranium eventually exceeded 90 percent.

In early 1944, X-10's fuel distribution was reconfigured to further enhance plutonium production. The standard configuration had been 459 channels loaded with 65 slugs each (about 40 t); this was changed to 709 channels with 44 slugs

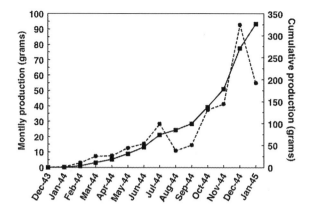

Fig. 5.9 Monthly (*dashed line, left scale*) and cumulative (*solid line, right scale*) production of plutonium from the X-10 reactor. Data from National Archives and Records Administration microfilm set A1218 (Manhattan Engineer District History), Roll 6 (Book IV—Pile Project X-10, Volume 2—Research, Part II—Clinton laboratories, "Top Secret" Appendix). By January, 1945, cumulative production reached 326.4 g

each. This did not significantly increase the amount of fuel in the pile, but reducing the amount of fuel in the center of the pile relative to that further out permitted operating at a higher power level without attaining too great a central temperature. Improved slug-canning techniques allowed even higher-temperature operation, to the point that, by May, 1944, the power level could be increased to 1,800 kW. In June and July of that year, installation of two 70,000-cfm fans allowed operation at an impressive 4,000 kW, four times the original design value. X-10 operated with remarkable reliability; the only real problem encountered was a bearing failure in one of the new fans, which necessitated temporary re-installation of one of the 30,000-cfm units during the summer of 1944.

Plutonium production began in December, 1943, with a mere 1.5 mg being isolated. By mid-1944, tens of grams were being turned out per month (Fig. 5.9), and by the time production ceased in January, 1945 (when the Hanford reactors were coming on-line), over three hundred grams had been extracted from 299 batches of slugs. It was X-10 plutonium that would lead to the discovery of the so-called spontaneous-fission crisis at Los Alamos in the summer of 1944 (Sect. 7.7). Had this discovery had to wait for Hanford-produced material, the Nagasaki *Fat Man* bomb would have been delayed by the better part of a year. An unanticipated bonus of X-10 operation was the production of quantities of radioactive lanthanum, which could be extracted from decaying barium, a direct fission product. As described in Sect. 7.11, this "radio-lanthanum" proved crucial in developing a diagnostic test of the plutonium implosion bomb. X-10 more than fulfilled its wartime mission.

5.3 Y-12: The Electromagnetic Separation Program

The rapid ascendance of Ernest Lawrence's electromagnetic "calutron" method to the top of the list of possible uranium enrichment techniques was described in Sect. 4.7. The electromagnetic method was the *only* enrichment method discussed at the September, 1942, Bohemian Grove meeting, where a recommendation was developed to have the Army enter into construction of a 100 g-per-day U-235 plant to be located in Tennessee. The late-1942 Lewis review committee concluded that an electromagnetic plant capable of producing one kilogram of U-235 per day would require at least 22,000 vacuum tanks; to achieve the same output with a diffusion plant would require a 4,600-stage installation. At its December 10, 1942 meeting, the Military Policy Committee opted to start with a more modest 500-tank electromagnetic plant. While neither the electromagnetic nor the diffusion approaches to enrichment would be easy, the advantage of the electromagnetic system was that the fundamental concepts were proven, and since it would operate in a "batch" mode, it could be built in sections, each of which could begin operating as soon as it was constructed. On the other hand, each section of the "continuous" diffusion plant would have to be operational before it could be connected to preceding and succeeding sections.

The Y-12 plant was located in an 825-acre tract within the Bear Creek Valley of the Clinton site (Fig. 4.14). It would become a mammoth undertaking. The second-most expensive facility of the entire Manhattan Project (about $478 million in construction and operating costs, in comparison to some $512 million for the gaseous diffusion plant), Y-12 would rank first if measured by number of personnel: a peak of nearly 22,500 employees in May, 1945. The complex would come to include nine main processing plants and over 200 auxiliary buildings, totaling altogether some 80 acres of floor space. The entire complex was surrounded by a 5.3-mile perimeter fence with 19 guard towers (Fig. 5.10).

To appreciate the accomplishments of Y-12, it is helpful to review the development of mass spectroscopy and the cyclotron presented in Sects. 2.1.4 and 2.1.8, in particular the idea how a stream of ionized atoms or molecules will naturally move in circular paths when directed into a magnetic field that lies perpendicular to their plane of motion.

Figure 5.11 shows a modified version of the two cyclotron Dees of Fig. 2.18. There is now no alternating-voltage supply, but a coil used to create the magnetic field is represented by the dashed circular outer line in the Figure; in effect, one has two copies of Fig. 2.7 placed back-to-back in order to double ion-separation production over what would be obtained if using just a single tank. Figure 5.12 shows how a single vacuum tank was represented in Manhattan District documents.

For access to the tanks for maintenance and to remove accumulated separated isotopes, it is convenient to place the tanks between two adjacent coils, as opposed to inside a coil. (Again, since complete separation is never achieved in practice, it is more correct to refer to *enrichment* than to *separation*.) This is illustrated

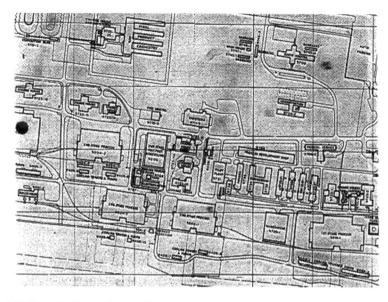

Fig. 5.10 Part of a layout diagram for the Y-12 complex. North is roughly to the *upper right*. The pilot-plant building discussed in the text, 9731, appears just to the *right of center*. Three 9204 Beta buildings (*left of center*, and *below center*) and one 9201 Alpha building (*lower right*) are visible. From *left* to *right*, the diagram covers about 2,900 feet. The two *grid lines* running vertically are 1,000 feet apart; those running horizontally are 500 feet apart. The entire complex measured approximately 8,500 feet end-to-end

Fig. 5.11 Schematic illustration of two back-to-back calutron "tanks" and a magnet coil (*circular dashed line*). As in Fig. 2.18, the magnetic field is perpendicular to the plane of the page

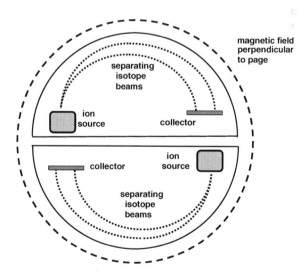

schematically in Fig. 5.13, where the tanks and coils are viewed from the side; the current which supplies the coils can run from one coil to the other through a connecting wire, which is not shown. If you are a physics student, you may

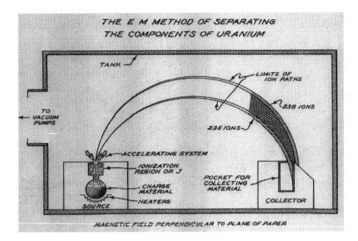

Fig. 5.12 Sketch of the electromagnetic separation method, reproduced from a Manhattan District History microfilm. *Source* A1218(9), image 831

Fig. 5.13 As Fig. 5.11, but seen from the side, with vacuum tanks sandwiched between two coils

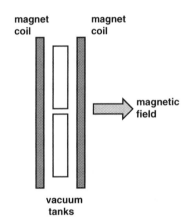

recognize such an arrangement to be a form of Helmholtz coils, which has the advantage that it creates a very uniform magnetic field between the two coils.

The limiting factor in calutron operation—and the reason Y-12 became such an enormous facility—is something known to cyclotron engineers as the "space-charge" problem. As the like-charged ion beams travel through the vacuum tank, they repel each other and so become disrupted from their ideal circular paths. This sets a practical limit on the strength of the beams, which is usually expressed as an equivalent electrical current. This in turn limits the mass of material that any one vacuum tank can theoretically separate per day. In the case of Y-12 calutrons, this capacity was about 100 mg of U-235 per day in the best of circumstances. To collect 50 kg from one tank (barely enough for a single bomb) would require 500,000 days of operation, or over 1,300 years. It was appreciated from the outset

Fig. 5.14 Schematic illustration of part of a calutron "track." In practice, a given track would include dozens of tanks

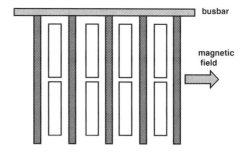

that only if one were willing to invest in a facility with at least hundreds of tanks might one have a chance of isolating enough material for a bomb in a year or two.

If hundreds of tanks are necessary, a convenient way to arrange them is to link together a number of copies of Fig. 5.13, connecting the coils with a current-carrying "busbar," as suggested in Fig. 5.14. Such an arrangement came to be known as a "track" of tanks and coils. Since the electrical current must run along a closed circuit, another refinement is to configure the track as a closed loop. Ernest Lawrence and his engineers conceived this idea early on, deciding on an oval-shaped arrangement. Such a real Y-12 track, known as an "Alpha racetrack," is shown in Fig. 5.15. This particular track contains 96 vacuum tanks placed as back-to-back pairs within 48 gaps between the magnet coils, which appear as rib-like structures. The number of gaps was chosen to be 48 because that number's large number of even divisors provided for greater flexibility in the use of power supply equipment. (The "Alpha" name emerged when later "Beta" tracks of a different design came along, as described below.) The linear structure running across the top of the photo holds the busbar, a square-foot solid-silver conductor that feeds

Fig. 5.15 *Left* A Y-12 alpha "racetrack." *Right* Workers tend to a *C-shaped* vacuum tank. *Sources* http://www.y12.doe.gov/about/history/getimages, http://commons.wikimedia.org/wiki/File:Alpha_calutron_tank.jpg

current to the coils. As shown in Fig. 5.15, the vacuum tanks in these tracks were C-shaped, and could be withdrawn on special gantries for material extraction and maintenance.

It is clear from Fig. 5.15 that the magnet coils involved in these units are much larger than would be the case for any laboratory mass spectrometer. The coils in the Alpha tracks were square (Fig. 5.21), and for the accelerating voltages and magnetic fields that could feasibly be provided, had to be of side length about 3 m. Even then, the separation of the U-235 and U-238 ion beams was only about one centimeter. The magnetic field utilized was of strength about 0.34 Tesla, some 7,000 times the average surface-level strength of the Earth's natural magnetic field.

By the fall of 1942, experiments with Lawrence's 184-in. cyclotron indicated that some 2,000 ion sources would be necessary to separate 100 g of U-235 per day, the goal set at the Bohemian Grove meeting. Stone and Webster conservatively assumed that no more than one ion source and collector could be fitted into each tank, and so began planning on as many as 2,000 tanks. From the capabilities of available electrical power-distribution equipment, it seemed feasible to assume that each production building could house two tracks, both containing about 100 tanks. If the gap between each successive pair of coils housed two tanks, each track would require 50 gaps. To provide for two thousand tanks, ten production buildings would consequently be required, as well as fabrication and maintenance shops, laboratories, and generating facilities. A particularly daunting aspect of the system was its vacuum requirements. The tanks would have to be pumped down to and maintained at pressures of about a hundred-millionth of standard atmospheric pressure. It was estimated that the vacuum volume for the plant would probably exceed by many orders of magnitude the entire evacuated space in the world at the time. Another consideration was facilities for chemical processing. The uranium oxide received from Mallinckrodt Chemical had to be transformed to uranium tetrachloride before being fed into the calutrons, and the processed material, which was often washed out of the tanks with acid, had to be collected and purified. Chemical operations alone at Y-12 employed several thousand people.

Through 1942, Lawrence and his engineers concentrated on refining the design of ion sources to incorporate multiple beams (Fig. 5.16). On November 18, he installed a new tank which contained a double source between the poles of his 184-in. cyclotron. Both sources were capable of producing two sets of beams, that is, there would altogether be four sets of U-235 and U-238 beams. The system was cantankerous, however; often only two beams could be kept in focus simultaneously. But even two sources per tank would be a major advance, as such a design would permit decreasing the total number of tanks to 1,000. In the meantime, Stone and Webster engineers had to begin designing buildings based on only very rough ideas as to the equipment they would contain; General Groves often invested enormous sums in construction before fully-workable enrichment systems had been developed.

Stone and Webster constructed Y-12, but the plant had then to be operated. For that task, Groves contracted with the Tennessee Eastman Corporation (a subsidiary

Fig. 5.16 Frank
Oppenheimer (*dark hair,
center*) and Robert Thornton
(*right*) examine a 4-source
Alpha-calutron ion emitter.
Source http://
commons.wikimedia.org/
wiki/File:Calutron_
emitter.jpg

of the Eastman Kodak Company), a firm to which he had entrusted construction of
an explosives plant in Kingsport, Tennessee. Tennessee Eastman's contract was on
a cost-plus-fee basis: a basic stipend of $22,500 per month plus $7,500 for each
track up to seven, plus an additional $4,000 for each track over that number.

Lawrence, Groves, and various industrial contractors met at Berkeley on
January 14, 1943, to begin planning the Y-12 project. The initial phase of
development called for five 96-tank tracks to be housed in three buildings; the
tracks themselves would be 122 feet long, 77 feet wide, and 15 feet high; the
buildings would require 6-foot foundations to support the weight of the magnets.
The center area of the tracks was large enough to be used as office space. Groves
wanted the first track in operation by July 1, 1943. The first floor of each building
(below ground level) would house vacuum pumps. The tracks would reside at
ground-level, and above them resided operating galleries from where employees,
mostly local young female high-school graduates, continuously monitored and
adjusted the ion beams in each tank (Fig. 5.17). The process was labor intensive,
requiring some 20 employees per operating separator. The magnetic fields had to
be kept extremely uniform; a deviation of only 0.6 % would result in collecting the
wrong isotope.

Design of the Y-12 facility and its equipment evolved continuously and
incrementally. First came a decision to use two-source ion emitters in each tank. In
early 1943, Edward Lofgren conceived the idea of building second-stage enrichers
which would be fed with slightly-enriched material (~ 15 % U-235) that emerged
from the first-stage tracks, and which would raise the enrichment level to 90 %.
Groves found the idea attractive, and authorized the first two such units on March
17. It was at this point that the original oval racetracks became known as "Alpha"
units, and the second-stage enrichers as "Beta" units. Beta tanks were half the
diameter of Alpha units, but operated at twice their magnetic field strength. Laid
out in a rectangular configuration of two parallel rows of 18 tanks, each Beta track
housed 36 tanks sandwiched between D-shaped magnet coils (Fig. 5.18). Beta
units also incorporated twin-source emitters.

Fig. 5.17 "Calutron girls" at their operating stations. Each operator monitored the performance of two vacuum tanks, but had no idea what was being produced. *Source* http://commons. wikimedia.org/wiki/ File:Y12_Calutron_ Operators.jpg

Fig. 5.18 This photograph, reproduced from Manhattan District History microfilms, shows two Beta tracks, one in the foreground and one in the background. *Source* A1218(10), image 0231

In July, 1943, Lawrence began advocating for multiple-beam sources within tanks, but Groves was reluctant to authorize changes that might delay plant completion. A compromise was struck: the first four Alpha and all Beta tracks would use two-beam sources, but the fifth Alpha track would use four-beam sources. The staff of the Radiation Laboratory was expanded to take on the additional design and engineering tasks; by mid-1944 it would reach 1,200 employees. Research alone for the electromagnetic program ran to about $20 million.

Well before design was complete, ground was broken for the first Alpha building, 9201-1, on February 18, 1943. Buildings containing Alpha tracks were known as "9201" buildings, while those containing Beta units were "9204" buildings. There would ultimately be five Alpha buildings housing nine tracks, plus four Beta buildings housing eight tracks. Altogether, these 17 tracks would contain 1,152 tanks, although not all came online until after the end of the war (Table 5.1). The first structure completed at the Y-12 site, and that which is

Table 5.1 Alpha and Beta calutron tracks

Building	Ion sources per tank x tanks per track	Tracks	Start date
9201-1	2 × 96 (Alpha I)	Alpha 1	13-Nov-43*
		Alpha-2	22-Jan-44
9201-2	2 × 96 (Alpha I)	Alpha 3	19-Mar-44
		Alpha 4	12-Apr-44
9201-3	4 × 96 (Alpha I)	Alpha 5	3-Jun-44
9201-4	4 × 96 (Alpha II)	Alpha 6	24-Jul-44
		Alpha 7	26-Aug-44
9201-5	4 × 96 (Alpha II)	Alpha 8	24-Sep-44
		Alpha 9	26-Oct-44
9204-1	2 × 36 (Beta)	Beta 1	15-Mar-44
		Beta 2	5-Jun-44
9204-2	2 × 36 (Beta)	Beta 3	12-Sep-44
		Beta 4	2-Nov-44
9204-3	2 × 36 (Beta)	Beta 5	30-Jan-45
		Beta 6	13-Dec-44
9204-4	2 × 36 (Beta)	Beta 7	1-Dec-45
		Beta 8	15-Nov-45

Track Alpha-1 was shut down shortly after the date shown; it was restarted on March 3, 1944

perhaps now most famous, was building 9731, the "pilot plant" building (Fig. 5.19). Completed in March, 1943, this building housed experimental Alpha and Beta units which were used for training operators. Designated as calutrons XAX and XBX, these units still stand in building 9731, and are now identified as Manhattan Project Signature Artifacts by the Department of Energy's Office of History and Heritage Resources. Building 9731 is open to tourists, and, unlike many Manhattan Project facilities, will not be demolished.

The experimental XAX Alpha unit was first successfully operated on August 17, 1943, by which time Groves was already considering further expansions. After reviewing design improvements at a meeting in Berkeley on September 2, he presented his plan to the MPC on September 9. Four additional 96-tank Alpha tracks, these with four ion sources per tank, would be constructed. Designated as Alpha II units, these tracks would reside in buildings 9201-4 and 9201-5. They would differ from the original oval configuration in being of rectangular layout (Fig. 5.20); tanks at the curved portions of the oval-shaped units, which would be re-named Alpha I units, proved difficult to regulate. Two more Beta tracks were also authorized at the same time. Production of the vacuum tanks, sources, and collectors was contracted to Westinghouse; General Electric took on responsibility for high-voltage electrical controls, and the magnet coils themselves were fabricated by the Allis-Chalmers company.

As mentioned in Chap. 4, one of the unique aspects of the electromagnetic program was its use of Treasury Department silver to make magnet coils. Normally, copper would have been used, but since that metal was used in shell

Fig. 5.19 Building 9731, the *light-colored*, flat-roofed building at *center left*, was the first building completed at the Y-12 complex. The large building is a Beta plant; compare Fig. 5.10. *Source* http://www.y12.doe.gov/about/history/getimages

Fig. 5.20 This photograph, reproduced from a Manhattan District History microfilm, shows an Alpha II racetrack. *Source* A1218(10), image 0214

casings it was a high-priority commodity during the war. Congress had authorized use of up to 86,000 t of Treasury Department silver for defense purposes; not having to divert mass amounts of copper was a huge boon for the Project's secrecy. Kenneth Nichols met with Undersecretary of the Treasury Daniel Bell on August 3, 1942, to inquire about borrowing 6,000 t of silver from the Treasury's vaults; Bell informed Nichols that the Treasury's preferred unit of measure was the troy ounce. [At 480 grains, a troy ounce is somewhat heavier than a common avoirdupois ounce, which weighs in at 437.5 grains; a troy ounce is equivalent to about 31.1 g]. Secretary of War Henry Stimson formally requested the silver in a letter to Treasury Secretary Henry Morgenthau on August 29, 1942. Stimson gave no indication what the silver was to be used for, stating only that the project was "a highly secret matter." His letter stipulated silver of purity 99.9 %, and assured Morgenthau that title to the silver would remain with the United States. The deadline for returning the silver was five years from its receipt, or upon written notice from the Treasury that all or any part of it was required for reasons connected with monetary requirements of the United States.

The War Department eventually withdrew more than 400,000 bullion bars of approximately 1,000 troy ounces each from the West Point Bullion Depository in West Point, New York. The first bars were withdrawn on October 30, 1942, and trucked about 70 miles south to a U.S. Metals Refining Company facility in Carteret, New Jersey, where they were cast into cylindrical billets weighing about 400 pounds each. By the time casting operations ceased in January, 1944, just over 75,000 billets weighing nearly 31 million pounds had been cast. Remarkably, this weight exceeded the 29.4 million pounds (about 14,700 t) withdrawn from the Treasury. Groves insisted on careful cleanup operations: workers coveralls were vacuumed clean, and machines, tools, furnaces, factory floors, and storage areas that had accumulated years of metal shards were dismantled and scraped clean. Armed guards observed every step in the processing to ensure that all trimmings were recovered. The recovery operation was so successful that more than 1.5 million pounds of silver were gained, versus less than 11,000 which were considered lost.

After being cast, the billets were trucked a few miles north to a Phelps Dodge Copper Products Company plant in Bayway, New Jersey. There they were heated and extruded into strips 3 in. wide by 5/8 in. thick by 40–50 feet long; if all of the Manhattan Project silver was shaped into one strip of that width and thickness, it would reach from Washington to outside Chicago. After being cooled, the strips were rolled to various thicknesses, depending on the particular magnet coils for which they were intended. They were then formed into tight coils (not yet the magnet coils) that were about the size of large automobile tires. Over 74,000 coils were produced, most of which were shipped to Wisconsin for magnet-coil fabrication (Fig. 5.21). In addition, some 268,000 pounds of silver were sent directly to Oak Ridge to be formed into busbar pieces. The coils shipped to Wisconsin went to the Allis-Chalmers Manufacturing Company in Milwaukee, where they were unwound, joined together with silver solder, and fed into a special machine that wound them around the steel bobbins of the magnet casings. Between February, 1943, and August 1944, 940 coils were wound, each containing on average about 14 t of silver. After fabrication they were shipped to Oak Ridge by rail.

By the summer of 1943, construction was in full swing at Y-12. Stone and Webster's construction payroll hit 10,000 by the first week of September, and would peak at about 20,000. Overall, the company would interview some 400,000 people for construction jobs; building the Y-12 complex would consume 67 million man-hours of labor. Tennessee Eastman began training operators; by November some 4,800 were ready. Ernest Lawrence, himself no stranger to large-scale operations, was awed by the size and complexity of Y-12, relating that

Fig. 5.21 This somewhat low-quality photograph, reproduced from a Manhattan District History microfilm, shows magnet coils being wound onto square bobbins, likely Alpha I coils. Note person in *lower right* foreground for scale. *Source* A1218(10), image 0443

"When you see the magnitude of that operation there, it sobers you up and makes you realize that whether we want to or no, that we've got to make things go and come through ... Just from the size of the thing, you can see that a thousand people would just be lost in this place, and we've got to make a definite attempt to just hire everybody in sight and somehow use them, because it's going to be an awful job to get those racetracks into operation on schedule. We must do it." Despite the pace of construction and operations at Clinton, the site's safety record was remarkable; Groves states that through December, 1946, only eight fatal accidents occurred: five by electrocution, one by gassing, one by burns, and one fall (Fig. 5.22).

Fig. 5.22 Construction at Y-12, 1944. *Source* http://www.y12.doe.gov/about/history/getimages.php

Problems began to emerge in the fall of 1943, however. Operators had trouble maintaining steady ion beams, and electrical failures, insulator burnouts, and vacuum leaks were endemic. Some of the steel tanks, which weighed about 14 t, were pulled several inches out of line by magnetic forces, putting tremendous stress on vacuum lines. The solution was to secure the tanks to the floor with steel straps. Worse, soon after the first Alpha track was started on November 13, it had to be shut down due to electrical shorts caused by coil windings being too close together and insulating oil being contaminated with rust, sediments, and organic materials. Furious, Groves arrived on December 15 to personally review the situation. The only option was to ship 80 coils back to Milwaukee for rebuilding, and modify designs to include oil-filtration systems. Refurbishing the coils cost over \$470,000.

As the magnets from the first Alpha track were being rebuilt, the second track entered service on January 22, 1944. Despite seemingly endless breakdowns, its performance gradually improved as experience was gained by maintenance and operating personnel. By the end of February it had enriched about 200 g of material to 12 % U-235; some of this went to Los Alamos, while the remainder was used as feed for beta calutrons. The rebuilt first alpha track re-entered service on March 3, and the first Beta unit began operation in mid-March. Buoyed by this growing success, Lawrence began advocating for another expansion, proposing that four new Alpha tracks be added to the nine already authorized. Groves did not authorize any additional Alpha tracks, but did decide to proceed with two more Beta buildings (tracks 5 through 8), in part to receive partially-enriched material from the gaseous diffusion plant (Sect. 5.4). Construction on the third Beta building began on May 22; the coils in these tracks were made with conventional copper windings.

During "routine" operation, Alpha tracks would be shut down about every tenth day to recover their uranium, and Beta tracks about every third day. It took a long time for productivity to settle into a routine basis, however. During the first months of 1944, not more than about 4 % of the U-235 in the Alpha sources was making its way to the receivers; for Beta stages the fraction was only about 5 %. Losses were due mostly to low ionization efficiency of the uranium tetrachloride feed material, and dissociative processes that yielded species other than just singly-ionized molecules. Much of the feed material ended up splattered around the insides of the vacuum tanks, which had to be scraped clean and washed over catchment sinks. Material that adhered to components that were too costly or awkward to pull out and clean was simply abandoned. More prosaic problems also cropped up. In one case, a mouse became trapped in a vacuum system, preventing proper pump-down. Several days of production were lost, as was the mouse. In another, what Groves described as a "suicidal" a bird perched on an insulator outside the building housing Alpha tracks 6 and 7, and caused a short. The bird received 13 kV, and the entire building was shut down.

Improvements accumulated through 1944. Between October 21 and November 19, U-235 production amounted to 1.5 kg, an amount nearly equal to that of all previous months combined (Fig. 5.23). By December 15, all nine Alpha tracks and

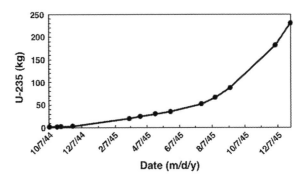

Fig. 5.23 Cumulative production of uranium-235 from Beta stages of the Y-12 plant through early 1946 (date format mm/dd/yy). Data from National Archives and Records Administration microfilm set A1218 (Manhattan Engineer District History), Roll 10 (Book V—Electromagnetic Plant, Volume 6—Operations, Top Secret Appendix, p. 4.)

Beta tracks 1, 2, and 3 were in operation, Beta tracks 4 and 6 were processing unenriched Alpha feed, and Beta 5 was being used for training. Y-12 operated on an around-the-clock basis.

In early 1945, an important evolution in the operation of the Clinton Engineer Works took place. With all uranium enrichments methods finally coming on-line, Groves began to think of harnessing them in series as opposed to treating them as competing in parallel. Following detailed calculations of how to optimize the rate of production of bomb-grade material, the decision was made on February 26 to begin the process by first feeding natural-abundance uranium hexafluoride to the S-50 thermal diffusion plant (Sect. 5.5), which would enrich the U-235 content from 0.72 to 0.86 %. This product would be fed to Alpha calutrons, but when the gaseous diffusion cascade had advanced to the stage of producing 1.1 %-enriched material, the S-50 product would be fed to it to be enriched to that level, after which material would go to Y-12 Alpha units. When enough diffusion stages were on-line to produce 20 % enriched material, the Alpha I units would be shut down, and K-25's product would be directed to Alpha II tracks, and, after that, to Beta units. Since the various plants at Clinton could achieve different but overlapping levels of enrichment, the sequence of feed steps was adjusted constantly as they came on-line. The S-50 plant could raise the enrichment level from 0.72 to 0.86 %, the Alpha stages of the Y-12 plant from 0.72 to about 20 %, Beta stages from 20 to 90 %, and K-25 from 0.72 to 36 %. Y-12's enriched uranium tetrachloride was converted to uranium hexafluoride for shipment to Los Alamos, with chemical processing carried out in gold trays to minimize contamination. The precious product, accounted for to fractions of a gram, was packed into gold-plated nickel cylinders about the size of coffee mugs, which were placed into cadmium-lined wooden boxes. The boxes were secured two at a time inside leather briefcases, which were chained to the wrists of armed Army couriers for a two-day train trip to New Mexico. By April, 1945, Y-12 had produced some 25 kg of bomb-grade

U-235; by mid-July the total would reach just over 50 kg. Every atom of uranium in the *Little Boy* bomb would pass through at least one stage of Ernest Lawrence's calutrons. At peak production, Alpha units were yielding a total of about 258 g per day of 10 %-enriched material, and Beta units were producing about 200 g per day of material enriched to at least 80 %, better output than the 100 g per day specified at the Bohemian Grove meeting.

The Clinton Engineer Works consumed an enormous amount of electricity. By mid-1945, transmission facilities at CEW could provide power at a peak rate of 310,000 kW, of which 200,000 were for Y-12 alone. Peak consumption, nearly 299,000 kW, occurred on September 1, 1945. At Y-12, electricity use began with consumption of 3.26 million kWh (MkWh) in November, 1943, fell to 0.28 MkWh during the depth of the magnet crisis in January, 1944, but grew steadily thereafter until peaking at 153 MkWh in July, 1945. The total amount of electrical energy consumed by Y-12 between November, 1943, through the end of July, 1945, totaled over 1.6 billion kWh, about 100 times the energy released by the U-235 *Little Boy* bomb. To put these numbers in perspective, this author's monthly household electricity consumption typically averages about 650 kWh; a million kWh would power my house for over 120 years. At its peak of operations in the summer of 1945, Clinton was consuming about one percent of the electrical power produced in the United States, much of it flowing through Lawrence's calutrons.

5.4   K-25: The Gaseous Diffusion Program

The K-25 gaseous diffusion enrichment complex was the single most expensive facility of the entire Manhattan Project, and also one of the most difficult to organize, design, engineer, and construct. While gaseous diffusion would eventually prove to be the most economical method of enriching uranium, it was nearly stillborn. In view of the cost and importance of K-25, you might expect that it is the topic of a vast literature, but this is not the case. The process for manufacturing the diffusion membrane is considered so highly-classified that gaseous diffusion receives virtually no mention in the publically-available version of the Manhattan District History; either a section devoted to K-25 was never prepared, or has been redacted. All that is openly known of the technique has to be gleaned from writers such as Hewlett & Anderson, Jones, and Smyth, all of whom had access to classified material and have published sanitized histories.

The fundamental principle of the gaseous diffusion method was described briefly in Chap. 1. The basic idea is that if a gas of mixed isotopic composition is pumped against a porous barrier containing millions of microscopic holes, atoms of lower mass will on average pass through slightly more frequently than those of higher mass. The result is a vey minute level of enrichment of the gas in the lighter-isotope component on the other side of the barrier. Since only a small enrichment factor can be achieved in any one step (see below), the slightly-enriched gas has to be pumped

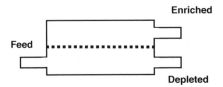

Enriched

Feed

Depleted

Fig. 5.24 Schematic illustration of one cell of a diffusion cascade. The porous membrane is represented by the *dashed line*. Feed material enters from the *left*. Gas enriched in the lighter isotope is pumped off to be sent to the next stage of the cascade, while gas depleted in the lighter isotope is sent back to the preceding stage for further processing

on to subsequent enriching stages. By linking together a number of processing "cells" in series in a cascade, bomb-grade material can eventually be isolated. The gas which emerges from each stage slightly "depleted" in the lighter isotope still contains atoms of that isotope, however, and so needs to be sent back "down" the cascade for additional processing. These ideas are indicated very schematically in Figs. 5.24 and 5.25.

Fig. 5.25 Schematic illustration of a diffusion cascade. The *circles* represent pumps. Gas enriched in the lighter isotope accumulates toward the *top* of the diagram, while that depleted in the lighter isotope accumulates toward the *bottom.* In reality, the cascade is *not* arranged vertically as this diagram suggests; in the K-25 plant, all cells were at ground level

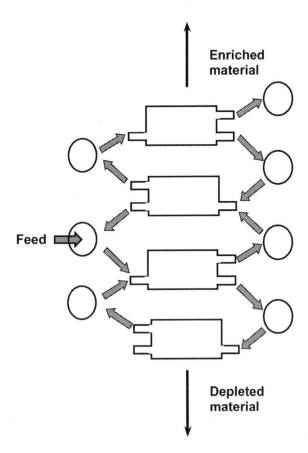

Enriched material

Feed

Depleted material

The enrichment capability of each cell is dictated by the statistical mechanics of diffusion. The theory is complex, but some of the basic results are straightforward. If the enriched gas from each stage is sent on to subsequent stages for a total of N stages, then the ratio of the number of lighter-isotope atoms to the number of heavier-isotope atoms in the emergent gas is given by

$$output\ ratio = (input - stage\ ratio)\left(\frac{mass\ of\ heavy\ isotope}{mass\ of\ light\ isotope}\right)^{N/2}. \qquad (5.1)$$

In the K-25 plant, the gas used was uranium hexafluoride, UF_6. Since fluorine has atomic weight 19, the heavy isotope ($U^{238}F_6$) has mass $238 + 6(19) = 352$, and the lighter one has mass 349 (there is only one stable isotope of fluorine). If the UF_6 input to the first-stage of the cascade has not been "pre-enriched," then the input abundance ratio will be the light-to-heavy abundance ratio of natural uranium, $0.0072/0.9928 = 0.00725$. Hence

$$output\ ratio \quad = \quad 0.00725\left(\frac{352}{349}\right)^{N/2} \quad = \quad 0.00725\,(1.0086)^{N/2}. \qquad (5.2)$$

To achieve a nine-to-one output ratio (90 % enrichment) requires $N \sim 1{,}664$. As Henry Smyth described the situation in his 1945 history of the Project, "many acres" of barrier are needed for a large-scale plant. If, however, you can start with 5 %-enriched material (input ratio $= 1/19$; why?), to get to the same 90 % enrichment will require about 1,200 stages—about 25 % fewer, but still a large number. The mathematics of diffusion is like that of compound interest: if you want to get to $1,000 at a fixed rate of interest (which plays the role of the mass ratio), you will achieve your goal much sooner if you start with $100 than with $10. In practice, the process is never as efficient as these numbers imply. Not all of the material that enters each stage undergoes diffusion, and some gas will naturally diffuse back through the barrier. In fact, detailed calculations indicated that the best arrangement from the point of view of plant size and power requirements is one in which only half of the gas that is pumped into each stage diffuses through the barrier, with the depleted half being returned to the preceding stage. In an actual plant, some 100,000 times the volume of gas that emerges from the end of the cascade may need to be fed to the input stage. The number of stages built below the feed point is dictated by a judgment of how economical it is considered to be to continue processing depleted material; in the K-25 plant, the feed point was about one-third of the way along the cascade.

The barrier must be robust and easy to manufacture, but the most important feature is the size of the diffusion holes. At atmospheric pressure, the mean free path of a molecule (the distance it will travel on average before colliding with another molecule) is on the order of 10^{-7} m, or about one ten-thousandth of a millimeter. To achieve true diffusion, the diameter of the holes in the barrier should be no more than about one-tenth of this figure, or about 100 Å (1 Å $= 10^{-10}$ m). For comparison, a typical human hair might have a diameter of about a million Ångstroms, although there is wide variation in hair sizes.

Diffusion processes were well-known to chemical engineers, so it is not surprising that this technique became the object of attention as a possible enrichment method once understanding of the crucial role of U-235 in the fission process became appreciated. In late 1940, John Dunning, Eugene Booth, Harold Urey, and mathematician Karl Cohen began research into the technique at Columbia University. Their first barrier material was partially fused glass, known as "fritted" glass, but that material could not stand up to the corrosive effects of uranium hexafluoride. The mid-1941 British MAUD report identified diffusion as the most promising enrichment technique; Franz Simon at Oxford was developing two 10-stage cascade models to test different pumping schemes. In November, 1941, when Vannevar Bush reorganized the project to appoint Program Chiefs, Urey was designated to lead diffusion work in America (Sect. 4.6). By that time, Dunning and Booth were experimenting with creating a porous metallic barrier by etching zinc from a sheet of brass, and had succeeded in enriching a small amount of uranium. (Brass is an amalgam of copper and zinc; etching away the zinc rendered the sheet porous).

At their May 25, 1942, meeting, S-1 administrators advocated proceeding with a diffusion pilot plant and engineering studies for a 1 kg/day full-scale plant, recommendations which Vannevar Bush took to President Roosevelt on June 17 (Sect. 4.9). By late October, Booth had a 12-stage demonstration system in operation at Columbia which achieved a small enrichment of uranium hexafluoride during a five-hour run. The Columbia system involved face-to-face cylinders about four inches in diameter, with dollar-coin-sized barrier samples placed between the faces. The entire assembly fit in a cabinet about eight feet square and three feet deep.

When the Columbia work came under Manhattan District auspices, it became christened as the Substitute (or, by some sources, Special) Alloy Materials (SAM) Laboratory. By the end of 1943, Urey had over 700 people working on gaseous diffusion problems at Columbia alone, plus several hundred more at other universities and industrial laboratories. As described in Sect. 4.10, the Lewis committee report of December, 1942, concluded that a $150-million, 4,600-stage diffusion plant would have the best chance of all methods of eventually producing about 25 kg of U-235 per month, and recommended proceeding with construction. At the December 10 Military Policy Committee meeting, it was decided to proceed, even though a 10-stage pilot plant under construction by the Kellogg Company (see below) would not be ready until June, 1943, and was not anticipated to be yielding any results until September of that year.

When the Kellogg Company took on the design and engineering of the diffusion plant in late 1942, no suitable barrier material had been developed. Ultimately, development of a useable barrier would prove one of the most difficult aspects of the entire Manhattan Project. The process material to be used in the plant, uranium hexafluoride, had the advantage that it could easily be made into a gas, but is extremely caustic. (Large-scale production of hexafluoride was pioneered by Philip Abelson, as described in the following Section.) The barrier would have to be strong enough to withstand both the corrosive effects of the gas and the high pressures under which it would operate. The only element that can withstand the caustic effects of UF_6 is nickel, and in late 1942 a Columbia group under the direction of Foster Nix (Bell Telephone Laboratories) turned their attention to experiments involving compressed nickel powders. These barriers proved sufficiently rugged, but insufficiently porous. In contrast, fine-enough holes could be realized with an electro-deposited mesh, but the mesh was not particularly strong. The mesh had been developed by Edward Norris, an interior decorator, as part of a paint sprayer that he had invented. Norris joined the Columbia group in late 1941, and by January, 1943, he and chemist Edward Adler had developed a material which looked to have the correct combination of porosity and strength. Construction of a six-stage pilot plant to hand-produce the Norris-Adler barrier was begun at Columbia in February; it would begin operating in July. For a full-scale plant, however, piecework would be impractical. Several million square feet of barrier would be required, which meant industrial-scale production. In April, 1943, the Houdaille-Hershey Corporation, a manufacturer of automobile accessories which had accumulated considerable experience in plating techniques, took on a contract to build and operate a $5-million barrier-production plant to be located in Decatur, Illinois, on the premise that the Norris-Adler barrier would prove amenable to mass production. The diffusion tanks themselves, some as large as 10,000 gallons, were manufactured by the Chrysler Corporation, which had to develop techniques for nickel-plating the insides of the tanks. Over the course of two years, Chrysler plated some 63 acres of steel surface.

Kellogg's December, 1942, contract with the Army was unusual. The firm was not required to make any guarantee that it could design, build, and get a plant into operation. Financial terms were left unspecified until the work was further developed; eventually the company accepted a fee of about $2.5 million for its efforts. A separate corporate entity, the Kellex Corporation, was set up to carry out the work. It has been suggested that the K-25 designation may be the only Clinton site code-name that had a meaning: K for Kellex and 25 for U-235; Y-12, X-10, and S-50 appear to be meaningless. As Henry Smyth described it, Kellex was a unique temporary cooperative of scientists, engineers, and administrators drawn from a number of schools and industries for the express purpose of carrying out one job. Percival Keith, a Kellogg Vice President and MIT chemical-engineering graduate, was designated to be in charge of the new corporation, which by 1944 would have some 3,700 employees. The firm also began undertaking its own barrier research, as well as construction of a 10-stage pilot plant in Jersey City, New Jersey, which would eventually be used to test full-size diffusion tanks under simulated operating conditions.

Diffusion-plant design was another area where American and British ideas conflicted. Karl Cohen's 4,600-stage analysis was predicated on a high-pressure, high-temperature single-cascade operation, whereas the British proposed a cascade-of-cascades arrangement which would operate at lower temperatures and pressures. The British approach would be more complex to engineer, but would place less stringent demands on the barrier material and would have the advantage of a shorter equilibrium time. On the other hand, a single-cascade design could be more easily configured to permit the system to be plumbed so that process material could be fed into or drawn off from any stage as desired. In his Appendix to Arthur Compton's third National Academy of Sciences report of November, 1941 (Sect. 4.5), Robert Mulliken estimated the equilibrium time for a low-pressure plant to be 5–12 days, as opposed to 100 days for a higher-pressure design.

The barrier was not the only issue Kellex faced. Since uranium hexafluoride reacts explosively with grease and moisture, neither could be allowed to mix with the process gas along the miles of pipes that a plant would require. How then, to lubricate the thousands of pumps and valves that would be involved? The solution turned out to be a water-resistant chemical known as polytetrafluorethane [PTFE; chemical formula $(C_2F_4)_n$], now commonly known as Teflon. Itself a fluorine compound, PTFE resists attack by fluoride compounds, and has one of the lowest coefficients of friction known of any solid material. Another problem was that the diffusion plant would require some 7,000 pumps. But when gases are compressed, they naturally heat up; the pumps had to be cooled as they operated, again with all seals vacuum-tight and non-leaking. K-25's pumps would be supplied by the Allis-Chalmers Company, which also manufactured the Y-12 magnet coils.

By mid-1943, as work was proceeding on the Decatur plant and surveying was underway for the K-25 plant, the barrier issue was approaching crisis proportions. The Norris-Adler barrier, for which the Decatur plant was being configured, was proving brittle, plagued with pinholes, and difficult to manufacture in uniform quality. A key advance was made in June of that year when Clarence Johnson, a Kellex engineer, developed a new barrier using a method coyly described in official histories as combining the techniques of Norris, Adler, and Nix. Significant contributions were also made by Hugh Taylor, a British-born Princeton University chemist who had been active in developing processes to isolate heavy water. A Military Policy Committee meeting held on August 13, 1943, concluded that a suitable barrier would probably be forthcoming, but research would have to continue on both the Columbia (Norris-Adler-Nix) and Kellex (Johnson) processes for the time being. Ultimately, many hands and minds would be involved in solving the barrier problem; the eventual success cannot be attributed to any one person.

By the end of 1943, the time to make a decision was approaching. At that point, Groves did something unusual in turning to British scientists to review the situation. On December 22, he met in New York with a 16-strong British contingent which included Franz Simon and Rudolf Peierls. The group was briefed by representatives from Kellex and Columbia, following which they adjourned to visit various laboratories before preparing their report, which was considered at a four-hour meeting at Kellex headquarters on January 5, 1944. The British felt that Johnson's barrier would be easier to manufacture and likely eventually prove superior to the Norris-Adler version, but, if time was the determining factor, the research already accumulated on the latter represented an important advantage. Houdaille-Hershey, however, was becoming pessimistic that they could produce the Norris-Adler barrier on a large scale. Kellex engineers countered that even with a switch to the Johnson barrier, they could have K-25 in operation by Groves' target date of July 1, 1945. Groves announced his decision at a visit to Decatur on January 16: the plant would be converted to fabricate the Johnson barrier.

As with the Y-12 complex, construction and operations were handled by two different contractors. Groves needed an operating contractor for K-25, and two days after his Decatur visit convinced the Carbide and Carbon Chemicals Corporation, a subsidiary of Union Carbide, to take on that task on a cost plus $75,000-per-month basis. The company appointed one of its vice-presidents, physical chemist and engineer George Felbeck, to be its K-25 project manager; Carbide also contributed to barrier research.

K-25 would require enormous amounts of steam to heat the process material and operate pumps. A concern in this regard was that a power interruption would not only delay production, but could set up pressure waves that could reverberate through the cascade and damage equipment. Fearing interruptions or sabotage, Groves did not want to rely on the TVA for electrical power, and decided that K-25 would have its own 238-MW steam-electric generating plant (enough to power a city the size of Boston), which would feed the main plant through protected underground cables. To construct the generating plant and the main K-25 plant itself, Groves chose the J. A. Jones construction company of Charlotte, North Carolina. He was familiar with the firm; Jones had built more Army camps than any other contractor in America. Jones would ultimately engage over sixty subcontractors in what was one of the largest construction projects in the world to its time. Work got underway in May, 1943, with a surveying party laying out a site for the generating plant on the bank of the Clinch river; surveying for the K-25 plant itself got underway later that month. When work on the power plant was begun, design of pumps for K-25 had not been settled; the power plant had to be designed to provide power at five separate frequencies. Despite this complication, it came online on Mach 1, 1944, only nine months after construction began.

Groves' original intent had been that K-25 would be capable of producing 90 % U-235, using, as Vincent Jones describes them, diffusion stages incorporating barriers in the form of annular bundles. However, detailed calculations indicated that available pumps and the tubular barriers would be most efficient up to an enrichment of 36.6 %, beyond which a different cell design and other pumps would be required. This prompted Groves, as soon as early 1943, to consider limiting K-25 to about 36 % enrichment, with its product to be fed to the Beta calutrons then being authorized. Groves formally announced the cutback at the August 13 MPC meeting, and asked Kellex to supply estimates on when 5, 15, 36.6, and 90 % plants might be expected to come into operation. (The 90 % figure would involve the later K-27 extension plant; see below.)

The K-25 complex was constructed in a 5,000-acre area in the northwest corner of the Clinton reservation, about 15 miles southwest of Oak Ridge (Fig. 4.14). To provide for a flat working area, almost 3 million cubic yards of earth were moved. Construction on the main building was begun on September 10, 1943, and the first concrete was poured on October 21. K-25's dedicated temporary construction camp was idyllically known as Happy Valley; its population would peak at about 17,000.

Fig. 5.26 An aerial view of the K-25 plant. *Source* http://commons.wikimedia.org/wiki/File:K-25_Aerial.jpg

The four-floor main process building, laid out in the shape of a giant letter U, was enormous (Fig. 5.26). Each side section was 2,450 feet long (just under a half-mile) by 450 feet wide; the total width exceeded 1,000 feet. Some 12,000 construction drawings would detail a facility with a total floor area of just over 5.5 million square feet, or about 120 acres—about 80 % of the floor area of the Pentagon. Some three million feet of pipes (over 500 miles) and a half-million valves would be involved, with the latter varying in size from 1/8 to 36 in. The construction force peaked at just over 19,600 in April, 1944. By June of that year, the plant was 37 % complete, and the estimated construction cost had escalated to $281 million. Kellex planned for a total of 2,892 diffusion stages. Ideally, as increasingly-enriched uranium accumulated toward the end-stages of the cascade (known as the "upper stages"), pumps and cells of steadily decreasing sizes could be used. As this would have involved complex and expensive manufacturing, Kellex decided on five sizes of pumps and four types of cells. The building itself comprised 54 sub-buildings linked together, and the cascade was divided into nine "sections," which, although they would normally operate as part of an overall cascade, could be operated individually. The fundamental operating entity was a "cell," a unit of six individual diffusion tanks.

The basement of the building housed lubricating, cooling, and electrical equipment. The diffusion tanks themselves resided on the ground floor, while the second aboveground floor served as a pipe gallery, and the top floor housed operating equipment. A central control room equipped with some 130,000 monitoring instruments was located on the top floor of the base of the U. Kellex divided its construction plan into five steps, designated as "Cases." Case I, to be completed by January 1, 1945, would see through to completion one cell for testing, then a building with a 54-stage pilot-plant, and finally enough functioning plant (402 stages) to produce 0.9 % U-235. Cases II, III, and IV would subsequently take the process to 5, 15, and 23 % enrichment by June 10, August 1, and September 13, respectively. Case V, to achieve 36 %, was to follow as soon as possible thereafter.

Cleanliness requirements during construction were practically at a surgical level. Workers wore special clothing and lintless gloves; even a thumbprint would leave enough moisture to be disastrous. Areas where process piping was being installed were equipped with pressurized ventilation and fed with filtered air. Some pieces of equipment required up to ten separate cleaning steps to remove all traces of dirt, grease, oxides, and moisture. Welding, which eventually involved 1,200 machines in simultaneous operation performing 14 specialized techniques, was done inside inflatable balloon enclosures. Since the entire plant would have to be constantly monitored for the presence of any leaks during operation, inert helium gas was fed into the piping system, and its presence sniffed for by sensitive portable mass spectrometers developed by Alfred Nier. Hundreds of Nier's devices were manufactured by General Electric and deployed throughout the K-25 plant; Nier was also involved in developing a system of over 50 fixed devices used to monitor the flow of various chemicals at locations throughout the building. These devices reported data back to the central control room, from where a single individual could monitor the entire plant. Another challenge in building K-25 was that projected nickel requirements for piping exceeded the entire world production of that metal. Again drawing on his knowledge of industrial firms, Groves contracted with Bart Laboratories in New Jersey, which specialized in electroplating oddly-shaped objects. Bart engineers were able to develop a method of electroplating the insides of pipes by using the pipe itself as the electroplating tank; rotating the pipe as current was passed through molten nickel ensured a uniform deposit of metal.

Progress with construction and operations at K-25 were detailed in monthly reports from Nichols to Groves. On April 17, 1944, the first six-stage cell was operated briefly as part of a preliminary mechanical test. By May, barrier of sufficient quality was beginning to become available; quantity production began in June. By August, operators could begin training at the 54-stage pilot plant located at the base of the U, using nitrogen in lieu of UF_6. On September 22, the first four diffusion tanks were received from Chrysler, but two were returned for tests of the effects of railroad handling. By November 9, the first dozen tanks were installed. A month later, Chrysler had shipped 324 tanks, of which about 200 were installed. By the end of 1944, the plant was 65 % complete, and 60 of the 402 stages of Case I were ready to be turned over to Carbide operators. By early January, 1945, all tanks necessary for Case I had been received, and Chrysler was producing 65–70 per week; by the end of the month the total number shipped would near 800.

After a period of leak testing and instrument calibration, the first process gas was introduced into the system on January 20, 1945. On March 10, Nichols reported that 102 of the 402 stages in Case I were in "direct recycle" operation, and that almost 1,100 tanks had been received. By March 12, two more buildings were connected to the system, and on the 24th, all of Case I went on-line. By early April just over half of the total 2,892 tanks had been received, and Cases I and II were producing 1.1 %-enriched U-235, which signaled that the facility could begin receiving its first slightly-enriched feed from the S-50 thermal diffusion plant. This occurred on April 28, by which time over 1,500 tanks were installed or ready for installation. By early June, all tanks had been shipped, nearly 1,500 were in operation, and K-25 was feeding 7 %-enriched product to Beta calutrons. On August 7, the day after the Hiroshima bombing, Nichols reported over 2,200 stages in operation. His report for August operations, dated September 6, indicted that all 2,892 stages were in operation by August 15, the day after the Japanese surrender. When the entire plant was operating, enrichment increased to 23 %.

Ultimately, K-25's product was not limited to the 36 % enrichment described above. In early 1945, Kellex developed plans for a 540-stage "extension" plant, which came to be known as K-27. By mixing waste output from the main K-25 cascade with natural uranium, K-27 produced a slightly enriched product which could be fed to the upper stages of K-25, increasing both its production and enrichment. Groves authorized construction of K-27 on March 31, 1945; it entered full operation in February, 1946, by which time all enrichment operations were being conducted by gaseous diffusion.

By any definition, K-25 was an outstanding engineering accomplishment. While the plant really came into its own only after the close of the war, Groves' gamble bequeathed America a means of enriching uranium that would operate flawlessly for years thereafter.

5.5 S-50: The Thermal Diffusion Program

The Manhattan Project's liquid thermal diffusion program has tended to be regarded with somewhat of a second-team status when compared to its much more gargantuan electromagnetic and gaseous-diffusion cousins. The S-50 plant was erected hastily, operated for a short time, and enriched uranium by only a small degree (from 0.72 to 0.86 % U-235), but its contribution was vital in giving the trouble-plagued electromagnetic separators a head-start on their efforts. Hewlett and Anderson have described S-50 as Groves' "last card" in his suite of options for securing fissile material.

The prime mover behind the thermal diffusion method was Ernest Lawrence's graduate student, Philip Abelson, who, among others, confirmed the discovery of fission at Berkeley in early 1939. Abelson formally received his Ph.D. in May, 1939, just a few months after his fission-confirmation work. He remained in Berkeley over the summer to complete some work on X-rays emitted during radioactive decay, and in September moved to Washington, D.C., to take up a position that Merle Tuve offered him at the Carnegie Institution of Washington (CIW). In the spring of 1940, Abelson took a brief leave to return to Berkeley to complete the neptunium-discovery work with Edwin McMillan (Sect. 3.8), efforts which directly motivated Glenn Seaborg to search for plutonium. After returning to Washington, Abelson began to consider possible approaches to enriching uranium isotopes, and after reviewing the research literature decided to explore the liquid-thermal–diffusion (LTD) method. Historian Joseph-James Ahern has suggested that Abelson's interest in LTD was triggered by a visit by Ross Gunn in July, 1940, who showed Abelson a copy of a paper by Harold Urey. As related in Chap. 4, Gunn was a member of Lyman Briggs' Uranium Committee, and had appreciated very early on the potential of nuclear fission as a power source for naval vessels.

As alluded to in Chap. 4, there was considerable political wrangling between the Army and the Navy over the LTD method. As we shall see, its development was begun by the Navy, but it was later appropriated essentially wholesale by the Manhattan District. District documents include a brief summary of the Navy work, but, after tracing its history to late 1942, jump abruptly to the S-50 project proper in mid-1944. However, there are now available a number of sources that fill in the gaps in this history. John Abelson, Philip Abelson's nephew, published a biography of his uncle based upon some autobiographical notes left by the latter, and writer Peter Vogel has prepared transcriptions of a number of letters and reports on the development of the LTD method that date from 1940 to 1944. The most significant sources, however, are two NRL reports, both of which list Abelson as first author. The first, dated January 4, 1943, describes progress up to that time, at which point the NRL had a small LTD pilot plant running. The second is dated September 10, 1946, and covers in detail the engineering theory of thermal diffusion plants and the full history of the method between 1940 and 1945.

The fundamental principle on which liquid thermal diffusion is based is that if a fluid (gas or liquid) comprising two isotopes of an element is subjected to a thermal gradient, the lighter isotope will move toward the hotter region, while the heavier one moves toward the cooler region. As a consequence, fluid containing the lighter isotope will be of lower density and will rise by convection, while that containing the heavier isotope will fall. Competition between this thermal diffusion and the ordinary diffusion of its isotopes through each other will lead, after some hours or days, to equilibrium between the two processes. The theory of thermal diffusion was first developed by David Enskog in Sweden (1911), and Sydney Chapman in England (1916). Its experimental proof was established by Chapman and F.W. Dootson in 1917. In Germany, Klaus Clusius and Gerhard Dickel first used a "column" approach in 1938 by placing a hot wire along the central axis of a vertical tube, and achieved a small enrichment of neon isotopes. Soon thereafter, Arthur Bramley and Keith Brewer of the U.S. Department of Agriculture conceived the idea of using two concentric tubes at different temperatures. Abelson adopted the Bramley and Brewer approach, using steam to heat the inner tube and water to cool the outer one, while injecting the process fluid into a narrow annular space between them.

Figure 5.27 shows a sketch of thermal diffusion "process column." The time for the column to achieve equilibrium depends upon the difference in temperature between the two tubes, their annular separation, and their lengths. The ultimate

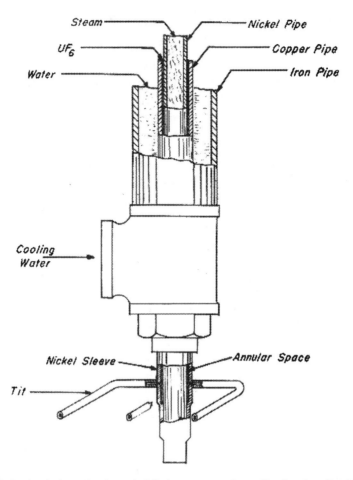

Fig. 5.27 Sectional view of a thermal diffusion process column. Uranium hexafluoride (UF_6) consisting of a mixture of light (U-235) and heavy (U-238) isotopes is driven into the narrow (0.25 mm) *annular space* between the *nickel* and *copper pipes*; the nickel pipes were 1.25 in. outside diameter. The desired lighter-isotope material is harvested from the *top* of the column. At the *top* and *bottom* of each tube, three small projecting "tits" provided access to the *annular space* for supply and withdrawal of material. From Reed (2011)

important characteristic of such a column is its so-called separation factor, which specifies its enrichment capability. For example, if a column has a separation factor of 1.2 (which was the case for the S-50 columns) and natural uranium is used, then, after processing, the percentage of U-235 will be 0.720 % (1.2) = 0.864 %.

Abelson's 1946 report indicates that his first column experiments were carried out at the CIW in July 1940; his goal was to repeat the German work by exploring diffusion of solutions of various potassium salts. Unfortunately, his attempt to use a solution of uranium salts produced what he called "an insoluble mess" at the bottom of the column. Merle Tuve became concerned that Abelson's experiments would produce radioactive contaminants, and began to look for another location for them. Tuve was a member of Briggs' Uranium Committee, and Briggs generously made space available to Abelson at the National Bureau of Standards (NBS). Abelson moved his experiments to the NBS in October, 1940, by which time the NRL had entered into a contract with the CIW to support the work; Abelson often advised Briggs on uranium matters during his subsequent nine-month stay at the Bureau. The NRL furnished Abelson's equipment, the CIW paid his salary, and the NBS provided laboratory space and an assistant chemist. Abelson's first order of business was to search for a suitable uranium compound to try in his columns, and he soon determined that one that might work was uranium hexafluoride, UF_6, commonly called "hex." Since, however, only a few grams of hex had ever been produced, he first had to develop a method for preparing it in kilogram quantities; he eventually obtained a patent for the process.

Between July 1, 1940, and June 1, 1941, Abelson constructed 11 columns of lengths between 2 and 12 feet, diameter 1.5 in., and annular separations between 0.5 and 2 mm. Experiments with water solutions of potassium salts showed that the equilibrium time and separation factor depended sensitively on the annular separation. A run with UF_6 in a 12-foot column in April 1941 yielded a small enrichment, but the measured value was only roughly equal to the probable error of measurement. On June 1, 1941, Abelson formally became an employee of the NRL, where a decision had been made to pursue study of LTD using 36-foot columns. These first NRL columns were collectively called the "experimental plant," to distinguish them from a later "pilot plant." Abelson achieved enrichment of chlorine isotopes with his first NRL column, but in November of that year it was ruined by decomposition products of carbon tetrachloride.

As related in Chap. 4, Arthur Compton's Committee on Atomic Fission was active during 1941. In an Appendix to the Committee's November 6, 1941, report, Robert Mulliken analyzed the feasibility and expected costs of various isotope-enrichment methods. He mentioned the LTD method in only a single brief paragraph, but did point out that "trials made recently with a solution using this method in an ingenious laboratory apparatus showed an astonishing rate and degree of separation of a dissolved salt from the water."

Between January and September of 1942, Abelson constructed five more experimental columns at the NRL using a hot-tube temperature of 286 °C. These were built with annular spacings of 0.53 mm, 0.65 mm, 0.38 mm, 0.2 mm, and 0.14 mm, and yielded separation factors of 2 % (January 1942), 1.4 % (March 1), 9.6 % (June 22), 21 % (July), and 12.6 % (September). Abelson regarded the 9.6 % result of June 22 as the first indisputably successful application of the method with uranium. Particularly encouraging was that the "pseudo-equilibrium time," the time for the column to produce a separation of one-half of the equilibrium value, was only eight hours. The optimal annular spacing appeared to be around 0.2 mm; a spacing of 0.25 mm would be used in the S-50 units. In July, the Navy, spurred by the success of the 36-foot columns, authorized the construction of a pilot plant with fourteen 48-foot columns with annular spacings of 0.25 mm to be built at the Anacostia Naval Station in Washington. However, as the NRL group gained experience through

1942, their fortunes within the formal Project administration were declining. According to Hewlett and Anderson, President Roosevelt made it clear to Vannevar Bush around March, 1942, that the Navy was to be excluded from S-1 affairs. Bush had evidently had a bumpy relationship with the Navy. Admiral Harold Bowen, who had been on the original Uranium Committee and was Director of the Navy's Bureau of Engineering, had criticized the OSRD for supplanting military-service laboratories and thus diverting funding away from the NRL. Admiral Alexander Van Keuren, who became Director of the NRL in 1942, was outraged by the Army's expenditure of what he described as "astronomical sums" of money on the uranium project.

Efforts to shut the Navy out of the work of the S-1 Committee were not entirely successful. In a letter to Conant on July 27, 1942, Harold Urey brought up Abelson's experiments, remarking that "This work has not been correlated with the other work of the Committee, for reasons that I do not understand, but efforts should be made ... to be sure that the work of that laboratory [NRL] ties in with the general purpose of this committee." Bush asked Briggs to get more information from Ross Gunn. In September, Briggs reported that Abelson was experimenting with 36-foot columns, and estimated that seven such columns in series would produce a doubling of the U-235 percentage. The catch was that the equilibrium time for such an arrangement, which was not specified, would be impracticably long. As related in Chap. 4, General Groves visited the Anacostia facility four days after his appointment as head of the Manhattan District, but was not favorably impressed.

By November 15, the Anacostia pilot plant was essentially complete, and by December 1 (the day before CP-1 went critical), five columns had been charged with material. The timing was propitious, as in late November the S-1 Executive Committee again reassessed enrichment methods, and decided to include the work at the NRL in its review despite its being officially orphaned. Consequently, Groves, Warren Lewis, and three DuPont employees visited the Anacostia plant on December 10, which for two full weeks between December 3 and 17 ran continuously with no shutdowns and a minimum of human intervention. On he 12th, Lewis wrote to Conant that the NRL work "is certainly of such interest that the development work ought to be continued intensively." He went on to report that the NRL workers expressed a desire for help by suitable experts, and told them that he would do anything he could to make "such men available through the NDRC." Conant replied on December 14, indicating that he would see if anything could be "done along these lines."

To work around the presidential injunction to exclude the Navy, Vannevar Bush wrote to Rear Admiral William Purnell (later a member of the Military Policy Committee) on December 31 to express "the hope that the work of the Naval Research Laboratory can be expedited so that a comparison can be made with other processes, and that ... the S-1 Executive Committee will do all it can to help." Noting that the Lewis committee felt that the NRL needed further facilities and manpower, Bush declared that "I would feel much gratified if you found it possible in some way to aid the [NRL]," and added that "Dr. Briggs has already undertaken to assure that any information that we have that can be of service to NRL … is made available to them." Purnell sent Abelson's reports to Conant, who had them reviewed by Briggs, Urey, and Eger Murphree, which group he also asked to visit the Anacostia facility. Accompanied by Lewis and chemical engineer William I. Thompson of Standard Oil, they did so, and submitted a report on January 23, 1943. Their assessment was that the NRL had made excellent progress, but they had concerns over a lack of solid production data: no appreciable amount of material had as yet been withdrawn from the columns. Thompson wrote an appendix to the report in which he analyzed possible large-scale plants of various configurations. The NRL group envisioned as most realistic a plant of 21,800 columns of length 36 feet, which would produce one kilogram per day of 90 % U-235. Individual columns would have a separation factor of 1,307, and their equilibrium time would be 625 days. Construction and operation costs for 625 days were estimated at some $72 million. As with the K-25 facility, an important requirement was that the heated inner tubes of the columns would have to be made of nickel. But even with appropriate strategic-materials priorities, product could not be expected until some 38 months following a decision to proceed, which would mean some time in early 1946.

On January 25, Murphree wrote to Briggs to emphasize the possibility that the thermal diffusion process could serve as an alternative to the initial stages of the K-25 plant. Briggs forwarded this idea to Conant, who, on the 27th, recommended that the NRL group should obtain more data and that an engineering group should study the process. Groves forwarded the documents to another review committee consisting of Lewis and several DuPont executives, including Crawford Greenewalt. They did not concur that thermal diffusion should become a substitute for gaseous diffusion, but did recommend continued research and preliminary engineering studies. The S-1 Executive Committee confirmed this conclusion on February 10. On the 19th, Murphree and Urey proposed a program of experiments which would include testing the reproducibility of results for different tubes, and drawing samples in order to quantify the approach to equilibrium. Briggs sent a copy of the proposal to Conant, and on 23rd followed up by suggesting to Conant that the S-1 Committee hoped that he would transmit the proposal to the Director of the NRL. Conant relayed this request to Groves the next day, who took no immediate action. That Groves was reluctant to pursue thermal diffusion at that time may have been due to having his hands full with getting construction of the Y-12 and K-25 plants underway, and with finding a site for the Los Alamos Laboratory. Also, his thinking at the time was directed to enrichment methods that would turn natural uranium into bomb-grade product in essentially one step as quickly as possible; the idea of using different enrichment methods in tandem had not yet emerged.

By the time Abelson, Gunn, and Van Keuren prepared their January 4, 1943 report, nine columns had been constructed at the Anacostia facility. Six were already operating, some for up to 500 hours. The earlier 36-foot "experimental" columns had been dismantled and checked for signs of corrosion; none were found. Between February and July, the NRL group constructed 18 columns, which operated for a cumulative total of 1,000 days. By September, they had produced some 236 pounds of slightly enriched hex, which they sent to the Metallurgical Laboratory in Chicago.

Groves did, however, keep himself informed of progress at Anacostia. On July 10, 1943, he wrote Conant that "progress at the Naval Research Laboratory ... has reached a point where it will be desirable to have this situation reviewed by the S-1 committee," and asked Conant "to take charge of this review and render a report." Conant notified Admiral Purnell that he proposed to appoint a committee consisting of Lewis, Urey, Murphree, and Briggs to again review the NRL work, expressing his hope that the NRL "would not regard such a visitation as an intrusion but rather as one more indication of the desire of the S-1 Committee to be of any assistance ... to the group which is doing such interesting and excellent work." On September 8, the committee conveyed to Conant the same concerns regarding cost, steam requirements, and long equilibrium times as they had in January, but did favor the S-1 Committee and the Manhattan District supporting work on improving the efficiency of the process. Apparently, such support never materialized.

Abelson and his group pressed on, proposing the development of a larger pilot plant or small production plant for the explicit purpose of "providing insurance against the complete failure of the Manhattan Project." Such a plant would require far more steam at higher pressures than was available at the Anacostia station, so they undertook a survey of other naval establishments. This quickly focused in on the Naval Boiler and Turbine Laboratory at the Philadelphia Naval Yard. Admiral Van Keuren, Abelson, and Gunn visited the site on July 24, 1943, and on November 17 formal orders were signed to authorize construction of a 300-column plant. They decided to first proceed with a 100-column installation (strictly, a "rack" of 102 columns) on the rationale that such a basic unit could be duplicated as desired if expansion was warranted. Construction of the Philadelphia plant began about January 1, 1944, with completion scheduled for July. Its 48-foot columns were to be operated as a cascade of seven stages, which was expected to deliver about 100 g of product per day at a concentration of 6 % U-235. The inner nickel tubes of the columns were formed from four 12-foot columns welded together, with nickel spacer buttons spot-welded to the tubes at 90 degree intervals at 6-in. spacings. Hung from steel racks, the tubes were heated by introducing condensing steam at the tops of their interiors; condensate was removed from the bottom for recirculation. The outer copper tubes were cooled by water flowing upward between them and external 4-in. iron tubes. When operating at a hot-wall temperature of 286 °C, about 1.6 kg of material resided within a single 48-foot column at any time. The power consumption for producing the steam was substantial: about 11.6 MW for one 102-column rack.

The circumstances of the resurrection of Manhattan District interest in thermal diffusion had an almost comedic flavor. In Abelson's telling, it began when Briggs obtained Gregory Breit as a new advisor on nuclear matters. Breit evidently knew that a high-ranking naval officer had been assigned to work at Los Alamos. One day in early 1944, Abelson received instructions to prepare a brief summary of the

Fig. 5.28 *Left* to *Right* Commander William Parsons, Rear Admiral William R. Purnell, and Brigadier General Thomas Farrell on Tinian island, August, 1945. *Source* http://commons. wikimedia.org/wiki/File:080125-f-3927s-040.jpg

NRL work and to appear at 8 p.m. on the balcony of the Warner Theatre in Washington, where he would encounter a naval officer who would whisper a code word. That officer was William S. Parsons, who was in charge of ordnance engineering for the uranium bomb at Los Alamos (Fig. 5.28; Chap. 7). In another version of the story, Hewlett and Anderson have it that Parsons visited the Philadelphia Navy Yard in the spring of 1944, and "discovered" that Abelson was building a thermal diffusion plant. Richard Rhodes has depicted the situation as more of a conspiracy between Abelson, Oppenheimer, and Parsons, with Abelson first making an effort to get information through to Los Alamos, and Oppenheimer and Parsons protecting the Navy by concocting the cover story that Parsons happened to learn of the NRL work on a visit to Philadelphia.

However covered, Oppenheimer wrote to Conant on March 4 to indicate that it seemed probable that some of the isotope-separation work being carried out at the NRL might be relevant to the purification of plutonium, and asked that Abelson's reports be sent to him. Conant cleared the request with Admiral Purnell, commenting "that the chances that they will find anything of use is slight, but I hesitate to turn down the request from that hard-pressed area." Conant forwarded the reports to Oppenheimer, who formally alerted Groves on April 28. Oppenheimer indicated that if the 100-column NRL plant were operated in parallel, it could theoretically produce 12 kg of material per day enriched to 1 % U-235, and that the LTD method might increase the electromagnetic-plant production by some 30–40 percent. Groves waited until May 31 before appointing Lewis, Murphree, and Richard Tolman to investigate the situation once again. The group visited the Philadelphia facility on May 31 and June 1, and turned in their report to Groves on June 3, just three days before the D-Day invasion of Europe. Work on the

100-column plant was well-advanced; it was expected to begin operation about July 15. The committee considered Oppenheimer's estimate of 12 kg per day of 1 % U-235 to be optimistic, but felt that 10 kg per day of 0.95 % U-235 was feasible.

Groves now moved with his typical dispatch. On June 5, he sent Lewis and Conant to Manhattan District Headquarters at Oak Ridge to confer with Colonel Nichols to discuss the feasibility of constructing a thermal diffusion plant there. They decided that the 238-MW powerhouse being constructed for the K-25 plant could provide sufficient steam for such a plant, at least until K-25 went into operation. At 11.6 MW per rack of columns, 238 MW could provide power for between 20 and 21 racks; 21 were built. Groves decided to proceed with construction of the S-50 plant on June 24. On June 26, Groves, Tolman, and Lieutenant Colonel Mark Fox, who had been appointed chief of the thermal diffusion project at Oak Ridge, visited the Philadelphia installation to inspect it and collect blueprints. The next day, Groves contracted with the H.K. Ferguson Company of Cleveland, Ohio, to construct the plant in 90 days. A second contract with Ferguson would follow for its operation. Groves initially demanded that the plant be in full operation in four months, with its first production unit operating 75 days after the beginning of construction. In a July 4 letter to Fox, he revised the schedule to demand that all units be in operation in 90 days, that is, by September 30. The 75-day requirement would be met, but not the 90-day one.

The main S-50 process building (Fig. 5.29) was 522 feet long, 82 feet wide, and 75 feet high. The most pressing initial problem for the project was to find contractors to mass-produce the large numbers of columns; twenty-one manufacturers

Fig. 5.29 The S-50 facility. The main process plant is the long, *dark* building to the *left of center*. The K-25 powerhouse (*three smokestacks*) is to its *right*, and a tank farm for supplying oil for the "new boiler plant" is to the *left*. The new boiler plant itself is between the main process building and the river. *Source* http://commons.wikimedia.org/wiki/File:S50plant.jpg

were consulted before the Grinnell Company of Providence, Rhode Island, and the Mehring and Hanson Company of Washington agreed to attempt the job. The outside diameter of the inner nickel tubes had to be maintained to tolerances of ±0.0003 in., and the clearance between the nickel and copper tubes to ±0.002 in. Since neither nickel nor copper tubes could be drawn in 48-foot lengths, shorter tubes had to be welded together. The first order for columns was placed with Mehring and Hanson on July 5.

The S-50 plant was designed as twenty-one copies of the 102-column Philadelphia installation, with the resulting 2,142 columns operated in parallel to provide a large quantity of slightly enriched U-235 as feed for the Y-12 and K-25 plants. Each rack was arranged as two rows of 51 columns, which for purposes of steam supply were divided into three groups of seven "sections." Columns could be isolated from each other for maintenance or product removal by "freeze-off" water directed through intercolumn connectors. Erected adjacent to the K-25 powerhouse on the bank of the Clinch River, the pace of construction of the S-50 plant was phenomenal. Ground was broken in early July, and foundations laid less than three weeks later. Installation of process equipment began on August 17, and the first columns were received from Grinnell on August 27. By September 16, sixty-nine days after the start of construction, 320 columns were on hand, one-third of the plant was complete, and preliminary operation of the first rack had begun. Of the twenty-one racks, number 21 was completed first, and was used for training operators. The first process material was introduced into that rack on October 18, and the first product was drawn off on October 30. Operation of S-50 was carried out on a cost plus $11,000 per month fee basis by the Fercleve Corporation (from *Fer*guson of *Cleve*land), a wholly-owned subsidiary of the Ferguson Company that was established to avoid the possibility of labor trouble when employing non-union laborers; Ferguson normally operated on a unionized basis.

Crucial to the operation of S-50 was the steam-supply system; the process required over 100,000 pounds of steam per hour for each rack. When power demand for the K-25 plant began to increase in early 1945, plans were made to construct a new boiler plant to service S-50. Construction began with site clearing on March 16; the boilers arrived on April 26, and steady operation was underway by July 13. Ironically, the plant was completed on August 15, the day after the Japanese surrender was announced.

Production from S-50 began in October 1944, with 10.5 pounds of output. During routine operation, enriched product was removed at two-to-four-hour intervals from the tops of columns by "milking" equipment. By mid-January, 1945, large-scale production was underway, with ten of the 21 racks producing, and construction of all racks nearly complete. By March 15, all 21 racks were yielding product, and in April, S-50 output began to go directly to the K-25 plant. Cumulative production amounted to nearly 45,000 pounds by the end of July, and just over 56,500 pounds by the end of September (Fig. 5.30). If all of this was of 0.86 % U-235 concentration, this would represent some 220 kg of U-235, enough for almost four Hiroshima *Little Boy* bombs. This productivity was less than the 10 kg per day of 90 % U-235 that the Lewis committee had estimated in June,

Fig. 5.30 S-50 production: monthly (*dashed line, right scale*); cumulative (*solid line, left scale*). *Source* MDH, Book VI—liquid thermal diffusion (S-50) project, *top secret supplement*

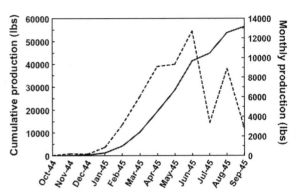

1944, because the S-50 columns were operated in parallel, not series. S-50's mission was to produce a large quantity of slightly-enriched material, as opposed to a small quantity of greatly-enriched material.

The cost of the S-50 plant was about $20 million, plus about $2 million in research costs borne by the Navy. This represented a mere 1 % of the total cost of the entire Manhattan Project, and less than one-twentieth the cost of either Y-12 or K-25. Nevertheless, Kenneth Nichols estimated that S-50 contributed to shortening the war by about nine days.

5.6 The Postwar Era at Clinton

By the end of the war, the continuous-feed gaseous diffusion process had proven itself more efficient at enriching uranium than the electromagnetic batch-feed method. Shutdown of Y-12's Alpha units began on September 4, 1945, and the last tank ceased operating on September 22. After the war, many calutrons were refitted with copper windings and operated to separate isotopes, which, after neutron bombardment in the X-10 pile, could be used as radioactive tracers for medical-imaging and cancer-treatment applications, an interesting example of wartime technology turned to humane use. In December, 1946, all but one Beta track was shut down, although the Alpha and Beta units in the pilot-plant building would be kept operating as part of a program to separate stable isotopes.

Beginning in 1958, Beta calutrons in building 9204-3 were used to produce medical isotopes. They continued in that role until they could no longer compete economically against overseas suppliers, and were finally shut down in 1998. The last of the Treasury silver was returned to West Point on June 1, 1970, just a few weeks before General Groves' death on July 13 of that year. Y-12 still operates as a Department of Energy "National Security Complex" under contract with the Babcock and Wilcox Company; an ultra-secure highly-enriched uranium materials facility was recently dedicated there as part of the site's mandate to retrieve and store nuclear materials. An April, 2010, federal Nuclear Posture Review advocated that a new uranium processing facility should be built at Y-12 to come on-line in 2021. According to a 2011 report by the International Panel on Fissile Materials, the United States produced a total of some 610 t of highly-enriched uranium between 1945 and 1995, about one metric ton of which was produced by the Y-12 plant.

The K-25 plant operated successfully for twenty years until it was shut down in 1964. The gaseous diffusion process proved so sound that, in the 1950s, three other diffusion plants were built at the Clinton site (K-29, K-31, and K-33), plus others in Kentucky and Ohio. The plant in Kentucky is still operating, although enrichment by more efficient centrifugation methods has been the preferred approach since the 1980s. K-25 went back into service in 1969 to produce low-enriched uranium for private customers, an arrangement that continued until it was shut down for good in 1985; all gaseous diffusion operations at the Clinton site ceased in 1987. Despite efforts to preserve a part of K-25 as a component of a Manhattan Project National Historic Park (Sect. 9.5), the last remnants of the structure were demolished on January 23, 2013. K-29 was demolished in 2006; K-31 and K-33 still stand, but are inactive.

The end of the war also brought operations at the S-50 plant to an abrupt halt. On September 4, orders were issued to terminate all operations at the earliest possible date and to place the plant in standby mode for possible future use. Columns were drained, washed, dried, capped, and employees given two weeks notice. In September, 1946, the decision was made to dispose of the S-50 plant, with its useful parts to be declared surplus and returned to the NRL or to be disposed of at sea. According to an Oak Ridge Associated Universities Dose Reconstruction Project published in 2006, disassembly of S-50 equipment was carried out in the late 1940s, with materials eventually being either salvaged or buried. Today, nothing remains of S-50 but for the concrete pad that it rested on.

The Clinton Laboratories eventually became Oak Ridge National Laboratory, which now has a staff of about 4,400 and is one of America's premier research facilities for neutron physics. In its postwar years, the X-10 pile would be used to synthesize medical isotopes which were distributed both domestically and internationally. X-10 was finally shut down in 1963 after operating for 20 years, and in 1966 was designated a National Historic Landmark. It is now accessible for public tours. At the University of Chicago, Stagg Field was torn down in 1957; a sculpture now marks the location where CP-1 stood. Both CP-2 and CP-3 remained in operation at Argonne (now Argonne National Laboratory) until 1954. After they were dismantled, CP-3's aluminum tank was filled with concrete, and contaminated hardware was dumped into the space between the tank and the biological shield, which was then also filled with concrete. The tank was tumbled into a 40-foot deep pit, covered with rubble, and capped with dirt. The area is now a public forest preserve, with a granite marker indicating the burial location (Fig. 5.31).

In addition to the Oak Ridge National Laboratory, Clinton's most enduring legacy is the town of Oak Ridge itself. At the time of the transfer of Manhattan District assets to the newly-formed Atomic Energy Commission at the beginning

Fig. 5.31 CP-2/CP-3 burial
marker. *Source* http://
commons.wikimedia.org/
wiki/File:Marker_
at_Site_A.jpg

of 1947 (Chap. 9), the town's population had declined to about 42,000, and
employment to about 29,000. The AEC operated and managed Oak Ridge through
an Oak Ridge Operations Office, but over the following years the city gradually
transitioned to normal municipal operations. To national fanfare, the town became
open to public access in March, 1949. The Atomic Energy Community Act of
1955 provided a legal basis for the establishment of local self-governance in Oak
Ridge, Los Alamos, and Hanford, and for disposal of federally-owned properties at
those sites; at Oak Ridge alone this would involve the appraisal and sale of nearly
6,500 pieces of real estate. A priority system was established for sale of homes to
residents, with the first sale occurring in September, 1956. By a 5,500-to-400 vote
in May, 1959, residents overwhelmingly approved incorporating the city. Following establishment of a city council and hiring of staff, the AEC turned over
operation of most municipal services to the new city on June 1,1960. Visitors to
Oak Ridge will find it an attractive, vibrant town with all of the amenities of
modern American city life.

The story of the Clinton Engineer Works during the Manhattan Project was one
of incremental improvements and problem-solving by a multitude of people who
together produced remarkable accomplishments in just 30 months. Without their
work, the Hiroshima *Little Boy* bomb would simply not have been, and the
Nagasaki bomb would have been seriously delayed. Clinton was central to the
success of the Manhattan Project.

Exercises

5.1. Consider an air-cooled reactor operating at a power output of 1 MW. The
    density and specific heat of air depend on temperature, but take as rough
    numbers 1 kg/m^3 and 1,000 J/(kg-K). If the airflow is 30,000 cubic feet per
    minute, what will be the temperature increase of the air? Recall $Q = mc\Delta T$
    from basic thermodynamics. 1 foot = 0.3048 m. [Ans: ~70 K].

5.2. Uranium fuel slugs for the X-10 reactor were in the shape of cylinders of diameter 1.1 in. and length 4.1 in. (1 in. = 2.54 cm). If each 24-foot long fuel channel was completely filled with slugs and the reactor contained 1,248 channels, what was the mass of a full fuel load? The density of uranium is 18.95 gr/cm^3. [Ans: About 105,700 kg, or 116 U. S. tons]

5.3. The density of graphite is about 2.15 gr/cm^3. If the graphite bricks in the CP-1 reactor were 16.5 in. long by 4.125 in. square in cross–section, what would be the total mass of the claimed ∼ 40,000 bricks in the entire assembly? Does your result accord roughly with the total mass quoted in the text? [Ans: About 395,700 kg, or 436 U. S. tons]

5.4. Suppose that you have available an enrichment process that increases the abundance of U-235 by 10 % at each stage, that is, by a factor of 1.1. If you begin with natural uranium (235 abundance fraction 0.0072), how many stages will you require in series to isolate bomb-grade U-235 of abundance fraction 0.9? [Ans: 51]

5.5. Verify the claim in the text that 1.6 billion kWh of energy is equivalent to about 100 times the energy released by the 13-kiloton *Little Boy* bomb. Explosion of one kiloton of TNT liberates 4.2×10^{12} J of energy.

5.6. Process columns in the S-50 thermal diffusion plant were 48 feet long. The outer diameter of the inner nickel pipe was 1.25 in., and the width of the annular space for the process fluid was 0.25 mm. If an operating column contained 1.6 kg of uranium hexafluoride, estimate the average density of that material during operation. [Ans: about 4.4 gr/cm^3]

5.7. For a gas of atoms or molecules at pressure P and absolute (Kelvin) temperature T, an approximate expression for the mean free path λ is

$$\lambda \sim \frac{kT}{\sqrt{2}\,\pi P d^2},$$

where k is Boltzmann's constant (1.38×10^{-23} J/K) and d is the effective diameter of the particles. Evaluate λ for standard atmospheric pressure $P = 101,300$ Pa, $T = 300$ K (room temperature), and $d = 3$ Å (O_2 molecules). How does your result compare to the ∼ 1,000 Å quoted in Sect. 5.4? [Ans: ∼ 1,022 Å]

Further Reading

Books, Journal Articles, and Reports

P. Abelson, R. Gunn, A. H. Van Keuren, *Progress report on liquid thermal diffusion research.* NRL report 0-1977, Jan 4, (1943)

P.H. Abelson, N. Rosen, J.I. Hoover, *Liquid thermal diffusion.* NRL report TID-5229, Sept 10, (1946). http://www.osti.gov/energycitations/product.biblio.jsp?osti_id=4311423

J. Abelson, P.H. Abelson, *Uncle Phil and the Atomic Bomb* (Roberts and Company, Greenwood Village, Colorado, 2008)

J.-J Ahern, We had the hose turned on us: Ross Gunn and the Naval Research Laboratory's early research into nuclear propulsion, 1939–1946. Hist. Stud. Phys. Biol. Sci. **33**, Part 2, 217–236 (2003)

H.L. Anderson, Fermi, Szilard and Trinity. Bull. At Sci. **30**(8), 40–47 (1974)

E. Booth, J. Dunning, W.L. Laurence, A.O. Nier, W. Zinn, *The Beginnings of the Nuclear Age* (Newcomen Society in North America, New York, 1969)

A. Bramley, A.K. Brewer, A thermal method for the separation of isotopes. J. Chem. Phys. **7**, 553–554 (1939)

R.P. Carlisle, J.M. Zenzen, *Supplying the Nuclear Arsenal: American Production Reactors, 1942–1992* (Johns Hopkins University Press, Baltimore, 1996)

S. Chapman, On the law of distribution of molecular velocities, and on the theory of viscosity and thermal conduction, in a non-uniform simple monatomic gas. Philos. Trans. R. Soc. Lond. **A216**, 279–348 (1916)

S. Chapman, F.W. Dootson, A note on thermal diffusion. The London, Edinburgh, and Dublin Philosophical Magazine and Journal of Science **33**, 248–253 (1917)

K. Clusius, G. Dickel, Neus Verfahren zur Gasentmischung und Isotopentrennung. Die Naturwissenschaften **26**, 546 (1938)

K.P. Cohen, S.K. Runcorn, H.E. Suess, H.G. Thode, Harold Clayton Urey, 29 April 1893—5 Jan 1981. Biographical Mem. Fellows Roy. Soc. **29**, 622–659 (1983)

A.L. Compere, W.L. Griffith, The U. S. Calutron Program for Uranium Enrichment: History, Technology, Operations, and Production. Oak Ridge National Laboratory report ORNL-5928 (1991)

A.H. Compton, *Atomic Quest* (Oxford University Press, New York, 1956)

D. Enskog, Über eine Verallgemeinerung der zweiten Maxwellschen Theorie der Gase. Physikalische Zeitschrift **12**, 56–60 (1911); Bemerkungen zu einer Fundamentalgleichung in der kinetischen Gastheorie, *ibid.*, 533–539

E. Fermi, Elementary theory of the chain-reacting pile. Science **105**(2751), 27–32 (1947)

E. Fermi, Experimental production of a divergent chain reaction. Am. J. Phys. **20**(9), 536–558 (1952)

L. Fine, J.A. Remington, *United States Army in World War II: The Technical Services—The Corps of Engineers: Construction in the United States* (Center of Military History, United States Army, Washington, 1989)

S. Groueff, Manhattan Project: The Untold Story of the Making of the Atomic Bomb. Little, Brown and Company, New York (1967). Reissued by Authors Guild Backprint.com (2000)

L.R. Groves, *Now It Can be Told: The Story of the Manhattan Project* (Da Capo Press, New York, 1983)

R.G. Hewlett, O.E., Jr. Anderson, *A History of the United States Atomic Energy Commission, Vol 1: The New World, 1939/1946* (Pennsylvania State University Press, University Park, PA, 1962)

L. Hoddeson, P.W. Henriksen, R.A. Meade, C. Westfall, *Critical Assembly: A Technical History of Los Alamos during the Oppenheimer Years* (Cambridge University Press, Cambridge, 1993)

V.C. Jones, *United States Army in World War II: Special Studies—Manhattan: The Army and the Atomic Bomb* (Center of Military History, United States Army, Washington, 1985)

R.L. Kathren, J.B. Gough, G.T. Benefiel, The Plutonium Story: The Journals of Professor Glenn T. Seaborg 1939–1946. Battelle Press, Columbus, Ohio (1994). An abbreviated version prepared by Seaborg is available at http://www.escholarship.org/uc/item/3hc273cb?display=all

P.C. Keith, The role of the process engineer in the atom bomb project. Chem. Eng. **53**, 112–122 (1964)

C.C. Kelly (ed.), *The Manhattan Project: The Birth of the Atomic Bomb in the Words of Its Creators, Eyewitnesses, and Historians* (Black Dog & Leventhal Press, New York, 2007)

C.C. Kelly, R.S. Norris, *A Guide to the Manhattan Project in Manhattan* (Atomic Heritage Foundation, Washington, 2012)

L.M. Libby, *The Uranium People* (Crane Russak, New York, 1979)

M. McBride, *55 Years That Changed History—A Manhattan Project Timeline, 1894 to 1949* (The Secret City Store, Oak Ridge, TN, 2010)

K.D. Nichols, *The Road to Trinity* (Morrow, New York, 1987)

A.O. Nier, Some reminiscences of mass spectroscopy and the Manhattan project. J. Chem. Educ. **66**(5), 385–388 (1989)

Norris, R. S.: Racing for the Bomb: General Leslie R. Groves, the Manhattan Project's Indispensable Man. Steerforth Press, South Royalton, VT (2002)

Oak Ridge Operations, U. S. Atomic Energy Commission: A City is Born: The History of Oak Ridge (1961). Reprinted by Oak Ridge Heritage and Preservation Association (2009)

W.E. Parkins, The uranium bomb, the calutron, and the space-charge problem. Phys. Today **58**(5), 45–51 (2005)

B.C. Reed, Liquid thermal diffusion during the Manhattan project. Phys. Perspect. **13**(2), 161–188 (2011)

B.C. Reed, From Treasury Vault to the Manhattan Project. Am. Sci. **99**(1), 40–47 (2011)

B.C. Reed, Bullion to B-fields: The silver program of the Manhattan Project. Michigan Academician **39**(3), 205–212 (2009)

R. Rhodes, *The Making of the Atomic Bomb* (Simon and Schuster, New York, 1986)

G.O. Robinson, *The Oak Ridge Story* (Southern Publishers, Kingsport, Tennessee, 1950)

E. Segrè, *Enrico Fermi Physicist* (University of Chicago Press, Chicago, 1970)

H.D. Smyth, *Atomic Energy for Military Purposes: The Official Report on the Development of the Atomic Bomb under the Auspices of the United States Government, 1940–1945* (Princeton University Press, Princeton, 1945)

Vogel, P.: The Last Wave from Port Chicago. Appendix A. Document transcriptions: The liquid thermal diffusion uranium isotope separation method (2001, 2009). http://www.petervogel.us/lastwave/chapterA.htm

A. Wattenberg, December 2, 1942: The event and the people. Bull. At. Sci. **38**(10), 22–32 (1982)

A. Wattenberg, The birth of the nuclear age. Phys. Today **46**(1), 44–51 (1993)

E.P. Wigner, *Symmetries and Reflections: Scientific Essays* (Ox Bow Press, Woodbridge, CT, 1979)

Wilcox, W. J.: The Role of Oak Ridge in the Manhattan Project. Privately published (2002)

W.J. Wilcox, *An Overview of the History of Y-12, 1942–1945*, 2nd edn. (The Secret City Store, Oak Ridge, TN, 2009)

A.L. Yergey, A.K. Yergey, Preparative scale mass spectrometry: A brief history of the Calutron. J. Am. Soc. Mass Spectrom. **8**(9), 943–953 (1997)

Websites and Web-based Documents

Argonne National Laboratory sites on Fermi and CP-1 witnesses: http://www.ne.anl.gov/About/legacy/unisci.shtml, http://www.ne.anl.gov/About/cp1-pioneers/index.html

Clinton Engineer Works history of diffusion operations: http://web.knoxnews.com/web/kns/bechteljacobs/role.html

Department of Energy: The First Reactor (1982): http://www.osti.gov/accomplishments/documents/fullText/ACC0044.pdf

Department of Energy: Radiological survey of CP-2/CP-3 burial area (1978): http://www.osti.gov/bridge/servlets/purl/6869820-y1102A/6869820.pdf

International Panel on Fissile Materials report on highly-enriched uranium and plutonium: http://fissilematerials.org/library/gfmr11.pdf

Kenneth Nichols on Clinton site codes: http://research.archives.gov/description/281585

Milwaukee Getaways Examiner article on CP-2/CP-3 burial area: http://www.examiner.com/article/the-first-nuclear-reactor

Oak Ridge Associated Universities dose reconstruction project: http://www.cdc.gov/niosh/ocas/pdfs/tbd/s50-r0.pdf

Oak Ridge National Laboratory report on X-10 graphite reactor: http://www.ornl.gov/info/reports/1957/3445605702068.pdf

Y-12 National security complex: http://www.y12.doe.gov/about/history/

Chapter 6
The Hanford Engineer Works

The Hanford Engineer Works represented even more of a gamble than its counterpart in Tennessee. The uranium enrichment facilities at Clinton were complex, plagued with difficulties, and subject to constant design changes, but at least involved processes that were in principle familiar to mechanical, electrical, and chemical engineers. At Hanford, the story was completely different. In 1943, there was no established nuclear industry, or even a discipline of nuclear engineering. Nobody had ever before designed or constructed a large-scale reactor, and there was no cadre of experienced operators ready to walk into a control room. Also, the dangers of this proposed new technology were immense. At Oak Ridge, an electrical short in a calutron or an explosion in a diffusion tank might set back production temporarily and endanger a small number of workers, but at Hanford an explosion in a reactor could potentially spread radioactive fission products over hundreds of square miles and endanger tens of thousands of lives. As General Groves wrote, the plutonium project was truly pioneering.

A few simple estimates will serve to give a sense of the magnitude of the potential radiological danger. As remarked in Chap. 5, a reactor fueled with natural uranium produces about 0.76 g of plutonium per day, per megawatt of operating power. To produce 10 kg of plutonium requires some 13,000 MW-days of trouble-free operation. If you desire to realize the 10 kg from 30 days of operation (let alone any time for processing the irradiated fuel), you will need about four reactors if each operates at 100 MW. At Hanford, three reactors were built, but each was designed to operate at 250 MW. This power level would theoretically yield about 190 g of plutonium per day per reactor, or about 18 days between 10 kg bombs once steady-state operation was achieved. The rate of fission-product generation from a 250 MW reactor is fantastic. At an energy release of 180 MeV per fission, 250 MW corresponds to about 8.7×10^{18} fissions per second. Each fission will give rise to two fission-product nuclei, most of which will have very short half lives. If we make the very rough assumption that the fission products decay at about the same rate as they are formed, we will have 1.7×10^{19} decays per second. As described in Chap. 2, the customary unit for rate of radioactivity is the Curie (Ci), where $1\ Ci = 3.7 \times 10^{10}$ decays per second. Our 250 MW corresponds to some 4.6×10^8 Curies. If a fuel slug remains in the reactor for

B. C. Reed, *The History and Science of the Manhattan Project*,
Undergraduate Lecture Notes in Physics, DOI: 10.1007/978-3-642-40297-5_6,
© Springer-Verlag Berlin Heidelberg 2014

100 days before being ejected for processing, then about 1 % of the fuel will be ejected on any given day, that is, some 4.6 *million* Curies worth of radioactive material will need to be handled safely *every day* from each reactor.

Many hundreds of different fission products with a wide spectrum of half-lives are produced in a reactor, so the forgoing is at best a very rough estimate. But the idea should be clear. Four million Curies corresponds to the radioactivity of 4,000 kg of pure radium. To complicate the issue further, only about one atom per 13,000 in a fuel slug will be transmuted into plutonium, and those transmuted atoms have to be extracted by first dissolving the slug in acid and then processing the resulting fluid with a complicated series of chemical reactions. The result is extracted plutonium plus a very daunting waste-disposal problem. It is no wonder that Groves insisted on a very isolated location for the Manhattan Project's reactors.

This Chapter describes the Hanford project. As with Chap. 5, this is done largely chronologically, beginning with considerations of contractor and site selection, and then working through pile design, construction, and operations. Section 6.6 briefly describes the postwar era at Hanford.

6.1 Contractor and Site Selection

The origins of the agreement of the DuPont corporation to design, build, and operate the plutonium production piles following General Groves' personal appeal to its Directors and the favorable report of the late-1942 Lewis review committee were described in Chap. 4. Following the Military Policy Committee meeting of December 10, 1942, which determined that production piles should be removed from the Clinton location, a site had to be procured for them.

On December 14, Colonel Nichols, Arthur Compton, and Lieutenant Colonel Franklin Matthias, Groves' Area Engineer for the plutonium project, traveled to Wilmington to discuss pile design and site selection with DuPont officials; Matthias had served as Groves' Deputy Manager of construction for the Pentagon project. Four helium-cooled 250 MW piles and two separation plants were planned for, with a goal of producing 600 g of plutonium per day. The piles, which would each require some 150 tons of uranium fuel and 350,000 cubic feet of helium coolant, were to be separated from each other by at least one mile, and the chemical separation plants from each other by four miles. Each pile was to be a self-contained unit, independent of the others in case of a disaster at any one of them. Laboratories would have to be at least eight miles from the separation plants, and a village for housing workers was to be at least 10 miles upwind from the nearest pile or separation plant; the village would eventually be located some 30 miles from the piles. To allow for the possibility of up to six piles, the site would require an area of about 15 by 15 miles.

Matthias reported back to Groves, who directed him to make inquiries as to locations where suitable electric power would be available. On December 16,

Matthias, Groves, and DuPont representatives met to draw up more specific criteria. Some 100,000 kW of power would have to be continuously available. While cooling the piles with circulating helium gas was the preferred design at that time, water-cooling was being given serious consideration, and Groves wanted to cover all bases: the site would require a water supply of 25,000 gallons per minute in case water-cooling was chosen (which it was). Level terrain with conditions suitable for heavy construction was desirable, with plenty of sand and gravel available for producing large quantities of concrete. Hanford would ultimately require over 780,000 cubic yards of concrete, enough for a 390-mile highway 20 feet wide by 6 in. thick. Overall, an area of close to 700 square miles was required, preferably in the form of a rectangle of about 24 by 28 miles which would completely enclose the 12 by 16-mile plant area. The setting should be remote, with no settlement of population greater than 1,000 within 20 miles. Consideration was given to locating the site within a 44 by 48-mile buffer area from which all residents would be removed, but this idea was eventually dropped.

Matthias and a group of DuPont representatives spent two weeks scouring the western United States in search of possible sites, looking at eleven altogether. Two sites in each of California and Washington looked promising. In California, these were near the Shasta Dam and the Hoover Dam on the California-Arizona border. In Washington, one site lay near the Grand Coulee Dam in the central-northeast part of the state, and the other, which had the advantage of access to the Bonneville Power Authority, was near the town of Hanford on the Columbia river in the south-central part of the state. Matthias reported back to Groves on December 31 (some sources say January 1) that the group was unanimously enthusiastic about the Hanford location. Groves inspected the site personally on January 16, 1943, and gave his approval. On February 9, Undersecretary of War Robert Patterson approved acquisition of more than 400,000 acres for the site of the Hanford Engineer Works (HEW). Hanford was the last site selected for the Manhattan Project.

The Hanford site comprised 670 square miles—about half the area of the entire state of Rhode Island—over a roughly circular area which extended 37 miles at its greatest north–south extent by 26 miles in maximum east–west breadth (Fig. 6.1). From the point of view of human habitation, the location was distinctly unappealing. Flat, semi-arid, and covered in grayish sand and gravel, the area could be swept by blinding sandstorms that lasted two or three days and which left everything coated in a layer of dust. Despite this, land acquisition proved to be a chronic bane for Groves and Matthias. About 88 % of the land was being used for grazing, 11 % was farmland, and less than 1 % was occupied by three small towns; Richland (population about 200), Hanford (about 100), and White Bluffs, which was about the same size as Richland. Acquisition was complicated by the presence of a number of interests: some 157,000 acres were owned by federal, state, or local governments; 225,000 by private individuals; 46,000 by railroads; and 6,000 by irrigation districts.

The first tract of land was acquired on March 10, but resistance soon arose on the parts of individuals and irrigation districts over what they thought to be low

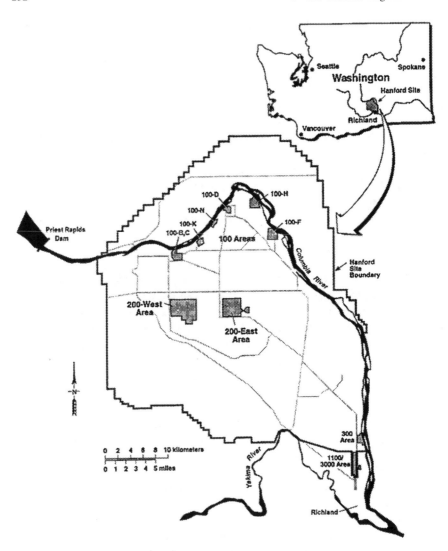

Fig. 6.1 The Hanford Engineer Works site. Piles were built at the 100-B, D, and F sites from west to east along the Columbia river. The 200-North site is not shown on this map; it was about 3 miles north of the 200-East site. The original village of Hanford was on the west bank of the Columbia, due east of the 200 area. *Source* HAER, Fig. 1

property valuations, inadequate advance-notice allowances (some residents were given as little as 48 hours to vacate), and insufficient compensation for crops. Rumors began circulating that the War Department was using the right of eminent domain to benefit DuPont. The issue reached a Military Policy Committee meeting on March 30, and then a cabinet meeting on June 17. At the latter, President Roosevelt, concerned with possible wartime food shortages, wondered if another site could be chosen. Groves had to explain to Henry Stimson that both DuPont and the Manhattan District had concluded that the Hanford site was the only one in the country where the work could be done. The acquisition process was not helped by faulty War Department appraisals. In the late spring of 1943, the Corps of Engineers Pacific District Real Estate Branch agreed to reappraise all tracts that had not yet been acquired. A number of cases went to trial, and settlements on over 1,200 tracts were averaging no more than seven cases per month until Groves requested more judges from the Department of Justice, and an end to the habit of juries inspecting the areas in question. Groves was particularly irritated with publicity in local newspapers, which was played up by Assistant Attorney General Norman Littell. Littell was responsible for prosecution of all cases of War Department condemnation procedures within the Lands Division of the Department of Justice, and had practiced law in Seattle before joining the Justice Department in 1940. The issue reached a head on November 18, 1943, when Attorney General Francis Biddle requested that Littell resign; Biddle and Littell had apparently been engaged in a long-standing feud between over administration of the Lands Division. Littell stalled until President Roosevelt removed him from office on November 26. Despite more expeditious proceedings, a number of landowners had to be evicted by court orders, and the acquisition program was still ongoing as of late 1946, when the official Manhattan District history was being prepared. By that time, $5 million had been spent on the acquisition program; Groves thought many of the settlements to be exorbitant.

6.2 Pile Design and Construction

Well before Groves took command of the Project or CP-1 had demonstrated the feasibility of a self-sustaining reaction, scientists and engineers at Arthur Compton's Metallurgical Laboratory were exploring possible configurations for production piles. The complexities were legion: design of reactors and separation plants, cooling and control systems, determining relevant chemical and metallurgical properties of uranium and plutonium, effects of reactor materials on the efficiency of the chain reaction, and ensuring human and environmental safety were but a few of the issues that occupied Met lab staff for months. In early 1942, one of Compton's first actions upon centralizing the pile program in Chicago was to establish an Engineering Council (later known as the Technical Council) to consider suggestions for pile design. As chief engineer, Compton chose Thomas V. Moore, a veteran of many years experience in the petroleum industry. A younger member of the group was John Wheeler of the Bohr and Wheeler fission theory. Initially, much of their effort focused on investigating helium-cooled, uranium-graphite configurations.

On June 18, 1942 (almost six months before CP-1 first achieved criticality), the Council gathered to consider designs that might be suitable for production-scale piles. Various suggestions were put forth. One plan was to use an actively-cooled

Fermi-type lattice, even though it would be necessary to dismantle the pile to retrieve the irradiated uranium. Walter Zinn suggested an arrangement of uranium in graphite cartridges that would move through a graphite block at about three feet per second, which would be fast enough to obviate the need for cooling the pile itself. John Wheeler proposed alternating layers of uranium and graphite, with the uranium-bearing layers connected to shafts to draw them out of the pile. Another concept, which was ultimately adopted, was to use cooled rods of uranium that extended through a large graphite block.

Following the Military Policy Committee decision of December, 1942, to proceed with full-scale piles, the most pressing issue was to decide which cooling method to adopt. If the piles were to be gas-cooled, two alternatives looked feasible. Air cooling was familiar to engineers, but would involve some neutron loss. On the other hand, helium cooling was attractive in view of its chemical inertness and that element's low neutron-capture cross-section. But all gases have relatively poor thermal properties, which would mean large volumes of gas would have to be pumped under high pressures, an issue which would complicate design of compressors and pumps. As for liquids, water-cooling was also familiar territory for engineers, but water captures neutrons and corrodes unprotected uranium metal. A number of Met Lab scientists favored heavy water, which could serve as both a coolant and a moderator, but that material was scarce. A drawback of any form of liquid-cooling was that a leak might render the pile inoperative, or cause an explosion if the coolant became vaporized under high pressure. During the summer of 1942, Moore and his team concentrated on designing a helium-cooled pile comprised of a block of graphite pierced by vertical holes in which graphite-uranium cartridges would be stacked and through which helium would be pumped. When Groves came into the project, Arthur Compton considered helium cooling to be the front-running possibility, but John Wheeler and Eugene Wigner continued to research the possibility of water cooling. At the same time as pile design went forward, there was also the issue of plutonium separation chemistry to consider. At one point, 12 alternate separation methods were under consideration; in May, 1943, DuPont officials decided on a bismuth-phosphate process for units at both Clinton and Hanford.

The successful operation of CP-1 indicated that water cooling would be feasible for large-scale piles. Within five weeks of CP-1's first criticality, Eugene Wigner and his group had developed a design for a 500-MW pile wherein a thin film of water would flow over aluminum-sheathed uranium slugs which would be contained within long aluminum tubes which ran through a graphite moderating structure. After being irradiated, slugs would be ejected from the back of the pile and collected in a pool of water to let their radioactivity die off before being transported away for chemical separation. Curiously, Wigner anticipated an operational lifetime for the reactor of only 100 days. This is the system that came to be used in the Hanford piles, although they ended up being operated for many years.

In mid-February, 1943, DuPont decided to terminate research on the helium-cooled design in favor of Wigner's water-cooled design. The decision to shift to water cooling was a major one, and involved a number of competing factors. Wigner had objected to helium on the grounds that the reactor would have to run at a very high temperature, perhaps 400–500 °C, which would mean serious material stress problems. Helium cooling would also require handling and purifying large volumes of gas, and maintaining a leakproof pressure enclosure for the pile. While it was expected that water cooling would reduce the reproduction factor by perhaps 3 %, DuPont engineers had become impressed by Wigner's design, and were confident that the chain reaction could be maintained. The decision in favor of water-cooling came *after* the Hanford site had been chosen, but the Columbia river was more than able to supply the requisite amount of water. As described below, however, helium was used to provide an inert operating atmosphere for the pile, which still meant providing a pressure enclosure. After the cooling decision was made, hard feelings persisted among some of the scientists at Chicago when they learned that DuPont planned to use its own staff for the detailed design work, consulting Chicago only occasionally. The Chicago group did not entirely lose control of their creation, however: Wigner reviewed all blueprints, and would eventually accumulate 37 patents on various kinds of reactors.

Like the Clinton facility, the Hanford Engineer Works was constructed from scratch. But in many ways construction at Hanford was far more challenging than at Clinton. With no cities nearby, DuPont had to plan from the outset for large on-site communities. They decided to build two: a construction camp at Hanford itself, and, more distant, a permanent housing area at Richland for employees and their families (Fig. 6.1). The construction camp was located about 6 miles from the nearest process area, and Richland village was about 25 miles from the piles. Planning for both began in early 1943. Nearly 400 miles of highways and over 150 miles of railroad track would also be constructed. Initial estimates called for being able to house a construction workforce of 25,000–28,000. Plans for the construction camp were deliberately made scalable, an approach which proved its worth: the construction force would grow to nearly twice the initial estimate.

For Richland, initial estimates projected a population of 6,500–7,500, but this was soon revised to 12,500, then to 16,000, and eventually to 17,500. As at Clinton, living conditions were rough-and-ready, with many families housed in prefabricated and portable residences. Once again, schools, stores, churches, recreational areas, hospitals, utilities, street maintenance, trash pickup, transit services, and fire and police forces had to be provided. Eventually, some 4,300 family dwelling units and 21 dormitories were put up. As at Oak Ridge, an extensive bus system was necessary; during the construction phase alone, some 340 million passenger-miles were driven.

If conditions at Richland were reminiscent of a boomtown, they must have seemed luxurious in comparison to those at the construction camp. Shops to machine graphite, fabricate concrete pipes, and prepare sections of steel plate and masonite panels for reactor shielding were interspersed among houses, heating and water plants, barracks, trailer courts, cafeterias, bars, administrative buildings, theaters, schools, hospitals, and libraries. The first DuPont employee arrived on February 28, 1943, and construction officially began on March 22 with the opening of an employment office in the city of Pasco, about 30 miles from the site. The construction camp began housing workers in April, although some workers spent their first six months living in tents. Between March, 1943, and August 1944, the local police force, the Hanford Site Patrol, recorded just over 8,000 "incidents," the vast majority of which involved intoxication and burglary, although the tally also included five violent deaths, 19 accidental deaths, and 88 cases of bootlegging.

Construction of the construction camp itself had to come first. Work on the first barracks began on April 6, and by September most people were working nine-hour days six days per week, with some laborers temporarily putting in ten-hour days seven days per week. In his diary, Matthias recorded on August 20 that the cafeteria was serving some 22,000 meals per day. By November, 5,300 workers were employed in erecting the construction camp alone, which by the end of the year boasted over 100 men's barracks, several dozen women's barracks, seven mess halls, and 1,200 trailers. On December 3, work went to two nine-hour shifts, and on January 1, 1944, a third shift was added. By July, 1944, when construction of the piles themselves was in full swing, the camp was home to 45,000 people. Total project man-hours at Hanford would run to over 126 million, with only about 15,000 lost due to labor disruptions. Walter Simon, DuPont's plant operations manager at Hanford, allegedly said that "Rome wasn't built in a day, but DuPont didn't have that job."

Isolation, sandstorms, and spousally-segregated living conditions made employee turnover an endemic problem. DuPont interviewed over 260,000 applicants and hired over 94,000 to maintain an average workforce of 22,500 over the life of project. By the summer of 1944, the turnover rate in construction personnel had reached 21 %. To raise morale, DuPont put up recreation halls, taverns, bowling alleys, tennis courts, baseball and softball fields, and brought in nationally-known entertainers. Groves directed that beer be could be sold in whatever quantities were needed. Unskilled laborers were attracted by an average daily pay of $8, twice the $3–$4 rate common in other parts of the country; for skilled laborers, the figures were $15 in comparison to $10. All employees signed a declaration of secrecy, which reminded them that violation of the national Espionage Act could result in 10 years in prison and fines of up to $10,000. Security agents would often pose as regular workers.

Three major types of working areas were laid out over the Hanford reservation. The piles themselves would be located in "100" areas: 100-B, 100-D, and 100-F, each about one mile square (as for the other letters of the alphabet, see below). The separation facilities were located about 10 miles south of the piles in "200" areas: 200-E, 200-W, and 200-N, for East, West, and North, respectively, with the 200-N area used as a storage area for irradiated fuel slugs. The 300 area, located just a few miles from Richland, was where uranium slug fabrication and testing took place. Each pile also required a plethora of support facilities: retention basins to hold spent cooling water until its radioactivity had declined to the point where it could be safely returned to the Columbia, water pumping and treatment plants, refrigeration and helium-purification facilities, fuel-storage areas, steam and electricity substations, and fire and first-aid stations. Equally monumental would be the three chemical separation plants. Colloquially known as "Queen Marys" after the famous ocean liner, each would be 800 feet long by 65 feet wide by 80 feet high (Fig. 6.2). Irradiated fuel from the piles made their journey to the Queen Marys in sealed casks aboard railroad cars.

Initial plans called for eight 100-MW piles laid out along the banks of the Columbia, designated as 100-A through 100-H. When Chicago scientists and DuPont engineers settled on a 250-MW water-cooled design, the decision was made in May, 1943, to cut the number of piles to three, to be located at the B, D, and F sites; the A and H sites were left vacant as safety areas. As shown in Fig. 6.1, the B-pile area was about 7 miles southwest of the D area, with F about 9.5 miles southeast of D. Various other reactors were built at Hanford after the war, but there never was an A-pile (Figs. 6.3 and 6.4).

The first pile built was the B-pile, and a particularly rich record on its construction and operation is available in a Department of Energy "Historic American Engineering Record" document (HAER; see Further Reading). This document is the source of many of the photographs appearing in this chapter, as well as of various facts and figures on design and operational details. Survey work for the B area was completed on April 15, 1943; ground was broken for a retention basin on August 27, and layout of the reactor building itself, the 105-B building, began on October 9.

The 105-B building had a footprint of 120 feet by 150 feet, and was 120 feet high. Including shielding, the outer dimensions of the pile itself were 37 feet from front to rear (roughly west to east), 46 feet from side to side (north–south), and 41 feet high. The graphite core for each pile measured 36 feet wide by 36 feet tall by 28 feet from front-to-rear.

Fig. 6.2 Queen Mary separation building *Source* http://commons.wikimedia.org/wiki/File:Queen MarysLarge.jpg

Fig. 6.3 The 100-B area, looking northwest, January, 1945. The Columbia river is in the background. The pile building itself is adjacent to the water tower. *Source* http://commons.wikimedia.org/wiki/File:Hanford_B_site_40s.jpg

Fig. 6.4 The B-pile building under construction. (HAER, Photo 3)

Figure 6.5 shows an overall view of how each pile was laid out. At the front face was the charging area, where slugs of uranium metal fuel were loaded into 2,004 aluminum process tubes, each of which was 44 feet long. The charging area was large enough to permit removal of fuel tubes for repairs if necessary. At the back of the pile was the discharge face, from which irradiated slugs would fall into a 20-foot deep pool for storage and transfer. The control room was situated on the left side of front face of the pile on the ground floor. Above the control room was a "rod room," from where nine 75-foot long control rods could be electrically or manually deployed. Exclusive of the pile itself, each pile building used 390 t of structural steel; 17,400 cubic yards of reinforced concrete; 50,000 concrete blocks; and 71,000 bricks. The piles themselves were welded to be gas-tight, and contained 2.5 million cubic feet of masonite; 4,415 t of steel plate; 1,093 t of cast iron; 2,200 t of graphite; 221,000 feet of copper tubing; 176,700 feet of plastic tubing; and some 86,000 feet of aluminum tubing. The total volume of land excavated at Hanford was equivalent to about 10 % of that of the Panama Canal. Organizing the construction was a mammoth task; over a two-year period, DuPont placed over 47,000 purchase orders and engaged 74 subcontracts with firms in 47 states. The firm's organizational charts ran to 24 feet in length. Despite the completely novel nature of Hanford, DuPont brought the project on-line a year ahead of schedule, and at a cost only about 10 % above that estimated in mid-1943.

The bottom-most layer of the pile structure was a 23-foot thick concrete footing, cast to accommodate instrument and gas-transfer ducts. Atop the footing lay a 1.5-inch steel baseplate. Each pile was surrounded on all sides by water-cooled cast-iron blocks which formed a thermal shield wall approximately 10 in. thick. The bottom layer of this shield served as a base for the graphite bricks of the pile, and absorbed about 99.6 % of the heat generated by the fission reactions. The cast-iron blocks were machined to accuracies of 0.003 in., and were interlocked to provide a radiation barrier. Holes bored through the shield for fuel-channel tubes had to match corresponding holes in moderator bricks to 1/64 of an inch. Working outward, the thermal shield was surrounded by a 4-foot thick biological shield comprised of over 350,000 blocks of alternating layers of steel and masonite, known as B-blocks. This layer reduced the ambient radiation by a factor of 10 billion; to achieve the same effect with concrete would have required a wall 15 feet thick. The entire assembly was then surrounded by a steel outer shell, which served as a containment structure for the pile's helium atmosphere.

As with the K-25 facility at Clinton, a particular issue in the construction of the piles was the quality of welding joints. Once a pile had been activated, it would be next to impossible to correct any internal problems; all joints had to be done properly the first time. Each pile required over 50,000 linear feet of welds, which had to be smooth to a tolerance of 0.015 in. This task was assigned to the highest-quality welders, who received a special pay grade and had to submit to background checks and periodic tests. Only about 18 % of such applicants qualified. Welds were inspected by use of X-rays or penetrating dyes; each weld was stamped with a welder's identification number.

Each pile comprised some 75,000 graphite moderating bricks, most being 4-3/16 in. square by 48 in. long. About one in five were bored lengthwise to accommodate fuel tubes spaced 8 and 3/8 in. on-center. The squareness tolerance of the bricks was held to ±0.004 in. to ensure snug fits, and their corners were

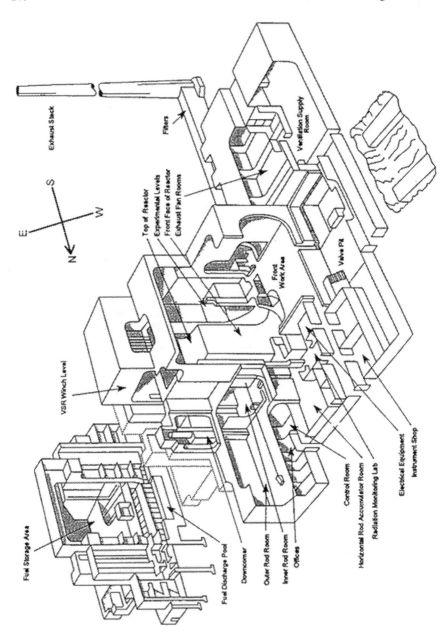

Fig. 6.5 Cutaway view of the Hanford B pile. (HAER, p. 133)

bevel-cut to provide passages for the helium atmosphere of the pile. Each brick weighed about 50–60 pounds, and their neutron-absorbing boron content was held to 0.5 parts per million. Bricks were milled a restricted-access building, and each was stamped with a quality code; the best-quality ones were used in the centers of the piles. A small test pile was built in the 300-area to check the fit of each brick, with the location of each recorded in order that layers could be correctly reconstructed in the real pile. After each layer of bricks was stacked, it was vacuumed to remove any contaminants. Milling of bricks for the B-pile began on December 10, 1943, and laying was finished on June 1, 1944, just a few days before the D-Day invasion of Europe. Graphite cleanliness was so critical that DuPont even had a laundry procedure which specified what soaps and detergents could be used to clean worker's clothes (Fig. 6.6).

6.3 Fuel and Cooling Systems

At the heart of each pile was its assembly of graphite moderator bricks, fuel channels, and fuel slugs. Eugene Wigner's early-1943 design for a 500-MW pile called for 1,500 process tubes piercing a graphite cylinder 28 feet in diameter by 28 feet deep. DuPont engineers modified the design by adding 500 fuel channels to make a roughly square-faced arrangement (Fig. 6.7).

The record as to who actually suggested the overdesign is unclear; many people were involved. Some sources indicate that Hood Worthington, the head of DuPont's design effort, followed what was normal chemical engineering practice at the time and invoked a one-third overcapacity margin. In his study of DuPont management practices at Hanford, Harry Thayer suggests that it was due to George

Fig. 6.6 Laying the graphite core of B-reactor. The rear face of the reactor is toward the *lower left*, and the inside of the front face to the *upper right*. (HAER, Photo 6)

Fig. 6.7 Front face of F-pile, February, 1945 (HAER, Photo 21)

Graves, the Assistant Manager of DuPont's TNX group. Other sources suggest that the idea was proposed by John Wheeler and Enrico Fermi, who were concerned about possible neutron-absorbing fission products poisoning the chain-reaction. While many physicists thought that the overdesign would make the piles more expensive than necessary to construct and operate, the conservatism would pay off. The additional tubes beyond Wigner's 1,500 contributed only about 10 % of the reactivity of the central ones, but would prove to be crucial to achieving the piles' design power ratings. One DuPont engineer estimated that had the additional tubes not been provided for from the start, eight to ten months would have been necessary to revise pile design and construction in response to the xenon-poisoning crisis described below (Sect. 6.5).

As constructed, each pile comprised a square central area of 42 tubes on a side, for a total of 1,764 tubes. To those were added 240 tubes arranged as two rows of 30 tubes each, centered on each of the four sides of the square. This gave a total of 2,004 tubes, each of which was uniquely numbered so that operators at the front and back faces of the pile could open the same tube simultaneously for refueling. Piles were shut down during re-fueling operations, during which tons of irradiated slugs would be discharged from the back face of the pile. The tubes, which had inside and outside diameters of 1.61 and 1.73 in., were developed by the Aluminum Company of America, which invested seven months of research in perfecting them (Fig. 6.8). During normal operation, each tube contained 32 active fuel slugs of outside diameter 1.44 in. (including an aluminum jacket 0.035 in.

thick) by 8.7 in. long. Relatively short slugs were used to minimize warpage due to thermal expansion. Each slug contained about eight pounds of natural uranium; with some 64,000 slugs inside the pile (2,004 tubes times 32 active slugs per tube), the usual fuel load was about 250 t. Fuel slugs were supported inside the tubes by two ribs which ran along the bottom of each tube, an arrangement which left an annular gap of only 0.086 in. for the flow of cooling water. With a flow speed of about 19.5 ft/sec, about 14 gallons of water would pass through each tube per minute. Nozzles at tube ends allowed for insertion and removal of fuel slugs and adjustment of the water flow rate. The piping was arranged such that if water flow to a tube was stopped for refueling or maintenance, the tube would remain full of water.

Early IBM computers were used to track the irradiation history of each fuel slug in order that the amount of plutonium production could be predicted. Dummy slugs, such as inert spacers and neutron absorbers, were used to help control the neutron flux within the pile; in routine operation a pile would contain almost as many dummy slugs as active ones. Dummy slugs could be reused, but since they too would become slightly radioactive, they also had to undergo a period of post-use thermal and radiological cooling before being re-inserted into a pile.

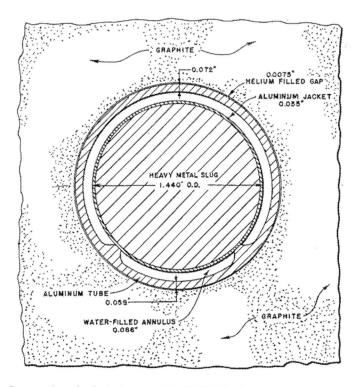

Fig. 6.8 Cross-section of a fuel tube assembly (HAER, Fig. 9, p. 143)

Because the Hanford piles operated at much greater power than the X-10 pile and involved potentially corrosive water cooling, requirements for the robustness of the "canning" of fuel slugs were much more demanding than at Clinton. Slug jackets had to be strong enough to withstand the thermal and neutron-bombardment environment within the pile without swelling or blistering (and hence releasing fission products), yet be easily dissolvable when the time came to process irradiated slugs to extract their plutonium. Finding a mass-production jacketing method proved to be so tricky that it almost derailed the plutonium project. At the Met Lab, researchers tried coating uranium slugs by various spraying, dipping, and canning methods, but to little avail. The Aluminum Company of America experimented with sealing the slugs in aluminum cans, but the process required welding on a cap without using any sort of soldering flux in order to maintain the purity of the slug. Aluminum is notoriously difficult to solder, and more often than not the result was cap failures. By October, 1943, Arthur Compton considered slug production to be the most critical job facing the project.

DuPont centralized much of the slug research to Hanford in March, 1944. Uranium arrived at Hanford in the form of billets, which would be extruded into rods with a 1,000-t press. From these rods, slugs were cut, machined smooth, and cleaned. The first experimental canning operations began later that month, but the number of acceptable slugs was limited to single-digits per day, a far cry from the thousands that would be required to fuel a pile. A critical breakthrough came with experimental determination of the correct temperature which would ensure proper bonding between a uranium slug and the aluminum can. First, a clean aluminum can was filled with a molten aluminum–silicon bonding material. After being cleaned, a slug would be dipped in a bath of bronze to prevent the uranium from alloying with the bonding material, into which it was then dipped. The slug would be quickly pressed into the can, and covered by an aluminum cap which would be welded into place; the process was called "underwater canning." Temperature control was crucial; the solder into which the slug was dipped melted at only a couple degrees below that of the aluminum can. After canning, slugs would spend 40 h in an autoclave to drive out any moisture. Each completed slug was inspected for blistering or distortions both visually and with X-rays; a flawed can could jam a process tube. The canning process was largely perfected by August, 1944, and, as experience was gained, the rejection rate fell to about 2 %.

Fueling the piles was accomplished by operators in a loading elevator who used a "charging machine" to push fuel slugs into process tubes, which simultaneously caused irradiated ones to emerge from the back of the pile. In operation, tubes typically averaged 59 fuel and dummy slugs. The rear face of the pile was sur-rounded by a 5-foot thick concrete wall, and workers would normally vacate the area after they had opened the discharge tubes but before pushing began. A discharge elevator on the rear face carried a cab which was shielded with 7 in. of lead, and was equipped with a periscope and power tools. The discharge system was a simple free-fall arrangement. After falling into the collecting pool, slugs would be sorted into buckets of active and dummy units. After an hour or two, their radioactivity would drop by a factor of 10, and then by another factor of 10 after 60 days.

Vital to the safe operation of each pile was its once-through cooling system. For all three piles, the total cooling water consumption would be equivalent to that of a city of about 1.3 million inhabitants. Some 30,000 gallons of water was pumped through each pile per minute, but only a small fraction of that would be inside the core at any moment (see Exercise 6.1). By using a single-pass arrangement, outlet temperatures could be kept at or below 65 °C. This ensured rapid heat dilution of the effluent water in the Columbia, which has a flow rate at Hanford on the order of 54 million gallons per minute. The cooling water would nevertheless become slightly radioactive from its single pass through the pile; to allow short-lived fission products to decay after discharge, effluent was held in a 7-million gallon retention basin for three to four hours before being returned to the river. Intake and discharge lines were guarded by grates to prevent fish from swimming up them. To monitor the health of fish, the University of Washington established an Applied Fisheries Laboratory at the site.

The cooling system contained multiple backups. Primary circulation was provided by electric pumps, with steam-driven pumps idling on the same lines in case of a power failure; the primary pumps were fitted with 4,600-pound flywheels so that they would keep running for 20–30 seconds until the steam pumps came up to full power. Each pile was also equipped with two elevated 300,000-gallon water tanks which could dump their contents into the piles by gravity feed.

In addition to fuel and cooling management, another concern was the operating environment of the piles. An ordinary air environment would not do. Nitrogen captures neutrons, and air also contains a small amount or argon, which becomes radioactive upon neutron capture. The solution was to weld each pile tightly closed and supply an inert helium atmosphere at a flow rate of about 2,600 cubic feet per minute; helium also had the advantage of being fairly thermally conductive for a gas. Pressure-tests of B-pile began on July 20, 1944, the same day as an unsuccessful assassination attempt against Adolf Hitler.

6.4   Control, Instrumentation, and Safety

Control of the Hanford piles was effected by a system of boron-steel control and backup rods similar to those used in the X-10 pile. At Hanford, nine 75-foot long, water-cooled horizontal control rods entered from the left side of the pile as seen from its front face. Hydraulically and electrically driven, these were arranged in three rows of three rods each, set five feet apart both vertically and horizontally. Seven were shim rods which controlled the bulk of the pile's reactivity. These could be moved at speeds of up to 30 in. per second, and could effect a complete shutdown of a pile unless a complete loss of cooling water occurred. The other two were regulating rods, which were used to handle finer minute-to-minute adjustments, and could be moved at speeds as slow as 0.01 in. per second. Above each pile resided 29 vertical safety rods. These were normally held in place by electric clutches which would release in the event of a power failure. Given that an earthquake or bombing could damage a pile in such a way as to prevent rods from deploying, a last-ditch safety system was mounted atop each pile: five, 105-gallon tanks filled with a boron solution, an arrangement reminiscent of the manual CP-1 "suicide squad." When released, the fluid would run into the vertical rod holes, but would ruin the pile in the process. In 1953, these were replaced with systems using boron-steel ball-bearings.

The safety systems received a real-world test on March 25, 1945, when a small explosive-laden Japanese balloon drifted into eastern Washington and struck an electrical transmission line that fed Hanford. The resulting two-minute power outage caused all 3 piles to scram automatically. Over the course of the war, the Japanese produced some 9,000 such balloons.

Operators constantly monitored the status of the piles through readouts from over 5,000 instruments. At a glance, they could determine the state of the water pressure at any tube inlet; the water temperature at all inlets and outlets; the water flow rate; the pressure of the helium gas; the temperature of the graphite moderator and the thermal shield; positions of all control rods; and monitor for the presence of any radiation leaks. Safety circuits were programmed to deploy control rods depending on the severity of a problem such as high or low water pressure, high radioactivity in the discharge water, overly high neutron flux, a power failure, or high effluent temperature. The power level was determined by the simple expedient of monitoring the temperature difference between inflowing and outflowing water, which in combination with the flow rate gave the heat generated by the pile.

Extensive efforts were made to ensure the safety of both workers and the environment. Radiation detectors monitored effluent water, retention basins, ventilating air, discharge areas, and control rooms. Workers who might be exposed to radiation always carried two personal dosimeters via which their daily and cumulative doses were tracked; they also wore protective clothing and face masks if needed. One type of dosimeter was a pocket ionization chamber called a "pencil." These would be electrostatically charged before being issued, and at the end of a shift the amount of discharge would indicate the amount of gamma-ray exposure sustained. The second system was a film badge housed in a worker's identification tag. The film would be fogged by beta or gamma radiation; exposure to radiation of different energies could be monitored by shielding different parts of the film.

Because plutonium tends to collect in bones, urine and blood samples were regularly collected and tested. A separate Health Instruments (HI) Division was responsible for setting radiation protection rules and standards, and for monitoring workers and the environment. The dose tolerance for workers was set to a very low level, 0.01 rems per day (see Sect. 5.2 for a brief discussion of rems). If workers had to enter a hazardous area, a HI monitor would first assess the area and set criteria for exposure time and distance from sources. HI Patrol Groups also routinely surveyed pile buildings and other areas to check for signs of contamination. As part of monitoring the external environment, Army guards would periodically shoot coyotes, whose thyroids would be examined for iodine, a characteristic fission product. Despite the pressure of wartime work, not a single serious case of radiation exposure occurred at either Oak Ridge or Hanford.

6.5 Operations and Plutonium Separation

The first fuel was loaded into B-reactor at 5:44 p.m. on September 13, 1944, by Enrico Fermi, giving the pile the so-called "blessing of the pope." During the initial loading phase, all control rods were inserted. The design of the pile was such that only a few hundred fully-loaded tubes would be needed to bring it to criticality, albeit at low power. Initially, only the central-most 1,595 tubes in B-pile were connected to the cooling system, and 895 of those were filled with aluminum dummy slugs.

The first operational benchmark was what reactor engineers term "dry criticality," which is when a pile achieves criticality with no coolant circulating. Given the poisoning effect of water, this is the smallest possible critical size of a pile. If cooling is then activated, criticality will be lost, and more tubes will need to be loaded to restore it. Dry-critical loading of B-pile began with a central area of 22 tubes on a side, and was achieved at 2:30 a.m. on September 15 with 400 tubes loaded. A period of control-rod tests and instrument calibrations followed, after which loading was resumed until 748 tubes were charged. At that point, the cooling system was activated, which, as expected, poisoned the reaction. Additional tubes were loaded, and "wet criticality" was achieved at 5:30 p.m. on Monday, September 18, with 838 tubes charged. Loading with control rods inserted continued into the early morning of September 19, by which time 903 tubes were charged, although two had to be shut down due to lack of water pressure. After further tests, control rods were withdrawn until wet-criticality was again achieved with the 901-tube loading. This occurred at 10:48 p.m. on Tuesday, September 26, 1944, and is regarded as the first official operation of the pile. By just after midnight, September 27, B-pile was operating at 200 kW, and by 1:40 a.m., 9 MW was achieved.

At first, everything seemed to be operating perfectly. But about an hour after reaching 9 MW, operators noticed that they were having to withdraw control rods to maintain power; the pile appeared to be dying. By 4:00 in the afternoon the power level had fallen to 4.5 MW, at which time it was intentionally reduced to 400 kW in an attempt to halt the decline. This proved unsuccessful, and by 6:30 p.m. the pile had shut itself down completely and was considered to be dead. There was no obvious problem: water flow and pressures were nominal, there was no evidence of any leaks or slug corrosion, and the helium atmosphere was normal.

Surprisingly, after a few hours of dormancy the pile spontaneously began coming back to life. The multiplication factor k rose back to greater than unity at about 1:00 a.m. the next morning, Thursday, September 28, and by 4:00 p.m. the power level could again be raised to 9 MW. But as soon as that level had been

reached, the multiplicity factor again began to decline. Sustained operation proved
to be impossible, and as Thursday became Friday, B pile was once again effectively dead (Fig. 6.9).

From the temporal pattern of the pile's on-again, off-again reactivity, Enrico
Fermi, John Wheeler, and DuPont chemical engineer Dale Babcock determined
that the problem was likely a fission product with a poisonously high neutron-
absorption cross-section. By monitoring the rate at which control rods had to be
withdrawn in order to hold the power steady at 9 MW, they determined that both
the parent isotope and its poisoning decay-product isotope both likely had half-
lives of the order of several hours. By Friday morning, enough data had been
gathered to indicate a half-life for the poison of about 9.7 hours. Examination of a
table of isotopes showed that the problem was likely an iodine-to-xenon decay
chain. The specific culprit was xenon-135, which arises from beta-decay of tel-
lurium-135, itself a direct fission product. Tellurium undergoes a 19 second beta
decay to iodine-135, which suffers a 6.6 hours beta decay to xenon-135, which has
a half-life 9.1 hours (modern value) before decaying to cesium and eventually to
barium. At over three *million* barns, Xe-135's neutron capture cross-section is the
largest known for any nuclide.

The only solution was to increase the amount of fuel in the reactor in order to
overcome the poisoning effect. This required plumbing in the initially unused fuel

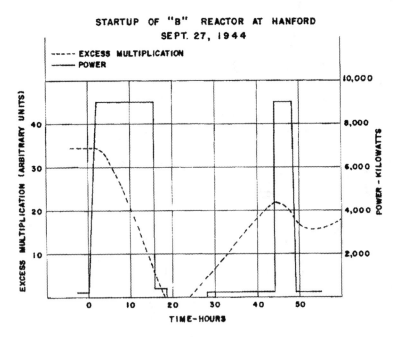

Fig. 6.9 Power output (*solid line, right scale*) and excess multiplication factor k-1 (*dashed line,
left scale*) for B-reactor startup. Time-zero corresponds to about midnight, September 26/27,
1944. Note how the excess multiplication drops as power is increased. From Babcock (1964)

tubes, which necessitated boring holes through the biological shield blocks. Compton presented the bad news to Groves in Chicago on October 3. Groves was highly critical of the scientists for not foreseeing the problem, and was not impressed by Compton's argument that a fundamental new discovery regarding the neutron properties of matter had been made. Compton then left for Hanford to review the situation personally.

Many accounts of the B-reactor startup present the xenon-poisoning episode as a completely unanticipated phenomenon, but this is far from the truth. As described by Babcock in an article published on the twentieth anniversary of the event, the possibility of a severely neutron-absorbing fission product had received considerable attention. The reproduction constant achieved in Fermi's CP-1 pile was only slightly greater than unity, and Wigner, Wheeler, and DuPont engineers were well aware that a production reactor would involve many materials not present in CP-1, particularly water and aluminum tubes. Wheeler carried out detailed calculations of how even slight changes in design specifications could affect the value of k, but the numbers were uncertain and not all fission products were known or could be predicted. As early as February, 1942, Wheeler had speculated on the possible effects of fission products, and in April of that year determined that a short-lived fission product could severely affect pile operation if it had a capture cross-section of about 100,000 barns or greater.

As design work progressed, each specification was assigned a plus-or-minus value for how it might affect k: the thickness of a fuel cladding or water jacket, the design of a control rod, the purity of graphite bricks, and so forth. The uncertainty of the situation is indicated by the fact that independent analyses by Sam Allison and John Wheeler in September, 1943, predicted "excess reactivity" values [that is, the value of $(k-1)$ expressed as a percentage] of +1.22, and −0.18 %, respectively. Wheeler's result led him to suggest adding the 504 additional fuel tubes at the periphery of the core, as well as slightly altering the diameter of the tubes. Wisely, DuPont accepted both suggestions. Wheeler also identified in advance some potential problematic fission products. One of particular concern was samarium-149, which is stable and has a thermal neutron capture cross-section of about 40,000 barns. However, the calculations were very sensitive to the fact that the distribution of fission products is by no means uniform with mass number (Fig. 3.7); the result could be very different if a different samarium isotope was preferentially created. It does seem to be true, however, that nobody anticipated a poison with a cross-section of millions of barns.

Work on charging an additional 102 tubes in B-reactor began on September 30, and was completed on October 3. With 1,003 tubes loaded, the pile was quickly brought back to criticality and taken to a power of 15 MW, where it was maintained until October 5. This did not overcome the poisoning, so the pile was shut down to load more tubes. Between October 12 and 15, the number of charged tubes was raised to 1,128, and the pile taken to a power 60 MW. Poisoning persisted; more fuel would be needed to get to the design power of 250 MW. Another shutdown on October 19 permitted raising the number of charged tubes to 1,300 and the power to 90 MW, and yet another on October 26 brought the number of operating tubes to 1,500. A power of 110 MW was achieved on November 3, but again could not be maintained, so operations were reduced to 90 MW on November 5. The next shutdown came on November 20, following which 1,595 tubes were made active.

B-pile achieved a power of 125 MW (half of its design capacity) on November 30, but it was clear that all tubes would be needed to get to 250 MW. Thus, the pile was again shut down on December 20 to install a full fuel load; extra reactivity was also obtained by replacing some dummy slugs with active ones. All 2,004 tubes were ready (less the two defective ones) by December 28. A power of 150 MW was achieved on the 29th, and 180 MW the next day. The full design rating of 250 MW was finally achieved on February 4, 1945, with about 1,950 tubes operating. With lessons learned from B pile, the D and F piles started life with full fuel loads. D went critical at 11:11 a.m. on December 17, 1944, with 2,000 tubes loaded, and F on February 25, 1945, with 1,994. Within a day, F-pile was operating at 100 MW, and by March 1 was running at 190 MW. Colonel Matthias recorded in his diary that on the morning of March 28, all three piles ran simultaneously for the first time at 250 MW. With all three piles operating at this power level, theoretical plutonium production would be about 17 kg per month, enough for almost three *Fat Man* bombs per month at about 6 kg per bomb. By May 3, some 1.6 kg of plutonium had been delivered to Los Alamos, and deliveries were taking only two days to get from Hanford to New Mexico. By June 1, Groves was ordering that production be maintained at five kilograms every 10 days. In early July, Matthias' diary makes frequent references to urgings from Groves and Oppenheimer to get material to Los Alamos as quickly as possible. D-pile was also used for polonium production: by May 4, four of its fuel channels had been loaded with 264 bismuth slugs.

Xenon poisoning was not the only operational concern. Another issue was graphite swelling, which had been anticipated by Eugene Wigner and is now known as "the Wigner disease." This is an effect where energetic neutrons knock carbon atoms out of their normal positions in graphite crystals, causing that material to expand. Matthias remarked on the effect in his diary entry for May 18, but investigating it was at that time a much lower priority than plutonium production. A year after startup, the graphite in the center of B-pile had expanded by about one inch, causing some of the tubes to warp. Curiously, cooler graphite would expand more than hot graphite under the same neutron flux, so the cooler edges of the pile actually expanded more than the central portion. B-pile was placed in standby mode on March 19, 1946, and power levels at the D and F piles were reduced to eliminate further expansion stresses. Full-power operation was resumed in July, 1948, when a solution to this effect was found: It was discovered that an annealing effect took place if the graphite blocks were operated at a temperature of about 250 °C as opposed to their usual 100 °C; the displaced carbon atoms would jump back into their crystalline planes. Operationally, this required changing to a helium-plus-carbon dioxide atmosphere. Another operational concern was the possibility of fuel slug failures; a ruptured or swollen slug could block cooling water or become stuck in a process tube. Although many slugs blistered and warped and some tubes had to be pulled, there were no total slug failures at Hanford during the war. The first actual rupture of a slug at Hanford did not occur until May, 1948, in F-pile.

For both the *Little Boy* and *Fat Man* bombs, expected availability of fissile material was always the pacing element of when they could be ready. Even while B-pile was undergoing its various reconfigurations in response to the xenon crisis, Groves began pressuring DuPont to find strategies to increase production. In October, 1944, DuPont estimated that production would begin with 200 g of plutonium in February, 1945, and increase to six kilograms per month by August, 1945. At this rate, after allowing time for material cool-down, processing, transport, and fabrication of bomb cores, the first plutonium test bomb would not have been ready until mid-October, 1945, and the first combat bomb not until a month or so later—after the proposed invasion of Japan had begun. Groves wanted five kilograms as soon as possible for a test device, and another five kilograms as soon as possible thereafter for a combat weapon.

There were three possible ways to increase production, and all were used in what came to be called the "speed-up program" or the "super acceleration" program: (1) operate piles at higher power levels, (2) push fuel slugs out of the piles sooner than normal (less plutonium per slug, but more slugs—and more waste), and (3) shorten the post-irradiation thermal and radiological cooling time for slugs before they were transported to the separation facilities. It had been intended that slugs should remain under water for about 120 days, but the time was first reduced to about 60 days, then to 30, and then, by mid-1945, to as little as 15 days. By March, 1945, Roger Williams (Sect. 4.10) advised Groves that DuPont should be able to deliver 5 kg by mid-June, and another five by mid-July. Groves pressed for even more efficiency, and the schedule was tightened to bring the delivery dates to June 1 and July 5. By Independence Day, 13.5 kg had been shipped, with another 1.1 ready to go. During the speed-up, B-pile remained at 250 MW, but by June, 1945, the D and F-piles would be operated at 280 MW and 265 MW, respectively.

Construction of the Queen Mary separation plants, also known as "canyon buildings," proceeded in tandem with that of the piles. Two Queen Marys, 221-T and 221-U, were completed by December 1944 in the 200-West area; a third reserve unit, 221-B, was constructed in the 200-East area and was completed in the spring of 1945. Essentially large concrete boxes, these huge buildings were divided internally into cells containing equipment for various stages of chemical processing. The cells were surrounded by seven-foot-thick concrete walls and covered with 35-t, six-foot-thick concrete lids which could be removed by an overhead crane which ran the length of the building. Each Queen Mary contained 40 cells, most of which measured about 15 feet square by 20 feet deep. Once operations started, the cells would become intensely radioactive; operators worked by remote control as they watched through periscopes and early television monitors. The separation plants, which were built to hundredth-of-an-inch tolerances, were largely designed using six-inch slide rules.

Part of the separation process involved centrifugal precipitation, a process akin to swirling a mixture in a flask. Leona Marshall (later Leona Libby), a 1943 University of Chicago Chemistry Ph.D. and the only female member of Fermi's CP-1 team (Fig. 3.5), became concerned that the swirling action might cause enough plutonium-bearing precipitate to collect that a low-grade chain reaction might occur. This proved not to be a problem at Hanford during the war, although a post-war accident was caused by this very effect.

Operators took over 221-T on October 9, 1944, and began test runs to become familiar with remote-operations procedures while B-reactor was going through its teething problems in November. The first test run of irradiated fuel was discharged from B-pile on November 6 (while it was being reconfigured); this was much sooner than the nominal irradiation time of 100 days, but slugs were desperately needed to test handling and separation processes. Spent fuel slugs were known colloquially as "lags." Aluminum cans were dissolved for the first time on November 25, and test runs using slugs rejected from the canning process started in early December. 221-T was ready for the first production-discharge run from B-pile on Christmas Day, and the first pure plutonium nitrate was produced before the end of January. The first Hanford plutonium to go to Los Alamos began its journey south on February 5, 1945. When operations had become routine by mid-1945, the average time for slugs to go from discharge to isolated plutonium was about 50 days, and the processing yield was up to 90 %. To receive the large volume of radioactive waste generated by the separation process (10,000 gallons per day per separation plant), 64 underground storage tanks were constructed, many as large as 500,000 gallons. Before disposal, the highly acidic waste was neutralized by addition of large quantities of sodium hydroxide. Many of the tanks began leaking in the 1950s.

By early 1945, all of General Groves' fissile-materials production programs were beginning to show results. The next task was for scientists and engineers at Los Alamos to turn fissile material into deliverable weapons. This work is the subject of the next chapter.

6.6 The Postwar Era at Hanford

Hanford continued to operate for many years after the end of the war. DuPont did not desire to remain in the nuclear business, and when its contract with the Army ended on September 1, 1946, General Electric became the operator of the facility; various operating contractors would follow in subsequent years. On being restarted in June, 1948, B-pile was taken to a power of 275 MW; by 1956 it was operating at 800 MW. In late 1956 it was shut down to install larger-capacity pumps for the cooling system, which by early 1958 permitted operation at 1,440 MW. This was increased to 1,900 MW a year later, and then to 2,090 MW in early 1961.

In the fall of 1948, Hanford acquired an important new project in addition to its role as a producer of plutonium: breeding tritium for use in fusion weapons

(Chap. 9). The process used was to seed fuel slugs with lithium, which would capture neutrons and produce tritium via the reaction $^{6}_{3}Li + ^{1}_{0}n \rightarrow ^{3}_{1}H + ^{4}_{2}He$. But there was a price to be paid for this: the neutron-capture cross-section of lithium-6 is so small that generating a single kilogram of tritium meant forgoing 80–100 kg of plutonium production; Hanford ultimately produced 10.6 kg of tritium. Further details on the fusion program are given in Chap. 9; we mention here only that the first American fusion device, "Mike," was tested at Enewetak Atoll in the Pacific Ocean on November 1, 1952. This device yielded an astonishing 10.4-Mt explosion, nearly 500 times the energy release of the plutonium-fueled *Trinity* and *Fat Man* devices.

The final shutdowns of the wartime F, D, and B piles came in June 1965, June 1967, and February 1968, respectively. Six other reactors were built at Hanford between 1949 and 1963; these operated at power levels of up to 4,000 MW. In 43 years of production, Hanford generated about 67,000 kg of plutonium, including over 15,000 kg from the B, D, and F piles. Between 1949 and 1964, the United States would build 11 more production reactors, which brought total U. S. plutonium production to 1994 to about 103,000 kg. At about 6 kg per *Fat Man* weapon, this represents enough for some 17,000 such devices; later improvements in bomb design decreased the amount of fissile material necessary per weapon. All piles constructed at Hanford were shut down by January, 1987.

In 1991, a group of local residents organized the B-Reactor Museum Association, a non-profit corporation dedicated to educating the public about the historical and technological significance of B-pile. In 1993, the Department of Energy issued a directive that the Hanford reactors be placed in "interim safe storage" for 75 years. This includes demolition of the reactor building down to the shield wall, and a "cocooning" process involving installation of an enclosing roof. Cocooning began in 1995, and is scheduled to be completed for all piles by 2015. In 1992, however, the National Park Service placed B-pile on the National Register of Historic Places, and in 2008 it became a National Historic Landmark. In response to community interest in preserving B-pile, the Department of Energy issued an alternative plan in 1999 that called for it to become a museum. B-pile may eventually become Hanford's component of a proposed Manhattan Project National Historical Park (Chap. 9).

The Hanford project exemplified all of the ingredients that made the Manhattan Project so successful: supremely competent and hard-driving leadership, lack of encumbering bureaucratic interference, outstanding contractors who insisted on rigorous quality control at every step of design, construction, and operation, and a remarkable dedication to safety and secrecy. General Groves' gamble more than paid off.

Exercises

6.1. From Fig. 6.8, the water annulus inside a Hanford fuel channel was of inner diameter 1.44 in. and thickness 0.086 in. If all 2,004 channels are in operation, compute the volume of water inside the 28-foot length of the channels that lay inside the core of the reactor at any moment. One U. S. fluid gallon has a volume of 231 cubic inches. [Ans: $\sim 1{,}202$ gallons].

6.2. A reactor operating at 250 MW is cooled by the flow of 30,000 gallons of water per minute. If the water makes a single pass through the reactor, by how much will its temperature increase? Density of water $= 1{,}000$ kg/m^3, specific heat of water $= 4{,}187$ J/(kg-K), one U. S. fluid gallon $= 3.786$ l. [Ans: ~ 32 K]

Further Reading

Books, Journal Articles, and Reports

D.F. Babcock, The discovery of Xenon-135 as a reactor poison. Nucl. News **7**, 38–42 (1964)

S.G. Bankoff, A. Weinberg, Notes on Hanford reactor startup. Phys. Today **57**(4), 17–19 (2004)

S. Cannon, The Hanford site historic district—Manhattan Project 1943–1946, Cold War Era 1947–1990. Pacific Northwest national Laboratory (2002). DOE/RL-97-1047 http://www.osti.gov/bridge/product.biblio.jsp?osti_id=807939, (2002)

R.P. Carlisle, J.M. Zenzen, *Supplying the Nuclear Arsenal: American Production Reactors, 1942-1992* (Johns Hopkins University Press, Baltimore, 1996)

A.H. Compton, *Atomic Quest* (Oxford University Press, New York, 1956)

L. Fine, J.A. Remington, *United States Army in World War II: The Technical Services – The Corps of Engineers: Construction in the United States Center of Military History* (United States Army, Washington, 1989)

M.S. Gerber, *The Plutonium Production Story at Hanford Site: Process and Facilities History*. Westinghouse Hanford Company, Richland, WA (1996). Report WHC-MR-0521, http://www.osti.gov/bridge/product.biblio.jsp?osti_id=664389

L.R. Groves, *Now It Can be Told: The Story of the Manhattan Project* (Da Capo Press, New York, 1983)

D. Harvey, *History of the Hanford Site, 1943-1990*. Pacific Northwest National Laboratory, http://ecology.pnnl.gov/library/History/Hanford-History-All.pdf

R.G. Hewlett, O.E. Anderson Jr., *A History of the United States Atomic Energy Commission, Vol. 1: The New World, 1939/1946* (Pennsylvania State University Press, University Park, PA, 1962)

L. Hoddeson, P.W. Henriksen, R.A. Meade, C. Westfall, *Critical Assembly: A Technical History of Los Alamos during the Oppenheimer Years* (Cambridge University Press, Cambridge, 1993)

V.C. Jones, *United States Army in World War II: Special Studies: Manhattan: The Army and the Atomic Bomb* (Center of Military History, United States Army, Washington, 1985)

C.C. Kelly (ed.), *The Manhattan Project: The Birth of the Atomic Bomb in the Words of Its Creators, Eyewitnesses, and Historians* (Black Dog and Leventhal Press, New York, 2007)

L.M. Libby, *The Uranium People* (Crane Russak, New York, 1979)

J.C. Marshall, Chronology of District "X" 17 June 1942: 28 October (1942)

National Nuclear Security Administration: The United States Plutonium Balance, 1944-2009. Report DOE/DP-0137 (1996; updated 2012). http://nnsa.energy.gov/sites/default/files/nnsa/06-12-inlinefiles/PU%20Report%20Revised%2006-26-2012%20%28UNC%29.pdf

R.F. Potter, Preserving the Hanford B-reactor: a monument to the dawn of the nuclear age. Phys. Soc. **39**(1), 16–19 (2010)
R. Rhodes, *The Making of the Atomic Bomb* (Simon and Schuster, New York, 1986)
A.H. Snell, Graveyard shift, Hanford, 28 September 1944: Henry W. Newson. Am. J. Phys. **50**(4), 343–348 (1982)
H. Thayer, Management of the Hanford engineer works in World War II. American society of civil engineers, New York (1996)
A.M. Weinberg, Eugene Wigner, Nuclear Engineer. Phys. Today **55**(10), 42–46 (2002)

Websites and Web-based Documents

B-Reactor Museum Association: http://www.b-reactor.org
Historic American Engineering Record: B Reactor (105-B) Building, HAER No. WA-164. DOE/ RL-2001-16. (United States Department of Energy, Richland, WA, 2001). http://wcpeace.org/ history/Hanford/HAER_WA-164_B-Reactor.pdf
Queen Mary buildings: http://www.atomicarchive.com/History/mp/p4s24.shtml

Chapter 7
Los Alamos, *Trinity*, and Tinian

The Los Alamos Laboratory was the intellectual center of the Manhattan Project, and the Laboratory's wartime Director, Dr. J. Robert Oppenheimer, is probably the most widely-recognized personality of the Project. Even decades later, the image of a collection of accomplished scientists and engineers shut away to labor for over two years in danger and secrecy under the direction of a brilliant, charismatic leader in a setting of spectacular natural beauty to produce a revolutionary new weapon still strikes a powerful emotional reaction.

Compared to the tasks faced by the organizers of the Clinton and Hanford Engineer Works, Los Alamos' mission of fashioning uranium and plutonium into deliverable weapons sounds straightforward. Arrange to bring enough fissile material together at the desired time inside a bomb casing, provide a source of neutrons for initiating the reaction, train a bomber crew to deliver the device, and the job is done. When Robert Oppenheimer took on the Directorship of Los Alamos in early 1943, he thought that he would require only a few dozen scientists, technicians, and engineers. But almost immediately, complexities in the nature of fissile materials and the engineering of bomb mechanics demanded expansions of the Laboratory staff. By mid-1945, Los Alamos employed over 2,000 people. Experimental physicists were needed to acquire measurements of nuclear parameters for various materials. Instruments had to be developed to measure such properties accurately and reproducibly. Employing numerical simulations of the time-evolution of a nuclear explosion over sub-microsecond time increments carried out with slide rules, mechanical calculators, and early computers, theoretical physicists worked to turn experimental results into predictions of critical masses to inform bomb design specifications. Chemists refined uranium and plutonium arriving from Oak Ridge and Hanford to purity levels of a few parts per million. After purification, the precious fissile materials were handed over to metallurgists, who worked to cast them into desired shapes, sometimes employing unusual alloying materials. Reactor-produced plutonium proved to have such a propensity to detonate too soon that ordnance experts had to develop a wholly-new high-speed triggering mechanism that had to operate within microsecond-level tolerances. Weapons engineers worked to integrate the fissile materials into practical bombs that could be carried by existing aircraft in combat conditions.

B. C. Reed, *The History and Science of the Manhattan Project*,
Undergraduate Lecture Notes in Physics, DOI: 10.1007/978-3-642-40297-5_7,
© Springer-Verlag Berlin Heidelberg 2014

Drop tests had to be conducted to refine bomb-casing designs to ensure stable flight characteristics, and reliable fuzing mechanisms had to be developed.

All of these tasks, as well as aircrew training, aircraft configuration, and preparations for overseas operations, were carried out against an ever-present deadline: when sufficient fissile material became available, a bomb had to be ready. Anticipated production schedules in Tennessee and Washington drove the pace of work at Los Alamos. Emilio Segrè wrote that genuine inventiveness was required; Los Alamos' products would be developed ab initio—literally, "from the beginning." That Oppenheimer and his staff accomplished their task in only 28 months is testimony to their brilliance and commitment. As Henry Smyth wrote, Los Alamos developed within less than three years into what was probably the best-equipped physics research laboratory in the world.

This chapter examines the work of the Los Alamos Laboratory from its beginnings in late 1942 through the *Trinity* test of July, 1945, and its involvement in preparations for the atomic missions against Japan. Target selection and the bombing missions themselves are described in Chap. 8.

7.1 Origins of the Laboratory

The idea of a centralized, secure laboratory under government control to coordinate fast-neutron research and bomb design was circulating well before the formal establishment of the Manhattan Engineer District. In the spring of 1942, the OSRD had contracts with no less than nine universities that had accelerators that could be used as neutron sources, but the work lacked overall coordination. Gregory Breit raised the issue of a centralized laboratory when he resigned from the project in May, 1942, and, a month later, Vannevar Bush and James Conant suggested in their report to Vice-President Wallace, Secretary of War Stimson, and General Marshall that a special committee take charge of all research and development on military uses of fissionable material. Immediately after the Bohemian Grove planning session described in Chap. 4, Oppenheimer, Fermi, Lawrence, Compton, Edwin McMillan, and others met in Chicago from September 19–23, 1942, to consider the notion of a bomb-design laboratory. These various ideas would become realized as Los Alamos and the Military Policy Committee (Sect. 4.10).

When General Groves was assigned to the project in September, 1942, his letter of appointment made no mention of a design laboratory. Groves began his new assignment with a familiarization tour of project sites, and met Robert Oppenheimer for the first time in Berkeley on October 8, at which time they discussed the concept of a centralized laboratory. Groves approved the idea on October 19, initially thinking that he would locate the facility near the production plants in Tennessee. In Manhattan District lingo, the bomb-design laboratory was known as Project Y.

Given Compton's involvement with the Project, his own University of Chicago Metallurgical Laboratory might have seemed a logical choice for a design center, or perhaps Ernest Lawrence's Radiation Laboratory at Berkeley. But Groves decided that a laboratory site would have to be isolated, relatively inaccessible, have a climate that would permit year-round construction and operations, be large enough to accommodate a testing area, and be sufficiently inland to be secure from enemy attack. None of Oak Ridge, Chicago, or Berkeley was sufficiently isolated, and the latter was also too vulnerable to Japanese attack. Groves assigned the problem of locating a site to Major John Dudley of the Corps of Engineers. After speaking with some of the scientists involved, Dudley estimated that a staff of some 265 would need to be accommodated. He investigated various locations in California, Nevada, Utah, Arizona, and New Mexico. One possibility near Los Angeles was rejected by Groves on security grounds, and another near Reno, Nevada, was discounted on the basis that heavy snowfalls would interfere with winter operations. Oak City, Utah, looked favorable, but would have required evicting several dozen families and taking a large amount of farm acreage out of production. The choices narrowed to two sites north of Albuquerque, New Mexico: one about 50 miles north of the city in the Jemez Springs area, and another about 25 miles northeast of Jemez near Los Alamos. "Jemez" is the Indian name for "Place of the Boiling Springs," and Los Alamos means "the poplars." The latter site, set on a mesa at an altitude of 7,300 feet, was then serving as the home of the Los Alamos Ranch School, a financially-troubled wilderness school for boys.

On November 16, 1942, Groves, Oppenheimer, Dudley, and McMillan set out on horseback to inspect the two sites. The Jemez Springs location proved to be in a valley prone to floods, and was deemed unsuitable. On the other hand, the Los Alamos mesa was surrounded by deep canyons which would be perfect for test sites. It also had the advantage of 54 ready-to-occupy buildings owned by the school, including 27 houses and dormitories. Oppenheimer owned a ranch not far from Los Alamos and was familiar with the area, having spent part of every summer there throughout the 1930s (Fig. 7.1).

Los Alamos was a bargain: just over 49,000 acres (about 75 square miles) were acquired at a cost of just under $415,000 (about $5.3 million 2013 dollars), a tiny fraction of the Manhattan Project's budget. (For perspective on this, the average list price of a house in Los Alamos in early 2013 was about $283,000.) The cost was modest as all but some 8,900 acres were federal lands under the jurisdiction of the Forest Service. Groves acquired right of entry to the lands and property of the school on November 23, obtained authority to acquire the site two days later, and authorized the Albuquerque District Engineer to proceed with construction on November 30, just two days before CP-1 went critical in Chicago. To allow students to complete their studies, the Ranch School was given until February 8 before it had to formally relinquish the site. Christmas vacation was cancelled, and the last four students were awarded their diplomas on January 21. One of those students, Stirling Colgate, would go on to earn a Ph.D. in nuclear physics at Cornell University, and later returned to Los Alamos to work on development of thermonuclear weapons. In March, 1943, Secretary of War Henry Stimson formally requested acquisition of the Forest Service lands from the Secretary of Agriculture "for the establishment of a demolition range." Agriculture Secretary Claude Wickward approved the request on April 8, by which time the work of the

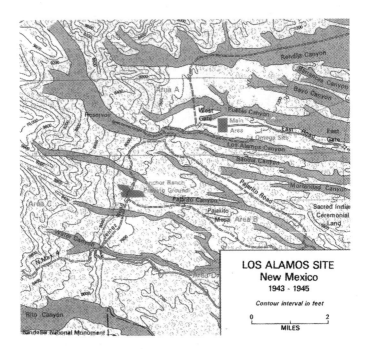

Fig. 7.1 The Los Alamos area. The "Main Area" is shown in more detail in Fig. 7.2. From V. C. Jones, United States Army in World War II: Special Studies—Manhattan: The Army and the Atomic Bomb. Courtesy Center of Military History, United States Army. See also Fig. 7.26

Laboratory was already getting underway. To its residents, Los Alamos became known as "The Hill." Oppenheimer biographers Kai Bird and Martin Sherwin have described the Laboratory as a combination army camp and mountain resort. The entire community would be fenced and guarded, and the Laboratory itself, known as the "Technical Area," would be built within an inner fenced area that had been the site of the school (Fig. 7.2); 25 outlying test sites were also eventually constructed. Construction costs at Los Alamos ran to some $26 million during the war.

Groves wrote in his memoirs that neither himself, Bush, or Conant felt committed to appointing Oppenheimer as Director. Indeed, he was seen to have a number of drawbacks. While regarded as brilliant and broadly-educated—he knew six languages—Oppenheimer was not an experimental physicist. A quintessential academic, he had no administrative experience such as being a department chair or Dean; his left-wing background was considered highly suspect, and, unlike Lawrence and Compton, he did not have a Nobel Prize. As experimental physicists, either Lawrence and Compton would have been naturals for the job, but neither could be spared from his own work. When it became apparent that no other candidates of Oppenheimer's quality were available, he was asked to take on the job. Lawrence had preferred the idea of McMillan as Director, and was apparently

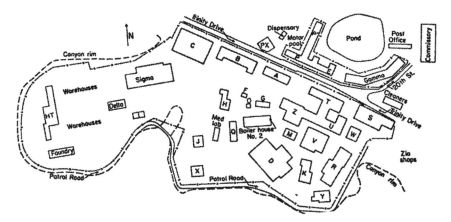

Fig. 7.2 Map of the main Los Alamos "Tech Area". The town proper and residential area were on the north side of Trinity Drive. *Source* Edith C. Truslow, *Manhattan District History: Nonscientific Aspects of Los Alamos Project Y 1942 through 1946.* Los Alamos report LA-5200, http://www.fas.org/sgp/othergov/doe/lanl/docs1/00321210.pdf

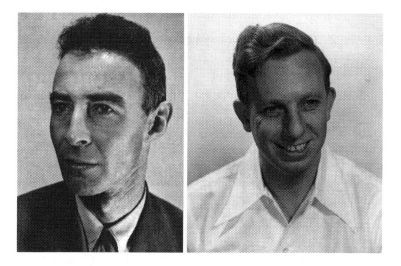

Fig. 7.3 *Left* Robert Oppenheimer (1904–1967), ca. 1944; *Right* John Manley (1907–1990) in 1957. *Sources* http://commons.wikimedia.org/wiki/File:JROppenheimer-LosAlamos.jpg; Los Alamos National Laboratory, courtesy AIP Emilio Segre Visual Archives, Physics Today Collection

outraged when Groves chose Oppenheimer; Luis Alvarez is said to have considered Oppenheimer incapable of running a hamburger stand. Security officers were so reluctant to clear Oppenheimer that Groves was forced to issue a direct order to them to do so. His July 20, 1943, directive to the District Engineer was that

In accordance with my verbal directions of July 15, it is desired that clearance be issued for the employment of Julius Robert Oppenheimer without delay, irrespective of the information which you have concerning Mr. Oppenheimer. He is absolutely essential to the project.

Oppenheimer's success at directing Los Alamos defied all expectations. Theoretical physicist Victor Weisskopf described Oppenheimer's managerial style: "He did not direct from the head office. He was intellectually and even physically present at each decisive step. He was present in the laboratory or in the seminar rooms, when a new effect was measured, when a new idea was conceived. It was not that he contributed so many ideas or suggestions; he did sometimes, but his main influence came from something else. It was his continuous and intense presence, which produced a sense of direct participation in all of us; it created that unique atmosphere of enthusiasm and challenge that pervaded the place throughout its time ... The location ... gave it a special character by its romantic isolation, in the midst of Indian culture. Living in this unusual landscape, separated from the rest of the world, in walking distance of the laboratories—all this created a community type of living, where work and leisure were not separated. But the special flavor came from the kind of people that were there. It was a large community of active scientists, many of them in their most vigorous and productive years." Another of Oppenheimer's biographers, Abraham Pais, described him with the words "In all my life I have never known a personality more complex than Robert Oppenheimer." Oppy, as he was known, would have to bring to bear all his abilities to his new task.

7.2 Organizing the Work: *The Los Alamos Primer*

Even before he was formally appointed as Director of Los Alamos, Robert Oppenheimer was delegated to recruit scientists to staff the new laboratory, and spent the latter part of 1942 and early 1943 traveling around the country doing so. The task was not easy. Oppenheimer could reveal very little of the Laboratory's ultimate purpose, and many leading scientists were already deeply involved in radar and other war work; some considered the bomb an improbable venture. As one history of Los Alamos put it, Oppenheimer had to recruit a staff for a purpose he could not disclose, at a place he could not specify, for a period he could not predict.

Oppenheimer particularly wished to recruit two outstanding physicists who were then working on radar at MIT, Robert Bacher and Isidor Rabi (Fig. 7.4). Both were crucial to the radar program, and initially refused to have any connection with a military-directed project. Rabi in particular was concerned that Los Alamos was planned to be a military installation, an arrangement squarely at odds with the scientific tradition of decentralized authority. In a letter to Conant on February 1, 1943, Oppenheimer related that following lengthy discussions with Rabi, McMillan, Bacher, and Alvarez, Rabi felt (and the others concurred) that an

Fig. 7.4 *Left to right* Robert Bacher (1905–2004); I. I. Rabi (1898–1988) in 1983; Kenneth Bainbridge (1904–1996) holding a photograph of the *Trinity* explosion, 1945. *Sources* http:// commons.wikimedia.org/wiki/File:Robert_F._Bacher.jpg; Photo by Sam Treiman, courtesy AIP Emilio Segre Visual Archives, Physics Today Collection; http://commons.wikimedia.org/wiki/ File:BainbridgeLarge.jpg

indispensable condition was that the Laboratory be demilitarized to avoid the possibility that scientific autonomy would lose out against having to follow military orders. In a heartfelt letter to Rabi on February 28, Oppenheimer stated that "I know that you have good personal reasons for not wanting to join the project, and I am not asking you to do so. Like Toscanini's violin, you do not like music." Oppenheimer went on to ask two things of Rabi, however: that he participate in an opening conference at the Laboratory to be held in April, and that he use his personal influence to persuade Hans Bethe (Cornell University, then also working on radar; Fig. 4.13) and Bacher to join the project, which they did. Rabi did not formally join Los Alamos, but visited frequently as a consultant.

Oppenheimer was formally appointed Director on February 25, 1943. As recorded in the appointment letter from Conant and Groves, a compromise had been found on the militarization issue. The Laboratory's work was to be divided into two periods. The first would involve "experimental studies in science, engineering, and ordnance," while the second would see "large-scale experiments involving difficult ordnance procedures and the handling of highly dangerous material." Los Alamos would operate on a strictly civilian basis during the first period, with personnel, purchasing, and business operations to be carried out under an operating contract with the University of California. But when the second part of the work was to be entered upon, which was anticipated as being no earlier than January 1, 1944, the scientific and engineering staff would become commissioned officers. Oppenheimer was authorized to show the letter to individuals whom he was trying to recruit.

Ultimately, Los Alamos functioned as a hybrid military-civilian-contractor organization with two heads. Formally, it was a military post with a Commanding Officer who reported to Groves, and who was responsible for maintenance of living conditions and the conduct of military personnel. All residents, civilian and military alike, were subject to military security and censorship regulations. Oppenheimer, as Director, was responsible for the technical, scientific, and security aspects of the program. Civilian employees never were commissioned, and remained employees of the University of California or other contractors. Los Alamos was formally activated as a military post on April 1, 1943, and the University of California contract became effective on April 20, retroactive to January 1.

Responsibility for overall direction of the Laboratory's scientific work lay in Oppenheimer's hands, but he was always assisted by a number of boards and committees. The first informal group, comprising Oppenheimer, Robert Wilson, Edwin McMillan, John Manley (Fig. 7.3), Robert Serber (a former postdoctoral student of Oppenheimer, then at the University of Illinois; Fig. 4.13) and Associate Director Edward Condon (Westinghouse Electric; Sect. 4.4), met on March 6, 1943, to begin considering when people and equipment would arrive and how the work would be organized. This initial group was superseded a few weeks later by a Planning Board, which met through early April to organize the laboratory's technical operations. The Planning Board was subsequently replaced by a more permanent Governing Board, which comprised Division leaders (see below), administrative officers, and individuals serving in technical liaison capacities.

The initial organizational structure of Los Alamos consisted of an Administrative Division and four Technical Divisions. The latter were Chemistry (later Chemistry and Metallurgy) under Glenn Seaborg's Berkeley colleague Joseph Kennedy, Ordnance and Engineering under Navy Commander William S. "Deak" Parsons (Sect. 5.5; Fig. 5.28), Experimental Physics under Robert Bacher, and Theoretical Physics under Hans Bethe. Within each Division were housed a number of individual research groups. While divisions, groups, and various oversight committees would come into and go out of existence as the work of the Laboratory evolved, the basic structure of groups operating within larger divisions remained, and is still in place today. Oppenheimer apparently considered that he would lead the Theoretical Division as well as serving as Director, but was dissuaded from that notion by Rabi.

The role of the Governing Board was to consider the work of the laboratory as a whole, and to relate it to progress in other parts of the Manhattan Project. Aside from technical issues, the Board also had its hands full with civic issues such as housing, construction priorities, water supply, recruitment, security restrictions, procurement bottlenecks, morale, and salary scales. Two later important appointments to the Board were George Kistiakowsky (Fig. 4.7) and Kenneth Bainbridge (Fig. 7.4), both of whom were recruited from Harvard University. Kistiakowsky was an expert on explosives, and would become intimately involved with the plutonium implosion bomb; Bainbridge, a physicist, would direct the *Trinity* test. The Board remained in place until mid-1944, when it was replaced by separate Administrative and Technical Boards during a reorganization of the Laboratory to deal with a crisis concerning plutonium.

Just as in a university or industrial laboratory setting, the work of the research groups required various support services such as a library, machine shops, photographic and drafting shops, optical shops, business offices, and safety and medical services. The ordnance program alone grew so extensive as to require its own machine shop, capable of handling some 2,000 man-hours of work per week; at one point, more than 500 machinists and toolmakers would come to be working at Los Alamos. By July, 1945, the library, which was organized by Robert Serber's wife, Charlotte, held some 3,000 books, copies of some 1,500 microfilmed reproductions of articles and parts of books, was receiving 160 journals per month, and served as a repository for some 6,000 internally-generated technical reports (over 200 per month). A Patent Office dealt with protection of government interests in any technology that might be developed; about 500 cases were reported to OSRD headquarters in Washington. The *Trinity* test represented for many inventions what patent attorneys refer to as their first "reduction to practice."

Directly reporting to Oppenheimer was the Health Group, which bore responsibility for setting health and safety standards and procedures for working with radioactive, explosive, and toxic materials. The work of the Health Group began to grow substantially in the spring of 1944 when the first significant quantities of plutonium began to arrive from Oak Ridge. Plutonium is not an external body hazard, but because it tends to collect in bones and kidneys and is only slowly eliminated from the body, the potential harmful dose was set at the very low level of 1 μg. Extremely sensitive tests had to be developed for detecting small quantities of plutonium in urine, about 10^{-10} μg/l. A sense of the scale of radiation safety operations can be gleaned from the statistics that in July, 1945, 630 respirators were decontaminated; 17,000 articles of clothing were laundered; and 3,550 rooms were being monitored. The Health Group at Los Alamos was but one of a number of such groups throughout the Manhattan Project. In early 1943, Groves appointed Dr. Stafford Warren of the University of Rochester to direct a research program on the biological effects of radiation. Warren effectively became the medical director for the Manhattan Project, and was commissioned as a Colonel in the Army Medical Corps. An extensive project was undertaken at Rochester in which radiation effects on hundreds of animals and over a quarter-million mice were studied. At Los Alamos, the Health Group was directed by Dr. Louis Hempelmann, a radiologist recruited from Washington University. No accidental occupational deaths occurred at Los Alamos during the war, but radiation overdoses did lead to two postwar deaths there (Sect. 7.11).

One of the first decisions made by the Planning Board was to sanction a series of orientation lectures for arriving scientific personnel. The lectures were delivered by Robert Serber on April 5, 7, 9, 12, and 14, 1943, and were recorded by the Laboratory's Deputy Director, Edward Condon. Condon's notes were printed up as a 24-page booklet titled *The Los Alamos Primer*. Designated as Los Alamos's first official technical report, only 36 copies were printed at the time. Declassified in 1965 and published in book form in 1992 with annotations by Serber, the *Primer* is now considered a foundational document in the history of nuclear weapons; a copy of the original typewritten report signed by Serber can be obtained from the Federation of American Scientists website. The lectures, which were attended by about 30 people, were held in a large library reading room, accompanied by background hammering as carpenters and electricians went about their work. At one point, a leg burst through the flimsy ceiling. In one annotation, Serber recalls that as he began lecturing and used the term "bomb," Oppenheimer, concerned that workmen would overhear, sent John Manley forward to tell Serber to use the term "gadget" instead. Edward Condon, upset with Groves' policy of compartmentalizing information, would resign from Los Alamos before the month of April was out.

The *Primer* still makes for fascinating reading. The first section, titled "Object," makes the situation clear: "The object of the project is to produce a *practical military weapon* in the form of a bomb in which the energy is released by a fast neutron chain reaction in one or more of the materials known to show nuclear fission." Subsequent sections touch on all major aspects of bomb design and operation: reaction cross-sections; the energy released in fission; how a chain reaction operates; the energy spectrum of fission neutrons; why natural uranium is safe against a fast-neutron chain reaction; the use of diffusion theory to estimate the critical mass; how a tamper can serve to lower the critical mass (Sect. 3.5); the expected efficiency of a nuclear weapon; the extent of damage expected from blast, thermal, and radiation effects; how a bomb could be triggered; and the probability of low-efficiency "fizzle" explosions arising from effects that could cause the weapon to detonate before the intended moment. Many experimental and theoretical details remained to be filled in, but the basic outline of an overall strategy for the development of fission bombs was fairly clear by the spring of 1943.

Immediately after Serber delivered his lectures, a series of conferences were organized to plan the Laboratory's research program. These were held from April 15 to May 6, during which time Los Alamos was visited by a special committee that had been appointed by Groves to review research and development plans. The chair of the committee was again Warren Lewis of MIT, who had been involved with the Compton committee in 1941 and with the DuPont-initiated review of the entire program in late 1942 (Sect. 4.10). The other members were Edwin L. Rose, an ordnance specialist and Director of Research for the Jones and Lamson Machine Company (a precision machine-tool company with ordnance contracts); theoretical physicist John van Vleck (also of the Compton committee); Harvard University physical chemist and explosives expert E. Bright Wilson; and Richard Tolman.

The committee submitted its report on May 10. They approved the Laboratory's proposed program of nuclear physics research, but recommended major changes in two areas. The first was that final purification of plutonium should be carried out at Los Alamos, rather than at the Metallurgical Laboratory in Chicago. The rationale for this was that since further purification would likely be required after experimental use of the material at Los Alamos, the purification might as well be done there. The other major recommendation was that ordnance development and engineering should be undertaken as soon as possible, and that such work should include the issues of safety, arming, firing and detonating devices, transport of the bomb by aircraft, and studies of bomb trajectories; the committee suggested that a Director of Ordnance and Engineering be appointed to coordinate these efforts. These proposals were estimated to require an increase in the number of chemists at the Laboratory by thirty, as well as a two-fold increase in the number of people working on ordnance issues. These expansions would prove to be merely the first steps in the growth of the Laboratory.

The appointment of an ordnance director resulted in a violation of President Roosevelt's admonition to Vannevar Bush to keep the Navy out of the Manhattan Project. At a Military Policy Committee meeting in May, 1943, Groves asked for advice in filling the position. His desire was to find an individual who possessed sound understanding of both the theory and practice of ordnance (high explosives, guns, and fusing mechanisms), but who also had a sufficiently strong scientific background to hold the respect of Los Alamos' professional scientists. (A technical comment: The term "high explosive" will appear occasionally throughout this Chapter. This term is properly used to designate an explosive such as TNT, as opposed to earlier powder-type explosives that date back hundred of years.) Since the appointee might well accompany the eventual bombs into combat, it was also desirable that he be a military officer. As Groves related the story, it was Bush himself who suggested Commander Parsons. Parsons had just completed several years of work on development and testing of proximity fuses, and had met Groves in the 1930s when he was working on radar development for the Navy and Groves was working on infrared technology for the Army. Chemist Joseph Hirschfelder, who worked closely with Parsons, considered him to be the "unsung hero" of Los Alamos.

To make estimates of critical masses, Los Alamos theoreticians needed accurate measurements of nuclear parameters such as cross-sections, secondary neutron numbers, and the energy spectrum of fission-generated neutrons. Setting up equipment to obtain such measurements became the first order of business for the Experimental Physics Division. Such a program required large-scale equipment such as particle accelerators, but there was no time to undertake the design and construction of such devices from scratch. As John Manley described it, "What we were trying to do was build a new laboratory in the wilds of New Mexico with no initial equipment except the library of Horatio Alger books or whatever it was that those boys in the Ranch School read, and the pack equipment that they used going horseback riding, none of which helped us very much in getting neutron-producing accelerators." To get work underway quickly, scientists' home universities sold or loaned the necessary equipment. A cyclotron from Harvard, two Van de Graaff generators from the University of Wisconsin, and a Cockcroft-Walton (deuteron–deuteron) accelerator from the University of Illinois made their way to Los Alamos. All were used to produce neutrons to bombard various materials; the energy ranges of the machines permitted experimenters to generate neutrons of energies from thermal to a few MeV. No one experimental method was ever relied upon for any particular energy; overlapping measurements were always conducted. The two Wisconsin machines were used to generate neutrons via proton bombardment of lithium ($_1^1H + _3^7Li \rightarrow _0^1n + _4^7Be$); together, they produced neutrons of energies from 20 keV to 2 MeV. The Cockcroft-Walton device generated neutrons up to 3 MeV via the reaction $_1^2H + _1^2H \rightarrow _0^1n + _2^3He$. The bottom pole-piece of the magnet for the Harvard cyclotron was laid on April 14 (the day of Serber's last lecture), and experiments with it began in July. Initially, the Laboratory possessed only about 1 g of U-235 and only micrograms of plutonium; scheduling of experiments and handoff of material between experimental groups had to be carefully monitored. Los Alamos' first experimental results emerged in mid-July, 1943: a measurement of the number of neutrons emitted in the slow-neutron fission of a 165-μg sample of plutonium. At 2.6 $\pm$ 0.2, this proved to be about 20 % greater than the corresponding number for uranium. Measurements of fission cross-sections for both elements began soon thereafter.

At the time the Lewis Committee was preparing its report, the Governing Board acted on a proposal that would indirectly lead to serious international repercussions years later. On May 6, Hans Bethe put forth a suggestion to hold a regular technical colloquium every week or two. Groves saw the idea as a potentially enormous risk to his policy of compartmentalization of information, wherein individuals were to have access only to what they strictly needed to know to do their job. Oppenheimer maintained that a colloquium would be the most efficient way to share information among individuals with legitimate need-to-know. Groves relented, although he had Oppenheimer agree to restrict the number of participants and to establish a vouching system. Groves' concern was such that he raised the issue at a meeting of the Military Policy Committee on June 24. The result, engineered by Vannevar Bush, was a June 29 letter to Oppenheimer from President Roosevelt. The President expressed his appreciation for the scientists' work on behalf of the war effort, but made clear the need for very strict secrecy. Groves wrote in his memoirs that he felt that the colloquium existed not so much to provide information as to support morale and a feeling of common purpose. However, his concern with security proved justified. One of the regular colloquium participants was theoretical physicist Klaus Fuchs (Fig. 7.5), a German-born member of the British Mission (Sect. 7.4) who later passed detailed design

Fig. 7.5 Klaus Fuchs (1911–1988), ca. 1940. *Source* http:// commons.wikimedia.org/ wiki/File:Klaus_Fuchs_- _police_photograph.jpg

information on the *Fat Man* implosion bomb to the Soviets. Fuchs' treachery was not discovered until after the war, at which time he was working for the British atomic energy program. In 1950, he was convicted of espionage and jailed. After his release in 1959, Fuchs emigrated to East Germany, and lived in Dresden until his death in 1988.

We cannot know what Fuchs might have passed to the Soviets had he *not* been privy to the colloquia, but attending them certainly gave him a synoptic view of the Laboratory's activities. As John Manley wrote, Fuchs didn't have to *penetrate* Los Alamos, he was an official member of the staff, and a very respected member of the Theoretical Division; by having himself appointed as liaison between the Theoretical Division and the (later) Explosives Division that did much of the work on the plutonium bomb, he gained an intimate working knowledge of that device. Two other Los Alamos employees, Theodore Hall and David Greenglass, also passed information to Soviet operatives.

Human nature being what it is, no security system will be perfect. While Groves' security was so tight that most scientist's wives had no idea what their husbands had been doing until after Hiroshima, some superiors did tell subordinates of the purpose of their work in order to boost morale and give them a sense of purpose. Groves' compartments were hardly hermetically sealed.

7.3 Life on the Hill

Oppenheimer's notion of running the Laboratory with a staff of a couple hundred soon ran up against the enormity of its task. On average, the working population of Los Alamos doubled about every nine months. By early June, 1943, "The Hill" was home to over 300 officers and enlisted personnel in addition to some 460 civilian employees. By the end of the year, the total was approaching 1,100. A census of personnel in May, 1945 counted 1,055 members of the military Special Engineer Detachment (below); 1,109 civilians, and 67 Women's Army Corps members, for a total of over 2,200. Like Oak Ridge, one product for which Los Alamos became known was babies. The most probable age of staff members was only 27. Many were recent college graduates starting families, and they wasted no time in doing so. During the war, 208 babies were born at Los Alamos (including Oppenheimer's daughter, Katherine, in December, 1944); nearly 1,000 would arrive between 1943 and 1949. All birth certificates listed addresses as Box 1663, Santa Fe, New Mexico, the Laboratory's official address. By June, 1944, one-fifth of all the married women at Los Alamos were in some stage of pregnancy, and approximately one-sixth of the population were children. The spate of fecundity prompted a poem:

> The General's in a stew
> He trusted you and you
> He thought you'd be scientific
> Instead you're just prolific
> And what is he to do?

By the time of the *Trinity* test in July, 1945, Los Alamos would boast a total population of just over 8,000. By the end of 1946, the number of housing and apartment units for families alone numbered 617, not including 16 ranch houses obtained from the original school, dozens of trailers, and 51 less ostentatious "winterized hutments". Thirty-six dormitories and 55 barracks provided living quarters for 2,700 single personnel. Fuller Lodge, one of the main ranch school buildings, served as a dining area; eventually it would serve some 13,000 meals per month. Because of wartime secrecy, no official census was attempted at Los Alamos until April, 1946, by which time it was a community of about 10,000.

Even more than Oak Ridge and Richland, Los Alamos was a frontier town. Despite the spectacularly beautiful natural surroundings and sense of companionship that veterans of The Hill would later speak of, life could be arduous. Living conditions for early arrivals were difficult, with several families often crowded together or housed in nearby guest ranches. Wartime construction restrictions dictated that new houses were to be equipped only with showers (as opposed to bathtubs), with the result that the only bathtub-equipped houses were a few which had served as residences for teachers at the Ranch School; this group of homes became known as "Bathtub Row." Transportation had to be arranged over roads that were primitive at best. Housing, water, milk, meat, and fresh vegetables were always in short supply. No sidewalks, garages, or paved roads were put in. All houses were painted Army green and referred to colloquially as greenhouses; Los Alamos was generally regarded as having the worst housing of the entire Manhattan Project. To conserve water, bathers were encouraged to limit showers to a minute or two. At the high altitude, simple meals could take hours to cook. Turning on a faucet might well yield algae, sediments, or worms. James Conant's granddaughter, Jennet Conant, has written that one GI named the place "Lost Almost." Ruth Marshak, wife of theoretical physicist Robert Marshak, described her feeling about the place as "akin to the pioneer women accompanying their husbands across uncharted plains westward, alert to dangers, resigned to the fact that they journeyed, for weal or woe, into the Unknown." For new arrivals, the first stop after a long, often dusty journey was an unassuming office at 109 East Palace Avenue in Santa Fe. There they would be met by Mrs. Dorothy McKibbin, who arranged for an even dustier ride northward to the mesa. McKibbin began work in March, 1943, and would remain on to manage the office until retiring in June, 1963.

A serious problem for all Project sites, particularly Los Alamos, was that of securing enough technically-trained personnel. This was addressed in two main ways. Many scientists' wives were pressed into service in technical/scientific, hospital, administrative, and school-system positions. By October, 1944, some 30 % of the Laboratory's 670 civilian employees were women. To prevent scientifically-educated individuals such as graduate students from being drafted and sent overseas or otherwise lost to the Project, the MED recruited these individuals into a so-called Special Engineer Detachment (SED), which was created on May 22, 1943, as the 9812th Technical Service Unit. By the end of 1943, nearly 475 SED's were present at Los Alamos. By August, 1944, they comprised almost one-third of the Laboratory's scientific staff, and their numbers reached some 1,800 by the end of the war. By the spring of 1945, about 29 % of SEDs posted to Los Alamos held college degrees, including a number of Masters and Doctorates. The transition from civilian to military life for SEDs was more than symbolic, however. Housing could not be provided for married enlisted men, and security regulations prohibited them from bringing their wives to Santa Fe or other nearby communities. Each man was allocated only 40 square feet in a military barracks, and not until the summer of 1944 were furlough regulations relaxed. SEDs were but one component of the Manhattan Project's complement of enlisted personnel, which by the fall of 1945 totaled about 5,000; at Los Alamos, some 42 % of the staff would be in uniform. Curiously, Groves does not mention the SEDs at all in his memoirs.

As at Oak Ridge and Richland, all of the services expected by a highly-educated population had to be provided. A Community Council was established (also known as the Town Council, proposed by Robert Wilson), with members elected by popular vote. Nursery and elementary schools had to be set up; by the end of 1946 the elementary school alone enrolled over 350 students. A high school, traffic laws, a court system, cafeterias, sewage systems, a fire department (chimney and brush fires were common; Los Alamos sported over 6,800 fire extinguishers), laundry services, a general store, a motor pool (hundreds of vehicles), an automobile repair garage, a cleaning and pressing shop, a post office, garbage collection, veterinary services (over 100 horses for the Military Police alone), dental services, and a hospital had to be organized. A policy for housing assignments and rental rates was established which took into account an employee's occupation, family status, and salary. Recreational activities included hiking, horseback riding, skiing, skating, numerous parties, visits to Indian pueblos, and a rough nine-hole golf course. For the first 18 months of the project, security regulations severely restricted personal off-site travel; Groves did not want anyone to feel the slightest desire to use any outside facilities. Edwin McMillan's wife, Elsie, who knew the purpose of the Laboratory, later remarked that "We had parties, yes, once in a while, and I've never drunk so much as there at the few parties, because you had to let off steam, you had to let off this feeling eating your soul, oh God are we doing right?"

Salary scales were a chronic area of discontent. Pay inequities often arose between academic scientists and technicians in that the latter received higher salaries, consistent with what they could command in the civilian world. Mail was subject to censorship; all letters had to be addressed to Box 1663. Letters could not contain last names or information which might provide a clue as to the Laboratory's location; the word "physicist" was strictly forbidden. Scientific workers were not allowed to maintain personal accounts in local banks; the Business Office would make up a monthly payroll and send it to Los Angeles, from where checks would be mailed out to banks designated by employees. Eventually, every resident over the age of six was issued a security pass. Even at the top administrative levels, Groves kept Los Alamos largely isolated from other branches of the Project. As he laid out in a memorandum in June, 1943, any liaisons with other sites or individuals had to be personally sanctioned by him, and discussions were to be limited to a list of approved topics. To the outside world, Los Alamos did not exist.

7.4 The British Mission

A group that made contributions to the Manhattan Project out of all proportion to its number was a contingent of British and European-born scientists known formally as the British Mission. The story of how the British Mission scientists came to America has as much if not more to do with politics as it did with physics and engineering, and merits a brief description.

It was described in Sect. 4.11 how the American assessment of cooperation with Britain on atomic matters cooled considerably between the fall of 1941 (when interchange on technical issues had been endorsed) and late 1942, when Vannevar Bush informed President Roosevelt that there would be no unfairness to the British if all interchange were to cease. The British, however, were not about to let the impact of their initial work slip away. Bush's counterpart in Britain was Sir John Anderson, a member of Churchill's War Cabinet. Anderson's involvement in wartime British administration ran deep: he would also serve as Chancellor of the Exchequer from September, 1943, to July, 1945. As a student, Anderson had studied geology, chemistry, and mathematics (he wrote a thesis on uranium), but he turned to a career in the civil service.

In March, 1942, Anderson wrote to Bush that he felt it desirable to continue complete collaboration. Bush later responded with a description of his June, 1942 rearrangement of the S-1 administrative structure (Sect. 4.9), but made no commitments. Soon thereafter, Churchill visited Roosevelt in America; they discussed the uranium issue on the afternoon of June 20. Churchill urged that Britain and America should pool their information, work as equal partners, and share whatever results might emerge, despite the fact that production plants would be located in the United States. Three weeks later, Roosevelt wrote to Churchill and Bush that he and the Prime Minister were "in complete accord," but no written agreement had been signed nor any details developed. On August 5, Anderson attempted to formalize the discussions in letters to Bush, suggesting that a British-designed diffusion plant be built in America, that a heavy-water pile program be transferred to Canada, that a common patent policy be developed, and that a joint nuclear energy commission be established. But Anderson's timing could hardly have been worse. The American program was in the middle of its transfer to military authority; for Groves and Bush, international negotiations could only be a hindrance. While the British had made some progress with diffusion, research on all of the other production methods—electromagnetic, piles, and centrifuges—were strictly American affairs. Bush informed Anderson of the evolving arrangements in America on October 1, and evasively referred to keeping up contact on how best to put the resources of both countries to work. Secretary of War Stimson discussed the issue with President Roosevelt on October 29, and suggested that matters be allowed to go along for the time being without sharing any more information than was necessary.

The rapidly-diverging viewpoints of British and American atomic-project leaders became clear on December 11, 1942, when James Conant conferred with

Directorate of Tube Alloys chief Wallace Akers (Sect. 4.10) in Washington. Conant presented the American perspective, which was that interchange should be restricted only to information that Britain could use during the war. Akers argued that Roosevelt and Churchill intended collaboration in both research and production, and felt that British scientists should have access to all large-scale American developments. Conant reported back to Bush the next day; four days thereafter, Bush carried to Roosevelt the 29-page December 15 MPC report which recommended no or only limited interchange. Other factors were in play, however. On September 29, Britain and Russia had concluded an agreement on exchange of new weapons which covered both those in use and any that might be developed. Roosevelt and Stimson had apparently known nothing of this until around December 26, when a copy of the agreement reached Stimson. Clearly, such an the agreement would cast into doubt the security of any atomic information passed on to the British. On December 28, Roosevelt initialed the MPC report, setting the policy at limited interchange: cooperation in the design and construction of the diffusion plant, research-level information interchange on plutonium and heavy water, and no sharing of information on the electromagnetic method or Los Alamos. Churchill brought up the issue with Roosevelt again when the two met at the Casablanca Conference in January, 1943, and further protested to Roosevelt aide Harry Hopkins in late February that interchange restrictions were contrary to the idea of a jointly-conducted effort. Hopkins took no action until prodded again by Churchill by cable on April 1. Roosevelt left it to Bush to develop a reply, which was that there was no reason to change the American position.

Churchill raised the issue yet again during a visit to Washington in late May, during which Bush was brought into discussions with Hopkins and British advisors. On the rationale that since a weapon might be developed in time for use in the war (in which case the "direct use" scenario would hold), Churchill departed with the understanding that he had secured from Roosevelt a private promise that the work was to be joint and that interchange would be resumed. Bush met with Roosevelt on June 24 to review the situation. Roosevelt had apparently not been apprised of Bush's discussion with Hopkins and the British advisors, and Bush left the meeting with the impression that Roosevelt had no intention of going beyond the standing limited-interchange policy. Churchill tenaciously raised the issue with Roosevelt again on July 9, and the President finally acquiesced: on the 20th he wrote to Bush to instruct him to renew full interchange with the British.

At the time of Roosevelt's July 20 directive, Bush was in London conferring with counterparts there on scientific aspects of the war. He met with Churchill on the 15th, and the Prime Minister expressed his frustration with the uranium issue in no uncertain terms. Unaware of the President's directive, Churchill, Hopkins, Bush, and Anderson met again on July 22, at which time Churchill offered a five-point proposition that would form the basis of the so-called Quebec Agreement that would be signed a month later. The essential points were that (1) the enterprise would be joint with free interchange; (2) neither government would employ nuclear weapons against the other; (3) neither would pass information to other countries without the consent of the other; (4) use of the bomb in war would require common consent; and (5) the President might limit commercial or industrial uses by Britain in such a manner as he considered fair in view of the expense being borne by the United States. Points (2)–(5) would go into the Quebec Agreement essentially unchanged, but the interchange issue would be referred to a joint committee. In Washington in early August, Bush and Conant met with Anderson at the British Embassy. Working from Churchill's draft proposal, Anderson added the establishment of a "Combined Policy Committee" to coordinate what work would be done in each country and to serve as a focal point for exchanging information. Interchange on scientific research and development was to be "full and effective," but interchange in the area of design, construction and operation of full-scale plants was left on an *ad hoc* basis to be decided by the Committee. Stimson, Bush, and Conant were specified as the American members of the Committee; the other members were two British military officers and the Canadian Minister of Munitions and Supply. The formal agreement was signed by Roosevelt and Churchill on August 19, 1943, during a meeting in Quebec City, and the Committee met for the first time in Washington on September 8.

As part of the interchange program, groups of British scientists, both native and newly-naturalized, came to America. In particular, they became involved with a pile-research program in Montreal, the diffusion and electromagnetic projects, and Los Alamos. James Chadwick headed the "British Scientific Mission in USA," and spent most of his time in Washington. Nineteen individuals would ultimately be appointed to work at Los Alamos, including Rudolf Peierls, Klaus Fuchs, and Otto Frisch; apparently General Groves was allowed no security vetting of these individuals. The first two members of the contingent, Frisch and Birmingham alumnus Ernest Titterton, arrived on December 13, 1943. As is described in the following sections, members of the British Mission at Los Alamos contributed a number of important theoretical and experimental insights. Hans Bethe was of the opinion that

> For the work of the Theoretical Division of the Los Alamos Project during the war the collaboration of the British Mission was absolutely essential. It is very difficult to say what would have happened under different conditions. However, at least, the work of the Theoretical Division would have been very much more difficult and very much less effective without the members of the British Mission, and it is not unlikely that our final weapon would have been considerably less efficient in this case.

General Groves' dismissive attitude toward British contributions to the Manhattan Project was probably driven by patriotic pride, but was unfair. Unfortunately, many Americans do not fully appreciate the contributions of the British Mission to the success of the Project.

7.5 The Physics of Criticality

This Section and the following two describe the physics underlying the concept of critical mass and bomb-core assembly that were central to the work of Los Alamos. Even if you do not wish to fully explore the technical details, it will be worth scanning these sections to get some understanding of the constraints that the scientists and engineers of Los Alamos faced.

As described in Sect. 3.5, the fundamental idea behind a critical mass is to assemble a great enough mass of fissile material such that once fissions have been initiated, more neutrons will cause subsequent fissions than will escape from the mass. The mass will eventually disrupt itself, but the goal is to obtain, at least for a while, a growing population of neutrons. The critical mass depends on the density of the material, the number of neutrons liberated per fission, and the cross-sections for fission and scattering. The most straightforward analytic way of determining the critical mass is by applying *diffusion theory* to the travel of neutrons from when they are created to when they encounter another nucleus. Derivations of the diffusion equation are available in a number of texts; only the essential expressions and results are discussed here. Stated more precisely, what diffusion theory provides is a way of calculating the critical *radius* for a given set of nuclear parameters. This can be transformed into an equivalent mass upon knowing the density of the material involved.

Central to the calculation of critical radius are the so-called fission and transport *mean free paths* for neutrons, respectively symbolized as λ_f and λ_t. These are given by

$$\lambda_f = \frac{1}{\sigma_f n} \tag{7.1}$$

and

$$\lambda_t = \frac{1}{\sigma_t n}. \tag{7.2}$$

σ_f is the fission cross-section, and σ_t is the so-called transport cross-section. If neutron scattering is isotropic, the transport cross-section is given by the sum of the fission and elastic-scattering cross-sections:

$$\sigma_t = \sigma_f + \sigma_{el}. \tag{7.3}$$

In words, the meanings of λ_f and λ_t can be expressed as "the average distance a neutron will travel before it is consumed in causing another fission," and "the average distance a neutron will travel before it is scattered or causes a fission." Recall from Chap. 3 that cross-sections are usually quoted in *barns* (bn);

1 bn $= 10^{-28}$ m^2. We do not consider here the role of *inelastic* scattering, which affects the situation only indirectly in that it lowers the mean neutron velocity.[1]

The symbol n in (7.1) and (7.2) represents the number density of nuclei, that is, the number of nuclei per cubic meter. If the fissile material has density ρ grams per cubic centimeter and atomic weight A grams per mole, then n (in nuclei per cubic meter) is given by

$$ n = 10^6 \left(\frac{\rho N_A}{A} \right), \tag{7.4} $$

where N_A is Avogadro's number, 6.022×10^{23}. The factor of 10^6 arises from converting cubic centimeters to cubic meters in the density.

Another important quantity is the average time that a neutron will travel before causing a fission, which is designated by the symbol τ. If neutrons have average speed v_{neut} and travel for an average distance λ_f before causing a fission, then it follows that

$$ \tau = \frac{\lambda_f}{v_{neut}}. \tag{7.5} $$

In the simplified case of an *untamped* spherical bomb core of radius R_{core}, that is, one that is not surrounded by any sort of enclosing jacket, diffusion theory shows that criticality will hold if the following transcendental equation is satisfied

$$ \left(\frac{R_{core}}{d} \right) \cot \left(\frac{R_{core}}{d} \right) + \frac{1}{\eta} \left(\frac{R_{core}}{d} \right) - 1 = 0. \tag{7.6} $$

In this expression, d is a measure of the characteristic size of the core, and is given by

$$ d = \sqrt{\frac{\lambda_f \, \lambda_t}{3 \left(-\alpha + \nu - 1 \right)}}, \tag{7.7} $$

---

[1] The neglect of inelastic scattering is not as drastic as it may seem, due to a combination of reasons. What matters to the growth of the neutron population is the time τ that a neutron will typically travel before causing another fission; see Eq. (7.5). But, if one averages through the many resonance spikes in Fig. 3.12, the fission cross-section for uranium-235 (and plutonium-239 as well) behaves approximately as $\sigma \sim 1/v_{neut}$. This means that the mean free path for fission, λ_f, is proportional to v_{neut} which, overall, makes τ independent of v_{neut}. This means that if a neutron has been either elastically or inelastically scattered, the time for which it will typically travel before causing a subsequent fission is largely independent of its speed. It would then seem that one should also add in the inelastic-scattering cross-section when forming the transport cross-section in Eq. (7.3). This is true, but another effect comes into play: elastic scattering is not isotropic. This has the effect of somewhat lowering the effective value of the elastic scattering cross-section. For elements like uranium and plutonium, the two effects largely cancel each other, with the net result that Eq. (7.3) is a quite reasonable approximation. Details are given in the Appendix to Serber's *Primer*; see also H. Soodak, M. R. Fleishman, I. Pullman and N. Tralli, *Reactor Handbook, Volume III Part A: Physics* (New York: Interscience Publishers, 1962), Chap. 3.

where v is the number of neutrons liberated per fission. The parameter α will be described presently. The quantity η in (7.6) is dimensionless, and is a measure of the ratio of the transport mean free path to the scale size:

$$\eta = \frac{2\lambda_t}{3d} = 2\sqrt{\frac{\lambda_t(-\alpha + v - 1)}{3\lambda_f}}. \tag{7.8}$$

Because of the presence of the cotangent, (7.6) cannot be solved analytically; it can only be solved by trial and error or by using, for example, the root-finding "Goal Seek" function in a spreadsheet.

The parameter α is involved with the time-growth of the number of neutrons in the bomb core. In working through the algebra of diffusion theory, it is actually easier to deal with the *density* of neutrons, that is, the number of neutrons per cubic meter. If the initial density of neutrons at the center of the core at the moment when fissions begin ("time zero") is N_o, then the central density at any later time t is given by

$$N_t(t) = N_o e^{(\alpha/\tau)t}. \tag{7.9}$$

The initial neutrons have to be provided by a suitable *initiator*; this is discussed in Sect. 7.7.1.

If $\alpha > 0$, the neutron density grows exponentially. In this case, one has a condition of *supercriticality*, and the energy liberated by fissions will also grow exponentially. If $\alpha < 0$, the reaction will quickly die out. If $\alpha = 0$, the neutron number density neither increases nor decreases once it has been established, in which case one has *threshold criticality*. To determine the so-called threshold critical radius, set $\alpha = 0$ in (7.7) and (7.8), and solve the criticality equation (7.6) for R_{core}. An untamped core is also known as a *bare* (or *naked*) core, and this value of R_{core} is consequently known as the *bare threshold critical radius*, designated R_{bare}. The corresponding bare threshold critical mass M_{bare} follows from $M_{bare} = 4\pi\rho R_{bare}^3/3$. It is this mass that is often referred to as *the* "critical mass," although, as explained below, this commonly-used term is not really uniquely defined.

Table 7.1 shows calculated bare critical radii and masses for uranium-235 and plutonium-239. Sources for the parameter values are cited in Reed, *The Physics of the Manhattan Project* (Springer, Berlin, 2011).

To put these numbers in some perspective, a regulation softball has a radius of about five centimeters and a mass of about 180 g. A threshold bare critical mass of plutonium is only slightly larger, but some 90 times heavier. Forty-six kilograms is equivalent to about 101 pounds, and 16.7 kg to about 37 pounds. We will see shortly, however, that these masses can be significantly reduced with use of a surrounding tamper (Fig. 7.6). One set of numbers to especially note are the neutron travel-times-to-fission, τ: they are on the order of a only few *nanoseconds*. Nuclear explosions are incredibly brief phenomena. Lest you think that openly publishing estimates of critical masses is to flirt with divulging classified data, put your mind at rest; such estimates have been available in the public domain for

Table 7.1 Parameter values and critical radii and masses for bare threshold criticality

Quantity	Unit	Physical meaning	^{235}U	^{239}Pu
A	g/mol	Atomic weight	235.04	239.04
ρ	g/cm^3	Density	18.71	15.6
σ_f	bn	Fission cross-section	1.235	1.800
σ_{el}	bn	Scattering cross-section	4.566	4.394
ν	–	Neutrons per fission	2.637	3.172
n	10^{28} nuclei/m^3	Nuclear number density	4.794	3.930
λ_f	cm	Fission mean free path	16.89	14.14
λ_t	cm	Transport mean free path	3.596	4.108
τ	10^{-9} s	Time between fissions	8.635	7.227
d	cm	Size scale factor; Eq. (7.7)	3.517	2.985
η	–	Equation (7.8)	0.6817	0.9174
R_{bare}	cm	Bare threshold critical radius	8.366	6.345
M_{bare}	kg	Bare threshold critical mass	45.9	16.7

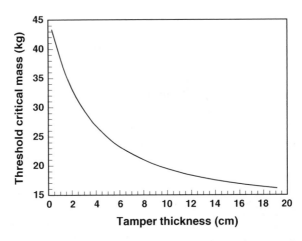

Fig. 7.6 Threshold critical mass of uranium-235 as a function of the thickness of a surrounding tungsten-carbide (steel) tamper. The tamper is presumed to fit snugly around the core with no gap between them. An example, based on the *Little Boy* bomb: If the tamper is assumed to have an outer radius of 18 cm, the core critical radius proves to be 6.17 cm. The core mass in this case is 18.4 kg; the tamper thickness will be 11.83 cm, and its mass will be about 350 kg (770 pounds). A critical mass of 18.4 kg represents a reduction of about 60 % from the untamped value of 45.9 kg. Adopted from Reed (2009)

decades. For example, a 1963 publication of the United States Atomic Energy Commission, "Reactor Physics Constants," a compilation of data for nuclear engineers, lists the *experimentally determined* bare critical mass for highly enriched uranium (93.9 % U-235) as 48.8 kg, and that for Pu-239 as 16.3 kg. Estimating a critical mass is one of the *least* difficult parts of making a nuclear weapon.

To students of physics and engineering, the forgoing equations will appear rather straightforward. It often comes as a surprise to non-scientists that calculations underlying nuclear weapons design are often no more complicated than those covered in many upper-level undergraduate physics courses. As Los Alamos mathematician Stan Ulam put it, "It is still an unending source of surprise for me to see how a few scribbles on a blackboard or on a sheet of paper could change the course of human affairs."

We come now to a very important consideration in fission-bomb design: Why it is desirable to assemble a core comprising more than one critical mass.

A bomb which contains only a single critical mass will not yield a very efficient explosion, as the core will rapidly expand and disperse itself. This typically happens over a timescale on the order of a single microsecond. To appreciate the effect of this expansion, look back to (7.8). The factor η appearing therein is independent of the density of the core material. Hence, for $\alpha = 0$, (7.6) will be satisfied by some unique value of R_{core}/d, which will be characteristic of the material being considered. Through the mean free paths and the number density n, the parameter d is proportional to the inverse of the density, $1/\rho$, so we can equivalently say that the solution of (7.6) demands a unique value of ρR_{core} for a given value of v and set of cross-sections. Stated more generally, this means that the condition for threshold criticality can be expressed as a constraint on the product ρR, where ρ is the mass density of the fissile material and R is the radius of the core.

Now suppose that we start with a core of more than one critical mass. This means that from the outset we will be specifying the value of R_{core} in (7.6), with $R_{core} > R_{bare}$. The value of ρR_{core} will then obviously be greater than that necessary for threshold criticality, ρR_{bare}. But if R_{core} is chosen in advance, of what use is equation (7.6)? The use is that there is still one variable: the time-growth parameter α. If R_{core} is specified, Eq. (7.6) can be solved for α, and it turns out that one will *always* find $\alpha > 0$ if $R_{core} > R_{bare}$. Hence, a way to get an exponentially *growing* supercritical reaction as opposed to a steady-state one is to start with more than one critical mass of material. For a core of two bare critical masses, α is typically on the order of 0.5.

Now consider what happens as the supercritical core begins expanding. The radius R will increase, but the density will drop. What happens to the product ρR? The mass M of the material is essentially fixed, and since density is equal to mass divided by volume, we will have $\rho \propto M/R^3$. This means that, at any time, $\rho R \propto M/R^2$. Consequently, it is inevitable that ρR will eventually fall below the threshold value $\rho_{original} R_{bare}$, at which time criticality will be lost. But if one started with only a single critical mass, critically would be lost as soon as the core begins expanding, which would be essentially immediately. It is to avoid this prompt shutdown that it is so important to assemble a multiple-critical-mass core.

At this point, it is instructive to return to the Frisch-Peierls formula quoted in Sect. 3.7 for the energy released by an exploding bomb. In terms of the present notation, this appears as

$$E \sim 0.2M \left(\frac{R_{core}}{\tau}\right)^2 \left(\sqrt{\frac{R_{core}}{R_{bare}}} - 1\right). \qquad (7.10)$$

As an example, consider 1.5 critical masses of U-235. From Table 7.1, this requires $M = 68.9$ kg. One and one-half critical masses will have $R_{core}/R_{bare} = 1.5^{1/3} = 1.145$, or $R_{core} = 9.577$ cm. With $\tau = 8.635 \times 10^{-9}$ s, we have, in MKS units (E will emerge in Joules)

$$E \sim 0.2(68.9) \left(\frac{0.09577}{8.635 \times 10^{-9}}\right)^2 \left(\sqrt{1.145} - 1\right) \sim 1.19 \times 10^{14} J. \qquad (7.11)$$

One kiloton is equivalent to 4.2×10^{12} J, so $E \sim 28$ kilotons, about the correct order of magnitude for a Manhattan Project fission bomb. Actually, this formula makes somewhat optimistic predictions. A detailed derivation shows that the factor of 0.2 in (7.10) is equal to α^2 [see Reed (2011), Sect. 2.4]. For 1.5 critical masses of U-235, $\alpha_{initial} \sim 0.3$, so the Frisch-Peierls factor of 0.2 is probably more on the order of $\sim 0.3^2 \sim 0.09$, which reduces E to ~ 13 kilotons, a figure very close to the yield of the Hiroshima *Little Boy* uranium bomb. This agreement is somewhat fortuitous, however, as the Frisch-Peierls formula does not account for two factors. The first is that, during the course of the explosion, α decreases from its initial value (given by solving 7.6) down to zero at the moment criticality is lost. The effective overall value of α^2 will thus be somewhat less than its initial value, and this will act to decrease the estimate of E. But countering this is a second effect: *Little Boy* was heavily tamped, which increased the yield by slowing the expansion and reflecting some neutrons back into the core (see below). The Frisch-Peierls formula is nevertheless very handy for getting an order-of-magnitude estimate of what one might expect.

If you are solving for α by trial and error using a calculator or spreadsheet, it is helpful to have an approximate starting value. To a rough approximation, a convenient expression for this is

$$\alpha \sim (\nu - 1)\left[1 - (R_{bare}/R_{core})^2\right]. \qquad (7.12)$$

For our example with 1.5 critical masses, this gives $\alpha \sim 0.39$, about 25 % high compared to the true value of 0.30.

From the above analysis, it would seem that you would have no hope of making an effective nuclear explosion if you have available less than one critical mass of material. Surprisingly, this is not the case. If you could crush the material from its normal density to a sufficiently great density, you could achieve a value of ρR which would exceed that for normal-density material, $\rho_{normal} R_{bare}$. If you have available a fraction f of a single bare critical mass, the condition for achieving criticality is to crush it to a density which satisfies $\rho_{compress} \geq \rho_{normal}/\sqrt{f}$. Engineering such an *implosion* is very difficult (Sect. 7.11), but offers the possibility of making a bomb with considerably less fissile material than would be the case for a non-implosion bomb. At Los Alamos, weapons engineers had to develop implosion for use in plutonium bombs because the properties of reactor-produced plutonium precluded use of a simpler triggering method which had been developed for the uranium bomb (Sects. 7.7, 7.11).

If the bomb core is non-spherical or is surrounded by a tamper (a metal jacket), the mathematics of criticality become more complicated; the geometry of the core and the thickness and properties of the tamper enter into the calculations, and there is no simple "ρR" measure of criticality to be had. In theory, an infinitely-thick tamper reduces the threshold critical radius by a factor of two, which reduces the critical mass by a factor of eight. As illustrated in Fig. 7.6, most of the critical mass reduction comes with the first few centimeters of tamper, so it is certainly worthwhile going to the trouble of providing one. The tamper also boosts the efficiency of the weapon by briefly retarding the core expansion, allowing criticality to persist a little longer. A weapon that makes use of both implosion and a tamper will be more efficient yet. A tamper is "dead weight" as far as transporting a bomb to a target goes, but is vital to its efficient functioning.

7.6 Critical Assemblies: The Gun and Implosion Methods

The question of how to assemble a supercritical mass was considered very early on in the Los Alamos project. In his *Primer*, the first and most straightforward system described by Robert Serber was the so-called "gun" method; he referred to it as "shooting." As ultimately realized in the Hiroshima *Little Boy* bomb and as sketched in Fig. 7.7, the concept is to begin with two sub-critical pieces of fissile material within the barrel of an artillery gun. A cylindrical "target" piece is held fixed at the nose end of the barrel, while a mating sleeve, the "projectile" piece, is fired toward the target piece from the tail end. When fully mated, the two comprise

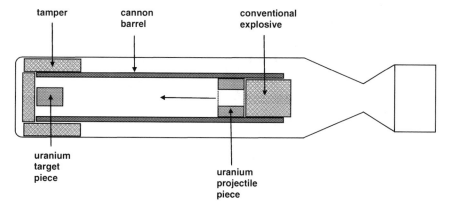

Fig. 7.7 Schematic illustration of a gun-type weapon. The uranium projectile is fired toward a mating target piece in the nose. See also Fig. 7.18

Fig. 7.8 Assembly process
for a gun-type fission weapon

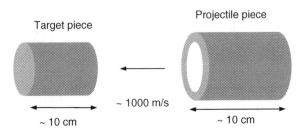

more than a critical mass of material. Since the *average* density of the projectile piece is fairly low because it is hollow, it can comprise by itself the equivalent of more than one "solid" critical mass, thereby giving the entire assembly over two critical masses. By surrounding the target piece with a tamper, the completed assembly can potentially comprise several tamped critical masses.

In World War II, the greatest muzzle velocity achievable with artillery pieces or naval cannons was about 1,000 m/s. Since the target and projectile pieces are on the order of 10 cm in size, an assembly speed of 1,000 m/s implies that about 100 μs will elapse between the time that the leading edge of the projectile piece first encounters the target piece, and when full assembly is achieved (Fig. 7.8). This 100-μs assembly timescale is *extremely* important, and we will return to it in the following section.

Another method described by Serber contains the genesis of what came to be known as the implosion technique. As described in the *Primer*, the idea was to mount pieces of fissile material on the inside of a ring, with explosive material distributed around the outside of the ring. When fired, the fissile pieces would be blown inward to form a cylinder or sphere, as suggested in Fig. 7.9.

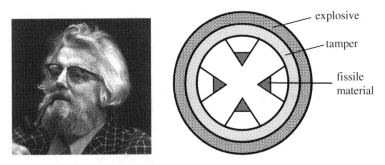

Fig. 7.9 Seth Neddermeyer (1907–1988) in his later years. *Source Photograph by David Azose, courtesy AIP Emilio Segre Visual Archives, Physics Today Collection.* Right: Sketch of early implosion concept adapted from Robert Serber's *Los Alamos Primer.* The *four triangular-shaped* wedges are pushed together by detonating an enclosing ring of explosives; the fissile material (*shaded*) consequently forms a cylindrical critical assembly. If the wedges are three-dimensional pyramids, a spherical assembly results. Serber's original sketch did not include the surrounding tamper

Conception of the implosion method is often attributed to physicist Seth Neddermeyer, a Caltech Ph.D. whom Oppenheimer had recruited from the National Bureau of Standards. However, Serber has dismissed this attribution as "television history," stating that the idea had been raised by Richard Tolman at the 1942 Berkeley summer conference. Serber and Tolman (and later Tolman alone) wrote memos on the subject which apparently went up to Bush and Conant. Tolman wrote Oppenheimer on March 27, 1943 (just before Serber's first orientation lecture), to describe how it might be possible to blow a shell of "active material" inward upon itself, beginning with an ordinary explosive. Neddermeyer apparently conceived of modifying the idea to surround a thick but initially centrally-hollow cylindrical or spherical core with a tamper, which itself would be surrounded by a layer of explosive. When detonated at many points simultaneously, the explosive would push inward at several kilometers per second, crushing the core to critical density in much less time than a gun mechanism could assemble subcritical pieces. However, the idea originated, Neddermeyer was struck by the concept, and the record shows that it was discussed at Los Alamos Planning Board meetings held on March 30 and April 2, 1943. By late April, Neddermeyer had developed calculations on the velocities that might be achieved, and was assigned by Oppenheimer to lead a small group devoted to implosion research within the Ordnance Division. While implosion was given low priority at first and considered a backup scheme in case the gun method failed to work, its priority was great enough that preliminary estimates of the size and weight of a spherical implosion weapon were generated in order to investigate how test-drop mockups would fare in comparison to models of the more conventionally-shaped gun bomb (Sects. 7.8 and 7.9). Neddermeyer carried out his first implosion test-shot on July 4, 1943, using tamped TNT surrounding hollow steel cylinders. The symmetry of the implosion was poor, but the shot did demonstrate the fundamental feasibility of using an explosion to crush something. In time, implosion would come to be crucial to the success of the plutonium bomb project.

7.7 Predetonation Physics

In the preceding section, it was remarked that the 100 μs assembly timescale for a gun-type bomb would prove to be of great significance. This timescale, which is purely mechanical in origin, is one of three timescales that are involved in the efficient functioning of a fission weapon. The other two involve the physics of the fissioning core, and also need to be appreciated in order to understand the importance of the 100-μs assembly time.

The first of the other two timescales involves how much time is required for the entire core to fission once the chain reaction has been initiated. In Sect. 7.5, it was described how once a neutron is emitted in a fission, it will travel for only about 10 ns before causing another fission. With such a small travel time between fissions, it will take only about 1 μs to fission the entire bomb core. This remarkably brief time can be understood with a simple estimate. Suppose that we have a core of mass M kilograms of fissile material of atomic weight A grams per mole. The number of nuclei N in the mass will be $N = 10^3 M N_A / A$, where N_A is Avogadro's number. If v neutrons are produced per generation, then the number of generations G that will be required to fission the entire mass will be $v^G = N$. At τ seconds per generation, the time to fission the entire mass will be $t_{fiss} \sim \tau\, G \sim \tau\, \ln(N)/\ln(v)$. For $M = 50$ kg of U-235 with $A = 235$ g/mol, $v = 2.6$, and $\tau \sim 8 \times 10^{-9}$ s, you should be able to show that $t_{fiss} \sim 0.5$ µs. Even if only half of the neutrons cause fissions ($v = 1.3$), $t_{fiss} \sim 2$ µs, still much less than the assembly timescale.

The second core-physics timescale concerns the amount of time required for the core to expand to the point where its decreasing density results in criticality shutdown. The time-evolution of the core to this condition, which is known technically as "second criticality" (first criticality is defined below), has to be determined via numerical simulations of exploding cores. The result is that this expansion takes about the same amount of time as required to fission the entire core once the chain reaction has started: a microsecond or two. The similarity of these timescales means that there is a very strong competition between the exponential growth of the explosion and the onset of second criticality. As Robert Serber wrote in *The Los Alamos Primer*, "Since only the last few generations will release enough energy to produce much expansion, it is just possible for the reaction to occur to an interesting extent before it is stopped by the spreading of the active material."

We can now understand the potential problem with the 100-µs assembly time. The difficulty is sketched in Fig. 7.10. At some point during the assembly of the core, a critical mass will come to be present in the partially-assembled system; this is known as "first criticality." If a stray neutron should initiate the first fission at some time after first criticality (which means that an exponential chain reaction will begin) but before the assembly is complete, the reaction may well reach second criticality before assembly can be completed. The result would be an explosion of much lower efficiency than what the weapon was designed to achieve on the presumption of the reaction not being initiated until assembly was fully completed. In general, because the chain reaction can begin at any time between first criticality and the fully-assembled state (the "supercritical period"), there will be a range of possible weapon efficiencies. The worst-case scenario is if the chain gets initiated just at the moment of first criticality, in which case there will likely

Fig. 7.10 Core achieves first criticality before assembly is completed

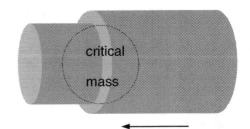

be little if any hope of completing the assembly before second criticality. Such an extreme predetonation is known to weapons engineers as a "fizzle."

There are two possible sources for stray neutrons, and their effects are additive. Both are controllable to some degree, albeit with difficulty. These sources are described in the following two sub-sections.

7.7.1 The (α, n) Problem

The first source of stray neutrons was described in Chap. 4: If the fissile material contains any light-element impurities, especially trace amounts of elements such as beryllium or aluminum, neutrons will be generated as a result of alpha-bombardments originating from alpha-decay of the fissile material. Chemical processing and purification will inevitably introduce some level of impurities, and since uranium and plutonium are both natural alpha-emitters, the problem is unavoidable: a stray neutron may initiate a premature reaction as soon as first criticality is achieved. The rates of alpha-emission per gram of material are fixed by nature, so minimizing the probability of a predetonation during core assembly means minimizing the level of impurities, and making the assembly time as short as possible. The probability can never be reduced to zero, but can be made acceptably small in most cases.

The first step in analyzing the (α, n) issue involves the half-life decay-rate formula of Chap. 2:

$$R_\alpha = 10^3 \left(\frac{N_A}{A}\right)\left(\frac{\ln 2}{t_{1/2}}\right) \text{ decays per kg per second,} \qquad (7.13)$$

where A is the atomic weight in g/mol, and $t_{1/2}$ is the alpha-decay half-life of the fissile material in seconds. We do not need to worry about the exponential decline in decay rate because the half-lives for U-235 and Pu-239 are so great that even if a bomb core has been sitting in storage for several decades, its activity will not have declined appreciably from when it was first manufactured.

Numbers for U-235 and Pu-239 appear in Table 7.2. The decay rates are large, but so far as the (α, n) problem goes, they are mitigated by two factors: the typical *yield* of such reactions, and the *range* of alpha particles within the fissile material.

The yield y of a reaction is a reflection of the fact that atoms are mostly empty space: not all alpha-particles will strike a light-element nucleus. In a postwar textbook on nuclear physics, Enrico Fermi gave some illustrative figures for two cases. The first is that 1 Ci of radium well-mixed with beryllium yields about 10–15×10^6 neutrons per second. The second case is that 1 Ci of polonium well-mixed with beryllium yields some 2.8×10^6 neutrons per second. (Both radium and polonium are natural alpha-emitters.) On recalling that 1 Ci is equivalent to 3.7×10^{10} decays per second, these figures give yields of 2.7–4.1×10^{-4} and 7.6×10^{-5} neutrons per alpha, respectively. Most light-element (α, n) reactions have yields on the order of $y \sim 10^{-4}$, which will be assumed here.

Table 7.2 Alpha-decay rates for bomb materials

Isotope	Half-life (years)	Alpha decay rate ($kg^{-1}\ s^{-1}$)
U-235	7.04×10^8	8.0×10^7
Pu-239	24,100	2.3×10^{12}

The range of a particle is a measure of how far it will travel through some material before coming to rest due to losing energy by causing successive ionizations in the material. Analyses of the rates of emission, yields, and ranges of alpha particles through samples of heavy elements leads to an expression for the average rate of neutron production R_{neut} (neutrons per second) in terms of the densities of the fissile material and the impurity. This is usually expressed as the ratio of the necessary number density of fissile nuclei to that of the light-element impurity in order to have the average rate of neutron production be *no more than* R_{neut}:

$$\left(\frac{n_{fissile}}{n_{light}}\right) > y\left(\frac{R_{alpha}}{R_{neut}}\right)\sqrt{\frac{A_{light}}{A_{fssile}}}, \tag{7.14}$$

where the A's again denote atomic weights.

Consider a 10-kg plutonium-239 core, which would have $R_{alpha} = 2.3 \times 10^{13}\ s^{-1}$. If we demand that R_{neut} be reduced to one neutron per 10,000 µs (=0.01 neutrons per 100 µs, or 100 neutron/s), take beryllium as the impurity ($A = 9$), and adopt $y = 10^{-4}$, we get

$$\left(\frac{n_{fissile}}{n_{light}}\right) > 10^{-4}\left(\frac{2.3 \times 10^{13}}{10^2}\right)\sqrt{\frac{9}{239}} \sim 4,460,000. \tag{7.15}$$

This means that no more than one atom in about 4.5 million can be one of beryllium! In a letter to James Conant on November 30, 1942, Robert Oppenheimer outlined fissile-material purity requirements, indicating a fraction by weight of beryllium in plutonium of no more than 10^{-7}; our result is of the correct order of magnitude. Oppenheimer similarly estimated the requirements for lithium and boron at a few times 10^{-7}. These are demanding, although not impossible levels of purity. Los Alamos chemists were able to reduce light-element impurities in plutonium to the level of a few parts per million, which was good enough in view of the second source of neutrons, which is described in the following subsection.

In the case of U-235, the impurity situation is much more forgiving. Suppose that we have a 50-kg core, again with beryllium as the contaminant:

$$\left(\frac{n_{fissile}}{n_{light}}\right) > 10^{-4}\left(\frac{4.0 \times 10^9}{10^2}\right)\sqrt{\frac{9}{235}} \sim 800. \tag{7.16}$$

One atom in a thousand is well within the limits of normal chemical purity.

It was appreciated from the outset at Los Alamos that the possibility of light-element-induced predetonation was going to be a much more demanding constraint for a plutonium bomb than for a uranium bomb. Impurities would have to be rigorously minimized, and the speed of assembly would have to be as great as possible. An artillery cannon capable of accelerating a projectile piece of plutonium to 1,000 m/s was anticipated to be about 17 feet long (Exercise 7.1). On the positive side, if the gun could be made to work for plutonium, it would surely work for uranium.

An issue related to alpha-particle bombardment yields is the question of how to initiate a nuclear explosion. At Los Alamos, this was accomplished by fabricating devices known, not surprisingly, as *initiators*; they were also known as "Urchins." Placed within the bomb core, these spheres, which were approximately the size of a golf ball, contained interior cavities lined with teeth which projected into a hollow center. Polonium and beryllium were deposited on opposite sides of the teeth. Upon being crushed by the incoming projectile piece of fissile material (uranium bomb) or by an implosion (plutonium bomb), the polonium and beryllium would mix; alphas from the polonium would strike beryllium nuclei, and liberate neutrons to trigger the detonation. The idea of such initiators was apparently conceived by Hans Bethe. Manhattan Project initiators used about 50 Ci of polonium-210 which was created by neutron bombardment of bismuth in the X-10 reactor at Oak Ridge and in the production reactors at Hanford via the reaction

$$\,^{1}_{0}n + \,^{209}_{83}Bi \rightarrow \,^{210}_{83}Bi \xrightarrow[5.0\,days]{\beta^{-}} \,^{210}_{84}Po. \tag{7.17}$$

50 Ci is equivalent to a mass of about 11 mg, and a rate of alpha emission of 1.85×10^{12} s^{-1}. If we suppose a yield of 10^{-4}, some 185 neutrons will be emitted if the initiator functions for 1 μs.

Initiator manufacture was a difficult business. Polonium is hazardous not only because of its high alpha-activity, but also because it is one of the most motile elements known; it is virtually impossible to work with and avoid its entrance into the human body. Fortunately, however, it does get eliminated rapidly and does not collect in bones as do radium and plutonium. Bismuth has a low thermal-neutron capture cross-section (0.01 barns), so large amounts of it had to be bombarded for a long time to make even a small amount of polonium. In a letter to General Groves on June 18, 1943, Oppenheimer related that one hundred pounds of bismuth placed near the center of the X-10 pile would create only 9 Ci of polonium every four months if the pile were operated at 20 kW/ton of fuel; a full fuel load for X-10 was about 120 t. Much greater supplies were anticipated from the Hanford piles when they went into operation; the same 100 pounds of bismuth would yield 4.5 Ci of polonium *per day*. Oppenheimer also stated that it would be desirable to have "a mean emission of 100 neutrons" during the operation of the initiator, a number of the same order as calculated above.

The polonium was separated from the parent bismuth material at a Monsanto Chemical Company facility located outside Dayton, Ohio, where Charles Thomas, Director of Research for Monsanto, set up a makeshift laboratory in the indoor tennis courts of the estate of his mother-in-law. The first batch of irradiated bismuth reached Dayton in January, 1944, and Los Alamos began receiving shipments of polonium by the end of March. By April, Monsanto was producing 2.5 Ci pet month; this rose to 6 Ci per month by summer, and by the time the Hanford piles came on-line in early 1945, Dayton was prepared to deliver 10 Ci per month. The uranium gun bomb used four initiators mounted within the projectile piece (Fig. 7.18); the implosion bomb used one at the very center of the core (Fig. 7.21). The first "production" Urchin unit was completed on June 21, 1945, only about three weeks before the *Trinity* test. Urchins were tested for resilience against leaks, vibrations, being dropped, and for resistance to water vapor—all circumstances they might experience in combat conditions.

To close this sub-section, we mention another aspect of the light-element impurity issue. This is that plutonium is rather brittle at room temperature, and is difficult to form into desired shapes unless alloyed with another metal. But common light alloying metals such as aluminum cannot be used because of the (α, n) problem; one has to use something heavier. Los Alamos metallurgists found that by alloying plutonium with 3 % gallium by weight, they could avoid the (α, n) problem while also depressing the melting point of the malleable δ-phase of plutonium (Sect. 3.8) sufficiently that it could be worked at room temperature. An advantage of this approach was that since the lower-density δ-phase transforms to the higher-density α-phase under compression, one realizes a gain in the sense that the critical mass of α-phase plutonium is less than that of the δ-phase material that one began with, leading to an efficiency enhancement.

In comparison to the high-profile work in physics and engineering carried out at Los Alamos, the work of the metallurgy group has tended to be overlooked. From a complement of about twenty in June, 1943, the staff of the Chemistry and Metallurgy Division would grow to number some 400, about one-sixth of the Laboratory personnel. Much of the research on the properties of plutonium was carried out by Charles Thomas and Cyril Smith (Fig. 7.11), a metallurgist employed with the American Brass Company who was working with the NDRC in Washington.

Some of the tasks faced by Los Alamos metallurgists were unusual. Uranium and plutonium will spontaneously ignite in air when powdered or thinly sliced, and so often had to be handled in an inert atmosphere. Plutonium is highly susceptible to corrosion; this was circumvented by plating bomb cores with thin coatings of silver. Other tasks included machining beryllium bricks for use in scattering and criticality experiments (Sect. 7.11), producing foils for nuclear-physics experiments, and developing crucibles for use in purification operations that did not themselves introduce further impurities. In a 1981 reminiscence, Smith put the importance of chemistry, engineering, and metallurgy at Los Alamos into perspective:

Fig. 7.11 Cyril Stanley Smith (1903–1992) in 1948. *Source* Allen M. Clary, Camera Portraits, courtesy AIP Emilio Segre Visual Archives

Of course the nuclear bomb was a physical concept, stemming from physical theory and experiment of the most magnificent kind, but the design would have been nothing without fantastic chemistry, without stupendous achievements in engineering both chemical and mechanical, or if the metallurgists had not been able to fabricate fantastic materials into many tricky shapes. Before any nuclear cross-section could be measured or before any critical assembly could be achieved, something had to be *made*.

7.7.2 The Spontaneous Fission Problem

The second source of possible pre-detonation-initiating neutrons arises from the fact that both uranium and plutonium suffer spontaneous fissions. This problem was nearly catastrophic for the plutonium-bomb program.

Because spontaneous fission (SF) is also a half-life phenomenon, the rate of such events can be computed with the formula for alpha-decays used above. Relevant numbers appear in Table 7.3. SF rates are listed in terms of both number of spontaneous fissions per kilogram per second, and number per gram per hour, which was the preferred unit in much Los Alamos technical documentation. Rates in both units are listed here for easy comparison with values to be cited from Los Alamos reports. The large number for Pu-240 is not a misprint.

Table 7.3 Spontaneous-fission rates for uranium and plutonium isotopes

Isotope	SF half-life (years)	SF rate ($kg^{-1} s^{-1}$)	SF rate ($g^{-1} h^{-1}$)
U-235	1×10^{19}	5.63×10^{-3}	0.02
U-238	8.2×10^{15}	6.78	24.4
Pu-239	8×10^{15}	6.92	24.9
Pu-240	1.14×10^{11}	483,000	1.74×10^6

Spontaneous fission differs from the light-element issue in that there is no mitigating yield factor involved: neutrons are emitted directly. Also, because neutrons are uncharged, they do not suffer any range limitation due to ionization-energy losses. Furthermore, since two to three neutrons are typically emitted per fission, the numbers in the last two columns of the Table should be multiplied by about 2.5 to get an idea of the rates of *neutron* emission.

In the case of either *pure* U-235 or *pure* Pu-239, spontaneous fission is not an overly serious concern for bomb engineers. Over 100 μs, a 10-kg core of pure Pu-239 will suffer on average only 0.007 spontaneous fissions, so the danger of a predetonation by this cause is quite low. If supercriticality for such a core lasts for a full 100 μs (that is, if 100 μs elapse between first criticality and when assembly is complete), the probability that the bomb will achieve its full design yield is greater than 99 %, *if* the separate issue of light-element impurity levels can be addressed. The situation is even more relaxed for a uranium core. The Hiroshima *Little Boy* bomb core had a mass of about 64 kg, about 20 % of which was U-238. But even with this level of contamination, the probability of a predetonation is less than 1 % for a 100-μs supercriticality time.

The problem with plutonium concerns the way it is produced. If an already-synthesized Pu-239 nucleus is stuck by a thermal neutron, it has about a one-in-four chance of capturing the neutron and so becoming Pu-240, as opposed to undergoing fission. This means that reactor-produced plutonium will *inevitably* contain some percentage of Pu-240. Leaving a fuel slug in the reactor for a longer time will give more Pu-239, but will also lead to more Pu-240 in the bargain. The problem is that because Pu 240 has such a spectacular SF rate, the presence of even a small amount of it can be a big problem. The *Trinity* and Nagasaki bombs each used about 6 kg of plutonium. If a 6-kg core is contaminated with even only 1 % Pu-240, it will suffer on average 2.9 SFs over the course of a 100-μs supercriticality time. As can be read from Fig. 7.12, the probability of *not* suffering a predetonation in this case is only about 10 %. There is no "correct" non-predetonation probability to aim for, but one would presumably want something in the vicinity of 90 % or better. The supercriticality period would have to be reduced to about 30 μs to have a 50 % chance of no predetonation, and to about 5 μs for a 90 % chance. Even if the Pu-240 contamination level is reduced to only 0.3 %, a 100-μs supercriticality timescale will still yield only a 50 % non-predetonation probability. Depending on such a low chance of success is not acceptable when one has invested hundreds of millions of dollars in synthesizing fissile material. Aside from the virtually impossible task of removing the offending Pu-240, the only option is to speed up the assembly process to on the order of a few microseconds. Unfortunately, this is impossible with any conceivable gun-type mechanism.

Figure 7.12 does not tell quite the entire story, however. The probability of predetonation can never be reduced strictly to zero, but its effect on the explosive yield of a bomb is another question altogether. If the chain-reaction is initiated when assembly is almost complete, the effect on the yield might be very slight. At the other extreme, if initiation occurs just at the time of first criticality, what

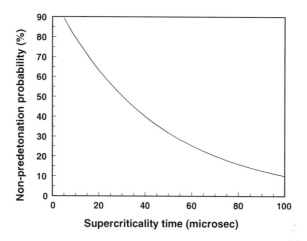

Fig. 7.12 Probability of no pre-detonation for a 6-kg Pu-239 core contaminated with 1 % Pu-240. From Reed (2009)

minimum "fizzle yield" might one expect? Would the explosion be violent enough to at least destroy the bomb, so that an enemy could not recover the fissile material? Los Alamos theoreticians devoted considerable effort to analyzing these issues. These questions are applicable to both the light-element and spontaneous-fission issues, but we will work with numbers assuming the latter problem, which was the more serious one for the plutonium bomb.

Answers to such questions are expressed with probabilities. It is impossible to make a statement to the effect of "Your bomb will achieve precisely percentage x of its design yield." Rather, one has to settle for an assessment such as "The chance of realizing at least percentage x of the design yield of a weapon of a specified core mass and percentage of spontaneously-fissioning contaminant is such-and-such." Figure 7.13 shows results of such calculations for a 6-kg Pu core contaminated with 1, 6, and 20 % Pu-240 by mass, assuming a supercriticality time before full assembly of 10 μs, a duration characteristic of an implosion weapon. The probability of achieving full yield with 1 % contamination is about 80 %, but falls to only about 27 % in the case of 6 % contamination. Manhattan Project plutonium was held to a Pu-240 contamination level of about 1 %.

The curve for 20 % contamination is included here as such a percentage is characteristic of the plutonium created in commercial power-producing reactors, so called "reactor-grade" plutonium. The Pu-240 fraction in such circumstances is very large because the fuel in commercial reactors typically remains in the reactor for many months. The chance of achieving any sensible fraction of the nominal design yield for a weapon constructed with such plutonium is abysmal. While this might seem comforting when considering the possibility of terrorists trying to develop a crude bomb based on plutonium extracted from spent fuel rods, bear in mind that a device that achieves even only a few percent of design yield would still create a devastating explosion and leave behind widely-scattered radioactive contamination.

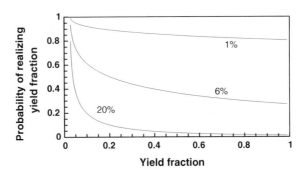

Fig. 7.13 Probability of achieving a given fraction of a fission weapon's design yield as a function of the desired fraction of the design yield for a 6-kg Pu-239 core contaminated with 1, 6, and 20 % Pu-240 for a supercriticality time of 10 µs. One has about a 50 % chance of obtaining at least 20 % of the design yield if the contamination level is 6 %, but only about a 30 % chance of achieving 80 % of the design yield for the same level of contamination. From Reed (2011)

Related calculations show that for a bomb of design yield 20 kilotons, one can expect a *minimum* energy release of about 500 t TNT equivalent in the worst-case scenario when the chain reaction starts at first criticality (supercriticality time 10 µs). This would be more than enough to destroy the weapon itself. (Since the calculation of the "fizzle yield" assumes that the chain reaction starts at the time of first criticality, the result is independent of the percentage of contaminating material.) As always, detailed results are dependent upon choices for various nuclear parameters.

The problem with plutonium spontaneous fission was not realized at Los Alamos until the summer of 1944. The circumstances of that discovery and the consequent reorientation of the Laboratory to deal with it are described in Sect. 7.10. In the meantime, however, to avoid the chronology of this chapter from becoming too scrambled, we return for the following two sections to 1943 to examine preparations made for testing the combat characteristics of bomb designs, and the configuration of the simpler gun-mechanism bomb.

7.8 The Delivery Program

Had the scientists and engineers of Los Alamos simply constructed and tested a nuclear weapon, they would have left their job only half-done. A laboratory experiment is one thing, a deliverable military weapon is quite another. To make a bomb ready for combat meant modifying aircraft to carry it, training crew members, designing a bomb casing that gave stable flight, and ensuring that electronic systems would function reliably in combat conditions and be immune from enemy interference.

Fig. 7.14 *Left Thin Man (front)* and *Fat Man* test bombs (*rear*) at Wendover Army Air Base (Utah); *Right* Norman Ramsey (1915–2011) *Sources*: http://commons.wikimedia.org/wiki/File:Thin_Man_plutonium_gun_bomb_casings.jpg; AIP Emilio Segre Visual Archives, W. F. Meggers Gallery of Nobel Laureates

William Parsons recognized these needs early on, and saw that an appropriate "delivery" program was organized as soon as possible. In October, 1943, a group within his Ordnance Division was charged with responsibility for integrating the design and delivery of weapons. This group was headed by physicist Norman Ramsey (Fig. 7.14), who possessed the ideal combination of familiarity with military operations and understanding of the science of the project. The son of an Army general, Ramsey had earned a doctorate from Columbia for work in the area of molecular-beam physics, and had been serving as a consultant to the Secretary of War on microwave radar when he was recruited to Los Alamos. In his postwar career at Harvard, Ramsey pioneered methods for precisely measuring the electron-transition frequencies of atoms and molecules, work which would lead to the development of atomic clocks and MRI scanners, and which earned him a share of the 1989 Nobel Prize for Physics.

The formal name of the program for preparation of combat bombs was Project Alberta, or simply Project A. Ramsey's first task was to undertake a survey of the sizes, shapes, and weights of bombs that could be carried by Army Air Forces aircraft. (The Air Force did not become a separate branch of the armed services until 1947.)

The gun bomb was anticipated to be 17 feet long by 23 in. in diameter. These dimensions dictated the necessary size of the bomb bay, and Ramsey soon zeroed in on the long-range B-29 "Superfortress" bomber, which was still undergoing flight tests at that time. With a gross takeoff weight (aircraft, fuel, bombs) of 70 t, the B-29 could carry a bomb load of 10 t for a combat range of nearly 1,600 miles. Powered by four 18-cylinder engines that each developed 2,200 horsepower, the B-29 would be the largest combat aircraft of World War II. The first production model came off the assembly line in July, 1943, and their first mission against the Japanese home islands occurred in June, 1944.

B-29's were equipped with two 150-in. long, 64-in. wide bomb bays, one forward and one aft of the wings. If the two bays could be joined together, they could accommodate the anticipated gun bomb under the main wing spar. On December 1, 1943, Army Air Forces Headquarters directed the Materiel Command at Wright Field in Dayton, Ohio, to undertake a high-priority modification of a B-29 bomber. This directive had as its name "Silver Plated Project," which eventually evolved to "Silver Plate," and then to "Silverplate." The first modified bomber was a prototype with a single 33-foot long bomb bay and release mechanisms for both *Thin Man* and *Fat Man* designs. A total of 46 Silverplate B-29's were produced by the end of 1945; the total would come to 65 by the time the project was terminated in December, 1947. The cost of each Silverplate aircraft has been estimated at about $815,000 in 1945 dollars (about $10 million in 2013 dollars).

Parsons arranged for Ramsey to supervise a drop-test program at the Dahlgren Naval Proving Ground in Virginia. To prepare a mockup 14/23-scale model of the gun bomb, Ramsey had a standard 23-in. diameter, 500-pound bomb split in half and the halves joined by a length of 14-in. diameter pipe. Known as the "sewer-pipe" bomb, the first drop test of the model was conducted on August 14, 1943, and proved, in Ramsey's words, "an ominous and spectacular failure. The bomb fell in a flat spin the like of which had rarely been seen before." (To imagine a flat spin, think of a spinning helicopter blade.) Adjustments to the tail-fin design and moving the bomb's center of gravity forward soon resulted in more stable flight. The eventual boxlike tail of the gun-type bomb was square-shaped and thirty inches on a side (Fig. 7.18).

By the fall of 1943, Ramsey was ready to begin tests with full-scale models. He and Parsons selected two external shapes and weights as representative of the bombs then under development: the 17-foot/23-in. gun model, and an ellipsoidally-shaped implosion model just over 9 feet long and 59 in. in diameter. Fifty-nine inches was the largest diameter that could be squeezed into a B-29 bomb bay, a constraint which set an absolute limit to *Fat Man's* girth. The nested structure of the implosion design (Fig. 7.21) meant that any change in the dimensions of any component propagated throughout the design: The diameter of the neutron-generating initiator at the center of the weapon dictated the dimensions of the fissile core, which dictated the dimensions of the surrounding tamper sphere and high-explosive assembly, all of which was contained within a spherical metal case held within an outer ballistic ellipsoid, with enough space to house arming and fusing circuits.

The origins of the names of the various Los Alamos bomb designs—*Thin Man, Fat Man,* and *Little Boy*—is a matter of debate. In a history of the delivery project prepared just after the end of the war, Ramsey asserted that it was Air Force representatives who coined the names *Thin Man* and *Fat Man*, the idea being to make telephone conversations sound as if aircraft were being modified to carry President Roosevelt (Thin Man) and Prime Minister Churchill (Fat Man). In a 1998 autobiography, Robert Serber claims to have named the bombs, with *Thin Man* being taken from the title of a 1934 detective novel by Dashiell Hammett, and

Fat Man referring to the role played by actor Sydney Greenstreet in the 1941 movie *The Maltese Falcon*, which starred Humphrey Bogart.

Tests of full-scale models for ballistic behavior and functioning of fusing and instrumentation circuits were begun in the spring of 1944 at Muroc Field in California. The site of a large dry lake bed, Muroc is now Edwards Air Force Base. The start of testing was delayed a month by torrential rain in what was supposed to be one of the driest places in the country. The prototype modified B-29 arrived on February 20, and the first drop test occurred in early March.

Parsons' conservatism in demanding an early start on the delivery program was well-founded. Fuses proved so unreliable that an investigation was begun of adapting a radar unit normally mounted on the tails of fighter aircraft as a substitute. High-speed photography revealed that Thin Man models proved to have very stable flight characteristics, but the *Fat Man* design wobbled violently, with its long axis departing up to 20° from the line of flight. Simply assembling the implosion bomb was arduous: one model required 1,500 bolts. (For the Nagasaki weapon, this number was cut to 90.) A release mechanism that worked properly for *Fat Man* failed completely for *Thin Man*, with several dangerous hang-ups occurring. In what would be the last test of this series on March 16, a *Thin Man* released prematurely and fell onto the bomb-bay doors, which had to be opened to release the bomb. The doors were seriously damaged, and this accident brought testing at Muroc to an abrupt if temporary halt. Later drop tests would prove equally harrowing. The first test with one of the more rugged replacement B-29's (see below) occurred on March 10, 1945, at the Salton Sea, but the bomb was released early and fell near a small town. At Wendover Field, Utah, on April 19, 1945, a bomb exploded just after clearing the bomb bay. Fortunately, it was a unit for testing the fusing mechanism and contained only one pound of explosive, just enough for observers to see whether it detonated at the desired height.

With the mid-1944 realization that the gun assembly method could not be used with plutonium (Sect. 7.10), the situation for the uranium gun bomb became much simpler. The assembly speed could be reduced to a leisurely 1,000 feet per second (~ 300 m/s), and the length of the bomb could be shortened to 10 feet, which meant that it could fit into a single B-29 bomb bay. The prototype bomber was reconfigured back to its original two-bay configuration, and the shortened gun bomb was dubbed "*Little Boy.*"

Tests at Muroc resumed in June, 1944. The new radar-driven fusing units functioned satisfactorily, but portly *Fat Man's* wobble proved more challenging to address. Replacing its parachute-like circular-shaped tail assembly with a square one helped to suppress but did not wholly eliminate the wobble. Ramsey had steel plates added to the tail assembly at 45° angles; this modification resulted in very stable flight and gave *Fat Man* its distinctive tail-end, which contributed 400 pounds to the weight of the bomb (Fig. 7.16). *Fat Man* test units were painted mustard yellow to make them easy to track, and they became known as "Pumpkin" units.

Tests of the gun-bomb firing mechanism were made at Wendover Army Air Base in Utah, one of the largest gunnery and bombing ranges in the world. In

Fig. 7.15 *Left* Colonel Paul Tibbets (1915–2007) waves from the cockpit of the *Enola Gay* shortly before takeoff for the Hiroshima mission. *Right* Frederick Ashworth (1912–2005) gives a talk at Los Alamos in his later years. *Sources* http://commons.wikimedia.org/wiki/File:Tibbets-wave.jpg; http://commons.wikimedia.org/wiki/File:Frederick_Ashworth.jpg

Fig. 7.16 *Left* assembled *Fat Man* bomb. *Note* signatures on tail. *Sources* http://commons.wikimedia.org/wiki/File:Fat_Man_on_Trailer.jpg. *Right* On Tinian island, *Fat Man* receives a coat of sealant. *Note* FM stencil on nose. *Source* http://commons.wikimedia.org/wiki/File:Fat_Man_on_Tinian.jpg

Manhattan Project lingo, Wendover was codenamed "Kingman," "Site K, and "W-47." The drop-test program was very successful from the outset; thirty-two tests involving natural uranium projectiles were conducted, and on only one occasion did the gun fail to fire, a consequence of a faulty electrical connection. These tests did result in one significant modification to the design of the breech of the gun bomb, however. The original design called for the bomb to be fully armed with conventional explosive upon aircraft take-off, but it was deemed desirable that it be possible to arm the bomb during flight lest a crash on take-off initiate a nuclear explosion. The breech was consequently modified to permit one person to be able to load or unload powder bags in the cramped space of the bomb bay. This would be the case in practice with the Hiroshima bomb, which Parsons himself armed in flight. No such arrangement was practical with the implosion weapon, which because of its enclosed design left the ground fully armed.

In parallel with technical refinements to bomb designs, air crews had to be selected and trained. On August 11, 1944, the Army Air Forces recommended freezing the design of the shapes of the bomb casings and starting crew training. Freezing the designs would permit modifications to a lot of B-29's to be started while a crew was being assembled. A special unit known as the 509th Composite Group, which was placed under the command of Colonel Paul Tibbets, would be responsible for dropping the combat bombs. The 509th trained at Wendover; Navy Commander Frederick Ashworth served as liaison between Los Alamos and Wendover, and would accompany the *Fat Man* bomb on its flight to Nagasaki (Fig. 7.15).

The first of 17 modified B-29's began arriving in October, 1944, and test flights began that month. Particular emphasis was put on training pilots to carry out unusual post-drop maneuvers designed to put the maximum possible distance between the aircraft and the bomb before the latter exploded. Drop-test bombs were filled with concrete to simulate the effect of how the plane would lurch upward after the bomb was released. This first group of modified aircraft proved to have poor flying qualities, however, so a new batch equipped with fuel-injected engines, variable-pitch propellers, and improved bomb-release mechanisms was obtained in the spring of 1945. Stripped of all of their guns and armor except for their tail turrets, aircraft of this second group were each 7,200 pounds lighter than normal B-29's. These modifications enabled them to fly above 30,000 feet at an average speed of 260 miles per hour, while carrying a payload of 10,000 pounds a distance of almost 2,000 miles. Another modification involved the addition of a position for an electronics test officer who would monitor the bomb's electrical circuits during flight.

Two of the aircraft in this second group would go down in history as the planes that carried the atomic bombs dropped on Japan. On May 9, 1945, the day after the German surrender was announced, Paul Tibbets was at the Martin Aircraft plant in Omaha, Nebraska, to pick out the bomber that he would use for the first atomic strike. B-29 production number B29-45-MO-44-86292 would be christened as *Enola Gay*, Tibbets' mother's maiden name. It was formally delivered to the Army Air Forces on May 18, flown by pilot Robert Lewis to Wendover on June 14, and then again by Lewis to Tinian island (see below), arriving on July 6. Serial number B29-35-MO-44-27297, *Bockscar*, was delivered on March 19, and arrived at Tinian on June 16. Named after its commander, Captain Frederick C. Bock, this aircraft is sometimes referred to with the two-word designation "*Bock's Car.*" *Enola Gay* and *Bockscar* are believed to be the only surviving Silverplate aircraft, and now reside at museums in Washington and Dayton, Ohio, respectively.

The pace of the 509th's training schedule was relentless, and went on right up to the time of the Hiroshima and Nagasaki missions. Between October, 1944, and mid-August, 1945, a total of 155 test bombs were dropped, a rate of nearly one per day. One of the most problematic issues encountered with the *Fat Man* bomb was a piece of equipment known as the X-unit, which was responsible for simultaneously triggering its network of spherically-distributed detonators (Sect. 7.11). This was a complex business. With redundant detonators for each of 32 implosion-lens segments, 64 cables were involved, all of which had to be the same length and have the same impedance. Not until late July, 1945, did a sufficient number of X-units begin to become available. The first drop test of a *Fat Man* with high explosives and an X-unit was not conducted until August 5, the day before the Hiroshima mission.

Combat models of the *Fat Man* bomb incorporated a number of safety features. Front-view photographs of the Nagasaki bomb show four cylindrical tubes about three inches in diameter protruding from the front of the bomb's outer casing (Fig. 7.16). These were contact fuses: if the bomb's fusing circuitry failed to trigger an "airburst," these would fire the detonating system when the bomb struck the ground. (*Little Boy* did not incorporate contact fuses as it would "self-assemble" upon striking the ground.) Another feature was that tempered-steel bomb casings proved vulnerable to 0.50-caliber machine gun fire which could cause internal damage. As a result, the entire *Fat Man* ballistic casing, plus cover plates on the *Little Boy* weapon and the rear tail covers on both bombs were to be made of hardened armor plate.

Provision also had to be made for an overseas combat base at which the bombs would be assembled, checked, and loaded onto aircraft. After surveying both Guam and Tinian islands and consulting with Groves and Admiral Chester Nimitz (Commander-in-Chief of both the United States Pacific Fleet and the Pacific Ocean Areas), Commander Ashworth selected Tinian: it was about 100 miles closer to Japan than Guam, had construction forces available, and its port facilities tended to be less overloaded than those at Guam. Tinian is a member of the Northern Mariana Islands chain, located just south of the island of Saipan (Fig. 7.17). Only about 12 miles long, the island had been taken by the Marines in July, 1944, and for a time was the site of the largest airport in the world: six runways each 8,500 feet long, which served as launching points for round-the-clock bombing raids against the Japanese home islands. It was not uncommon for 400 aircraft to leave the field in less than two hours. Tinian's Manhattan codename was "Destination."

Ashworth oversaw construction of 509th facilities on Tinian. Air-conditioned assembly buildings were erected, along with warehouses and shops. Special pits were constructed for hydraulically loading bombs into aircraft from underneath; there was otherwise insufficient clearance between the bodies of the aircraft and the ground to accommodate the weapons. If bottlenecks in construction or transportation arose, Ashworth needed only to invoke the code word "Silverplate," which came to designate all atomic-bomb related activities within the military, and which required instant cooperation from all personnel. The 509th moved its operations to Tinian in late June, 1945, to undertake practice missions in advance of their "hot runs" against Hiroshima and Nagasaki.

Fig. 7.17 Map of Tinian and Saipan. One minute of latitude corresponds to a distance of about 1.15 miles. See also Fig. 8.3. *Source* http://commons.wikimedia. org/wiki/File:Map_Saipan_ Tinian_islands_closer.jpg

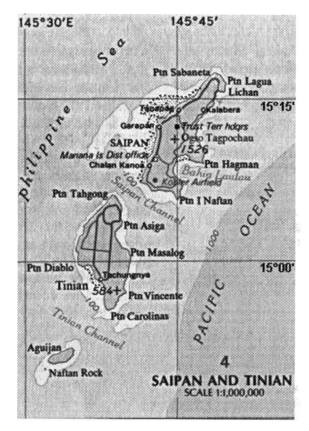

At this point, it is appropriate to return to Los Alamos to explore the development of the gun bomb, the emergence of the spontaneous fission crisis, and the development and testing of the implosion weapon. Delivery of bomb components to Tinian, and bomb assembly and testing are described in Sect. 7.14; further details on the selection and training of 509th personnel are described in Chap. 8.

7.9 The Gun Bomb: *Little Boy*

During the first six months of Los Alamos' existence, only the gun method of fissile-material assembly was considered sound enough to warrant an extensive engineering program. Responsibility for the design, engineering, drop tests, and assembly of the uranium gun bomb lay with the Gun Group of the Ordnance Division, which was directed by Commander Albert Francis Birch, a Harvard University geophysicist and Navy Commander who had an extensive background in physics, electronics, and mechanical design.

While the gun method was straightforward in principle, it faced a number of unique engineering issues which bore little resemblance to typical military ordnance problems. Standard naval cannons had integer calibers ranging from 4 to 16 in. If it were decided that the nuclear guns should have a non-standard caliber, additional design work would be required. This proved to be the case; the *Little Boy* gun ultimately had a 6.5-in. bore. The neutron-reflecting properties of the steel used in the guns had to be determined; if it should prove reflective, it could contribute to lowering the critical mass. But the most unusual aspect of the design was that rather than exiting the gun as would a normal shell, the projectile piece was to be *stopped* by the tamper (also known as the "target case") after mating with the target piece, so that the chain reaction could proceed for enough time to give reasonable efficiency. Fortunately, U-235 proved to be strong enough to be able to withstand such deceleration without disintegrating.

When the *Thin Man* configuration was under consideration, established ordnance practice indicated that a gun designed to achieve a 3,000 foot-per-second muzzle velocity would weigh five tons and have to be able to withstand a breech pressure of some 75,000 pounds per square inch from the chemical explosive used to propel the projectile piece. These requirements created a potentially serious problem for delivering such a weapon. The payload of a B-29 bomber depended on the duration of the mission: up to 10 t could be carried to combat radii of about 1,600 miles, which was about the distance from Tinian to Tokyo. But massive artillery cannons were not normally carried as bombs. How then could the weight of *Thin Man*, with its cannon, tamper, casing, and instrumentation, be reduced to a safely deliverable level?

The resolution of these conflicting demands came with Edwin Rose's realization that regular artillery pieces are designed to withstand the stresses of thousands of firings, but that such a requirement was entirely inessential for a weapon which would be fired only a few times in tests, and only once in combat. By sacrificing durability, the otherwise prohibitive weight of a gun bomb could be reduced to a point where it could be configured as a practical weapon.

Los Alamos established a gun testing area, the Anchor Ranch Proving Ground, about three miles from the main laboratory area (Fig. 7.1). The first true experimental gun units did not arrive until March, 1944, but the first test shots were fired on September 17, 1943. In a memoir published in 1980, Edwin McMillan remarked that he considered the September 17 shot to mark the transition from the "early" to the "late" history of Los Alamos. All Los Alamos guns were fabricated by the Naval Gun Factory at the Washington Navy Yard. The first two guns delivered from Washington were 3,000 foot-per-second prototypes, but they arrived just as the plutonium spontaneous fission crisis was beginning to emerge, and were abandoned unused. Three new *Little Boy* guns designed for 1,000 foot-per-second operation were promptly ordered. Since the Gun Group could not do any test-firings using "active" U-235 components, they had to find a substitute material whose mechanical properties mimicked that of U-235. Natural uranium proved adequate for this purpose. Since gun tubes designed for lightness could not be repetitively test-fired, proof-testing consisted of a few instrumented firings at

1,000 feet per second with a 200-pound projectile, after which the guns were greased and stored for future use. In addition to such "live" guns, a number of dummy models made from discarded naval guns were used in drop tests of simulated assembled bombs; these units were not intended for test firing.

An important element in both the gun and implosion designs was the choice of tamper material. The best option would be a heavy metal which elastically scattered neutrons. Responsibility for investigating tampers was assigned to the Radioactivity Group of the Experimental Physics Division. By October, 1943, the list of possible tamper materials had been narrowed to tungsten carbide (steel), natural uranium, beryllium oxide, iron, and lead, although measurements would be made on over two dozen elements, including gold and platinum. As Robert Serber wrote in a *Primer* annotation, "The active material seemed so precious that everything else in contrast seemed cheap. The notion of vaporizing a few hundred pounds of gold in the explosion did not strike us as odd." Ironically, beryllium generates neutrons when struck by alpha particles, but is otherwise an excellent reflector of neutrons and would have made an ideal tamper material but for the fact that such use would have virtually exhausted the country's supply of the metal at that time. Beryllium was, however, used as a reflective tamper in so-called "criticality" experiments (Sect. 7.11).

Tungsten carbide was chosen as the tamper for the *Little Boy* gun bomb on the basis of its high elastic-scattering cross-section, while natural uranium was used in the *Fat Man* design in view of its inertial and nuclear properties. Remarkably, the first gun-bomb target case to be test-fired proved to be one of the best made. Known as "old faithful," it was tested four times at Anchor Ranch, and was incorporated into the bomb dropped at Hiroshima. *Little Boy's* 28-in. diameter target case was three feet long and weighed over 5,000 pounds. Within the target case resided a 13-in. diameter tungsten-carbide liner (the tamper material proper), which surrounded the 6.5-in. diameter gun tube (Fig. 7.18). The chemical symbol for tungsten carbide, WC, led to its becoming known as "Watercress."

The altitude at which combat bombs would be detonated was also given careful consideration. In addition to liberating great quantities of electromagnetic radiation and billions of Curies of radioactivity, a nuclear explosion differs from a conventional one of the same energy in that pressures generated are higher at closer distances. Based on the results of the *Trinity* test, the detonation heights for the Hiroshima and Nagasaki bombs were set at 1,850 feet. This was chosen to maximize destruction by the shock wave created by the bombs, while minimizing the amount of fallout that would be created if they were otherwise detonated near ground level and irradiated tons of dirt and debris. The Ordnance Division's concern with the altitude issue was that most combat bombs detonate near ground level; little thought had ever been given to mechanisms designed for high-altitude operation. Extreme reliability was the paramount consideration. In a conventional mission where thousands of bombs might be dropped, the failure of a few percent will likely not affect the outcome of the operation. But any type of fuse that failed even 1 % of the time would be unacceptable for a single bomb whose development had consumed hundreds of millions of dollars. As a result, fuse specifications

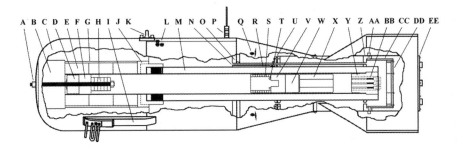

Fig. 7.18 Cross-section drawing of Y-1852 *Little Boy* showing major components. Not shown are radar units, clock box with pullout wires, barometric switches and tubing, batteries, and electrical wiring. *Numbers in parentheses* indicate quantity of identical components. Drawing is to scale. Copyright by and used with kind permission of John Coster-Mullen. (a) Front nose elastic locknut attached to 1-in. diameter Cd-plated draw bolt, (b) 15.125-in. diameter forged steel nose nut, (c) 28-in. diameter forged steel target case, (d) Impact-absorbing anvil with shim, (e) 13-in. diameter 3-piece WC tamper liner assembly with 6.5-in. bore, (f) 6.5-in. diameter WC tamper insert base, (g) 14-in. diameter K-46 steel WC tamper liner sleeve, (h) 4-in. diameter U-235 target insert discs (6), (i) Yagi antenna assemblies (4), (j) Target-case to gun-tube adapter with 4 vent slots and 6.5-in. hole, (k) Lift lug, (l) Safing/arming plugs (3), (m) 6.5-in. bore gun, (n) 0.75-in. diameter armored tubes containing priming wiring (3), (o) 27.25-in. diameter bulkhead plate, (p) Electrical plugs (3), (q) Barometric ports (8), (r) 1-in. diameter rear alignment rods (3), (s) 6.25-in. diameter U-235 projectile rings (9), (t) Polonium-beryllium initiators (4), (u) Tail tube forward plate, (v) Projectile WC filler plug, (w) Projectile steel back, (x) 2-pound Cordite powder bags (4), (y) Gun breech with removable inner breech plug and stationary outer bushing, (z) Tail tube aft plate, (aa) 2.25-in. long 5/8-18 socket-head tail tube bolts (4), (bb) Mark-15 Mod 1 electric gun primers with AN-3102-20AN receptacles (3), (cc) 15-in. diameter armored inner tail tube, (dd) Inner armor plate bolted to 15-in. diameter armored tube, (ee) Rear plate with smoke puff tubes bolted to 17-in. diameter tail tube

called for a less than one in ten-thousand chance of the bomb failing to fire within about 100 feet of the desired altitude.

Two major lines of fuse development were investigated. One was to use barometric switches which would be sensitive to air pressure as a function of altitude. The other, mentioned earlier, was to adapt electronic techniques such as proximity fuses or fighter-plane tail-warning radar sets for use with the weapons, presuming that a reliable signal could be obtained with a falling bomb. For both *Little Boy* and *Fat Man*, a redundant series–parallel system comprising clocks, barometers, and four modified tail-warning radars known as "Archies" was adopted. The first stage in the firing process was that when the bombs were released, pullout switches activated timers that counted off a 15-s delay before the arming system became activated; this was to ensure a safe separation distance from the aircraft (in 15 s, a bomb will free-fall about 1,100 m; an aircraft flying at 300 miles per hour will travel about 2,200 m in the same time). Following this, barometric switches activated the radar units at an altitude of 17,000 feet; these were designed to close a relay at a predetermined altitude when any two of them detected the desired firing altitude. To lessen the possibility of failure due to Japanese jamming, each radar operated on a slightly different frequency.

The final *Little Boy* bomb, sketched in Fig. 7.18, was ten feet long, 28 in. in diameter, and weighed about 9,700 pounds. The gun barrel itself was six feet long and weighed 1,000 pounds. The target and projectile pieces were not cast as solid wholes; rather, they each comprised a number of washer-like rings that were cast as uranium became available from Oak Ridge. The projectile was made up of nine rings totaling 7 in. in length, with inside and outside diameters of 4 and 6.25 in. Because the amount of uranium received from Oak Ridge varied from shipment to shipment, none of the individual rings were of the same thickness (nor, likely, of exactly the same enrichment). The projectile had a volume of 126.8 in.^{3}, or 2,078 cm^3. At a density for pure U-235 of 18.71 g/cm^3, the assembled projectile rings totaled 38.9 kg. The target consisted of six rings, also of 7 in. total length, but with inside and outside diameters of one and four inches for a volume of 82.4 in.^{3} (1,351 cm^3) and a mass of 25.3 kg. The assembled core totaled just over 64 kg, about 60 % of which resided in the projectile. The projectile piece traveled about 52 in. ($\sim$130 cm) before meeting the target piece, which resided about 20 in. (half a meter) to the rear of the nose of the target case. The target assembly and tamper liner were secured to the front of the bomb with a nut which itself weighed several hundred pounds.

By December, 1944, General Groves was confident enough of anticipated uranium production schedules that he ordered all research and development on the gun bomb to be complete by July 1, 1945. The design was frozen in February, 1945, and *Little Boy* was ready for combat by May, 1945. Deployment awaited only enough U-235, which was expected to be ready about August 1. The gun bomb, Robert Serber's "shooting" concept, would be the first nuclear weapon used in combat.

7.10 The Spontaneous Fission Crisis: Reorganizing the Laboratory

The potential for a problem with spontaneous fission (SF)-induced predetonation was not wholly unappreciated when Los Alamos was established. Robert Serber discussed the issue in his *Primer*, but the only SF data then available pertained to natural uranium. SF in natural uranium had been discovered by Flerov and Pet-rzhak in the Soviet Union in 1940 (and openly published in the *Physical Review*), and it was certainly anticipated that plutonium would likely suffer the same effect.

In the United States, a group at Berkeley led by Emilio Segrè began SF research around late 1941/early 1942. Using plutonium created by deuteron bombardment of U-238 in Ernest Lawrence's 60-in. cyclotron, they determined, by June, 1943, a SF rate for the new element of 18 per gram per hour. From Table 7.3, they must have been working primarily with Pu-239 created via the reaction

$$^2_1H + ^{238}_{92}U \rightarrow ^1_0n + ^{239}_{93}Np \xrightarrow[2.356\ days]{\beta^-} ^{239}_{94}Pu. \tag{7.18}$$

As a humorous side note, Robert Serber relates in the *Primer* (published 1992) that the last time he saw Segrè in Berkeley, the latter was driving a beat-up old car with a bumper sticker which read "My Owner has A Nobel Prize"; Segrè had shared the 1959 Nobel Prize for Physics for the discovery of the antiproton.

The consequences of creating highly spontaneously-fissile Pu-240 in a reactor along with the desired Pu-239 were also anticipated early on. In his diary entry for March 18, 1943, Glenn Seaborg wrote that "The possibility of an appreciable yield from the (n, gamma) reaction on $^{239}94$ seems rather remote; however, a cross-section 1 % of the fission cross-section would result in enough $^{240}94$ to complicate the purity problem ... If the spontaneous rate of $^{240}94$ is high, e.g., a half-life of less than 10^{10} years, it might be serious." By (n, gamma), Seaborg means a reaction where a Pu-239 nucleus absorbs a neutron to become Pu-240, and then sheds excess energy by emission of a gamma-ray. The modern value for the thermal-neutron radiative-capture cross-section of Pu-239 is 271 barns, which is about one-third of the fission cross-section of 750 barns. This is over 30 times greater than the 1 % Seaborg specified as having the potential to create complications. Nature was somewhat more on his side in the case of the SF rate, but not enough to offset the large capture cross-section: the spontaneous-fission half-life for Pu-240 is 1.1×10^{11} years, only about 10 times greater than his "serious" threshold of 10^{10} years. (A *longer* half-life is preferred, as it results in fewer spontaneous fissions per second from a given mass of material.)

Oppenheimer invited Segrè to move his work to Los Alamos, which he did in June, 1943. Because spontaneous fission counts are low and can be confounded by small fluctuations in background radiation, a special remote field station for SF measurements was set up in a Forest Service cabin at Pajarito Canyon, a 14-mile drive from the main area of the Laboratory. Segrè described the cabin-laboratory as being in one of the most picturesque settings one could dream of; Fermi was fond of visiting the site. Segrè and his group were set up by August, working with five 20-μg samples of ^{239}Pu that had been prepared at Berkeley. These proved too small for any reliable determination of the SF rate, but the group was able to measure the number of neutrons per spontaneous fission as $v = 2.3$. From the data in Table 7.3, 20 μg of pure Pu-239 would yield on average only about 0.36 spontaneous fissions per *month*. The 20-μg samples slowly accumulated spontaneous-fission counts, a grand total of six over the course of five months to January 31, 1944. However, small-number statistics are inevitably subject to great uncertainty; real confidence would come only with the arrival of pile-produced plutonium from Oak Ridge.

Soon after getting set up, the group noticed a curious effect with spontaneous fission of uranium. This was that the rate for ^{238}U agreed with what they had measured in Berkeley, but the rate for ^{235}U was *higher* at Los Alamos than at Berkeley. Since the ^{235}U itself should not have changed in any way, the cause of the increased rate must presumably be something external. It was soon determined that at the higher altitude of Los Alamos, cosmic rays were inducing more fissions than they could at sea level in Berkeley, where many of them were absorbed by the thicker intervening atmosphere. The cosmic rays were not energetic enough at either location to induce fission in ^{238}U, and their effect on ^{235}U at sea level had gone unappreciated. The rate of natural sea-level spontaneous fissions in ^{238}U greatly exceeded the rate of cosmic-ray induced fissions, which led to the realization that the majority of spontaneous fissions in uranium arises from the heavier isotope. The very low SF rate deduced for ^{235}U in combination with the relatively gentle purity requirements for that isotope led to the relaxation of the assembly velocity requirement for the uranium bomb, as described in the preceding section.

The first X-10 plutonium arrived at Los Alamos in the spring of 1944, and was placed in detection chambers on April 5. More arrived over the following week, and it soon became clear that a problem was developing. During the first three days of observations, the pile-produced material exhibited a rate of spontaneous fissions five times that of samples which had been prepared with the cyclotron. On April 15, Segrè reported a tentative rate of 200 spontaneous fissions per gram per hour, or eight times the rate of pure Pu-239. By May 9, the estimate had risen to 261 per gram per hour, which corresponds to a Pu-240 contamination level of 0.01 %. This may not sound drastic, but for a 10-kg core that is supercritical for 100 µs before assembly becomes complete, a contamination level only ten times greater (0.1 %) would reduce the probability of achieving the weapon's full design yield to only about 67 %. Hanford-produced plutonium would contain far more than 0.1 % Pu-240 due to the much greater neutron flux of the 250-MW production-scale reactors, unless the fuel was withdrawn much earlier than planned. Robert Bacher reported the news to Arthur Compton during an early-June visit to Chicago, and related that Compton turned as white as a sheet of paper.

Oppenheimer presented the evidence at a Laboratory colloquium on July 4, with James Conant in attendance. It was clear that the gun-assembly mechanism for the plutonium bomb would have to be abandoned. On the 17th, a conference was held in Chicago with Compton, Oppenheimer, Charles Thomas, and Conant present; Fermi, Groves, and Kenneth Nichols attended another meeting of the same group later that evening. Excerpts from a handwritten summary prepared by Conant give a sense of the severity of the situation (slightly edited):

> The disquieting prospect first discussed with Conant by Oppenheimer on the visit to L. A. on July 4 … was considered. It was concluded that the evidence was now so clear that "49" prepared at Hanford could not be used in the gun method of assembly …. Dr. Oppenheimer was not very optimistic about a speedy solution of the implosion method which is now left as the only hopeful way of using 49.

The next day, Oppenheimer summarized the situation in a letter to Groves: "At the present time the method to which an over-riding priority must be assigned is the method of implosion." The depth of Conant's reaction is captured in a handwritten note scribbled on a June 13 letter from Charles Thomas which, ironically, had reported encouraging progress in purifying plutonium (Fig. 7.19). John Manley summed up the situation in one sentence: "The choice was to junk the whole discovery of the chain reaction that produced plutonium, and all of the investment in time and effort of the Hanford plant, unless somebody could come up with a way of assembling the plutonium material into a weapon that would explode."

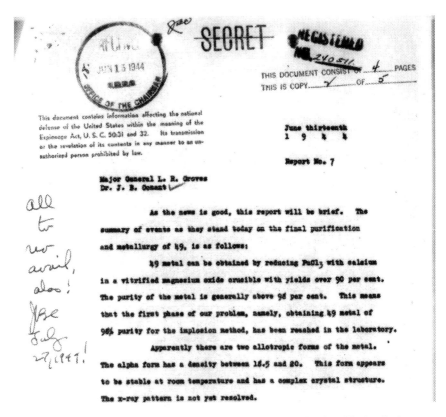

Fig. 7.19 James Conant's gloomy assessment of the plutonium situation. The handwritten note reads "All to no avail, alas! JBC July 27, 1944". *Source* M1392(1), 0900.jpg

Oppenheimer promptly developed a plan to reorganize the Laboratory to deal with the crisis. The reorganization was approved at an Administrative Board meeting on July 20, and was formally put into place on August 14. The shakeup was extensive. The Governing Board was abolished, replaced by separate Administrative and Technical Boards. The plutonium program was removed from the original Ordnance Division, and divided between two new Divisions. The first of these, X Division ("Explosives"), headed by George Kistiakowsky, would be concerned with experimentation involving explosives; methods of initiation; development, fabrication and testing of implosion systems; and developing a suitable design for assembly of the explosives and the initiating system. X Division absorbed several groups which had formerly resided within the Ordnance Division. The other new arrival was G Division ("Gadget"; also known as the Weapon Physics Division), which would be under the leadership of Robert Bacher. G Division was to be responsible for developing methods for investigating the hydrodynamics of implosion, with particular emphasis on symmetry, compression, behavior of materials, and developing design specifications for the tamper, active-material core, and neutron-initiating source. G-Division absorbed several groups which had been part of the Experimental Physics Division, which was re-named R (Research) Division; this was led by Robert Wilson of Princeton University. R-Division performed criticality experiments (Sect. 7.11), carried out work to measure nuclear parameters such as cross-sections and spontaneous fission rates, and was also involved in developing instrumentation for the *Trinity* test. The Ordnance Division (now O-Division) retained responsibility for the uranium gun bomb, and remained under William Parsons' leadership; Kistiakowsky and Bacher were to keep each other and Parsons closely informed of their work.

Implosion experimentalists were aided by a Theoretical Division group under the direction of Edward Teller. This group had actually been established in January, 1944, to address analyses of estimating the properties of imploded metals under millions of atmospheres of pressure. To numerically integrate equations describing implosion hydrodynamics, these efforts utilized early computers fed information via punch-cards. With this work, the Manhattan Project became the first major scientific endeavor where large-scale numerical simulations complemented experiment and theory, formally establishing simulation as a "third leg" of physics research. Unfortunately, Teller remained so distracted by the idea of the fusion "super" bomb that Oppenheimer (at Hans Bethe's request) replaced him with Rudolf Peierls in June, 1944. Overall, implosion would come to be the concern of over 14 groups within the T, G, and X-Divisions. Other reorganizations also came into effect at the same time. Enrico Fermi, who had frequently consulted at Los Alamos, arrived on a full-time basis to head the new F-Division after completing his work at Hanford. Named after him, this division included in its responsibilities problems that did not fit into the work of other divisions, including investigation of the hydrogen bomb; Fermi and Parsons also became Associate Directors of the Laboratory.

As the implosion program grew in complexity over subsequent months, other committees arose. The most important of these were the Intermediate Scheduling Conference (ISC; under Parsons), the Technical and Scheduling Conference (TSC), and the "Cowpuncher" Committee. Both of the latter were under the leadership of Fermi's Chicago colleague Samuel Allison, who arrived in November, 1944. The ISC was responsible for coordinating aspects of the "packaging" of the gun and implosion bombs for testing and eventual delivery to their combat bases, while the TSC took on responsibility for scheduling experiments, shop time, and the use of fissile material. The Cowpuncher committee came into existence in March 1945; its responsibilities are described in Sect. 7.12.

7.11   The Implosion Bomb: *Fat Man*

Despite starting out with priority lower than the gun-bomb project, the implosion program under Seth Neddermeyer (Sect. 7.6) had enjoyed an increasing measure of attention and resources from the fall of 1943 onward. Neddermeyer made some progress, but achieved only very rough implosion symmetry due to the presence of "jets" of material which traveled ahead of the main mass of compressed material. Such asymmetries promised to render the method too inefficient for a practical weapon. But with Los Alamos' mid-1944 reorganization, implosion began to take center stage on the mesa.

The jet problem appeared insuperable until John von Neumann of the Institute for Advanced Study in Princeton, New Jersey, visited Los Alamos in late September, 1943. A brilliant mathematician, von Neumann had been studying shock waves for the NDRC, and had considerable experience in analyzing shaped explosive charges used in armor-piercing projectiles. His work had convinced him that more symmetric implosions could be obtained if higher material velocities than what Neddermeyer had been working with could be achieved. Neddermeyer's superior, William Parsons, saw the advantage of von Neumann's approach, and the decision was made at a Governing Board meeting on October 28, 1943, to strengthen the implosion program. The higher priority was ratified by Conant and Groves at a Military Policy Committee meeting on November 9, well over half a year before the spontaneous fission crisis emerged.

Unfortunately, Neddermeyer and Parsons were of almost completely opposite personalities, and found it difficult to establish an effective working relationship. Neddermeyer preferred the academic tradition of working alone or in a small group, and chafed under Parsons' more rigorous military approach. It soon became clear that some change would be necessary if implosion research were to be effectively pursued. Oppenheimer's solution was to bring in Harvard University explosives expert George Kistiakowsky (Fig. 4.7) to oversee the work. Kistiakowsky had visited Los Alamos as a consultant while serving as chief of the NDRC Explosives Division, and joined the Laboratory full time in February, 1944, to serve as Parsons' deputy. This position made him Neddermeyer's superior; as a scientist, Kistiakowsky served as an effective buffer between Neddermeyer and Parsons. Oppenheimer formally relieved Neddermeyer of his leadership of the Implosion Experimentation group on June 15, 1944, but Neddermeyer did remain on as a technical advisor and as a member of an implosion steering committee. Parallel to Kistiakowsky in the hierarchy of the Ordnance Division was Edwin McMillan, who took on directorship of the gun-bomb program. Another valuable recruit to the Laboratory at the time of the reorganization was Lieutenant Commander Norris Bradbury, a Stanford physicist and naval reserve officer (Fig. 7.31). Bradbury had been carrying out research in projectile ballistics at the Dahlgren Proving Ground, and was brought to Los Alamos to assist with "implosion lens" research; he also headed the implosion field-test program.

It is difficult to convey a sense of the state of the implosion program in the spring and summer of 1944. In an official history of the Los Alamos Project, David Hawkins summarized the situation as: "at that time there was not a single experimental result that gave good reason to believe that a plutonium bomb could be made at all." In a report prepared in the spring of that year, Kistiakowsky outlined work to be carried out during the last quarter of the year, and summarized

Fig. 7.20 Schematic
illustration of a binary-
explosive implosion lens
segment. Not to scale

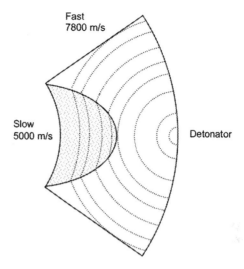

Fast
7800 m/s

Slow
5000 m/s

Detonator

his pessimism with a prediction for November and December: "the test of the
gadget failed … Kistiakowsky goes nuts and is locked up."

Implosion research took another significant step forward with a suggestion by
British Mission member James Tuck. His idea was to modify the shaped-charge
concept into a system of three-dimensional implosion "lenses." In combination
with the use of electric detonators, this concept was key to the eventual success of
the implosion bomb. Tuck, Neddermeyer, and von Neumann subsequently filed for
a patent on the concept, which has never been made public.

The fundamental idea of an implosion lens is sketched in Fig. 7.20, which
shows a single lens in side-view cross-section. To extend the concept to three
dimensions, imagine a somewhat pyramidal-shaped five or six-sided block about a
foot across and a foot and a half from end-to-end (left to right in the Figure, which
is not to scale). Each block comprises two castings of different explosives that fit
together very precisely, and which interlocks with neighboring blocks to form a
complete sphere. The outer casting of each block is a fast-burning explosive
known as "Composition B" (Comp B), which had been developed by Kistiakowsky. The inner lens-shaped casting is a slower-burning material known as
Baratol, a mixture of barium nitrate and TNT. A detonator at the outer edge of the
block of Comp B triggers an outward-expanding detonation wave, which progresses to the left in the Figure. When the detonation wave hits the Baratol, it too
begins exploding. If the interface between the two materials is of just the right
shape, the two waves can be arranged to combine as they progress along the
interface to create an inwardly-directed *converging* burn wave in the Baratol. The
right-to-left progression of the implosion is indicated schematically by the dashed
lines in Fig. 7.20.

In the *Trinity* and *Fat Man* devices, 32 such "binary explosive" assemblies
interlocked to create a complete sphere, as indicated in Fig. 7.21. The full sphere

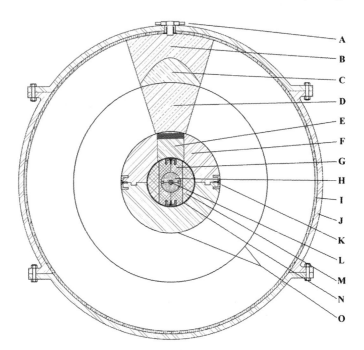

Fig. 7.21 Cross-section drawing of the Y-1561 *Fat Man* implosion sphere showing major components. Only one set of 32 lenses, inner charges, and detonators is depicted. *Numbers in parentheses* indicate quantity of identical components. Drawing is to scale. Copyright by and used with kind permission of John Coster-Mullen. (a) 1773 Electronic Bridge Wire detonator inserted into brass chimney sleeve (32), (b) Comp B component of outer polygonal lens (32), (c) Cone-shaped Baratol component of outer polygonal lens (32), (d) Comp B inner polygonal charge (32), (e) Removable aluminum pusher trap-door plug screwed into upper pusher hemisphere, (f) 18.5-in. diameter aluminum pusher hemispheres (2), (g) 5-in. diameter U-238 two-piece tamper plug, (h) 3.62-in. diameter Pu-239 hemisphere with 2.75-in. diameter jet ring, (i) 0.5-in. thick cork lining, (j) 7-piece Y-1561 Duralumin sphere, (k) Aluminum cup holding pusher hemispheres together (4), (l) 0.8-in. diameter Polonium-beryllium initiator, (m) 8.75-in. diameter U-238 tamper sphere, (n) 9-in. diameter boron plastic shell, (o) Felt padding layer under lenses and inner charges

surrounds an inner spherical assembly of 32 blocks of Comp B (item D in Fig. 7.21), which surrounds the tamper/core assembly. The choice of 32 assemblies was dictated by the fact that this is the number of pentagonal and hexagonal-shaped blocks that can be fitted together to give nearly regular outer faces; think of the patches on a soccer ball. The *Trinity* and Nagasaki weapons used 12 pentagonal and 20 hexagonal sections, which respectively weighed about 47 and 31 pounds each.

The purpose of the inner layer of Comp B blocks, which are detonated by the imploding Baratol lenses, is to achieve a high-speed symmetric crushing of the tamper and core. The higher speed achievable with Comp B was essential in lowering the compression timescale to a few microseconds in order to beat the spontaneous-fission predetonation problem. A trap-door arrangement with a plug of tamper material (item E in Fig. 7.21) allowed for insertion of the core while the bomb was being assembled. As might be imagined, assembling the HE configuration is difficult and time consuming: one must literally hand-assemble a three-dimensional jigsaw puzzle with explosive pieces that fit together very precisely. The total weight of the high-explosive assembly alone was about 5,300 pounds, just over half the bomb's total weight of about 10,200 pounds. The 1-in. thick outer casing alone contributed 1,100 pounds.

One of the most serious complications in attempting to estimate the yield of the *Fat Man* design arose from the nested aluminum/uranium tamper-spheres configuration. Geoffrey Taylor, another member of the British Mission, had determined that when a heavy metal is accelerated against a light metal, the interaction is stable. But if the acceleration is done with the light metal into a heavy metal, the interface becomes unstable, and gives rise to jets of light material spurting ahead of the main mass of that material, as Neddermeyer had discovered. The effect has been compared to fresh paint dripping from a ceiling. This jetting effect is now known as a Rayleigh–Taylor instability; to avoid it, the implosion must be extremely symmetric. Given this, one might wonder why the lighter aluminum shell in Fig. 7.21 (item F) was imploded into the heavier uranium shell (item M). The reason was explosion efficiency: it has been estimated that some 20 % of *Fat Man's* yield was due to fast neutrons fissioning U-238 in the tamper sphere.

The dynamics of implosion are extremely complicated. The pressure created in the bomb core was estimated to be similar to that at the center of the earth, and the properties of materials in such circumstances were not well known. Detonation waves can interfere with each other unless they are arranged to be perfectly converging, which requires simultaneous multi-point triggering; variations in the velocity of the implosion must be held to less than about 5 %. In the original conception of the implosion scheme, the jetting problem was aggravated by the intent of trying to compress a thin shell of fissile material to many times its normal density; there was little confidence that the necessary symmetry could be maintained. In September, 1944, Robert Christy (Fig. 7.22), a former student of Oppenheimer's and one of the first persons that had been recruited to Los Alamos, proposed a configuration with a core which was solid except for a small central void to hold the initiator. Christy's design came to be known as the "Christy core," and was adopted for the *Trinity* and Nagasaki bombs. As Christy described it:

> Earlier designs of the implosion bomb had been a relatively thin shell of plutonium, which would then be blown in by the implosion. It was assembled in the center with ideally very high density and spherical shape. But, there were constant worries at the time that, because of irregularities in the explosive, it would end up in a totally unacceptable form. They were worried it wouldn't be spherical and that it might end up with jets coming in and it wouldn't even go off. These worries were very real. They wanted to be sure it would not fail. It would be a very bad thing if they had a failure. So I suggested if they took the hole out of the middle, and just made it solid, it couldn't very well be made non-spherical. There was a very small hole for the initiator that was required.

Responsibility for developing the explosive components of the implosion bomb lay with George Kistiakowsky's X-Division, which eventually came to have a staff of some 600. This meant investigating methods of detonating high-explosive (HE) components, improving the quality of castings, developing and testing the lens system, and fabricating explosive charges. Kistiakowsky organized an extensive series of test shots to investigate the best number of detonation points, the types and arrangement of explosives, and the material to be collapsed. The level of activity of the testing program can be judged by the fact that some 20,000 castings of acceptable quality were created over a period of 18 months, while many more than that were rejected. Some 100,000 pounds of HE were used per month. Casting operations became so extensive that a separate site (Sawmill, or "S" site) was set up for that purpose, staffed largely by SEDs. It began to come into operation in May, 1944. McAllister Hull, a 21-year old SED, arrived at Los Alamos in the fall of 1944 to, as he put it, "figure out how to cast the lenses to the specifications required." Hull had worked at an ordnance plant where TNT was cast into shells, and had much practical experience with such operations. The lens castings were done as slurries in modified commercial candy-making machines (Fig. 7.23).

Three main problems cropped up with casting operations, however. As cast explosives cooled, internal voids would tend to form, and surfaces tended to have bubbles. Also, the chemicals from which the explosives were made tended to separate during cooling. Hull and his group managed to lick each problem in turn. The molds for the lenses had double walls, which housed cooling-water coils. By pumping hot water through the coils as the molten explosive was poured into the mold, formation of surface bubbles could be eliminated by gradually lowering the water temperature. The void and separation problems took longer to solve. As a student, Hull had worked for a time as a waiter, where he had gained experience in producing smooth, well-blended milkshakes. Working from that experience, he

Fig. 7.23 An original
implosion-lens casting
machine on display at
National Museum of Nuclear
Science and History in
Albuquerque, NM. Photo by
author

developed a protocol for placing a stirrer in the setting explosive and withdrawing
it vertically just ahead of the solidification line. This prevented void formation
while keeping the chemicals well mixed. By late 1944, a fairly reliable if labor-
intensive casting system was in place, with men usually working three shifts per
day, even (against regulations) during thunderstorms. As Hull related:

> I used a stirrer, gradually pulled up as the casting cooled from the outside, to keep the
> Baratol mixture uniform and the interior cavity free. I determined the rate at which to pull
> up the stirrer by casting 10 inner lenses simultaneously, then sawing them in half at five-
> minute intervals to see where the solid line had reached. The stirrer was pulled up at a rate
> to keep the blades ahead of the solidification curve inside the lens.

In his memoirs, Hull relates a story involving General Groves that would be
unbelievable if it were not true. Groves and Oppenheimer came by one day to
witness the casting operation, and Groves accidentally stepped on a water line. The
line popped away from its wall connection, and a jet of near-boiling water struck
Groves on his backside. Hull, in uniform, suppressed his laughter while turning off
the faucet, but broke up when Oppenheimer said "It just goes to show the
incompressibility of water."

Because the explosives would tend to stick to the molds, ordinary Vaseline was used as a releasing compound. After being removed from their molds, all castings were checked for uniformity by X-raying them. Those which contained voids were repaired by drilling into them with non-conducting tools to get to the void, and then pouring in molten explosive, rather like a dentist filling a cavity. George Kistiakowsky put this operation in perspective with the remark that 1 g of such explosives could finish off a hand. After repairs, castings were machined to remove any flashings or roughness; thousands of machining operations were conducted without a single accidental detonation. Design and manufacture of molds and producing enough castings of acceptable quality were always pacing elements in implosion testing.

An enormously challenging but critically important area of work for the implosion program was development of suitable diagnostic routines. Essentially, the problem was to obtain information on events *inside* an explosion to time resolutions on the order of a microsecond. The seven complementary and over-lapping methods developed are a testament to the creativity available at Los Alamos and the dedication with which the project was pursued.

The most direct diagnostic technique was what were called "terminal observations": examining the remains of detonated explosives. This was actually much more refined than it sounds. A flat mold for a two-dimensional cross-section of an implosion lens would be created, explosives cast into it, and a detonator placed at the "top." (Many diagnostic experiments used two-dimensional lenses because they were easy to make by cutting or casting the explosives.) A steel plate in the shape of the lens would be cut out and the casting placed atop it. Upon detonating the explosive, the burn wave would leave an imprint on the plate, from which the symmetry of the detonation could be measured. A variety of pairs of explosives corresponding to different "indexes of refraction" were tried in an effort to get the correct ratio of detonation speeds.

In the optical realm, cameras were developed where a shutter remained open while film was advanced at high speed on a rotating drum, or where an image was scanned along a fixed film by means of rotating mirror. Such cameras could obtain images with sub-microsecond time resolution. Another technique was to image blocks of imploding explosive with brief but intense bursts of X-rays detected by banks of Geiger counters. The so-called "magnetic method" took advantage of the fact that the motion of a metal within a magnetic field alters the field. A changing magnetic field will induce a current in a wire, and the time-evolution of the current can be analyzed to provide information of the progress of the explosion. This method proved excellent for giving information on the velocity of the external surface of a metallic imploding sphere, and was first tried on January 4, 1944. Surprisingly, the magnetic field required was only about 10 Gauss, much less than that of a modern-day refrigerator magnet. The magnetic method was unique in that it was the only diagnostic technique that could be applied to a full-scale implosion assembly. Complementing the magnetic method was the "electric method," which involved recording responses from electrical contacts formed between an imploding sphere and a network of prearranged pins. This method was especially valuable in that it gave three-dimensional information on velocity asymmetries during implosions.

One of the most innovative diagnostics was the "radiolanthanum" or "RaLa" method. Conceived by Robert Serber in November, 1943, the implementation of this method was carried out under the direction of Bruno Rossi, who, like Enrico Fermi, had fled Italy for America. This method was predicated on including a strong gamma-ray emitter within an imploding sphere, and monitoring the intensity of gamma rays as a function of time to follow the changing density of the sphere. Serber estimated that the strength of the gamma-ray source would have to be on the order of 100 Ci, a number which eventually had to be increased. As described in Sect. 5.2, the gamma-ray source, radioactive lanthanum-140 (half-life 40 h) was obtained by way of beta-decay of barium-140, a direct fission product extracted from the Oak Ridge X-10 reactor. A single batch of radiolanthanum could contain up to 2,300 Ci of radioactivity, an extremely dangerous amount. A special extraction laboratory was established in Tennessee, from whence the material was shipped in lead-lined containers for its 1,400-mile journey to Los Alamos. The first RaLa feasibility-study shot was fired on September 22, 1944, using a mockup core made of iron and a source of only 40 Ci. A second shot followed on October 4 with a 130-Ci source, and a third on October 14. On December 14, a test showed encouraging evidence of compression of a Christy-core assembly. Subsequent solid-core tests, conducted on February 7 and 14, 1945, used new electric detonators and gave even more encouraging results. The month of March, 1945, saw little RaLa testing because of a shortage of radiolanthanum, but the first use of implosion lenses with a RaLa shot was carried out on April 1.

The final diagnostic technique, the "betatron" method, was proposed by Seth Neddermeyer and Donald Kerst in August, 1944. This method also employed gamma rays, but in a manner that complemented the RaLa technique. A betatron is a machine for accelerating electrons to high speeds. When accelerated, the electrons emit gamma-rays, and such gamma-rays could be directed to pass through an implosion assembly from the outside, as opposed to originating from within as in the RaLa method. The gamma-rays would be affected by the changing density of the assembly, and were detected by a large ionization chamber located opposite the betatron on the other side of the explosion. Both the betatron and the detection chamber had to be located behind protective concrete walls several feet thick to shield them from the explosions. Within days of receiving Neddermeyer and Kerst's proposal, Oppenheimer had located the only suitable betatron unit in the country, which was being made for the Army and undergoing testing at the University of Illinois. Oppenheimer's priority request for its transfer to Los Alamos was granted, and the 6-t unit arrived in mid-December. By mid-January, it was producing images. (The images were recorded on films; there were no digital cameras in 1945!)

In Robert Bacher's G-Division, implosion work centered on the problem of developing detonators that would fire with sufficiently great simultaneity to initiate a highly symmetric implosion. The simultaneity required was far beyond that available in any commercial detonators, which were normally required to trigger

Fig. 7.24 *Left* Luis Alvarez's (1911–1988) Los Alamos ID-badge photo. *Right* Donald Hornig (1920–2013) in 1964. *Sources* http://commons.wikimedia.org/wiki/File:Luis_Alvarez_ID_badge. png; AIP Emilio Segrè Visual Archives, Physics Today Collection

only one explosion at a time. Much of the work on detonator design fell to Luis Alvarez and Donald Hornig (Fig. 7.24), who took an Edison-like trail and error approach. At Harvard University, Hornig had written a thesis on shock waves produced by explosions, and had come to Los Alamos from the Underwater Explosives Laboratory of the Woods Hole Oceanographic Institute in Massachusetts. Hornig worked with "spark-gap" switches, wherein a high voltage causes a spark to jump between two pieces of metal placed a small distance apart within an explosive. Curiously, the explosive triggered by the detonator did not directly initiate an implosion lens, but rather drove a copper "slapper" plate into the lens, which triggered the implosion via a high-pressure pulse. The first test of multiple electric detonators was carried out in May, 1944, and by late that year Horning had managed reduce the timing spread for firing down to several hundredths of a microsecond. Detonator production always lagged behind schedule, however; refinements in their design continued right up to the time of the *Trinity* test.

The most dramatic work of the R and G-Divisions involved so-called criticality experiments. These were assemblies of varying amounts of U-235 or Pu-239 arranged to approximate critical masses. Short of a real explosion, there was no way to determine the extent of supercriticality that would be achieved with a full-scale gun or implosion assembly, but data from subcritical and barely-critical experiments could be extrapolated to give checks on theoretical estimates. Initially, criticality experiments involved assembling blocks of uranium hydride, on the premise that the hydrogen would slow down neutrons and hence give researchers experience with slower reactions before moving to fast-neutron configurations. By surrounding a subcritical assembly of hydride blocks with neutron-reflective beryllium tamper blocks, the number of fissions could be enhanced; such experiments were known as "Godiva" assemblies, where an otherwise bare core would be "clothed" by the tamper blocks. Some hydride assemblies were so near-critical that the neutron-reflecting effect of the body of a person hovering over the assembly could make it supercritical; the experimenter would hop away just as criticality was reached.

By September, 1944, enough pure uranium metal was becoming available to begin criticality experiments without hydration. The first such experiments used a 1.5-in. diameter sphere (two hemispheres) of uranium enriched to 70 % U-235; later experiments involved spheres up to 4.5 in. diameter of 73 %-enriched material. (The bare critical diameter for pure U-235 is about 6.6 in.). When a neutron source was placed within the sphere, the number of neutrons emerging from the sphere would be greater than from the neutron source alone due to the effect of induced fissions; by extrapolating to infinite neutron multiplication, the critical mass could be determined. By March, 1945, enough uranium had been accumulated to make a tamped critical mass, and, on April 4, a combination of 4.5-in. hemispheres and tamper cubes was brought to within 1 % of criticality. The first critical plutonium assembly was achieved in April, 1945, using a plutonium-water solution with a beryllium tamper.

Criticality experiments resulted two postwar fatalities at Los Alamos. On the night of August 21, 1945, Harry Daghlian was working alone (against regulations) with a plutonium sphere and tamper blocks when a block slipped out of his hand and caused a brief chain reaction. Daghlian had to partially disassemble the pile to halt the reaction, but received a radiation dose estimated at 500 rems. Such a dose is usually considered to be the single-shot dose that will cause 50 % of exposed individuals to die within 30 days. (Details of damage caused by various doses are discussed in Sect. 7.13). Daghlian died 25 days later, on September 15. His hands, which had been the closest parts of his body to the assembly, became gangrenous, and his kidneys eventually became unable to remove decomposition products from his blood. A similar accident took the life of Louis Slotin on May 21, 1946. Slotin was demonstrating how to make criticality measurements using the same spheres Daghlian had used; they would become known as the "demon core." Slotin was gradually decreasing the separation between the hemispheres with a screwdriver, but the screwdriver slipped and the spheres came together. Thermal expansion quickly halted the reaction, but Slotin received a radiation dose estimated at over 2,000 rems, and died nine days later. Seven other people were in the room at the time; two suffered acute radiation symptoms, but recovered. The Slotin accident permanently ended all hands-on criticality work at Los Alamos.

Less hands-on but also potentially dangerous were experiments that came to be known as "Dragon drops." In October, 1944, Otto Frisch proposed constructing a device where a slug of uranium hydride would be dropped through the center of an almost-critical assembly of the same material (Fig. 7.25). When the slug passed through, the assembly would become supercritical for a brief time. Richard Feynman, a Los Alamos theoretician and future Nobel Laureate, described this as "tickling the dragon's tail," and Frisch's machine became known as the Dragon

machine. Frisch likened it to the curiosity of an explorer who has climbed a
volcano and wants to take one step nearer to look into the crater, but not fall in.
Given the nature of the setup, Frisch was surprised when the Coordinating Council
deemed the experiment worth pursuing.

As realized, the Dragon machine stood about 6 m high. Designed to be operated
largely by remote control, the operator could not activate the so-called "Here We
Go" button until various safety interlocks had been activated. A steel box which
contained uranium hydride rode on guide wires and was dropped from the top of
the device, which looked like an oil-well derrick. The box would pass through a
lower table on which had been mounted more hydride, producing, for about 0.01 s,
a very slightly supercritical assembly. Frisch estimated that even if the box became
stuck, the resulting explosion would be equivalent to only a few ounces of high
explosive.

Frisch was ready by mid-December, and began with tests using dummy materials before moving to active material. On January 20, 1945, the Dragon machine
produced the world's first fast-neutron chain reaction. The reactions were brief, but
bursts of up to 10^{15} neutrons were created, accompanied by power releases of up to
20 million Watts and temperature increases in the hydride of up to 2 °C/ms over
about 3 ms; there was not a single accident or instance of material hanging up in
the drop mechanism. Because other experimental groups needed the hydride,
experiments ceased in February, and the machine was subsequently dismantled.
Dragon experiments contributed data on such parameters as the generation time
between fissions, and the exponential growth rate of the chain reaction.

Prospects for implosion slowly improved through the latter half of 1944 and the
first half of 1945. For James Conant, his July, 1944, pessimism began to give way

to guarded optimism. By October, he was giving a lensed device a 50:50 chance of working for a test on May 1, 1945, and three-to-one odds for a test on July 1. Conant was visiting the Laboratory at the time of the December 14 test referred to above, and concluded that while the method had the possibility of giving relatively high efficiency (a few percent), it still faced enormous difficulties: "Further experiments which may be completed by March 1 will show the chances of doing this in 1945. My own bets are very much against it." He judged that an implosion bomb would likely yield less than 850 t TNT equivalent, and perhaps only 500 t.

By early 1945, progress was such a schedule for working toward to a full-scale test was developed at a Technical and Scheduling Conference held on February 17. Full-scale lens molds were to be available for casting by April 2, and full-scale lens shots to test the timing of multi-point detonations were to be ready by April 15. By April 25, shots with hemispheres of explosives were to be ready. Detonators should come into routine production between March 15 and April 15. A full-scale test of implosion without fissile material but using the magnetic diagnostic method should be made between April 15 and May 1. Between May 15 and June 15, plutonium spheres had to be fabricated and tested for criticality. Fabrication of implosion lenses for the full-scale test were to be underway by June 4, and fabrication and assembly of the implosive sphere should begin by July 4. The target date for the test itself was set as July 20.

On February 28, just eleven days after the TSC meeting, Oppenheimer and Groves decided provisionally on the Christy-core design with explosive lenses made of Comp B and Baratol. Characteristic of so many decisions in the Manhattan Project, their choice was a gamble: few implosion lenses had by then been tested in the diagnostic program.

7.12 Trinity

The most important source of information on the *Trinity* test is a publically-available report assembled by the test's Director, Kenneth Bainbridge. Prepared soon after the test and augmented in 1946 with information acquired from two tests conduced at Bikini Atoll in the Pacific, the report was cleared for public release as Los Alamos report LA-6300-H in 1976, and is now readily available online. It is required reading for any serious student of the Manhattan Project.

Given the uncertainties with the implosion method, the idea of a full-scale test was circulating well before the spontaneous fission crisis emerged in mid-1944. A test was considered essential because of the enormous leap from laboratory experiments and theory to a practical "gadget"; no one wanted the first test of a *Fat Man* weapon to be over enemy territory, where, if it failed, an enemy might be able recover a large amount of fissile material. General Groves saw the idea of a full-scale test as a waste of fissile material, and proposed that any test device contain only enough to just start a chain reaction. Oppenheimer objected to this on the rationale that it would be practically impossible to specify the precise amount of material necessary to achieve such a circumstance. On February 16, 1944, he wrote Groves to emphasize that the "implosion gadget must be tested in a range where the energy release is comparable with that contemplated for final use." Groves relented, and preparations for a full-scale test began in March, 1944, when Oppenheimer appointed Bainbridge to oversee the operation.

The first issue was to locate a suitable site. Criteria included flatness in order to facilitate measurements, favorable weather, wind patters that would not expose populated areas to excessive fallout, and proximity to Los Alamos to simplify travel. The Secretary of the Interior wanted no Indians to be displaced for the test, and Groves added that stipulation into the mix as well. Four sites were considered in New Mexico, including the Jornada del Muerto ("Journey of death") desert east of the Rio Grande; one in Colorado; two in California including near the town of Rice in the Mojave desert in the eastern part of the state; and sand bars off the coat of Texas. The choice came down to the Jornada and Rice locations, with Jornada winning out. Proximity to Los Alamos was likely a factor, although it has been claimed that Groves rejected the Rice location because it was in use by General George Patton, whom Groves refused to approach regarding its use. One source quotes Groves as saying that Patton was "the most disagreeable man I ever met."

Located about 160 miles south of Los Alamos, the Jornada site comprised an 18 by 24-mile tract in the northern portion of the Alamogordo Army Air Field. The town of Alamogordo (2008 population about 36,000) is located about 60 miles southeast of the point where the bomb was detonated; Socorro (presently about 9,000 inhabitants) lies about 35 miles to the northwest. Summertime temperatures in the area routinely reach over 100 °F. At the time of the test, the nearest habitation was about 12 miles distant. Before the war, the had land supported some cattle grazing, but in 1942 the Army appropriated the four-room ranch house of the family of George McDonald to serve as a part of the Alamogordo Bombing and Gunnery Range. The house was used as the assembly station for the *Trinity* bomb; while it was somewhat damaged by the explosion, it still stands about two miles southeast of ground zero (Figs. 7.26, 7.27 and 7.28). Now restored to the way it appeared in 1945, the house is accessible to tourists during the two weekends per year that the site is normally open to visitors.

An enduring mystery is how the name *Trinity*, which served to designate both the site and the test, came to be chosen. Oppenheimer claims to have suggested it, and a common speculation is that his love of the poetry of John Donne may have been involved. The first four lines of Donne's devotional poem "Batter My Heart" read

> Batter my heart, three-person'd God, for you
> As yet but knock, breathe, shine, and seek to mend;
> That I may rise and stand, o'erthrow me, and bend
> Your force to break, blow, burn, and make me new.

Donne is alluding to the Christian notion of deity as Father, Son, and Holy Ghost. Another speculation derives from Oppenheimer's interest in Hindu culture, where the concept of Trinity involves three gods: Brahma the Creator, Vishnu the

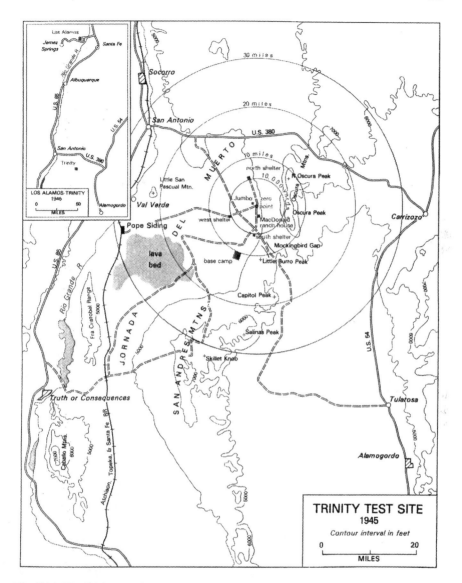

Fig. 7.26 The Trinity test site. From V. C. Jones, *United States Army in World War II: Special studies—Manhattan: The army and the Atomic Bomb.* Courtesy Center of Military History, United States Army

Preserver, and Shiva the Destroyer. In this faith, whatever exists in the Universe is never destroyed but rather transformed, appropriate imagery for a nuclear explosion.

Except for the ranch house, the site was completely undeveloped. A Base Camp of barracks, officers quarters, warehouses, repair shops, bomb-proof structures,

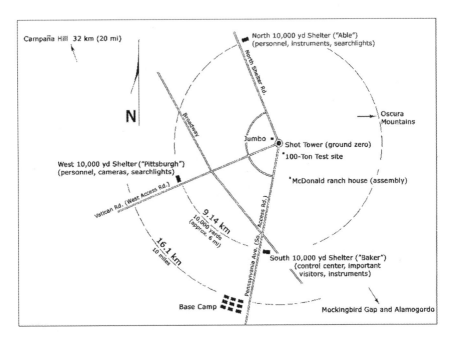

Campaña Hill 32 km (20 mi)

North 10,000 yd Shelter ("Able")
(personnel, instruments, searchlights)

N

Broadway

North Shelter Rd.

Oscura
Mountains

Jumbo

Shot Tower (ground zero)

100-Ton Test site

West 10,000 yd Shelter ("Pittsburgh")
(personnel, cameras, searchlights)

McDonald ranch house (assembly)

Vatican Rd. (West Access Rd.)

9.14 km
10,000 yards
(approx. 6 mi)

16.1 km
10 miles

Pennsylvania Ave. (So. Access Rd.)

South 10,000 yd Shelter ("Baker")
(control center, important
visitors, instruments)

Base Camp

Mockingbird Gap and Alamogordo

Fig. 7.27 Detail map of Ground Zero area. *Source* http://www.lahdra.org/pubs/reports/
In%20Pieces/Chapter%2010-%20Trinity%20Test.pdf, based on Lamont (1965)

Fig. 7.28 The author at the
McDonald Ranch House,
October 2004

technical facilities, a mess hall, and other support facilities had to be constructed to
serve the needs of a staff that would grow to number over 250. Over 20 miles of
blacktopped roads and 200 miles of telephone lines would have to be provided,
along with a fleet over 100 vehicles. Oppenheimer approved the construction plans
on October 27, 1944, and the first residents, a detachment of Military Police under
the command of Lieutenant Harold Bush, arrived to take up their duty in late

Fig. 7.29 Trinity Base camp. *Source* http://commons.wikimedia.org/wiki/File:Trinitybase_camp.jpg

December, 1944. The Base Camp was located about 17,000 yards (9.6 miles) south of "ground-zero," the location of the test itself. The *Trinity* bomb would be mounted atop a 100-foot high surplus steel Forest Service fire-watch tower whose concrete footings were sunk some 20 feet into the earth. Groves witnessed the explosion from Base Camp along with various distinguished visitors, including Bush, Conant, and Fermi (Fig. 7.29).

Within an area of about 100 square miles centered on ground zero were placed three instrument stations, roughly to the North, West, and South, all 10,000 yards from the test site (Fig. 7.27). The South station also served as the control center where the final switches to activate an automatic firing sequence would be thrown; Oppenheimer witnessed the test from that point. At the time of the test, all shelters were under the supervision of a scientist until the bomb detonated, at which time command passed to a medical doctor who was authorized to order evacuation if necessary. The scientists in charge at the North, West, and South shelters were respectively Robert Wilson, John Manley, and Robert Oppenheimer's brother, Frank. Personnel who had participated in the development of the bomb but who were not needed at the control station during the countdown witnessed the spectacle from a vantage point on Campaña Hill, some 20 miles to the northwest. This group included such notables as Hans Bethe, James Chadwick, Ernest Lawrence, Edward Teller, and Robert Serber.

The precise number of people that witnessed the test was not documented, but film-badge counts indicate that some 350 people were at the site sometime during July 16, 1945 (the day of the test). One of the major players of the Manhattan Project who would *not* witness *Trinity* was Arthur Compton. Oppenheimer had sent him an invitation reading "Anytime after the 15th would be a good time for our fishing trip." Compton decided not to attend so as not to raise questions at the

Met Lab, but after the test Oppenheimer called him to report that "You'll be interested to know that we caught a very big fish."

By early 1945, preparations for *Trinity* were becoming so complex that Oppenheimer appointed the "Cowpuncher Committee" to provide executive direction for the implosion program—to "ride herd" on it. Cowpuncher comprised the Laboratory's top scientific and administrative personnel: Oppenheimer, Bainbridge, Bethe, Kistiakowsky, Parsons, Bacher, Allison, and Cyril Smith. The committee first met on March 3, and assigned highest priority to initiator development, detonators, and procuring lens molds.

To test-run procedures and calibrate instruments in advance of the full-scale test, a rehearsal test was conducted at about 4:30 a.m. on May 7, 1945. This involved detonating 108 t of high explosive mounted atop a 20-foot high tower located about 800 yards southeast of where the *Trinity* tower would be erected. The height of this explosion was not arbitrary. At that time, the best prediction for the *Trinity* yield was about 5,000 t TNT equivalent. Theoretical analysis indicated that for an observer at distance d from a nuclear explosion of yield E, the air pressure behind the initial shock wave would be proportional to $E^{2/3}/d^2$, so the center of gravity of the 108-t stack was placed at 28 feet above the ground to scale to *Trinity's* planned 100-foot high detonation and anticipated yield. (For comments on the *peak* pressure during the passage of the shock wave, see Sect. 7.13.) *Trinity's* yield would prove to be much more than 5 kilotons, however, which resulted in many recording instruments being overwhelmed in the real test. Monitoring instruments were also deployed at scaled distances. To create a low-level simulation of the fallout pattern to be expected from a nuclear explosion, the TNT was seeded with tubes containing fission products from a Hanford fuel slug. These were sufficient to supply 1,000 Ci of beta-activity, and 400 of gamma-activity. The TNT shot proved a valuable test of procedures, and revealed a number of issues that needed to be resolved before the real test. Some of these were technical, such as interference on instrument cables, while others were more prosaic, such as failure to provide enough batteries to power all of the instruments that had been deployed. Probably the most important lesson was that there should be a cutoff date beyond which no further apparatus would be introduced into the experimental area. Another was that there should be no kibitzing (horseplay) at the tower during bomb assembly.

One of the most curious aspects of the *Trinity* test was the "Jumbo" program. When the chances for implosion looked slim, it was thought that it would be wise to set off the explosion within some sort of vessel that could contain the force of the high explosive, so that, in the event of a *nuclear* fizzle, the plutonium could be recovered. The pressure requirement was estimated to be 60,000 pounds per square inch, or about 4,000 atmospheres. One scheme considered was to suspend the bomb in a tank of water of weight 50–100 times that of the high-explosive. A drawback of this scheme was that plutonium dispersed in the condensed steam that would be created by the explosion would be supercritical if the container held, unless neutron-absorbing boron were added to quench the reaction. The only option that looked feasible was to set the bomb off within a strong containing vessel. This led to the design and procurement of Jumbo, a massive steel cylinder within which the bomb would be placed; the ends would then be closed off. As related by Kenneth Bainbridge, "Jumbo represented to many of us the physical manifestation of the lowest point in the Laboratory's hopes for the success of an implosion bomb. It was a very weighty albatross around our necks."

The design of Jumbo fell to the Engineering Group of Kistiakowsky's X-Division. As early as May, 1944, scale-model "Jumbinos" were undergoing feasibility tests. In its final incarnation, Jumbo weighed in at 214 t, was 28 feet long, 10 feet in inside diameter, had a shell 14 in. thick, and cost $12 million (about $150 million in 2013 dollars). Manufactured by the Babcock and Wilcox Corporation in Ohio, the giant vessel was carried 1,500 miles by rail on a special flatcar (which itself weighed 157 t) over a circuitous route that included travel down the Mississippi river to New Orleans. Jumbo's rail journey ended at a siding 30 miles from ground zero. From there it was hauled to the test site at three miles per hour on a 73-t, 64-wheel trailer. By the time of the test, however, confidence in a successful implosion was much greater, and the anticipated need for Jumbo had diminished. Also, experimenters were concerned that the vessel would interfere with monitoring instruments. The plan was abandoned, and Jumbo was erected on a tower some 800 yards northwest of the explosion. The tower was vaporized, but Jumbo survived. Had it been used, the result would have been tons of radioactive fallout in the New Mexico sky, and chunks of shrapnel hurled to great distances. Easily large enough for more than one person to stand inside, the remaining 100-t body of Jumbo, less its ends (blown off, according to some sources), now lies where it was on the morning of July 16 (Fig. 7.30). One of the ends now serves as a tourist attraction in Socorro.

The *Trinity* test was probably the most monitored and photographed scientific experiment in history to its time. Physicists proposed no end of experiments, but as shop time was at a premium in the weeks leading up to the test, all proposals had to be submitted to a review committee for classification as essential (efficiency, blast pressure, detonator performance), desirable (fireball photography and analysis,

Fig. 7.30 *Left* Jumbo, 1945. *Source* http://commons.wikimedia.org/wiki/File:Trinity_Jumbo.jpg . *Right* The author (*light-colored shirt* and *hat*) inside the 100-t body of Jumbo, 800 yards from *Trinity* ground zero

motion of the surrounding earth), or unnecessary. No experiment could affect the operation of the bomb, and no experiments were allowed to be installed within four weeks of the test date in order to leave time for set-ups, rehearsals, and debugging. Proposers were required to submit answers to over a dozen questions, including estimated manpower requirements, calibrations, signal line needs, actuation mechanisms, and shop time.

Six chief groups of experiments were arranged: implosion diagnostics; energy release measurements; damage, blast, and shock; general phenomena; radiation measurements; and meteorology. Within these groups were deployed dozens of individual experiments designed to measure every conceivable aspect of the explosion. An incomplete list includes detonator simultaneity; shock wave transmission through the imploding high-explosive; fission-rate growth; gamma rays; neutrons; fission products; atmospheric pressure effects; seismic disturbances; earth displacement; and ignition of structural materials. Over 50 types of cameras were used, from simple pinhole models to motor-driven units capable of exposing up to 10,000 frames per second; some 100,000 individual exposures were obtained. Spectrographic cameras recorded light of various wavelengths emitted by the fireball. Gold foils placed in protective tubes spread around the site would become radioactive due to neutron bombardment, and so reveal the strength of the neutron flux. Fission fragments in the soil would be collected from a lead-lined tank with a trap-door in the bottom; such fragments were a valuable source of information on the efficiency of the bomb. Pressure gauges were deployed to measure the energy released by the explosion. Some pieces of equipment would knowingly be destroyed by the shock wave created by the bomb, and had to be designed to transmit their data between the time of the explosion and their destruction. In all, some 500 miles of wires and cables were installed for the test.

Groves paid particular attention to obtaining shock measurements from both airborne and ground-level sensors. Barographs were deployed at distances of 800, 1,500 and 10,000 yards, and from 50 to 100 miles from the site of the explosion. These units served two purposes: their data would bear on setting the detonation heights of combat weapons, and Groves wanted evidence in the event of any damage lawsuits arising from the test. To obtain radiation-exposure records, films were mailed to dummy addresses through local post offices, and picked up later by intelligence officers. Groves also deployed a security contingent of 160 men north of the test area, lest it prove necessary to evacuate ranches and towns at the last moment.

Workdays at the site often stretched to 18 hours. On June 9, the Cowpuncher Committee set Friday, July 13, as the earliest possible test date, with the 23rd as a probable date. On June 30, the earliest possible date was revised to Monday, July 16. For political reasons (see below), Groves wanted the test as soon as possible. Oppenheimer thought the 14th possible, but settled on the 16th. On July 2, the plutonium core hemispheres for the *Trinity* device were completed, and on the fourth a mockup device was assembled and checked for criticality. On the 6th, *Trinity's* uranium tamper was machined, and on the 10th, the best available lens castings were selected.

Meteorological conditions were of particular concern when setting the test schedule, and the story of the work of the project's weather forecaster makes for an interesting example of how the same circumstances can be related very differently by different observers. The Manhattan Project's meteorology supervisor was Jack Hubbard, who had been obtained from the California Institute of Technology. Equipped with portable weather stations, field radar sets, devices which gave temperature and humidity readings at different altitudes, balloons, and local and national records, one of Hubbard's first responsibilities had been to choose a date for the 100-t test, and he identified April 27 and May 7 as the optimum dates. The latter was chosen, and his forecast proved accurate; Bainbridge described the meteorological service for the test as excellent. July weather in the southwest can be more unstable than May weather, however.

For the *Trinity* test, physics, meteorology and politics collided incompatibly, and different accounts offer conflicting assessments of Hubbard's work. The demands of the various experimental groups were practically impossible to reconcile. For some groups, rain before the test might not be a concern, but for others an instrumentation cable might be rendered useless if it had not had time to dry out. Hubbard's first choice of dates was July 18–21, with the 12th to the 14th as second choice, and the 16th as only a possible date. The 16th was favored, however, because that would be the earliest date for which the bomb would be ready, and Groves was under intense pressure to carry out the test as soon as possible. President Truman would be in Germany for the Potsdam Conference from July 16 to August 2, negotiating with Winston Churchill and Josef Stalin regarding post-war occupation arrangements in Europe and the prosecution of the war against Japan. The conference had originally been set to begin on July 6, but Truman had asked for a postponement to the 15th to give Los Alamos more time. The British and Canadians were informed at a July 4 meeting of the Combined Policy Committee that America intended to use the bomb.

The strategic situation was complex and fluid. Planning for a November 1 American-British invasion of the southern island of Japan was already very advanced. The Soviet Union had committed to enter the war against Japan within three months after the defeat of Germany, which had been declared on May 8. If the Soviets did declare war on Japan, they would insist on territorial claims, a situation American and British leaders preferred to avoid. A successful test would strengthen the hand of American and British negotiators, and could be parlayed into an ultimatum to Japan to surrender. As events played out, the Soviets would honor their commitment on the last possible date, August 8, between the bombings of Hiroshima and Nagasaki; Soviet territorial claims ended up being restricted to some smaller Japanese islands.

Setting the test date was out of Hubbard's hands, but he did his best. From June 25 onwards, hourly observations were recorded by weather stations at Base Camp and Ground Zero. On July 6, Hubbard predicted that the area would be dominated by a stagnant tropical air mass, which proved to be partly true. On learning that the test had been set for the 16th, he recorded in his diary: "Right in the middle of a period of thunderstorms, what son-of-a-bitch could have done this?"

Out of concern that radioactive fallout could be carried over populated areas, the most pressing weather considerations were wind and rain. South-southwest winds were preferred in order to blow fallout to the northeast. On the morning of the 15th, Hubbard predicted that the next day would see light and variable winds from east to west below 14,000 feet, and west-southwest winds above 15,000 feet. What he apparently did not predict were the strong localized thunderstorms that moved into the area about 2:00 a.m. on the 16th, two hours before the scheduled test time.

Robert Norris has presented a different view of the Hubbard story. This is that when Hubbard was acquired, the purpose of the work was not revealed, and Cal Tech assigned one of its "lesser-qualified" staff. In this version of the story, Groves apparently began to appreciate this as the test neared, and brought in Air Force meteorologist Colonel Ben Holzman, who had participated in the selection of the date for the D-Day landings in Normandy. Groves wrote in his memoirs that Hubbard had been making accurate long-range predictions, but the only time he was not right was "on the one day that counted." Groves states that he dismissed the forecasters in the hours before the test, deciding to rely on his own predictions.

Rehearsal tests were conducted on July 8, 12, 13, and 14. *Trinity's* plutonium hemispheres were conveyed to the site by car from Los Alamos on the 11th, and initiators arrived the next day. Final assembly of the high-explosive components was carried at one of the outlying sites as Los Alamos on the 13th, and they were brought down to the site by truck later that day. The mood of the Laboratory soured when a magnetic-method test shot carried out on the 14th seemed to indicate that the bomb would not function efficiently, but Hans Bethe saved the day by demonstrating that the analysis was flawed and that acceptable detonator symmetry had in fact been achieved. A stanza of poetry caught the sense of the time:

> From this crude lab that spawned a dud
> Their necks to Truman's axe uncurled
> Lo, the embattled savants stood
> And fired the flop heard round the world.

Final assembly of the *Trinity* device began at one p.m. on Friday, July 13, within a tent at the base of the 100-foot tower. The date and time were chosen by George Kistiakowsky in the hope that they would bring good luck. Just after three p.m., the core assembly was ready for insertion within the high explosive, but a hitch arose. The metallic plutonium core, warm from its own internally-generated alpha-decay heat and the desert climate, did not fit into the cooler high-explosive assembly. Leaving them in contact for a couple minutes brought them to thermal equilibrium, and the core slipped into place. Assembly of the bomb's innards was complete by 5:45 p.m., and it was raised to the top of the tower in preparation for installation of detonators and firing circuitry the next day. As the bomb was raised, a protective bed of mattresses was placed under it. Sunday, July 15, was reserved for final inspections (Fig. 7.31).

Fig. 7.31 The *Trinity* device atop its test tower on July 15, 1945, with Norris Bradbury (1909-1997). The cables feeding from the box halfway up the device go to the implosion-lens detonators discussed in the text. *Source* http://commons.wikimedia. org/wiki/File:Trinity_ Gadget_002.jpg

Bainbridge's report contains a copy of a detailed schedule for the test. Among minutiae regarding the precise placement and handling of components, one finds more prosaic matters such as "Light must be available to work in tent at night," and "Bring up G-Engineer footstool." For Sunday, July 15, the entirety of the schedule read "Look for rabbit's feet and four-leaved clovers. Should we have the chaplain down here? Period for inspection available from 0900 to 1000." The entry for July 16 reads only as "Monday, 16 July, 0400 Bang!"

The last group of people to attend the bomb was an arming party headed by Bainbridge, which set out at about 10:00 p.m. on the night of the 15th to activate timing and arming switches so that the bomb could be triggered from the South-10,000 station, and to collect Donald Hornig, who had earlier ascended the tower to switch out a practice detonating circuit for the operational one and to stand guard over the bomb. In Hornig's words:

Oppenheimer was really terribly worried ... that it would be easy to sabotage. So he thought someone had better baby sit it right up until the moment it was fired. They asked for volunteers and as the youngest guy present, I was selected. I don't know if it was that or that I was most expendable or best able to climb a 100-foot tower! By then there was a violent thunder and lightning storm. I climbed up there, took along a book, *Desert Island Decameron*, and climbed the tower on top of which there was the bomb, all wired up and ready to go. Little metal shack, open on one side, no windows on the other three, and a 60-Watt bulb and just a folding chair for me to sit beside the bomb, and there I was! All I had was a telephone. I wasn't equipped to defend myself, I don't know what I was supposed to do. There were no instructions! The possibility of lightning striking the tower was very much on my mind. But it was very wet and the odds were the tower would act like a giant lightning rod and the electricity would just go straight down to the wet desert. In that case, nothing would have happened. The other case was that it would set the bomb off. And in that case, I'd never know about it! So I read my book.

By the time of the test, Hubbard had not slept in over two days. At a weather conference held at 2:00 a.m. on the 16th, he predicted that conditions would become acceptable at dawn. Holzman apparently agreed, and the shot was set for

5:30. Groves demanded that Hubbard sign his forecast, stating that he had better be right, "or I will hang you." Groves then placed a call to the Governor of New Mexico to inform him that it might be necessary to declare martial law throughout the central part of the state. Importantly, however, the winds for the test were as desired.

At Base Camp, Enrico Fermi occupied himself by offering to take wagers on whether or not the bomb would ignite the atmosphere and, if so, would it destroy only New Mexico or the entire world; he guessed that if nitrogen in the air were ignited it would go only about 35 miles. He added that it would not make any difference whether the bomb went off or not, as it would still have been a worthwhile experiment. Groves was not amused by Fermi's diversions, but the latter was not the only one in a wagering mood. Physicists established a pool on the yield of the bomb, with an ante of $1 each. Edward Teller optimistically bet on 45 kilotons; Hans Bethe opted for 8 kilotons. Oppenheimer picked 200 t, and had a side bet with George Kistiakowsky of $10 against a month of Kistiakowsky's salary that the bomb wouldn't work at all. The pool winner was I. I. Rabi, who arrived too late to choose a low number, and had to settle for 18 kilotons; he took home $102. Others had different concerns. Kenneth Bainbridge later wrote that "My personal nightmare was knowing that if the bomb didn't go off or hangfired [a delay between triggering and detonation], I, as head of the test, would have to go to the tower first and seek to find out what had gone wrong." For his part, Oppenheimer was practically a nervous wreck: he had suffered a bout of chicken pox and had lost 30 pounds; despite standing at over six feet, his weight was only about 115 pounds.

In the control bunker at S-10,000, the tension was palpable. As described by Brigadier General Thomas Farrell, Groves' deputy:

> The scene inside the shelter was dramatic beyond words. In and around the shelter were some twenty-odd people concerned with last minute arrangements … For some hectic two hours preceding the blast, General Groves stayed with the Director, walking with him and steadying his tense excitement. Every time the Director would be about to explode because of some untoward happening, General Groves would take him off and walk with him in the rain, counseling with him and reassuring him that everything would be all right.

Groves departed for Base Camp 20 minutes before the detonation. He had dictated that he and Farrell were not to be together in situations where there was an element of danger, which arguably existed at both locations.

The final countdown began at 5:10 a.m., and was conducted by Samuel Allison. At T-minus 45 seconds, arming-party physicist Joseph McKibben threw a final switch that activated a timing apparatus with a rotating drum and pin-actuated switches to trigger time-sensitive instruments. Donald Hornig manned a final cutoff switch that was the only way the test could have been stopped.

The exact time of the *Trinity* detonation is only approximately known because of difficulty in picking up a national time-service radio broadcast at the shelter. Bainbridge's report gives as a best estimate 5:29:15 a.m., plus 20 seconds or minus 5 seconds. Witnesses at Base Camp were instructed to lie flat on the ground, face away from the tower, and not to rise until after the blast wave had passed. From Farrell's description:

> As the time interval grew smaller ... the tension increased by leaps and bounds. Dr. Oppenheimer, on whom had rested a very heavy burden, grew tenser as the last seconds ticked off. He scarcely breathed. He held on to a post to steady himself. For the last few seconds he stared directly ahead and then when the announcer [Allison] shouted 'Now!' and there came this tremendous burst of light followed shortly thereafter by the deep growling roar of the explosion, his face relaxed into an expression of tremendous relief. Several of the observers standing back of the shelter to watch the lighting effects were knocked flat by the blast.

> The tension in the room let up and all started congratulating each other ... Dr. Kistia-kowsky ... threw his arms around Dr. Oppenheimer and embraced him with shouts of glee.

A number of descriptions of the explosion have been published, a few of which are reproduced here. One of the most striking was provided by Farrell, a devout Catholic, in his subsequent report to Groves:

> The effects could well be called unprecedented, magnificent, beautiful, stupendous and terrifying. No man-made phenomenon of such tremendous power had ever occurred before. The lighting effects beggared description. The whole country was lighted by a searing light with the intensity many times that of the midday sun. It was golden, purple, violet, gray and blue. It lighted every peak, crevasse and ridge of the nearby mountain range with a clarity and beauty that cannot be described but must be seen to be imagined. It was that beauty the great poets dream about but describe most poorly and inadequately. Thirty seconds after the explosion came, first, the air blast pressing hard against the people and things, to be followed almost immediately by the strong, sustained, awesome roar which warned of doomsday and made us feel that we puny things were blasphemous to dare tamper with the forces heretofore reserved to The Almighty. Words are inadequate tools for the job of acquainting those not present with the physical, mental, and psychological effects. It had to be witnessed to be realized.

Farrell commented to Groves immediately after the test that "The war is over." "Yes," was Groves' reply, "just as soon as we drop one or two of these things on Japan" (Fig. 7.32).

At Base Camp, Enrico Fermi estimated the strength of the blast by an elegantly simple experiment:

> The explosion took place at about 5:30 a.m. I had my face protected by a large board in which a piece of dark welding glass had been inserted. My first impression of the explosion was the very intense flash of light, and a sensation of heat on the parts of my body that were exposed. Although I did not look directly towards the object, I had the impression that suddenly the countryside became brighter than in full daylight. I subsequently looked in the direction of the explosion through the dark glass and could see something that looked like a conglomeration of flames that promptly started rising. After a few seconds the rising flames lost their brightness and appeared as a huge pillar of smoke with an expanded head like a gigantic mushroom that rose rapidly beyond the clouds probably to a height of the order of 30,000 feet. After reaching its full height, the smoke stayed stationary for a while before the wind started dispersing it.
>
> About 40 seconds after the explosion the air blast reached me. I tried to estimate its strength by dropping from about six feet small pieces of paper before, during and after the

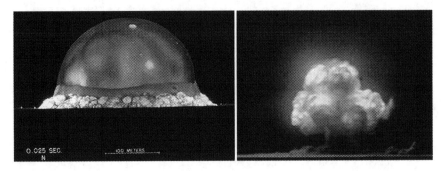

Fig. 7.32 *Left* The *Trinity* fireball at 25 ms into the nuclear age. *Right* The *Trinity* mushroom cloud a few seconds later. *Sources* http://commons.wikimedia.org/wiki/File:Trinity_Test_Fireball_25ms.jpg; http://commons.wikimedia.org/wiki/File:Trinity_shot_color.jpg

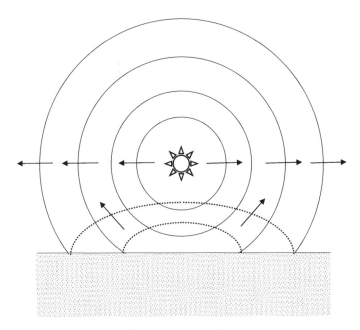

Fig. 7.33 Schematic illustration of formation of a reflected shock wave. Adopted from Glasstone and Dolan (1977)

passage of the blast wave. Since at the time, there was no wind I could observe very distinctly and actually measure the displacement of the pieces of paper that were in the process of falling while the blast was passing. The shift was about 2 1/2 m, which, at the time, I estimated to correspond to the blast that would be produced by ten thousand tons of T.N.T.

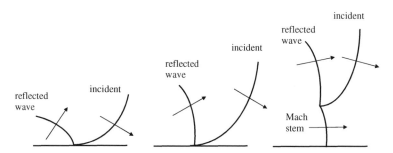

Fig. 7.34 Schematic illustration of formation of the Mach stem. Adopted from Glasstone and Dolan (1977)

Kenneth Bainbridge described the test as "a foul and awesome display." After the shock wave passed, Bainbridge congratulated Oppenheimer and said to him: "Now we are all sons of bitches." In a 1975 reminiscence, Bainbridge related that in 1966, Oppenheimer told Bainbridge's daughter that her father's assessment was the best thing anyone said after the test.

Hans Bethe on Campañia Hill: "it looked like a giant magnesium flare which kept on for what seemed a whole minute but was actually only 1 or 2 seconds. The white ball grew and after a few seconds became clouded with dust whipped up by the explosion from the ground and rose and left behind a black trail of dust particles. The rise, though it seemed slow, took place at a velocity of 120 m/s. After more than half a minute the flame died down and the ball, which had been a brilliant white became a dull purple. It continued to rise and spread at the same time, and finally broke through and rose above the clouds which were 15,000 feet above the ground. It could be distinguished from the clouds by its color and could be followed to a height of 40,000 feet above the ground."

James Conant at Base Camp: "Then came a burst of white light that seemed to fill the sky and seemed to last for seconds. I had expected a relatively quick and bright flash. The enormity of the light and its length quite stunned me. My instantaneous reaction was that something had gone wrong and that the thermal nuclear transformation of the atmosphere, once discussed as a possibility and only jokingly referred to a few minutes earlier, had actually occurred."

I. I. Rabi at Base Camp: "We were lying there, very tense, in the early dawn, and there were just a few streaks of gold in the east; you could see your neighbor very dimly. Those 10 seconds were the longest 10 seconds that I have ever experienced. Suddenly, there was an enormous flash of light, the brightest light I have ever seen or that I think anyone has ever seen. It blasted; it pounced; it bored its way right through you. It was a vision that was seen with more than the eye. It was seen to last forever. You would wish it would stop; although it lasted about 2 seconds. Finally it was over, diminishing, and we looked toward the place where the bomb had been; there was an enormous ball of fire which grew and grew and it rolled as it grew; it went up into the air, in yellow flashes and into scarlet and green. It looked menacing. It seemed to come toward one. A new thing had just been born; a new control; a new understanding of man, which man had acquired over nature."

Emilio Segrè at Base Camp: "We saw the whole sky flash with unbelievable brightness in spite of the very dark glasses we wore … In a fraction of a second, at our distance, one received enough light to produce a sunburn." In his later biography of Fermi, Segrè wrote that "Even though the purpose was grim and terrifying, it was one of the greatest physics experiments of all time ... The feat will stand as a great monument of human endeavor for a long time to come."

Norris Bradbury at the Control Shelter: "The shot was truly awe-inspiring. Most experiences in life can be comprehended by prior experiences but the atom bomb did not fit into any preconception possessed by anybody. The most startling feature was the intense light."

Robert Christy on Campañia Hill: "It was awe-inspiring. It just grew bigger and bigger, and it turned purple ... The debris was intensely radioactive, and it was sending out beta particles and gamma rays in all directions, and those ionized the air. So the air around this ball emitted a bluish glow ... It was most fantastic, to see this thing going up and swirling around and eventually cooling off to the point where it was no longer visible."

George Kistiakowsky put his arms around Oppenheimer and said "Oppie, you owe me $10." In a 1980 reminiscence, Kistiakowsky claimed to still have the $10 bill.

Oppenheimer's reaction to the test is a matter of debate. His brother, Frank, when interviewed for the 1980 documentary *The Day After Trinity*, stated that he thought all his brother had said was "It worked!" In postwar years, Oppenheimer uttered a number of dramatic, quasi-philosophical statements on his reaction to the test. A 1947 lecture on "Physics in the Contemporary World" at MIT included the following frequently-quoted passage:

Despite the vision and the farseeing wisdom of our wartime heads of state, the physicists felt a peculiar intimate responsibility for suggesting, for supporting, and in the end, in large measure, for achieving, the realization of atomic weapons. Nor can we forget that these weapons, as they were in fact used, dramatized so mercilessly the inhumanity and evil of modem war. In some sort of crude sense which no vulgarity, no humor, no overstatement can quite extinguish, the physicists have known sin; and this is knowledge which they cannot lose.

A number of physicists were offended by this statement. Freeman Dyson, a physicist at Cornell University in the years after the war, put it this way:

Most of the Los Alamos people at Cornell repudiated Oppy's remark indignantly. They felt no sense of sin. They had done a difficult and necessary job to help win the war. They felt it was unfair of Oppy to weep in public over their guilt when anybody who built any kind of lethal weapons for use in war was equally guilty. I understood the anger of the Los Alamos people, but I agreed with Oppy. The sin of the physicists at Los Alamos did not lie in their having built a lethal weapon. To have built the bomb, when their country was engaged in a desperate war against Hitler's Germany, was morally justifiable. But they did not just build the bomb. The enjoyed building it. They had the best time of their lives while building it. That, I believe, is what Oppy had in mind when he said they had sinned. And he was right.

In a 1965 interview for a television documentary, *The Decision to Drop the Bomb*, Oppenheimer gave this reaction to *Trinity*:

We knew the world would not be the same. Few people laughed, few people cried, most people were silent. I remembered the line from the Hindu scripture, the *Bhagavad-Gita*. Vishnu is trying to persuade the Prince that he should do his duty and to impress him takes on his multi-armed form and says, "Now I am become Death, the destroyer of worlds." I suppose we all thought that, one way or another.

Groves permitted access to the Manhattan project to a single journalist, William L. Laurence, a science reporter with *The New York Times* (Sect. 3.6). Laurence witnessed the *Trinity* explosion from Campañia Hill. In the first of many articles on the Project published in the *Times*, Laurence gave a dramatic description of the explosion on the front-page of the September 26, 1945, edition (excerpted):

> "At that great moment in history, ranking with the moment in the long ago when man first put fire to work for him and started on his march to civilization, the vast energy locked within the hearts of the atoms of matter was released for the first time in a burst of flame such as never before been seen on this planet, illuminating earth and sky for a brief span that seemed eternal with the light of many super-suns. ... It was like the grand finale of a mighty symphony of the elements, fascinating and terrifying, uplifting and crushing, ominous and devastating, full of great promise and great forebodings. ... And just at that instant there rose from the bowels of the earth a light not of this world, the light of many suns in one. ... On that moment hung eternity. Time stood still. Space contracted into a pinpoint. ... The thunder reverberated all through the desert, bounced back and forth from the Sierra Oscuros, echo upon echo. The ground trembled under our feet as in an earthquake."

In the same article, Laurence quoted George Kistiakowsky as saying "I am sure that at the end of the world—in the last milli-second of the earth's existence—the last man will see what we saw." The best assessment of the significance of the *Trinity* test may be that by novelist Joseph Kanon: "July 1945 at Alamagordo is the hinge of the century. Nothing after would ever be the same."

Trinity's most dramatic visual manifestation was its enormous ball of fire. Much of the immediate energy from a nuclear explosion is in the form of X-rays and ultraviolet light, and since cold air is opaque to radiation at these wavelengths, the air surrounding the weapon absorbs the energy and heats up dramatically, to a temperature of about $1,000,000°$ out to a radius of a few feet. Because this bubble of hot, incandescent air emits energy in the X-ray and ultraviolet regions of the electromagnetic spectrum, it will be *invisible* to an outside observer. But the bubble is surrounded by a cooler envelope, which, although incredibly hot by everyday standards, will be visible to observers at a distance. The temperature of this surrounding air, however, has little physical significance as far as measuring the energy release of the bomb is concerned. As the fireball increases in size, its total light emission increases, up to a first maximum (Fig. 7.35; Stefan's law of thermal radiation indicates that emission is proportional to surface area times the fourth power of the temperature), after which it begins cooling due to the growing mass of accreted air. Like a hot-air balloon, the fireball will also rise. The temperature within the fireball is so great that all of the weapon residues will be in the form of vapor, including the fission products. As the fireball expands and cools, these vapors condense to form a cloud of solid debris particles; the fireball also picks up water from the atmosphere. All of this material will eventually become fallout, sometimes in the form of radioactive rain. As the fireball ascends, cooling of its outside and air drag often creates a toroidal (doughnut-like) shape. At this stage, the cloud will often have a reddish appearance due to the presence of nitrogen-oxide compounds at its surface.

The air inside the fireball cools by successive radiation and re-absorption of X-rays. When the air has cooled to a temperature of about $300,000°$, a "hydrodynamic shock" forms, a so-called "front" of compressed air. The shock front travels faster than energy can be transported by successive absorption and re-emission of radiation, so it "decouples" from the hot sphere and moves out ahead of the latter, leaving behind a region of relatively cool air which "eats into" the central hot sphere. For outside observers, visible radiation comes from the shock wave. As the shock front cools, its observable temperature bottoms out at a minimum of about 2,000°. The shock front also becomes transparent; an observer, if he or she still has eyes and sentience, can now look into higher-temperature air, which results in a second brightness maximum. This "double maximum" in the time-evolution of visible radiation is uniquely characteristic of a nuclear explosion (Fig. 7.35). During this time, however, the central fireball is still hot enough to be essentially opaque, and hence invisible. As the shock front progresses outwards, there soon comes a time when, for a while, the air pressure behind the front is actually lower than ambient atmospheric pressure, a so-called "negative pressure" region. In this phase, air rushes inward to the site of the explosion, an "afterwind."

As Hans Bethe and Robert Christy wrote in an undated memorandum (presumably summer 1945), "the ball of fire will rise to the stratosphere (about 15 km height) in about two or three minutes …. The flash of light obtained in the first instant will be as bright as the sun at a distance of about 100 km from the explosion …. At a time when it reaches the stratosphere it will still appear as bright as the moon at a distance of about 250 km. The radioactive materials are expected to be near the center of the ball of fire and rise with that ball of fire to the stratosphere. Presumably the ball of fire will rise to a very considerable height (100 km or more) before its rise is stopped by either diffusion or cooling. If the radioactive material ever comes down again it will certainly be spread out over a radius of at least 100 km and probably very much more and will, therefore, be completely harmless".

It has been estimated that *Trinity* released about a *trillion* Curies of radioactivity.

The second iconic image of a nuclear detonation is the characteristic mushroom-shape cloud that forms after the explosion. This happens for so-called "airburst" weapons, that is, ones detonated above the ground. (To a weapons strategist, an airburst is technically an explosion at a height such that the fireball does not touch the ground when its luminosity is at the second maximum described above. An "optimum-height" airburst is one which maximizes the blast damage area.) The "stem" of the mushroom is formed when the initial blast wave reflects from the ground. The reflected wave, however, will be traveling through air that has already been heated and compressed by the passage of the initial wave, and so moves *faster* than the initial wave. As sketched in Figs. 7.33 and 7.34, the reflected wave catches up to the initial wave, forming the stem. In technical parlance, the stem is known as a "Mach stem."

The incredible temperatures created in the fireball can be estimated from simple thermodynamics. Fission of a uranium nucleus releases about 200 MeV of energy, most of which goes into the kinetic energy of the fission fragments. From kinetic theory, the kinetic energy of a particle is equivalent to an absolute temperature

T given by $3kT/2$, where k is Boltzmann's constant, 1.38×10^{-23} J/K. A fission fragment of 100 MeV kinetic energy (1.6×10^{-11} J) therefore has a temperature equivalent of about 8×10^{11} K. The fragments will be quickly slowed down by collisions with air molecules, but the result is still impressive.

About one-third of the total energy liberated by a fission weapon is in the form of ultraviolet, visible, and infrared light. The rate of delivery of this energy is so prompt that combustible materials such as paper, wood, and fabrics will be charred or burst into flame out to great distances. Such materials can be ignited by the prompt delivery of 10 physical calories of radiant energy per square centimeter; a 20-kt explosion delivers this much energy to a radius of 6,000 feet. At *Trinity*, some fir timbers were slightly scorched to this distance; such charring requires a temperature of about 400 °C. For human beings, moderate burns to unprotected skin can be produced by deposit of about 3 calories per square centimeter. For a 20-kt explosion, the radius for this effect is about 10,000 feet; at Nagasaki, skin burns were reported to 14,000 feet. *Trinity's* radiant energy output (i.e., heat) alone was estimated at 3 kilotons TNT equivalent.

Bainbridge's report on the test includes a graph of the brightness of the explosion as a function of time (Fig. 7.35). Brightness here is measured in "Suns" equivalent at a distance of 10,000 yards from the explosion. At $t = 10^{-4}$ s, the illumination was approximately 80 Suns; it dropped to about 0.1 Suns at $t \sim 0.04$ s, rose back to about 2 Suns at $t = 0.4$ s, and then declined to about 0.4 Suns at $t \sim 10$ seconds. At a brightness of 80 Suns and neglecting any effects due to atmospheric absorption and cloud cover, *Trinity* would momentarily have appeared over 30 times brighter than Venus to an observer located on the moon, and would have been visible to observers on Mercury, Venus, and Mars. Not until the fireball cooled to ~ 2 Suns equivalent a few tenths of a second after the explosion would it have diminished to the brightness of Venus for a lunar observer, and even after 10 seconds would still have outshone Jupiter for such an observer. (On the day of the test, the moon was at first-quarter phase and had set about 1 a.m. New Mexico time, some four and one-half hours before the detonation. Only Venus and Mars were above the horizon at the time of the test.)

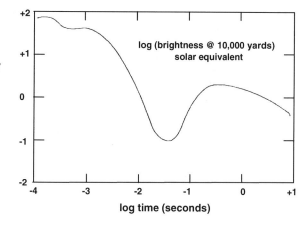

Fig. 7.35 Brightness of the *Trinity* explosion as a function of time. Scales are logarithmic. This figure was produced by scanning a copy of Fig. 7 of Los Alamos report LA-6300. From Reed (2006)

High-speed photography of the *Trinity* fireball showed that it struck the ground about 0.65 ms after the detonation. For a detonation height of 100 feet, this corresponds to an average expansion speed of about 46 km/s; for comparison, the speed of sound is only about 340 m/s. Some quarter-million square meters (70 acres) of surrounding desert sand was fused to a depth of about half an inch into a fragile, greenish, glassy material that came to be known as *Trinitite*. The greenish color is due to the presence of iron in the sand; a small sample owned by this author is still very slightly radioactive.

Groves was anxious to get word to Secretary of War Stimson in Potsdam, and called his secretary in Washington, Jean O'Leary, about ninety minutes after the test (about 9:00 a.m. Washington time). O'Leary proceeded to the Pentagon office of Stimson advisor George Harrison, where they drafted a brief coded cable:

> Operated on this morning. Diagnosis not yet complete but results are satisfactory and already exceed expectations. Local press release necessary as interest extends a great distance. Dr. Groves pleased. He returns tomorrow. I will keep you posted.

Stimson received the cable at 7:30 p.m. Potsdam time (1:30 p.m. Washington time, six hours after the test), and immediately relayed it to President Truman. In his diary for July 18, Truman remarked that at a lunch alone with Churchill he "Discussed Manhattan (it is a success)." He also recorded that "Believe Japs will fold up before Russia comes in. I am sure they will when Manhattan appears over their homeland. I shall inform Stalin about it at an opportune time."

Groves hastened back to Washington, arriving in his office about 2:00 p.m. the day after the test. That evening, he prepared a lengthier memorandum to Stimson. Completed early in the morning of Wednesday, July 18, it was sent by courier to Potsdam, where it was handed to Stimson at 11:35 a.m. on Saturday morning, July 21. A few passages drawn from the memo testify to the enormity of the blast:

> The light from the explosion was clearly seen at Albuquerque, Santa Fe, El Paso, and other points generally to about 180 miles away. The sound was heard ... generally to 100 miles. Only a few windows were broken, although one was some 125 miles away. A crater from which all vegetation had vanished, with a diameter of 1,200 feet ... in the center was a shallow bowl 130 feet in diameter and 6 feet in depth ... The steel from the tower was evaporated ... I no longer consider the Pentagon a safe shelter from such a bomb ... Radioactive material in small quantities was located as much as 120 miles away ... My liaison officer at the Alamogordo Air Base, sixty miles away [reported] a blinding flash of light that lighted the entire northwestern sky.

Upon receiving the report, Stimson took it to General Marshall and President Truman; Churchill was also informed. It took Stimson the better part of an hour to read the report. Curiously, the version of the report reproduced in Groves' memoirs does not include a statement included in the original version: "It resulted from the atomic fission of about 13½ pounds of plutonium which was compressed by the detonation of a surrounding sphere of some 5,000 pounds of high explosive."

On the morning of the 24th, another cable from Harrison informed Stimson that "operation may be possible any time from August 1 depending on state of preparation of patient and condition of atmosphere." Later that morning, a combined American and British Chiefs of Staff meeting convened with Churchill and Truman; Truman biographer David McCullough pinpoints this meeting as critical in the decision to use the bomb. That evening, Truman approached Josef Stalin to let him know that America had developed a new weapon "of unusual destructive force." Stalin, who was probably well-briefed on the project, apparently showed no special interest, replying only that he hoped that America would make "good use of it against the Japanese." In their analysis of Soviet nuclear espionage, *Bombshell: The Secret Story of America's Unknown Atomic Spy Conspiracy*, Joseph Albright and Marcia Kunstel write that on February 28, 1945 (the day on which Los Alamos settled on the Christy core for the implosion bomb), the NKGB in Moscow ("State Security People's Commisariat") finished a comprehensive report on atomic intelligence which would go to Lavrenti Beria, the People's Commisar for Internal Affairs. The Soviets knew of the main features of the implosion weapon five months before the *Trinity* test.

Groves had prepared a number of press releases written to accommodate a range of test outcomes. The story he went with was that a remotely-located ammunition magazine containing "a considerable amount of high explosives and pyrotechnics" had exploded on the grounds of the Alamogordo Army Air Base, but that there had been no loss of life or any injuries. The story was widely reported in the area and along the west coast, but received no exposure on the east coast except for a few lines in the early edition of a Washington paper.

Because the yield of *Trinity* was some three times greater than predicted, many instruments were overwhelmed by the explosion. No blast-measuring device within 200 feet of the tower survived, although one located at 208 feet gave a pressure reading of nearly 5 t per square inch, almost 700 atmospheres. Most γ-ray and neutron measurements were overloaded. Diaphragm gauges designed to measure the peak blast pressure gave a result of 9.9 kilotons, but radiochemical analyses of soil samples indicated nearly twice that figure, 18.6 kilotons. A 20-kiloton explosion would have been equivalent to an efficiency of about 18 %. One immediate effect of *Trinity's* unexpectedly great yield was that Oppenheimer proposed to Groves on July 19 that the U-235 that had been accumulated for the *Little Boy* gun bomb be used instead to make composite uranium–plutonium cores. Groves sensibly preferred to go with existing plans and vetoed the idea, but composite cores were incorporated into postwar weapons.

A number of re-evaluations of *Trinity* measurements have been carried out in light of information subsequently gathered from the Hiroshima and Nagasaki bombings, as well as various postwar atomic tests. Based on data from a 1946 test, a 1952 analysis calibrated *Trinity* as 23.8 kilotons. A December, 2000, Department of Energy report on all United States nuclear tests lists an official yield of 21 kilotons. This is roughly equivalent to 2,100 fully-loaded B-29 bombers dropping 84,000 five-hundred pound bombs simultaneously. The *Trinity* explosion was the largest man-made explosion in history to its time; the previous record, estimated to be 2.9 kilotons, had been set by the accidental explosion of a munitions ship in Halifax harbor in Nova Scotia in 1917.

Because wind patterns at the time of the test were favorable, there was no serious fallout from *Trinity*. Nevertheless, there were consequences. In addition to creating direct fission (and subsequent decay) products, the explosion vaporized an estimated 100–250 t of sand, much of which would have been rendered radioactive by neutron bombardment (as was additional soil that did not get lofted into the atmosphere.) The radioactive cloud split into three parts, with the majority moving northeast and dropping radioactivity over an area of about 100 miles long by 30 miles wide. Readings of about 3 rems/h (R/h) in the affected area were not uncommon; the present-day (2012) standard for maximum exposure for people who work with radioactive materials is 5 rems *per year*. The lead-lined tanks that had been prepared to retrieve soil samples could make only brief passes through the crater itself, where soil samples registered initial activities of 600–700 R/h. Exposure limits which would trigger evacuation of shelters and surrounding areas were not rigidly defined, although 10 R/h was loosely accepted as the threshold of concern, with a recommendation that no person "of his own will" receive more than 5 rems at one exposure. The North-10,000 shelter was evacuated about twenty minutes after the explosion when 10 R/h was recorded, but it is suspected that this may have been a mis-read.

The most seriously-affected radiation victims were likely animals, particularly grazing Hereford cattle at local ranches. A few weeks after the test, several cows began losing hair, which grew back in white as opposed to its normal reddish tint; Louis Hempelmann's Health Group bought four cows and brought them to Los Alamos for study. Because the breed purity of discolored cattle would be questioned, ranchers faced a cut in price, and in December, 1945, Los Alamos bought some 75 animals that were most heavily damaged. None of them died of unexplained causes, and they reproduced normally, with several being eaten. Some of the more seriously exposed ones did eventually develop skin cancers on their backs, but the overall conclusion was that there was no gross differences between the exposed cattle and their offspring when compared with an unexposed control group.

One effect of *Trinity* fallout turned up far from the site. In the fall of 1945, the Eastman Kodak Corporation in Rochester, New York, found that several batches of industrial X-ray film were flecked with spot-like imperfections. The film itself was fine, but strawboard liners used to separate the films in their cartons were embedded with radioactive particles. The strawboard had been prepared by paper mills in Iowa and Indiana in the weeks following *Trinity*, and it is thought that rain washed fallout into rivers which were used as water sources during the paper processing. One of the culprit fallout products was Cerium-141, which has about a 32-day beta-decay half-life.

In the aftermath of the bombings of Hiroshima and Nagasaki, some alarmist commentators asserted that both cities were uninhabitable, an assessment which would have been a surprise to surviving residents. To help quell concern over radiation effects, Groves arranged what would now be called a "media day" for reporters and photographers at the *Trinity* site on September 9, with everyone wearing protective booties (Figs. 7.36 and 7.37). Radioactivity was measured at 12 R/h, and the visit was kept brief. Ironically, because *Trinity* was detonated so close to the ground, the site was radiologically hotter then either Hiroshima or Nagasaki. Systematic studies of long-term effects at *Trinity* began in 1947, and were carried out periodically thereafter. In the 1947 survey, plutonium was found in the soil and on plants at locations up to 85 miles from the detonation site, and some birds, rodents, and insects were malformed, had eye cataracts, or unusual spottings. Another study a year later found no damaged birds or rodents, indicating that effects were not genetically passed on. Trinitite proved to be insoluble in water, so it could not easily enter plants or animals.

In the years following the war, some efforts were launched to try to make the *Trinity* site into a national monument. Various studies to this effect were carried out, but competing interests of using the land for grazing and the impact of what became the White Sands Missile Range doomed such ideas. In the 1950s, the

Fig. 7.36 *Trinity* ground zero, September 1945. Oppenheimer (*center, hat*), Groves, and others look at the remains of the 100-foot tower. *Source* http://commons.wikimedia.org/wiki/File:Trinity_Test_-_Oppenheimer_and_Groves_at_Ground_Zero_001.jpg

Fig. 7.37 Aerial view of the aftermath of the *Trinity* test. The 0.1 kt test crater is from the 100-t TNT test. The area covered by the image is about 1,550 m wide by 1,400 m tall. *Source* http://commons.wikimedia.org/wiki/File:Trinity_crater_(annotated)_2.jpg

Fig. 7.38 *Left* The author, *second* from *left*, at the *Trinity* ground-zero monument, October, 2004. *Right* Monument plaque

Trinitite was packed into barrels and buried; a 1967 study calculated that a person would have to eat some 100,000 kg of the material to ingest the maximum permissible body burden of 4 nCi of plutonium-239, although only 10 kg would have to be consumed to reach maximum permissible beta and gamma-ray exposure from fission products. In 1965, the National Park Service declared the site a National Historic Landmark, and erected a monument (Fig. 7.38); in 1975, the location was designated a National Historic Site. The Army donated Jumbo to the city of Socorro, but no means could be found to remove it from the site.

The *Trinity* site is now open to tourists two days per year, the first Saturdays of April and October, depending on security conditions at the White Sands Missile Range. This author has visited the site, and found it an unusual experience. When approaching many places of historic significance, one is often struck by a sense of awe before actually arriving, but that is not the case with *Trinity*. After crossing miles of desert, one's first indication that something happened is to see the corpse of Jumbo. Actually *standing* at ground zero or inside the McDonald ranch house is another matter, but the approach to the site itself is not at all a memorable experience. Tourists need have no concern about visiting the site as far as residual radioactivity is concerned: a 1985 Los Alamos report on a radiological survey of the area concluded that exposure during public visits to the ground-zero area amounts to less than 0.2 % of Department of Energy Radiation Protection Standards for members of the public (Fig. 7.39).

With the successful completion of the *Trinity* test, the stage was set for combat use of nuclear weapons. Before proceeding to a discussion of overseas preparations for the combat missions, however, it seems appropriate to briefly describe some of the destructive effects of nuclear weapons. This is the topic of the following section, after which we will return to Tinian island. The following section

Fig. 7.39 Author at the
West-10,000 instrument
bunker

is mildly technical, but non-technical readers might wish to scan it to gain at least a qualitative sense of the destructiveness of nuclear weapons.

7.13 A Brief Tutorial on Bomb Effects

The three main damaging effects of nuclear weapons on people and structures are pressure ("blast"), thermal radiation (heat), and fallout. Because these effects are contingent on factors such as weapon yield, explosion height, shielding due to structures and terrain, and weather conditions such as haze or fog, there are no simple general formulae that can be deployed to estimate effects in all circumstances. Professional weapons engineers often make use of test data that have been distilled into approximate formulae and graphical summaries that appear in volumes such as that prepared by Glasstone and Dolan. For pedagogical purposes, however, we can use some approximate relations to make order-of-magnitude estimates, assuming clear skies, an airburst weapon, and flat terrain.

Rather confusingly, some of the units involved with these expressions are American (miles, pounds per square inch) while others are MKS (calories, kilotons). This reflects that fact that much of the available information on weapons effects derives from postwar American weapons tests, when customary United States units were the norm. We look at each of the three major effects in turn.

7.13.1 Blast Pressure

The majority of the physical destruction caused by nuclear weapons is due to the high-pressure shock wave that races out from the fireball. Normal atmospheric pressure is 14.7 pounds per square inch (psi). Weapons effects are usually stated in terms of the *overpressure* created, which is the number of psi generated in excess of this ambient value. Seemingly small overpressures can have devastating effects. An overpressure of 1 psi is sufficient to break ordinary glass windows. Wood frame homes are destroyed under the action of a 5 psi overpressure, which is also about the threshold for human eardrum rupture. Massive multistory buildings will sustain moderate damage at 6–7 psi overpressure and be demolished at 20 psi, which corresponds to a wind of speed 500 miles per hour. Eight to ten psi overpressure is sufficient to destroy brick houses and collapse factories and commercial buildings. Even if you are in no danger of being trapped within a collapsing structure, you are not necessarily safe: the threshold for human death from compressive effects sets in at about 40 psi.

The overpressure that an observer or structure experiences depends on the yield of the weapon and the "slant range" to the explosion—the direct line-of-sight distance between the explosion and the observer. In the case of an optimum-height airburst weapon, if the yield of a weapon is Y kilotons and the slant range is R miles, the *maximum* overpressure in psi is given approximately by the formula

$$P_{max} \sim 1.4 \frac{\sqrt{Y}}{R^{3/2}}. \qquad (7.19)$$

For example, at a slant range of 2 miles from a 20-kiloton yield, $P_{max} \sim 2.2$ psi. Your house will be damaged, but likely survive—as will you, if you can avoid flying debris, fallout, and thermal burns. But bear in mind that weapons technology has advanced considerably since 1945; yields of several hundred kilotons are now not uncommon (Chap. 9). At two miles, a 400-kiloton yield will give an overpressure of nearly 10 psi.

7.13.2 Thermal Burns

The harmful effects of prompt exposure to thermal radiation on humans are usually divided into two categories: "flash" burns caused by direct skin exposure, and "contact" burns caused by ignited clothing or a fire otherwise initiated by the explosion. Even the color of clothing a person is wearing can be important: black fabric will absorb more thermal radiation than white fabric, and hence more readily burst into flame. The effects of flash burns are easier to quantify than those of contact burns, but they too depend on unpredictable factors such as exposure duration and individual skin pigmentation.

The unit of measure used to quantify flash burns is the number of calories of energy deposited per square centimeter of skin (cal/cm^2). The resulting burns themselves are classified as first, second, or third degree. First-degree burns are the mildest, from which recovery without scarring can be expected. A bad sunburn is a classic example of a first-degree burn, and prompt exposure of 2–3 cal/cm^2 will cause such burns for most people. Second-degree burns ($\sim$4–5 cal/cm^2) will develop scabs, but normally heal in a week or two unless an infection sets in. Third-degree burns ($>\sim$6 cal/cm^2) are the most harmful: burnt areas are so damaged that they cannot transmit pain impulses, and so pain is felt only from surrounding areas. With such burns, skin grafts will be necessary to prevent scarring. To put these numbers in perspective, some 10–15 cal/cm^2 are required to char pine, redwood and maple trees; clothing and upholstery fabrics will typically ignite on exposure to 20–25 cal/cm^2.

Burn effects are very dependent on atmospheric conditions, so only an approximate expression for thermal exposure can be offered. The symbol used to designate thermal exposure is Q, and the formula is

$$Q \sim 1.1 \left(\frac{\tau Y}{R^2} \right) \ (cal/cm^2), \qquad (7.20)$$

where Y is again the weapon yield in kilotons and R the slant range in miles. The factor τ is known as the "transmittance," and is a measure of the attenuating effects of the atmosphere. For fairly low-altitude airbursts (within a few miles of the earth's surface) and distances within a few miles of the detonation, a sensible value is $\tau \sim 0.7$. For a 20-kt bomb at $R = 2$ miles and $\tau \sim 0.7$, $Q \sim 3.9$ cal/cm^2, enough for about a second-degree burn. If you are actually *looking* at the fireball, be advised that the focusing effect of your eyes can lead to serious retinal burns. It has been estimated that at Hiroshima, some two-thirds of those who died in the first day after the bombing were badly burned. A 400-kiloton bomb at two miles will be fatal; you will literally be burnt alive.

7.13.3 Radiation

For many people, the most feared consequence of a nuclear explosion is exposure to radioactivity. In reality, however, for most victims of a nuclear attack, the radiation exposure will likely pale in comparison to pressure and heat effects: if you are near enough to suffer acute radiation exposure, you have probably been blasted or burnt to death. It is perhaps because radiation is invisible and presents no symptoms in low doses that it has become imbued with such fear.

Weapons analysts divide radiation effects into two categories: initial, or "prompt" exposure, and long-term or "residual" exposure. The demarcation time between the two is not defined in any hard-and-fast way, but one minute after the explosion is usually taken as a working definition. The most damaging prompt

Table 7.4 Effects of acute radiation exposure. After Sartori (1983) and Glasstone and Dolan (1977)

Dose (rems)	Symptoms, treatments, prognosis
0–100	Few or no visible symptoms. No treatment required; excellent prognosis
100–200	Vomiting, headache, dizziness; some loss of white blood cells. No hospitalization required; full recovery in a few weeks
200–600	Severe loss of white blood cells, internal bleeding, ulceration, hemorrhage, hair loss at ~ 300 rems, danger of infection. Treat with blood transfusions and antibiotics. Guarded prognosis at low end of dose range, but probability of death ~ 90 % at high end of dose range. Cause of death: hemorrhage, infection
600–1000	As 200–600 but more severe. Treatment via bone marrow transplant, but probability of death 90–100 %
1000–5000	Diarrhea, fever. Treat to maintain electrolyte balance; death in 2 days—2 weeks due to circulatory collapse
>5000	Immediate onset convulsions and tremors. Treat with sedatives. Death in no more than 1–2 days due to respiratory failure and brain tissue swelling

radiations are neutrons and gamma rays emitted directly by the explosion and as a consequence of neutron-capture by nitrogen molecules in the surrounding air, which creates gamma-rays. This latter effect, while strictly secondary to the explosion, happens so quickly as to qualify as a source of prompt radiation.

As with blast and thermal effects, an individual's exposure to (and reaction to) radioactivity is dependent on factors such as weather conditions and shielding offered by surrounding structures. While an approximate exposure formula for *prompt* radiation exposure for unprotected individuals has been developed (below), it is essentially impossible to do so for the residual effects, as so many contingencies come into play: Do winds transport much of the fallout to distant locations? Have food and water supplies become contaminated? Can air be filtered? Is medical treatment available? We will look at the prompt dose issue, and the probability of eventually contracting a long-term cancer from the dose received.

The "rem" unit of radiation dose was introduced in Sect. 5.2. For an unprotected person a distance R miles from a warhead of yield Y kilotons, the prompt dose received, in rems, is given very roughly by the expression

$$D_{prompt} \sim \frac{6Y}{R^{7.6}}. \tag{7.21}$$

For our 20-kiloton bomb at 2 miles, $D_{prompt} \sim 0.6$ rems, an almost harmless amount; recall that a single-shot lethal dose is ~ 500 rems. Table 7.4 summarizes effects of various acute radiation doses.

Even if you do not receive an acutely harmful dose of radiation, there is a statistical chance that you will in the long-term die from a radiation-induced cancer. In the medical community, this would be counted as an *excess cancer death*. The reason for this terminology is that statistics show that some 20 % of the population will die of cancer even if they have never been exposed to any human-caused radiation. (The percentage varies by location and sub-populations, but adopting an average of 20 % will serve for our purposes.) Thus, in a population of 100,000, we can expect that some 20,000 people will die of cancer. What, then, is an individual's *excess* probability of cancer death if he or she has been exposed to some man-made radiation? The effects of ionizing radiation on humans and animals have been extensively studied, and a definitive publication in this regard, "The Biological Effects of Ionizing Radiation," has been prepared by the United States National Academy of Sciences. While there is some "noise" in the statistics, the overall result can be summarized with the rule of thumb that for every 100 rems worth of radiation dosage, your chance of dying by cancer increases to about 24 %, that is, a 100-rem dose increases you chance of dying due to cancer from 20 to 24 %. If the entire population of a city of 100,000 acquired 100-rem doses (which would be a *lot* of exposure), then some 24,000 people can be expected to die of cancer, which corresponds to 4,000 *excess* deaths. We can express this as

$$excess\ deaths \sim \frac{0.04\ (population\ exposed)(dose\ in\ rems)}{(100\ rems)}. \qquad (7.22)$$

For the 0.6-rem dose calculated above, this model predicts 24 excess deaths for a population of 100,000 so exposed. Of course, it would be impossible to determine which *individual* deaths out of the (nominal) 20,024 were actually caused by the exposure. Note that these calculations do not include any other causes of death, such as accidents, murders, falls, other medical conditions, etc. It has been estimated that the roughly 100,000 survivors of Hiroshima and Nagasaki received average radiation doses of 20 rems, which implies some 800 excess deaths. In comparison, the number killed by blast, burns, and acute radiation was on the order of 100,000, with many of those suffering injuries from multiple causes.

In the United States, the annual average per-person radiation dose is about 0.6 rems, with about 0.3 rems arising from each of background radiation and medical procedures. The Nuclear Regulatory Commission requires that its licensees limit maximum additional annual dosages to members of the public to 0.1 rems; for adults who work with radioactive materials, the limit is 5 rems. Exposing a population of 300 million to a 0.1-rem dose could be expected to lead to some 12,000 excess deaths. In comparison, some 30,000 people die in traffic accidents annually in the United States, plus about the same number from gunshot wounds.

A parenthetical comment on this 4 % per 100 rems model: By this rationale, a dose of 2500 rems would give a 100 % chance of an excess cancer. This is true, but Table 7.4 tells you that you would die of much more unpleasant effects long before you have a chance to develop a cancer.

7.14   Project A: Preparation of Combat Bombs

As preparations for the *Trinity* test proceeded, a parallel set of preparations for combat use of atomic bombs was also underway. Some of the preparations for the development and deployment of combat bombs were described in Sect. 7.8. This section describes delivery of bomb components to Tinian island, and practice missions carried out there in advance of the Hiroshima and Nagasaki bombings.

The first Los Alamos bomb-preparation personnel departed for Tinian on June 18, 1945, nearly a month before bomb components began to arrive. In July, the uranium for the *Little Boy* gun bomb was delivered to Tinian in two shipments, one by sea and one by air. On Saturday, July 14, the projectile rings, encased in a lead-lined cylinder, departed Los Alamos for Kirtland Field in Albuquerque. The cylinder was attached to a parachute, loaded aboard a DC-3 transport plane, and flown to just outside San Francisco. From there it was convoyed to Hunters Point Naval Shipyard, where it resided for the 14th and the 15th until being loaded onto the fast heavy cruiser USS *Indianapolis*, and bolted to the deck. The *Indianapolis* also carried the "inert" parts of *Little Boy*, which weighed about 10,000 pounds. *Indianapolis* departed San Francisco at 8:00 a.m. on Monday, July, 16, just three and one-half hours after the *Trinity* test, and arrived at Tinian on Saturday, July 28. The six *Little Boy* target rings, cast later, arrived by C-54 transport aircraft with two rings as the sole cargo aboard each of three planes. The C-54s departed Kirtland on the afternoon of July 26, and also arrived at Tinian on the 28th.

The projectile and target pieces and initiators were loaded inside the bomb on July 30. With the installation of radar altimeters and barometric switches the next day, *Little Boy* was ready for combat, awaiting only weather good enough for a visual bombing run. Back at Los Alamos, the Theoretical Division's most recent predicted yield was 13.4 kilotons, which would prove to be remarkably accurate.

In Potsdam, on the evening of July 26 the governments of America, China, and Great Britain (Russia was not yet at war with Japan) issued the joint Potsdam Declaration, which called on Japan to surrender unconditionally or face "prompt and utter destruction." With summer time in effect, Potsdam was eight hours earlier than Tinian, which put the time of the declaration as the very early hours of July 27 on Tinian, the day before the arrival of *Little Boy's* projectile and target rings. Physics and politics were again crossing paths.

The declaration was broadcast to Japan by radio, and leaflets describing it were dropped from American bombers. Japan is one time zone west of Tinian and 7 hours ahead of Germany; the broadcasts were picked up in Tokyo at 7:00 a.m. on the morning of Friday, July 27. Japanese government officials spent all of that day debating the ultimatum, but Prime Minister Kantaro Suzuki concluded that the only recourse was for Japan to fight on and treat the declaration with what historians have characterized as "silent contempt." On Saturday afternoon in Tokyo (Saturday morning in Potsdam), the same day as *Little Boy's* target rings arrived at Tinian, Suzuki related the official response to a press conference. Radio Tokyo began broadcasting Suzuki's statement on Sunday afternoon in Japan (Sunday morning in Potsdam). Japan's atomic fate was sealed two days before the completion of *Little Boy*.

Delivery of *Fat Man* components to Tinian went on in parallel with the preparations for *Little Boy*. At the same time as *Little Boy's* target rings departed Kirtland field, two other C-54s carrying the *Fat Man* plutonium core and initiator also departed, and likewise arrived at Tinian on the 28th. On the morning of July

Fig. 7.40 *Little Boy* test units, and the Nagasaki F31 *Fat Man* plutonium implosion weapon shortly before its mission. Courtesy of the Los Alamos National Laboratory Archives

28, three B-29 bombers, each carrying a high-explosive implosion preassembly, departed from Kirtland. These arrived at Tinian about midday on August 2 (late evening August 1 in Washington).

These various lots of components were not simply spares. In addition to bombing runs with weapons of the same shapes and weights as "active" *Little Boy* and *Fat Man* units, a number of tests were run to check various systems using both inert bombs and ones loaded with conventional explosives. *Little Boy* test bombs were known as "L" units, and *Fat Man* ones as "F" units. July 23 saw the dropping of unit L1, which was fired in the air by radar fusing (Fig. 7.40).

Units L2 and L3 followed on July 24 and 25. On July 29, unit L6 was used to test the procedure for emergency reloading of the bomb into another aircraft at Iwo Jima. The same unit was used on July 31 in a test where the bomber flew to Iwo Jima accompanied by two observation planes, rendezvoused, and returned to Tinian to complete a drop test; this was essentially a dress rehearsal for the Hiroshima mission to follow in a few days. Following this test, all rehearsals preparatory to a combat delivery of a *Little Boy* with active material were complete; unit L11 was designated for the Hiroshima bombing.

The first *Fat Man* test, with unit F13, was made on August 1. This unit was used to test the fusing and detonating circuits, and was "inert" in that it used cast plaster blocks in place of high explosives. Unit F18 was dropped *unsuccessfully* (the firing mechanism did not operate properly) on August 5, the day before the Hiroshima mission. Unit F33, a fully-functioning model except for having an inert core, was dropped on August 8; this was a rehearsal for the Nagasaki mission the following day. The "active" Nagasaki *Fat Man* was unit F31 (Fig. 7.40). Another unit, F32, was held at Tinian in case a third combat drop was to be made, but its fissile material never left the United States.

Training for 509th crews consisted of much more than simulated *Little Boy* and *Fat Man* drops, however. Between June 30 and July 18, they flew seven training and orientation missions comprising 27 sorties (a "sortie" means the flight of an individual aircraft, whether alone or as part of a group); bombs were not carried on these missions. Between July 1 and August 2, 15 practice-bombing missions totaling 89 sorties were conducted. These used conventional 500 and 1,000-pound bombs dropped on nearby lightly-defended Japanese-held islands. Curiously, these were not considered to count as "combat" missions. What did count as combat operations were 16 "Pumpkin" missions (51 sorties), where *Fat Man*-shaped 10,000-pound bombs containing 6,300 pounds of high-explosive were dropped from altitudes of about 30,000 feet over various cities in Japan proper (see Sect. 7.8 for the origin of the Pumpkin terminology.) Two of these sorties had to be aborted, with the result that only 49 Pumpkins were dropped; in one case the bomb was jettisoned, and in the other it was returned safely to Tinian. Pumpkin missions extended from July 20 right up to the day of the Japanese surrender, August 14. It is not generally appreciated that 509th missions continued for almost a week after the Hiroshima and Nagasaki bombings.

After unloading its cargo at Tinian, the *Indianapolis* sailed to Guam, about 130 miles to the south, and then proceeded toward Leyte Island in the Philippines. There its crew of 1,196 were to join a Task Force in preparation for the scheduled November 1 invasion of Kyushu, the southernmost main island of Japan. But just before midnight on Sunday, July 29, the ship was torpedoed by the Japanese submarine I-58. The *Indianapolis* sank within 12 min; some 850 men managed to escape into the sea. A distress call was sent, but it is not clear if any transmitting power remained. Not until Thursday morning, August 2, were 316 survivors inadvertently discovered. Many of those who survived the sinking had succumbed to shark attacks. The loss of the *Indianapolis* represented the greatest single loss of life at sea in the history of the Navy, and has been called the worst naval disaster in American history. The *New York Times* reported the story at the bottom of its front page on Wednesday, August 15, the same day that the headlines reported that Japan had decided to surrender. The *Indianapolis'* Captain, Charles McVay, survived, but was court-martialed and found guilty for failing to steer a zigzag course to avoid torpedoes, even though he had not been explicitly ordered to do so. In recognition of McVay's bravery in combat before the sinking, however, the Secretary of the Navy lifted the sentence. McVay was promoted to Rear Admiral upon his retirement in 1949, but tragically committed suicide in 1968. In July, 2001, the Navy announced that McVay's record had been amended to exonerate him for the loss of the *Indianapolis* and her crew.

As technical preparation of bombs was underway, legal groundwork for their use was being finalized. On July 22, General Marshall, in Potsdam, directed his acting Chief of Staff in Washington, General Thomas Handy, to prepare a directive for submission to himself and Stimson. Groves prepared the orders on the 23rd, and relayed them back to Marshall through Handy. Marshall informed Handy on the 25th that Truman and Stimson had approved them. The text of the orders read:

25 July 1945
TO: General Carl Spaatz
Commanding General
United States Army Strategic Air Forces

1. The 509 Composite Group, 20th Air Force, will deliver its first special bomb as soon as weather will permit visual bombing after about 3 August 1945 on one of the targets: Hiroshima, Kokura, Niigata and Nagasaki. To carry military and civilian scientific personnel from the War Department to observe and record the effects of the explosion of the bomb, additional aircraft will accompany the airplane carrying the bomb. The observing planes will stay several miles distant from the point of impact of the bomb.
2. Additional bombs will be delivered on the above targets as soon as made ready by the project staff. Further instructions will be issued concerning targets other than those listed above.
3. Discussion of any and all information concerning the use of the weapon against Japan is reserved to the Secretary of War and the President of the United States. No communiques on the subject or releases of information will be issued by Commanders in the field without specific prior authority. Any news stories will be sent to the War Department for specific clearance.
4. The foregoing directive is issued to you by direction and with the approval of the Secretary of War and of the Chief of Staff, USA. It is desired that you personally deliver one copy of this directive to General MacArthur and one copy to Admiral Nimitz for their information.

(Sgd) THOS. T. HANDY
THOS. T. HANDY
General, G.S.C.
Acting Chief of Staff

Effectively, these orders made the decision to use the bombs the responsibility of commanders in the field; no further authorization from higher-ups would be necessary. Groves also sent Marshall a memorandum describing operational plans. Attached to the memo was a small map of Japan cut out from a *National Geographic* map, accompanied by descriptions of each of the four target cities listed in Handy's orders. All of these cities except for Nagasaki had been specifically "reserved" against bombing to provide virgin targets for the new weapons; Groves also included a draft of the necessary orders for the Joint Chiefs of Staff to release them to General Spaatz for attack. (Target selection is discussed in more detail in Chap. 8.) Groves also outlined a schedule for anticipated future bomb availability:

The second implosion bomb should be ready 24 August … Additional bombs will be ready for delivery at an accelerating rate, increasing from about three in September to possibly seven in December, with a sharp increase in production expected early in 1946.

An excerpt from President Truman's personal diary for July 25, the day before the Potsdam Declaration was issued, offered a somewhat apocalyptic perspective:

We have discovered the most terrible bomb in the history of the world. It may be the fire destruction prophesied in the Euphrates Valley Era, after Noah and his fabulous Ark. Anyway we think we have found the way to cause a disintegration of the atom. An experiment in the New Mexico desert was startling—to put it mildly. Thirteen pounds of the explosive caused the complete disintegration of a steel tower 60 feet high, created a crater 6 feet deep and 1,200 feet in diameter, knocked over a steel tower 1/2 mile away and knocked men down 10,000 yards away. The explosion was visible for more than 200 miles and audible for 40 miles and more.

This weapon is to be used against Japan between now and August 10th. I have told the Sec. of War, Mr. Stimson, to use it so that military objectives and soldiers and sailors are the target and not women and children. Even if the Japs are savages, ruthless, merciless and fanatic, we as the leader of the world for the common welfare cannot drop that terrible bomb on the old capital or the new. He and I are in accord. The target will be a purely military one and we will issue a warning statement asking the Japs to surrender and save lives. I'm sure they will not do that, but we will have given them the chance. It is certainly a good thing for the world that Hitler's crowd or Stalin's did not discover this atomic bomb. It seems to be the most terrible thing ever discovered, but it can be made the most useful…

The "old" capital referred to by Truman is the city of Kyoto, the historic capital of Japan. Groves wanted Kyoto on the target list, but, as is described in Chap. 8, Stimson deleted it. Truman's belief that he had ordered the bomb to be used against purely military targets was illusory. Hiroshima and Nagasaki were indeed sites of important Japanese military bases, but the world was about to learn that nuclear weapons are of power sufficient to obliterate entire cities at one blow.

In Washington, Stimson's office was busy drafting statements and press releases in preparation for when the bombings would be reported to the public. The pace of preparations in the Pacific proceeded so rapidly that on July 30, Stimson, by then returned to Washington, had to send an urgent cable to Truman (still in Potsdam) with proposed revisions to the statements, noting that "The time schedule on Groves' project is progressing so rapidly that it is now essential that statement for release by you be available not later than Wednesday, 1 August." Truman received the message early in the morning on the 31st, and wrote in pencil on its reverse that any release be held until at least August 2, by which time he would be at sea on his way home (Fig. 7.41).

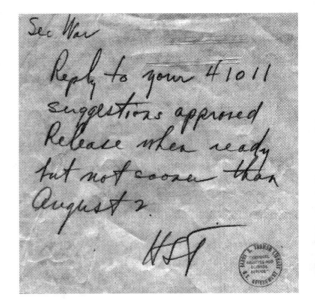

Fig. 7.41 President Truman to secretary of war Stimson, July 31, 1945. *Source* http://www.trumanlibrary.org/oralhist/arnimage2.htm# transcription

President Truman presumably meant August 2, Washington time (August 2 on Tinian would correspond to August 1 in America). By late July, 1945, the use of Los Alamos' bombs was essentially a foregone conclusion, awaiting only final delivery of fissile material to the Pacific and acceptable weather.

Exercises

7.1. Consider a gun-type bomb with a *solid cylindrical* projectile piece of radius r and mass m which is fired under breech pressure P toward a mating target piece a distance x away. For simplicity, assume that the pressure P maintains its value as the projectile piece moves along the gun barrel. Using simple force and kinematic concepts $[F = ma; \; v^2 = 2ax]$, develop an expression for the velocity that the projectile will have after traveling down the barrel. Apply your result to a projectile with $r = 3$ in., $P = 75,000$ pounds per square inch, $m = 50$ kg, and $x = 17$ feet. Be careful with conversion factors; 1 in. $= 2.54$ cm; 1 pound per square inch $= 6895$ Pascals. Does you result accord approximately with the figures given in this chapter? [Ans: 1398 m/s]

7.2. Working from the parameter values cited in Table 7.1, verify the critical masses for U-235 and Pu-239 given in the Table; convince yourself that you understand how to solve the criticality equation (7.6). Now consider U-233: $A = 233.04$ g/mol, $\rho = 18.55$ g/cm^3, $\sigma_f = 1.946$ bn, $\sigma_{el} = 4.447$ bn, $v = 2.755$ neutrons per fission. What is the critical mass? [Ans: 14.2 kg]

7.3. From the decay-rate formula and spontaneous fission data of Sect. 7.7, compute the expected number of spontaneous fissions from 20 μg of Pu-239 over the course of 30 days. Does your result agree with the value of ~ 0.36 cited in the text?

7.4. Suppose you have 1 g of plutonium that is 99.99 % Pu-239 by weight, with the remaining 0.01 % being Pu-240. Compute the hourly spontaneous fission rate of your gram of plutonium. [Ans: 199]

7.5. It is remarked in the text that the shock-wave pressure created by a nuclear explosion is proportional to $E^{2/3}/d^2$, where E is the energy liberated by the explosion and d is distance. If it had been predicted that the *Trinity* test would liberate energy equivalent to 20,000 t of TNT, how many tons of TNT should have been used in the May, 1945, calibration test to produce the same pressure at ground zero if the *Trinity* and test shots were at elevations of 100 and 28 feet, respectively? [Ans: 439 t]

7.6. From the dimensions given in Fig. 7.21, compute the masses of the aluminum and natural uranium tamper spheres in the *Fat Man* bomb; neglect the effect of the "trap-door" access. Take the densities of aluminum and natural uranium to be 2.699 and 18.95 g/cm^3, respectively. [Ans: aluminum 131 kg = 289 lb; uranium 101 kg = 223 lb]

7.7. In the study of thermodynamic properties of materials, the following simple differential equation is used to model the change in volume dV of a sample of material of volume V when it is subjected to a change in pressure dP:

$$\frac{dV}{dP} = -\frac{V}{B}.$$

B is the *bulk modulus* of the material, a measure of its compressibility; a material of higher-B is more difficult to compress than one of lower B. The bulk modulus of plutonium is about 30 GPa (assume constant). If an implosion bomb subjects a plutonium core to a pressure increase of one million atmospheres (1 atm $\sim$ 10^5 Pa), integrate the differential equation to estimate the ratio of the final volume of the plutonium to its initial volume. [Ans: $V_{final}/V_{initial} \sim 0.036$]

7.8. To minimize fallout created by an air-burst nuclear weapon, the weapon should be detonated at a height such that the fireball, at its maximum size, does not touch the ground. An approximate expression for the maximum radius in miles of the fireball created by an air-burst weapon of yield Y kilotons is $R \sim 0.041 Y^{0.4}$. If a 200-kiloton weapon is detonated at this height, to what distance from ground zero will the maximum overpressure exceed 5 psi? Use Eq. (7.19) for the maximum overpressure. [Ans: 2.48 miles]

7.9. The purpose of this problem is to make a very crude estimate of the radioactivity produced by a fission weapon. Suppose that fission of ^{235}U happens exclusively by the reaction

$$^{235}_{92}\text{U} + ^{1}_{0}\text{n} \rightarrow ^{141}_{56}\text{Ba} + ^{92}_{36}\text{Kr} + 3\left(^{1}_{0}\text{n}\right)$$

Assume that 1 kg of ^{235}U is fissioned in this way. ^{141}Ba and ^{92}Kr then both subsequently decay by beta-decay with half-lives of 18 min and 1.8 s, respectively. Use the decay-rate expression of this chapter to estimate the "immediate" beta-radioactivity so generated; for simplicity, ignore the neutrons released in the reaction. If this radioactivity falls out over an area of 10 square miles, what will be the resulting immediate radioactivity in Curies per square meter? [Ans: Appx. 2.7×10^{13} Ci; 1.0×10^6 Ci/m^2]

7.10. Due to a reactor accident, it is predicted that a city of population 200,000 will be exposed to radiation doses averaging 3 rems per person. You are the civic official responsible for deciding whether or not to evacuate the city. The Chief of Police tells you that the chaos to be expected in an evacuation will probably result in about 300 deaths due to traffic accidents, heart attacks, and other such causes. Compare the number of expected radiation-induced excess cancer deaths to the number of deaths expected to be caused by the evacuation. What would you do? [Ans: Appx. 240 excess deaths]

Further Reading

Books, Journal Articles, and Reports

J. Albright, M. Kunstel, *Bombshell: The Secret Story of America's Unknown Atomic Spy Conspiracy* (Times Books, New York, 1997)

L. Badash, J.O. Hirschfelder, H.P. Broida (eds.), *Reminiscences of Los Alamos 1943–1945* (Reidel, Dordrecht, 1980)

K.T. Bainbridge, Trinity. Los Alamos report LA-6300-H, http://library.lanl.gov/cgi-bin/getfile?00317133.pdf

K.T. Bainbridge, Orchestrating the test, in *All In Our Time: The Reminiscences of Twelve Nuclear Pioneers*, ed. by J. Wilson (The Bulletin of the Atomic Scientists, Chicago, 1975)

B. Bederson, SEDs at Los Alamos: A personal Memoir. Phys. Perspect. **3**(1), 52–75 (2001)

J. Bernstein, *Plutonium: A History of the World's Most Dangerous Element* (Joseph Henry Press, Washington, 2007)

H.A. Bethe, Theory of the Fireball. Los Alamos report LA-3064 (1964), http://www.fas.org/sgp/othergov/doe/lanl/docs1/00367118.pdf

H.A. Bethe, R.F. Christy, Memorandum on the Immediate After Effects of the Gadget, http://www.lanl.gov/history/admin/files/Memorandum_on_the_Immediate_After_Effects_of_the_Gadget_Hans_Bethe_and_Robert_Christy.pdf

K. Bird, M.J. Sherwin, *American Prometheus: The Triumph and Tragedy of J. Robert Oppenheimer* (Knopf, New York, 2005)

A.A. Broyles, Nuclear explosions. Am. J. Phys. **50**(7), 586–594 (1982)

R.H. Campbell, *The Silverplate Bombers* (McFarland & Co., Jefferson, North Carolina, 2005)

R.P. Carlisle, J.M. Zenzen, *Supplying the Nuclear Arsenal: American Production Reactors, 1942–1992* (Johns Hopkins University Press, Baltimore, 1996)

M.B. Chambers, Technically Sweet Los Alamos: The Development of a Federally Sponsored Scientific Community. Ph.D. thesis, University of New Mexico (1974)

J. Conant, *109 East Palace: Robert Oppenheimer and the Secret City of Los Alamos* (Simon and Schuster, New York, 2005)

R.H. Condit, Plutonium: An Introduction. Lawrence Livermore National Laboratory report UCRL-JC-115357 (1993), www.osti.gov/bridge/product.biblio.jsp?osti_id=10133699

J. Coster-Mullen, *Atom Bombs: The Top Secret Inside Story of Little Boy and Fat Man* (Coster-Mullen, Waukesha, WI, 2010)

Department of Energy Nevada Operations Office: United States Nuclear Tests July 1945 through September 1992. (Report DOE/NV-209 REV 15), http://www.nv.doe.gov/library/publications/historical/DOENV_209_REV15.pdf

D.F. Dvorak, The other atomic bomb commander: Colonel Cliff Heflin and his "Special" 216th AAF Base Unit. Air Power Hist. **59**(4), 14–27 (2012)

F. Dyson, *Disturbing the Universe* (Harper and Row, New York, 1979)

D.C. Fakley, The British Mission. Los Alamos Sci. **4**(7), 186–189 (1983)

E. Fermi, *Nuclear Physics* (University of Chicago Press, Chicago, 1949)

R.H. Ferrell, *Harry S. Truman and the Bomb*. High Plains Publishing Co., Worland, WY (1996)

F.L. Fey, Health Physics Survey of Trinity Site. Los Alamos report LA-3719 (June, 1967), http://library.lanl.gov/cgi-bin/getfile?00314894.pdf

G.N. Flerov, K.A. Petrzhak, Spontaneous fission of uranium. Phys. Rev. **58**, 89 (1940)

S. Glasstone, P.J. Dolan, *The Effects of Nuclear Weapons* (United States Department of Defense and Energy Research and Development Agency, Washington, 1977)

P. Goodchild, *J. Robert Oppenheimer: Shatterer of Worlds*. British Broadcasting Corporation, London (1980)

L.R. Groves, *Now It Can be Told: The Story of the Manhattan Project* (Da Capo Press, New York, 1983)

W.R. Hansen, J.C. Rodgers, Radiological Survey and Evaluation of the Fallout Area from the Trinity Test. Los Alamos report LA-10256-MS (June 1985)

D. Hawkins, Manhattan District History. Project Y: The Los Alamos Project. Volume I: Inception until August 1945. Los Alamos, NM: Los Alamos Scientific Laboratory (1947). Los Alamos publication LAMS-2532, available online at http://library.lanl.gov/cgi-bin/getfile?LAMS-2532.htm

R.G. Hewlett, O.E. Anderson, Jr., *A History of the United States Atomic Energy Commission, Vol. 1: The New World, 1939/1946*. Pennsylvania State University Press, University Park, PA: (1962)

L. Hoddeson, P.W. Henriksen, R.A. Meade, C. Westfall, *Critical Assembly: A Technical History of Los Alamos during the Oppenheimer Years* (Cambridge University Press, Cambridge, 1993)

M. Hull, A. Bianco, *Rider of the Pale Horse: A Memoir of Los Alamos and Beyond* (University of New Mexico Press, Albuquerque, 2005)

J. Hunner, *Inventing Los Alamos: The Growth of an Atomic Community* (University of Oklahoma Press, Norman, OK, 2004)

V.C. Jones, *United States Army in World War II: Special Studies—Manhattan: The Army and the Atomic Bomb* (Center of Military History, United States Army, Washington, 1985)

R.L. Kathren, J.B. Gough, G.T. Benefiel, *The Plutonium Story: The Journals of Professor Glenn T. Seaborg* 1939–1946 Battelle Press, Columbus, Ohio (1994). An abbreviated version prepared by Seaborg is available at http://www.escholarship.org/uc/item/3hc273cb?display=all

C.C. Kelly (ed.), *Oppenheimer and the Manhattan Project* (World Scientific Publishing, Singapore, 2006)

C.C. Kelly (ed.), *The Manhattan Project: The Birth of the Atomic Bomb in the Words of Its Creators, Eyewitnesses, and Historians* (Black Dog & Leventhal Press, New York, 2007)

C.C. Kelly, R.S. Norris, *A Guide to the Manhattan Project in Manhattan* (Atomic Heritage Foundation, Washington, 2012)

G.B. Kistiakowsky, Trinity—a reminiscence. Bull. Atomic Scientists **36**(6), 19–22 (1980)

L. Lamont, *Day of Trinity* (Antheneum, New York, 1965)

W.L. Laurence, Drama of the Atomic Bomb Found Climax in July 16 Test. The New York Times, 26 Sept 1945, pp. 1, 16

L.M. Libby, *The Uranium People* (Crane Russak, New York, 1979)

S. L. Lippincott, A Conversation with Robert F. Christy—Part I. Phys. Perspective **8**(3), 282–317 (2006). A Conversation with Robert F. Christy—Part II. Phys. Perspective **8**(4), 408–450 (2006)

Los Alamos: Beginning of an Era 1943–1945. Los Alamos Historical Society, Los Alamos (2002), http://www.atomicarchive.com/Docs/ManhattanProject/la_index.shtml

R.E. Malenfant, Experiments with the Dragon Machine. Los Alamos publication LA-14241-H (2005), http://www.osti.gov/energycitations/purl.cover.jsp?purl=/876514-I1Txj9/

D. McCullough, *Truman* (Simon and Schuster, New York, 1992)

T. Merlan, *Life at Trinity Base Camp* (Human Systems Research, Las Cruces, NM, 2001)

R.A. Muller, *Physics and Technology for Future Presidents: The Science Behind the Headlines* (Norton, New York, 2008)

D.E. Neuenschwander, Jumbo: Silent Partner in the Trinity Test. Radiations. Fall 2004, 12–14, http://www.spsnational.org/radiations/2004/neuenschwander.pdf

K.D. Nichols, *The Road to Trinity* (Morrow, New York, 1987)

R.S. Norris, *Racing for the Bomb: General Leslie R. Groves, the Manhattan Project's Indispensable Man*. Steerforth Press, South Royalton, VT (2002)

J.R. Oppenheimer, Physics in the contemporary world. Bull. Atomic Scientists **4**(3), 65–68, 85–86 (1947)

A. Pais, R.P. Crease, *J. Robert Oppenheimer: A Life*. Oxford University Press, Oxford (2006)

N. Polmar, *The Enola Gay: The B-29 that dropped the Atomic Bomb on Hiroshima* (Brassey's, Inc., Dulles, VA, 2004)

I.I. Rabi, *Science: the Center of Culture* (World Publishing, New York, 1970)

N. Ramsey, History of Project A, http://www.alternatewars.com/WW2/WW2_Documents/War_Department/MED/History_of_Project_A.htm

F. Reines, Yield of the Hiroshima Bomb. Los Alamos report LA-1398, 18 April 1952, http://www.fas.org/sgp/othergov/doe/lanl/la-1398.pdf

B.C. Reed, Seeing the light: Visibility of the July 1945 *Trinity* atomic bomb test from the inner solar system. Phys. Teacher **44**(9), 604–606 (2006)

B.C. Reed, A brief primer on tamped fission-bomb cores. Am. J. Phys. **77**(8), 730–733 (2009)

B.C. Reed, Predetonation probability of a fission-bomb core. Am. J. Phys. **78**(8), 804–808 (2010)

B.C. Reed, Fission fizzles: Estimating the yield of a predetonated nuclear weapon. Am. J. Phys. **79**(7), 769–773 (2011a)

B.C. Reed, *The Physics of the Manhattan Project* (Springer, Berlin, 2011b)

R. Rhodes, *The Making of the Atomic Bomb* (Simon and Schuster, New York, 1986)

H. Russ, *Project Alberta: The Preparation of Atomic Bombs for use in World War II* (Exceptional Books, Los Alamos, 1984)

L. Sartori, Effects of nuclear weapons. Phys. Today **36**(3), 32–41 (1983)

E. Segrè, *Enrico Fermi* (Physicist University of Chicago Press, Chicago, 1970)

R.W. Seidel, *Los Alamos and the development of the Atomic Bomb* (Otowi Crossing Press, Los Alamos, 1995)

R. Serber, *The Los Alamos Primer: The First Lectures on How to Build an Atomic Bomb* (University of California Press, Berkeley, 1992)

R. Serber, R.P Crease, *Peace and War: Reminiscences of a Life on the Frontiers of Science*. Columbia University Press, New York (1998)

C.S. Smith, Some Recollections of Metallurgy at Los Alamos, 1943–45. J. Nucl. Mater. **100**(1–3), 3–10 (1981)

H.D. Smyth, *Atomic Energy for Military Purposes: The Official Report on the Development of the Atomic Bomb under the Auspices of the United States Government, 1940–1945* (Princeton University Press, Princeton, 1945)

H. Soodak, M.R. Fleishman, I. Pullman, N. Tralli, *Reactor Handbook, Volume III Part A: Physics* (Interscience, New York, 1962)

M.B. Stoff, J.F. Fanton, R.H. Williams, *The Manhattan Project: A Documentary Introduction to the Atomic Age* (McGraw-Hill, New York, 1991)

C. Sublette, Nuclear Weapons Frequently Asked Questions, http://nuclearweaponarchive.org/Nwfaq/Nfaq8.html

F.M. Szasz, *The Day the Sun Rose Twice* (University of New Mexico Press, Albuquerque, 1984)

F.M. Szasz, *British Scientists and the Manhattan Project: The Los Alamos Years* (Palgrave McMillan, London, 1992)

R.F. Taschek, J.H. Williams, Measurements on $\sigma_f(49)/\sigma_f(25)$ and the value of $\sigma_f(49)$ as a function of neutron energy. Los Alamos report LA-28, 4 Oct 1943

E.C. Truslow, Manhattan District History: Nonscientific Aspects of Los Alamos Project Y 1942 through 1946. Los Alamos Scientific Laboratory, Los Alamos, NM (1973). Available online at http://www.fas.org/sgp/othergov/doe/lanl/docs1/00321210.pdf

S.M. Ulam, *Adventures of a Mathematician* (Charles Scribner's Sons, New York, 1976)

United States National Academy of Sciences, Health Risks from Exposure to Low Levels of Ionizing Radiation: BEIR VII Phase 2, Washington, D. C. (2006)

S.L. Warren, The role of radiology in the development of the atomic bomb, in Radiology in World War II: Clinical Series, ed. by K.D.A. Allen, Office of the Surgeon General, Department of the Army, Washington (1966)

S. Weintraub, *The Last Great Victory: The End of World War II July/August 1945* (Dutton, New York, 1995)

V.F. Weisskopf, The Los Alamos Years. Phys. Today **20**(10), 39–42 (1967)

J.H. Williams, Measurements of ν_{49}/ν_{25}. Los Alamos report LA-25, 21 Sept 1943

J. Wilson, *All in Our Time: The Reminiscences of Twelve Nuclear Pioneers* (Bulletin of the Atomic Scientists, Chicago, 1975)

Websites and Web-based Documents

American Institute of Physics Array of Contemporary American Physicists website on Manhattan Project, http://www.aip.org/history/acap/institutions/manhattan.jsp#losalamos

British Mission to Los Alamos, http://www.lanl.gov/history/wartime/britishmission.shtml

Documents on Los Alamos land acquisition, http://www.lanl.gov/history/road/pdf/Acquisition%20of%20Land%20for%20Demolition%20Range,%20November%2025,%201942.pdf; http://www.lanl.gov/history/road/pdf/Secretary%20of%20War%20Requisitioning%20Lands,%20March%2022,%201943.pdf; http://www.lanl.gov/history/road/pdf/Secretary%20of%20Agriculture%20Granting%20Use%20of%20Land%20for%20Demolition%20Range,%20April%208,%201943.pdf

Federation of American Scientists index to Los Alamos reports, http://www.fas.org/sgp/othergov/doe/lanl/index1.html

Fermi's description of Trinity test, http://www.lanl.gov/history/story.php?story_id=13; http://www.atomicarchive.com/Docs/Trinity/Fermi.shtml

Harry Daghlian and Louis Slotin, http://en.wikipedia.org/wiki/Harry_K._Daghlian,_Jr; http://en.wikipedia.org/wiki/Louis_Slotin

James Tuck, http://bayesrules.net/JamesTuckVitaeAndBiography.pdf

Los Alamos National Laboratory History site, http://www.lanl.gov/about/history-innovation/history-site.php

Los Alamos Primer, http://www.fas.org/sgp/othergov/doe/lanl/docs1/00349710.pdf

McDonald Ranch House, http://en.wikipedia.org/wiki/McDonald_Ranch_House

Monsanto Corporation Dayton operations, http://moundmuseum.com/

Oppenheimer appointment letter as Director of Los Alamos, www.lanl.gov/history/road/pdf/Conant-Groves.pdf

Oppenheimer quotes, http://en.wikiquote.org/wiki/Robert_Oppenheimer

Oppenheimer memoranda on Gadget and Explosives Divisions, http://www.lanl.gov/history/road/pdf/Organization%20of%20Explosives%20Division,%20August%2014,%201944.pdf; http://www.lanl.gov/history/road/pdf/Organization%20of%20Gadget%20Division,%20August%2014,%201944.pdf

Roosevelt to Oppenheimer, June 29, 1943, http://lcweb2.loc.gov/mss/mcc/083/0001.jpg and http://lcweb2.loc.gov/mss/mcc/083/0002.jpg

Sir John Anderson, http://en.wikipedia.org/wiki/John_Anderson,_1st_Viscount_Waverley

Tinian Island, http://www.globalsecurity.org/military/facility/tinian.htm

Trinity eyewitness accounts, http://www.dannen.com/decision/trin-eye.html

Chapter 8
Hiroshima and Nagasaki

The atomic bombings of Hiroshima Nagasaki were the culminating events of the Manhattan Project. As described in Chap. 7, training of crews to deliver the bombs began in the fall of 1944. Planning for the eventual use of the bombs in the sense of choosing targets and considering wartime postwar strategic implications of such a radical new weapon also began to come under consideration by scientists, politicians, government advisors, and military officials at about the same time.

This chapter examines the preparations for the bombing missions, debates that were conducted as to whether the bombs should be used directly or be demonstrated first, the missions themselves, the effects of the bombs, reactions to their use, and the still-debated role of the bombings in the circumstances of the Japanese surrender in August, 1945.

8.1 The 509th Composite Group: Training and Targets

As described to in Sect. 7.8, selection and training of crews to drop the bombs was an integral part of the delivery program. In this section, the history of this unique group is briefly related.

In consultation with General Henry Arnold (Commander of the Army Air Forces), General Groves decided—again for reasons of security and compartmentalization—to organize a self-sustaining Air Force unit to deal with bomb delivery. During the summer and fall of 1944, Air Force and Manhattan Project personnel screened possible candidates to command the new unit, settling on Lieutenant Colonel Paul W. Tibbets. A superb combat pilot, Tibbets had flown the first B-17 bomber across the English Channel on a bombing mission in World War II, and later led the first American raid on North Africa. After more than twenty-five combat missions, he returned to the United States to become involved with flight-testing the B-29 bomber. On September 1, 1944, Tibbets underwent a final security grilling at the Colorado Springs headquarters of General Uzal Ent, Commanding General of the Second Air Force. After answering questions to the satisfaction of Groves' security chief, Colonel John Lansdale (Sect. 4.10), Tibbets

B. C. Reed, *The History and Science of the Manhattan Project*,
Undergraduate Lecture Notes in Physics, DOI: 10.1007/978-3-642-40297-5_8,
© Springer-Verlag Berlin Heidelberg 2014

Fig. 8.1 *Left* Partial crew of the *Enola Gay*: Standing (l-r): John Porter (ground maintenance officer), Theodore Van Kirk, Thomas Ferebee, Paul Tibbets, Robert Lewis, Jacob Beser; kneeling (l-r): Joseph Stiborik, Robert Caron, Richard Nelson, Robert Shumard, Wyatt Duzenbury. Not present: William Parsons, Morris Jeppson. Photo courtesy John Coster-Mullen. *Right* Morris Jeppson. *Source commons.wikimedia.org/wiki/File:Morris_Jeppson.jpg*

was ushered into Ent's office. There he was introduced to William Parsons and Norman Ramsey, who briefed him on his new assignment.

Arnold gave Tibbets wide liberty in his choice of personnel to staff his new command, but Tibbetts could tell selectees nothing of their ultimate mission. Familiar with some of the best pilots, navigators, and bombardiers of the war, Tibbets wasted no time in recruiting them. Among his earliest acquisitions were two personal friends with whom he had flown a number of missions: bombardier Major Thomas Ferebee, and navigator Theodore "Dutch" Van Kirk (Fig. 8.1), veterans of 63 and 58 missions, respectively. Both would fly with Tibbets on the Hiroshima mission. At this writing (early 2013), Van Kirk is the only surviving crewmember of the two atomic bombardment planes. First Lieutenant Jacob Beser, the 509th's radar officer, would be the only crew member to fly in both the Hiroshima and Nagasaki "strike" aircraft, the ones that carried the bombs. Beser would be responsible for monitoring Japanese radar to determine if they were trying to jam the bomb's firing mechanisms, or perhaps even cause a premature detonation.

What made the 509th unique was that it was a "composite" group. Air Force squadrons were normally single-purpose entities: maintenance, bombardment, engineering, transport, and the like. The 509th drew together a number of separate units to form a self-sustaining whole: the 393rd Heavy Bombardment Group (which comprised 15 bomber crews); the 320th Troop Carrier Squadron; the 390th Air Service Group; the 603rd Air Engineering Squadron; the 1,027th Air Material Squadron; the 1st Special Ordnance Squadron (Aviation); the 1,395th Military Police Company (which included some 50 Manhattan Project agents); and the 1st Technical Detachment, War Department Miscellaneous Group, a catch-all unit of civilian and military scientists and technicians. The 509th was authorized to a

Table 8.1 Hiroshima, Nagasaki, and Tokyo Bombings

Statistic	Hiroshima	Nagasaki	Tokyo
Planes	1	1	279
Bombs	1 atomic	1 atomic	1,667 t
Population per square mile	46,000	65,000	130,000
Square miles destroyed	4.7	1.8	15.8
Killed and missing (thousands)	70–80	35–40	83.6
Injured (thousands)	70	40	102
Mortality (thousands/ square mile)	15	20	5.3

complement of 225 officers and 1,542 enlisted men; the addition of some 50 members of Project Alberta brought the total to over 1,800.

The 509th's first B-29 flight at Wendover Field occurred on October 21, 1944, with pilot Robert Lewis at the controls. In addition to test drops at Wendover, practice bombing runs involved flying from Wendover almost 600 miles due south to the Salton Sea in southern California, dropping a single "blockbuster" bomb (often filled with concrete), and then executing a 155° diving escape turn designed to put about eight miles between the bomber and the eventual nuclear explosion. From the formula given in Chap. 7 for blast overpressure, a 20-kiloton bomb at 8 miles would create an overpressure of about 0.3 psi, which the bomber was expected to have no trouble surviving. The 509th's bombers would have no fighter escorts during their missions in order to avoid drawing the attention of Japanese defenders; also, to survive the shock wave, fighter aircraft would have to be so far from the bomber that they could provide no real protection. Colonel Tibbets knew that B-29's stripped of their guns and armor could fly as high as 34,000 feet, well out of the range of anti-aircraft guns and above the ceiling of Japanese "Zero" fighters. The 393rd bombardment group received its fifteenth stripped-down B-29 on November 24, 1944, bringing it to full strength. On December 17, the 41st anniversary of the Wright brothers first flight, the 509th was formally activated. Ferebee and Van Kirk were appointed as Group Bombardier and Group Navigator.

In the spring of 1945, the pace of training for the 509th built toward deployment to the Pacific. On May 19, the first members of the group arrived on Tinian; others would follow until all personnel and aircraft were present by early August. Technically, the 509th lay in the theatre of operations of General Curtis LeMay, who had taken command of the Twenty-First bomber command of the Twentieth Air Force in January, 1945. LeMay kept his headquarters on the island of Guam, about 130 miles south of Tinian.

To cripple Japanese industry, LeMay had decided upon a strategy of nighttime low-level ($\sim$5,000 feet altitude) incendiary bombing. On the night of March 9–10, 1945, a fleet of nearly 300 B-29 s firebombed Tokyo, dropping some sixteen hundred tons of incendiary bombs (Table 8.1). Individual fires coalesced into a firestorm, with the result that some 16 square miles of the city were burnt out (about 25 % of the city). Some one million people were rendered homeless, and 84,000 were killed (some estimates claim over 100,000), a toll greater than the number of immediate deaths that would occur at either Hiroshima or Nagasaki; only 14 aircraft were lost. The air was heated so much that B-29s at 6,000 feet experienced serious turbulence; crew members could smell the burning flesh of victims below. Similar raids followed against the cities of Nagoya, Osaka, and Kobe, and another raid on April 13 burnt out eleven more square miles of Tokyo.

As the above suggests, the ferocity of World War II in the Pacific is almost beyond comprehension. Some 5,000 Americans and many more Japanese were dying each week as American forces advanced through Japanese-held islands. A mid-June, 1945, Joint Chiefs of Staff planning document for the proposed invasion of Japan summarized some casualty statistics. To take the islands of Leyte (late 1944), Luzon (early 1945), Iwo Jima (February–March 1945), and Okinawa (April–June 1945), United States casualties totaled some 110,000. An incomplete tally of Japanese killed and taken prisoner totaled over 300,000. Some 140,000 civilians on Okinawa alone are estimated to have been killed or committed suicide. If an invasion of the home islands of Japan went ahead and the Japanese kept up such a fanatic level of resistance, casualties could be astronomical. During the entire war, no Japanese unit had ever surrendered.

The proposed invasion of Japan comprised two elements. The southern island of Kyushu (home to Nagasaki) was to be the target of Operation Olympic, scheduled to begin on November 1, 1945 (Figs. 8.2 and 8.3). This would involve some 760,000 ground forces and support from three naval fleets. For comparison, the number of troops landed during the June 6, 1944, D-Day invasion of Normandy was about 156,000. The plan was for forces to advance about one-third of the way along the island (the dashed line in Fig. 8.2), setting up air bases to support an invasion of the area around Tokyo (on the main island of Honshu) which was scheduled for March 1, 1946. The Honshu invasion was planned to involve just over one million ground forces. In the meantime, LeMay's bombers would continue laying waste to Japanese cities, anticipating delivery of 100,000 t of bombs per month by the end of 1945, and 220,000 t per month by March, 1946. General Douglas MacArthur is said to have expressed the fear that once American forces had established themselves on Kyushu, they might face a guerrilla war which could go on for 10 years. One American intelligence estimate indicated some 560,000 Japanese troops stationed in Kyushu as of August, 1945.

In the early spring of 1945, General Groves turned his attention to the issue of selecting targets. Groves met with General Marshall (Groves's office diary records meetings with both Marshall and Stimson on March 7), and asked Marshall to designate a contact within the Army's Operations Planning Division. Marshall was reluctant to bring any more people into the issue than necessary, and directed Groves to see to targeting himself. For Groves, his target criteria were, as he put it, "places the bombing of which would most adversely affect the will of the Japanese people to continue the war." Beyond that, targets should be military in nature: headquarters, troop concentrations, and centers of production. Groves contacted General Lauris Norstad, Chief of Staff of the Army Strategic Air Force, to

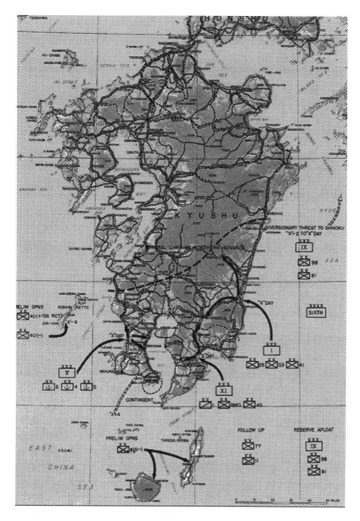

Fig. 8.2 Map showing invasion locations for southern Kyushu. The scale bar at the lower right is 50 miles long. Compare Fig. 8.14. *Source* http://commons.wikimedia.org/wiki/File:Operation_Olympic.jpg

establish a committee to make target recommendations. Chosen were three staff members from General Arnold's office, plus three scientists from Los Alamos: John von Neumann, Robert Wilson, and William Penney of the British Mission; the latter had carried out extensive analyses of what levels of damage to Japanese cities might result from bombs of various yields detonated at various heights. The committee's charge was to develop a list of four previously unbombed cities, chosen such that three could be available for each mission, with weather predicted to be good enough for visual bombing.

The first of three meetings of the Target Committee took place at General Norstad's office in the Pentagon on Friday, April 27. Groves opened the meeting with a short briefing, after which he left General Farrell in charge. Much of the discussion in this first meeting concerned the dismal prospects for acceptable weather over Japan in the summer months. Experience indicated that June would be the worst month. In July, seven good days (defined as 3/10 or less cloud cover) could be expected, only six in August, and even fewer in September. Only once in 5 years had there been two successive good visual bombing days for Tokyo. January would be the best month, but there was no question of waiting that long. After lunch, the discussion turned to possible targets. Colonel William Fisher of the Air Force summarized ongoing operations. The Twenty-First bomber command of the 20th Air Force had 33 primary targets on its priority list. As the minutes of the meeting recorded, "the 20th Air Force is operating primarily to lay waste all the main Japanese cities, and they do not propose to save some important primary target for us if it interferes with the operation of the war from their point of view. Their existing procedure has been to bomb the hell out of Tokyo, bomb the aircraft manufacturing and assembly plants, engine plants and in general paralyze the aircraft industry". Then followed a list of eight cities, including Tokyo and Nagasaki, which were being bombed "with the prime purpose in mind of not leaving one stone lying on another."

As to possible Manhattan District targets, four were specifically discussed (Fig. 8.3): Hiroshima was the largest untouched target not on the Twenty-First's priority list, and the site of the Japanese Second Army Headquarters (from which the defense of Kyushu would be directed); Yawata, not far from Osaka, was the site of a steel industry, Yokohama (on Tokyo Bay, south of Tokyo); and Tokyo itself. However, Tokyo was not considered a high priority as it was "practically all rubble with only the palace grounds left standing." (The Palace and Emperor Hirohito had deliberately been spared destruction.) The meeting adjourned at 4:00 p.m. with a list of 17 target areas identified as needing further research regarding damage already inflicted, weather data, amount of damage expected from the new weapons, and "the ultimate distance at which people will be killed." Particular consideration was to be given to large urban areas not less than three miles in diameter which were sited within larger populated areas.

The second meeting of the Target Committee was held in Oppenheimer's office at Los Alamos over May 10–11, just after the 100-ton test at the *Trinity* site. The agenda was extensive, and included topics as diverse as optimum bomb detonation heights, weather reports, procedures for a bomber having to jettison a bomb or return to base with a non-released one, status of targets, expected psychological and radiological effects, rehearsals, and coordination with the Twenty-First's regular bombing campaigns. Detonation heights which would yield 5-psi over-pressures were desired, but corresponding damage radii were not specified in the record of the meeting. In spite of a lack of firm bomb-yield estimates, considerable latitude was available in the detonation heights; it was predicted that the bombs could be detonated as much as 40 % below optimum height or 15 % above optimum height with only a 25 % loss in the damage area. For *Little Boy*,

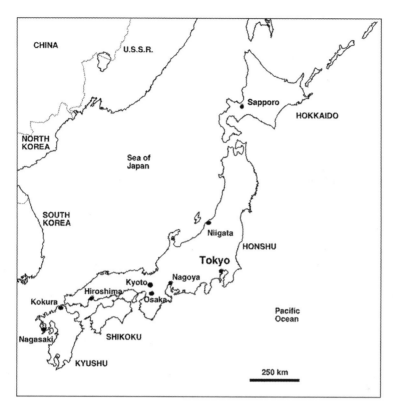

Fig. 8.3 Map of Japan, showing main islands and major cities. A number of smaller islands are omitted. Adapted from http://www.hist-geo.co.uk/japan/outline/japan-cities-1.php

detonation heights of 1,550 and 2,400 feet were considered appropriate for yields of 5 and 15 kilotons, respectively. The yield outlook for *Fat Man* was still pessimistic, with heights being estimated for yields of only 0.7, 2, and 5 kilotons. In view of the uncertainties, it was decided that four different fuse-height settings should be available: 1,000, 1,400, 2,000, and 2,400 feet, with 1,400 feet likely to be used for both bombs.

By the time of the second meeting, the Air Force had relented from its position at the April 27 meeting and was willing to "reserve" (leave unbombed) five targets for Manhattan consideration. First on the list was Kyoto, the historic capital and cultural center of Japan, with a population of about one million; industries were being moved there as other cities were being destroyed. It was pointed out that "Kyoto is an intellectual center for Japan and the people there are more apt to appreciate the significance of such a weapon". Second on the list was Hiroshima. Yokohama remained in third place, although considered disadvantageous in view of its heavy concentration of anti-aircraft defenses. In fourth place, and new on the list, was Kokura, the site of one of the largest arsenals in Japan. Bringing up the rear came Niigata, north of Tokyo on the western side of Honshu, a port of embarkation that was also the site of machine-tool industries and oil refining. After some discussion, the first four were recommended (in the order described here) for target status. Nagasaki seems not to have been discussed during this meeting.

The third and last meeting of the target committee was held in the Pentagon on May 28, with Parsons, Ramsey, Ashworth, and Tibbets in attendance. After a brief discussion of revisions to detonation-height settings—now five options between 1,100 and 2,500 feet—Tibbets gave a detailed description of his crews' training regimens. Each of 15 bombardiers had accrued at least 50 releases at altitude, with most having performed 80–100. Drops conducted with radar-based bombing runs at 20,000 feet altitude were averaging within 1,000 feet of targets; visual runs were achieving 50 % success within 500 feet. Round-trip flights up to 4,300 miles with 10,000-pound bomb loads had been conducted, with plenty of fuel to spare. Parsons reported that 19 Pumpkins had been shipped from Wendover, and that it looked feasible to have 25–30 at Tinian by July 15, with production reaching 75 per month by mid-June. 509th ground echelons were already in place on Tinian; the entire group would arrive by mid-July. A 37-man bomb-assembly field crew comprising both civilian and military personnel had been designated. The civilians would hold assimilated military ranks; Robert Serber, for example, became an instant Colonel. The list of reserved targets had shrunk to three: Kyoto, Hiroshima, and Niigata; no reason was recorded as to why Yokohama and Kokura had been dropped. The overall conclusion was that activities were solidly on track for *Little Boy* to be ready by August 1. *Fat Man* was not discussed in detail, pending results of the *Trinity* test, still 6 weeks in the future.

Groves' personal preferred target was Kyoto, in view of its having a large enough area to gain maximum knowledge of the bomb's effects. However, that city was spared by the personal intervention of Secretary of War Henry Stimson. On May 30, 2 days after the target committee meeting, Groves was conferring with Stimson when the latter asked about the status of targets. Stimson, who had personally visited Kyoto on two occasions, immediately objected to targeting the city on the grounds of its historical, cultural, and religious significance to the Japanese; he wished to spare it on humanitarian grounds.

On June 14, Groves forwarded to General Marshall a revised list of Kokura, Hiroshima, and Niigata, but was not about to give up on his preference. Through June and July, he attempted, on up to perhaps a dozen or so occasions, to get Kyoto back on the list, even after Stimson had departed for the Potsdam conference. In a cable to Stimson on July 21, George Harrison (Sect. 7.12) stated that "All your local military advisors engaged in preparations definitely favor your pet city and would like to feel free to use it as first choice". Stimson consulted with Truman, who concurred with his Secretary of War; Stimson replied that he was aware of no factors to change his decision. Kyoto's reprieve was Nagasaki's doom: in the Groves-Handy orders of July 25 (Sect. 7.14), Nagasaki had replaced the historic capital, with Kokura listed afresh. In his memoirs, Groves takes credit for sparing Kyoto, claiming that that he prevailed upon General Arnold to keep it on the reserved list when he realized that the Air Force might delete it from the list after

Stimson's refusal to approve it. The fates of Hiroshima and Nagasaki were cast weeks before the bombing missions.

8.2 Fall 1944: Postwar Planning Begins

That the development and use of nuclear weapons would radically alter the balance of power in the world and could potentially precipitate a dangerous arms race was evident to many of the leading figures of the Manhattan Project. Despite the pressure of the war, consideration had to be given at the highest levels to issues that would come to the fore as soon as the existence of the bomb was revealed. A number of possibilities were on the table: Should it be used against an enemy without warning, or should a demonstration be arranged first? After the war, would atomic energy come under civilian or military control? What could be publicly revealed of the work of the Project without violating security concerns? What legislation and Congressional oversight would need to be established? What should be the role of the government in supporting research and regulating private nuclear industries? To forestall an arms race, would some sort of international control be necessary, with knowledge being shared among different countries?

Some of the first to raise these issues were scientists at Arthur Compton's Metallurgical Laboratory in Chicago. By the summer of 1944, the X-10 reactor was functioning and construction at Hanford was well underway. Technical work at Chicago was beginning to wind down, and attention began to turn to questions of possible use of bombs, long-term prospects for the Laboratory, and the wider ramifications of atomic energy. Compton asked Isaiah "Zay" Jeffries, a metallurgist and General Electric executive whom he had brought into the Met Lab as a consultant, to head a committee to prepare a "Prospectus on Nucleonics," the latter being the term Met Lab scientists applied to what they foresaw as a vast postwar research and industrial field. Other members of the committee included Robert Mulliken and Enrico Fermi. Their report, submitted to Compton on November 18, 1944, contained seven sections. The first five reviewed the history of nuclear physics and potential peacetime applications in areas as diverse as commercial power production, naval propulsion, medicine, agriculture, and industry. Potential research applications for radioactive tracer isotopes were numerous; one specifically mentioned was tracking metabolic and photosynthetic pathways. More speculative possibilities involved using nuclear explosives in immense construction projects, or to divert hurricanes. It is for its last two sections, however, that the Jeffries Report is now remembered. In particular, section six, "The Impact of Nucleonics on International Relations and the Social Order," was remarkably prophetic in its vision of possible future events. A copy of the report can be found in Martin Sherwin's *A World Destroyed*.

Knowing that the laws of physics are universal and that any industrially advanced country could harness nuclear energy, the report cautioned that America could not secure lasting security by simply attempting to stay ahead of other nations in nucleonics research and development; breakthroughs could happen anywhere. Anticipating much of the future Cold War and current-day concerns with nuclear proliferation and terrorism, the report stated that

> Nuclear weapons might be produced in small hidden locations in countries not normally associated with a large scale armament industry ... A nation, or even a political group ... will be able to unleash a "blitzkrieg" infinitely more terrifying than that of 1939–1940 ... The weight of the weapons of destruction required to deliver this blow will be infinitesimal compared to that used up in a present day heavy bombing raid, and they could easily be smuggled in by commercial aircraft or even deposited in advance by agents of the aggressor.

To forestall such a destabilizing situation, the committee advocated that a central international authority would have to be established to exercise control over nuclear power, supervision of associated materials, and making available such materials for legitimate research needs. In unknowing anticipation of what would come to be called the strategy of "mutually assured destruction," the group felt that until such an authority was established, "The most that an independent American nucleonic re-armament can achieve is the certainty that a sudden total devastation of New York or Chicago can be answered the next day by an even more extensive devastation of the cities of the aggressor, and the hope that the fear of such a retaliation will paralyze the aggressor." The report also addressed the need for broad public education on nuclear issues, believing such to be the only way to assure the "moral development necessary to prevent the misuse of nuclear energy".

Turning to the role of a free-market economy, the last section of the report used the alcohol industry as an example to make the point that there need be no inherent conflict between the ideas of a regulating authority and the usual operation of private enterprise: production and sales could in the hands of private industries, but conducted under government oversight. But a vigorous nucleonics industry would not by itself be sufficient. Since most private industries were not set up for long-term research, the group felt it vital that government-supported nucleonics laboratories having "ample facilities for both fundamental and applied research" be established. Much of what the Jeffries Report predicted in the industrial and government-activities areas came to pass. Civilian nuclear power under government licensing and regulation was in place by the 1950s, along with a complex of national laboratories. But effective international control of nuclear materials would become a political quagmire, and the anticipated arms race ensued.

About the same time that the Jeffries committee was formed, forward planning also began to garner more attention at the upper administrative levels of the Manhattan Project. The vast complex of production facilities and laboratories constructed for the Project would not simply vanish the day after the war ended. Should they come under civilian or military jurisdiction? How would they be funded and operated? In August, 1944, the Military Policy Committee authorized Richard Tolman to head a Committee on Postwar Policy to study the relation of atomic energy to national security. Tolman's small group (himself, Warren Lewis, Henry Smyth of Princeton, and Rear Admiral Earle Mills of the Navy) conducted interviews with over 40 Project scientists and also received written submissions. Their December 28 report to Groves emphasized that nuclear power for propelling naval vessels should be developed immediately, and that, within bounds dictated by security considerations, a nucleonics industry should be strongly encouraged. Also, wide dissemination of knowledge would be essential to encourage a level of post-war progress in the field necessary to maintain national security. Perhaps most importantly, they envisioned a national authority which would distribute research and development funds among military, civilian, academic, and industrial laboratories. International relations lay outside the committee's charge, and it ventured no opinions in that area.

Vannevar Bush and James Conant had their own ideas as well. On September 19, 1944, they wrote to Henry Stimson to point out that the time would soon come to consider how to release basic scientific information and enact legislation for domestic control of nuclear power. On September 30, they followed up with a more extensive memorandum titled "Salient Points Concerning Future International Handling of Subject of Atomic Bombs." This 3-page document would prove just as prophetic as the Jeffries report. In six brief paragraphs, Bush and Conant laid out what was virtually a script for the Cold War. Their first paragraph set out the baseline scenario: "There is every reason to believe that before August 1, 1945, atomic bombs will have been demonstrated and ... be the equivalent of 1,000 to 10,000 t of high explosive ... one B-29 bomber could accomplish with such a bomb the same damage against weak industrial and civilian targets as 100 to 1,000 B-29 bombers." The second paragraph pointed to a possibility for how fission bombs might be used to trigger even more violent explosions: "It is believed that such energy can be used as a detonator for setting off ... the transformation of heavy hydrogen atoms into helium. If this can be done a factor of a thousand or more would be introduced into the amount of energy released ... That such a situation presents a new challenge to the world is evident." Paragraph three predicted that the then-current advantage of the United States and Great Britain with respect to such weapons would surely be temporary; any nation with good technical and scientific resources could reach the U.S.-Britain position in 3 or 4 years, an estimate which would prove very accurate. Paragraph four argued that since the Manhattan Project was so vast, information regarding various aspects of it was actually quite widespread, so plans should be made for disclosure of the history and development of the Project "as soon as the first bomb has been demonstrated. This demonstration might be over enemy territory, or in our own country, with subsequent notice to Japan that the materials would be used against the Japanese unless surrender was forthcoming."

Paragraph five of Bush and Conant's memo advised that it would be extremely dangerous for the United States and Britain to attempt to carry on further development in complete secrecy, for such would undoubtedly motivate Russia to do the same; if another country were to develop fusion bombs first, the United States would find itself "in a terrifying situation." To counter this, they proposed that an international system of free exchange of all scientific information on the subject be set up, to be established under the auspices of an international office "deriving its power from whatever association of nations is developed at the close of the present war." They further suggested that the technical staff of the supervising office be given "free access in all countries not only to the scientific laboratories where such work is contained, but to the military establishments as well." While acknowledging the naivety of this idea, they closed with the warning that "the hazards to the future of the world are sufficiently great to warrant this attempt ... Under these conditions there is reason to hope that the weapons would never be employed and indeed that the existence of these weapons might decrease the chance of another major war."

The arguments common to the Jeffries, Tolman, and Bush-Conant reports are striking. Such recommendations needed to be considered at the highest levels, but the day-to-day pressures of the war naturally intervened. In late October, 1944, Bush suggested to Stimson that one approach might be to establish an advisory group that would report directly to the President. Stimson and Groves updated President Roosevelt on the status of the Project on December 30, but they did not discuss postwar planning; Stimson apparently felt that the time was not yet appropriate to broach the idea of an advisory committee. As the calendar turned to 1945, planning was relegated to official limbo, where would remain until Harry Truman assumed the Presidency in April of that year. Stimson did raise the issue with President Roosevelt in their last conversation together on March 15, 1945, but nothing came of it at the time.

8.3 President Truman Learns of the Manhattan Project

On the afternoon of April 12, 1945, President Franklin Roosevelt died of a cerebral hemorrhage at the age of 63, having served just over 12 years as Chief Executive. Vice President Harry Truman, who knew of the existence of the Manhattan Project but knew almost nothing of its details, was sworn in that evening at the White House. Truman had officially met with Roosevelt only eight times since becoming the Vice-Presidential candidate in the 1944 election (Fig. 8.4).

After a brief Cabinet meeting following his swearing-in, Truman was approached by Stimson, who related that he wished to inform the new President "about an immense project that was underway—a project looking to the development of a new explosive of almost unbelievable destructive power." The next afternoon, James Byrnes, head of the Office of War Mobilization (and soon to be Truman's Secretary of State), dramatically told the President that "we are perfecting an explosive great enough to destroy the whole world. It might well put us in a position to dictate our own terms at the end of the war." In the pressure of adjusting to his new job, almost 2 weeks would elapse before Truman received a full briefing on the Project. As Truman biographer David McCullough has written, the bomb project was a Roosevelt legacy inherited by Truman with no written guidance save Roosevelt's Quebec City agreement with Churchill (Sect. 7.4). Without Roosevelt's personal backing, the project would

Fig. 8.4 *Left* President Harry S. Truman (1884–1972). *Right* Truman, Secretary of State James Byrnes, and Ambassador to Belgium Charles Sawyer in Antwerp, Belgium, July 15, 1945. *Sources* http://commons.wikimedia.org/wiki/File:Harry_S._Truman_-_NARA_-_530677.jpg; *NARA-198780.tif*

never have obtained the priority it needed to succeed, but the results fell squarely in Truman's lap.

At noon on Wednesday, April 25, Stimson and Groves briefed the new President on the Manhattan Project. The President's schedule was crowded, but Stimson felt that a full briefing could not be put off any longer. That evening, the opening conference of the United Nations would take place in San Francisco. Truman was to address the delegates by radio, and Stimson felt that it would be inappropriate for him to do so without an appreciation of the potentialities of the new weapon. Two days earlier, Groves had submitted to Stimson a background memorandum to be given to the President. Essentially a primer on the entire Project, this memorandum, titled "Atomic Fission Bombs," ran to only 24 double-spaced pages, but managed to cover every aspect of the Project from the idea of uranium fission up to the prospects for fusion weapons.

Stimson arrived at the Oval Office a few minutes before Groves, and had prepared a two-page covering memorandum of his own. The first sentence read: "Within 4 months we shall in all probability have completed the most terrible weapon even known in human history, one bomb of which could destroy a whole city." Echoing the Jeffries report, Stimson expressed the fear that the future could see a time when such a weapon could be constructed in secret and used suddenly with devastating power against an unsuspecting nation or group, unless some system of control could be developed. Such a system, however, would "undoubtedly be a matter of the greatest difficulty and would involve such thorough-going rights of inspection and internal controls as we have never heretofore contemplated." The development of this weapon, he felt, "has placed a certain moral responsibility upon us which we cannot shirk without very serious responsibility for any disaster to civilization which it would further." After Stimson had finished reading his memo regarding postwar responsibilities,

Groves entered the meeting, and the three men went through his longer document in detail.

Groves' memorandum is a model of how to prepare an effective summary document. The essential facts on the expected power of such bombs are laid out in the first few pages. The opening sentences could not have failed to catch Truman's attention: "The successful development of the Atomic Fission Bomb will provide the United States with a weapon of tremendous power which should be a decisive factor in winning the present war more quickly with a saving in American lives and treasure. If the United States continues to lead in the development of atomic energy weapons, its future will be much safer and the chances of preserving world peace greatly increased. Each bomb is estimated to have the equivalent effect of from 5,000 to 20,000 t of TNT now, and ultimately, possibly as much as 100,000 t." The balance of the report includes discussions of the history of the discovery of fission; the fissile properties of uranium and plutonium; establishment of the Briggs committee; the scale of work going on at Oak Ridge, Hanford, and Los Alamos; the concept of a graphite pile; the notion of critical mass; gun and implosion bombs; anticipated operational plans; collaboration with the British; a summary of what foreign countries might be up to; and the necessity for postwar planning. Groves related that the gun bomb was expected to yield between 8 and 20 kilotons, and the implosion device between 4 and 6 kilotons. The first gun bomb, which was not expected to require a full-scale test, was predicted to be ready by August 1; a second one should be ready by the end of the year, with subsequent ones to follow at about 60-day intervals thereafter. A test of the implosion device should be possible by the early part of July. A second test of the implosion bomb, if necessary, should be ready by the first of August, with bombs themselves ready in quantity—about one every 10 days—by the latter part of August. The target, wrote Groves, "is and was always expected to be Japan." Costs of construction and operations to March 31, 1945, had accumulated to nearly $1.5 billion, a figure which was expected to grow to nearly $2 billion by the end of June. After outlining issues that would have to be addressed in the postwar period, Groves closed by remarking that George Harrison (Sect. 7.12) had suggested setting up a committee to develop recommendations for consideration by the executive and legislative branches of the government for the time when secrecy was no longer in effect.

In a summary of the meeting for his own files, Groves remarked that the President did not show any concern over the amount of money being spent, but made it clear that he was "in entire agreement with the necessity for the project." Truman approved the idea of a committee to begin developing policy proposals; Stimson was to recruit members.

One cannot help but imagine that Truman must have felt that he had glimpsed merely the tip of an iceberg of staggering complexity that had been developed in secrecy so extreme that even as a Senator and subsequently as Vice-President he had picked up only an inkling of its true magnitude. One cannot also help but wonder if he had some sense that if the project were successful, it could represent an incredible deliverance from the war that he had inherited.

8.4 Advice and Dissent: The Interim Committee, the Scientific Panel, and the Franck Report

Henry Stimson wasted no time in pulling together his advisory committee. On May 2, he was back at the White House with a proposed list of eight members: himself, his aide George Harrison (who would serve as alternate Chair when Stimson could not attend), Undersecretary of the Navy Ralph Bard, Vannevar Bush, James Conant, Karl Compton, Assistant Secretary of State William Clayton, and, as the President's personal representative, soon-to-be Secretary of State James Byrnes. In recognition of the fact that Congress would presumably establish a permanent body to supervise and regulate atomic energy, this group was known as the "Interim Committee." Their first meeting took place on May 9, with Stimson making opening remarks: "Gentlemen, it is our responsibility to recommend action that may turn the course of civilization." The committee's charge was to "to study and report on the entire problem of temporary war-time controls and later publicity, and to survey and make recommendations on post-war research, development, and control, and on legislation necessary for these purposes." For background, the group reviewed Groves' April 23 report.

When Stimson first approached Conant to serve on the committee, the latter suggested that it might be valuable to invite some of the leading scientists to present their views on international relations in the context of the bomb. This suggestion was the first item on the agenda for the committee's second meeting, which was held on May 14. It was agreed to appoint a Scientific Panel whose members were Arthur Compton, Ernest Lawrence, Robert Oppenheimer, and Enrico Fermi. The Panel would be free to advise not only on technical matters, "but also to present to the Committee their views concerning the political aspects of the problem." The balance of the second meeting was taken up with consideration of relations with the British, and development of statements to be made public following the *Trinity* test and eventual use of the bomb; William Laurence of the New York Times (Sects. 3.6 and 7.12) was assigned to work up draft statements. At its third meeting on May 18, the group decided to invite the Scientific Panel to meet with the committee on May 31, 3 days after the last meeting of the Target Committee.

Including a 1-hour lunch break, the May 31 meeting ran from 10:00 a.m. to 4:15 p.m., and was pivotal in the sense of arriving at a "decision" as to how the bomb would be used. The entire committee plus the Scientific Panel was present, as were Generals Groves and Marshall as invited guests. Stimson opened with a statement as to how he viewed the significance of the Project:

> The Secretary expressed the view, a view shared by General Marshall, that this project should not be considered simply in terms of military weapons, but as a new relationship of man to the universe. This discovery might be compared to the discoveries of the Copernican theory and of the laws of gravity, but far more important than these in its effect on the lives of men. While the advances in the field to date had been fostered by the needs of war, it was important to realize that the implications of the project went far beyond the needs of the present war. It must be controlled if possible to make it an assurance of future peace rather than a menace to civilization.

Arthur Compton reviewed the development of the Project, after which the discussion turned to domestic issues. Lawrence felt that research "had to go on unceasingly," that plant expansion had to be pursued, and that a stockpile of bombs and material needed to be built up. All of the members of the Scientific Panel spoke up on the importance of a vigorous post-war research program. As to the issue of controls and inspections, Oppenheimer felt that knowledge of the subject was so widespread that steps should be taken to make American developments known to the world, and that it might be wise for the United States to offer to the world free interchange of information with particular emphasis on the development of peace-time uses. Stimson wondered what kind of inspections might be effective, and what would be the positions of democratic governments versus those of totalitarian regimes under a program of international control coupled with scientific freedom? Vannevar Bush was of the opinion that it would be hard for America to remain permanently ahead if results of research were to be turned over to the Russians with no reciprocal exchange. General Marshall cautioned against putting too much faith in the effectiveness of an inspection proposal, but did suggest that it might be desirable to invite two Russian scientists to witness the *Trinity* test; this idea was apparently not pursued.

The group broke for lunch at 1:15 p.m. No record of the lunch-time conversation was kept, but the idea of giving the Japanese a demonstration of the bomb's power before deploying it in a way that would cause any loss of life was evidently raised; perhaps a test on a remote island might do. This idea has been attributed to both Arthur Compton and Ernest Lawrence; James Byrnes apparently asked for elaboration. In the discussion that followed, it seems that nobody was able to conceive of a demonstration powerful enough to convince the Japanese that continued resistance would be pointless. Other objections were that America would look ridiculous if a demonstration proved to be a dud, and that the Japanese might bring prisoners of war into the demonstration area. In his memoirs, Arthur Compton wrote that "Throughout the morning's discussions it seemed to be a foregone conclusion that the bomb would be used." Discussion on how to use the bomb resumed after lunch; Marshall did not attend the afternoon session. As the minutes recorded (emphasis as in original):

> After much discussion concerning various types of targets and the effects to be produced, *the Secretary expressed the conclusion, on which there was general agreement, that we could not give the Japanese any warning; that we could not concentrate on a civilian area; but that we should seek to make a profound psychological impression on as many of the inhabitants as possible. At the suggestion of Dr. Conant the Secretary agreed that the most desirable target would be a vital war plant employing a large number of workers and closely surrounded by worker's houses.*

As the meeting drew to a close, Harrison remarked that the Scientific Panel was a continuing group that should feel free to present its views to the Committee at any time. In particular, the Committee wished to hear Panel member's thoughts as to what sort of controlling organization should be established. The question arose as to what Panel members were at liberty tell their subordinates about the Committee. It was agreed that they should feel free to relate that the Committee had been appointed by Stimson, and that they (the Panel) had been given complete freedom to present their views on any phase of the subject. The Scientific Panel agreed to meet again at Los Alamos on June 16. Byrnes went directly to the White House to brief President Truman on the Committee's deliberations, and Stimson further discussed the matter with the President on June 6.

Arthur Compton took to heart the notion of soliciting the views of his subordinates. After returning to the Metallurgical Laboratory, he met with a group of senior scientists on June 2, and asked them for input. Various committees were established to consider issues such as research, education, and controls and organization, but it was a group headed by James Franck that was to have the most impact. Franck had shared the 1925 Nobel Prize for Physics with Gustav Hertz, and had emigrated to the United States from Germany in the mid-1930s, settling at the University of Chicago. In the summer of 1945, he was Director of the Met Lab's Chemistry Division.

Franck's committee, which included Glenn Seaborg and Leo Szilard, was to prepare a report on "Political and Social Problems" associated with the bomb. Working over the week of June 4–11, they drafted a document known as the Franck Report, which is now widely acknowledged to be a founding manifesto of the nuclear non-proliferation movement.

While the Franck Report echoed many of the points already made in the Jeffries Report, it added some arguments with a tone of high morality. A few excerpts drawn from the Preamble give the idea:

> The scientists on this Project do not presume to speak authoritatively on the problem of national and international policy. However, we found ourselves, by force of events during the last 5 years, in the position of a small group of citizens cognizant of a grave danger for the safety of this country as well as for the future of all the other nations, of which the rest of mankind is unaware … All of us, familiar with the present state of nucleonics, live with the vision before our eyes of sudden destruction visited on our own country, of a Pearl Harbor disaster repeated in thousand-fold magnification in every one of our major cities.

The next section, "Prospects of Armaments Race," reiterated the conclusion of the Jeffries report and Oppenheimer's argument to the Interim Committee: that knowledge of the fundamental scientific facts of nucleonics was so widespread that it would be foolish to hope that secrecy could protect America for more than a few years. Also, America would be at a significant disadvantage if an arms race did develop, as its population centers and industries tend to be very centralized, as opposed to those in possible enemy countries such as Russia. Not anticipating the yield of the soon-to-be-tested *Trinity* device, they posited that "Ten years hence, it may be that atomic bombs containing perhaps 20 kg of active material can be detonated at 6 % efficiency, and thus each have an effect equal to that of 20,000 t of TNT."

Having developed the argument that nuclear weapons could not be kept secret for long and that an arms race could potentially be disastrous, the group proceeded to their central thesis: "From this point of view, the way in which the nuclear weapons now being secretly developed in this country are first revealed to the world appears to be of great, perhaps fateful importance." Given that the Japanese were still fighting on after many of their cities had been reduced to rubble, the authors felt it doubtful that the first available bombs would be sufficient to break Japan's will to resist. On the other hand, if one were to look forward to an international agreement on the prevention of nuclear warfare, "the military advantages and the saving of American lives achieved by the sudden use of atomic bombs against Japan may be outweighed by the ensuing loss of confidence and by a wave of horror and repulsion sweeping over the rest of the world and perhaps even dividing public opinion at home. *From this point of view, a demonstration of the new weapon might best be made, before the eyes of representatives of all the United Nations, on the desert or a barren island.* ... After such a demonstration the weapon might perhaps be used against Japan if the sanction of the United Nations (and of public opinion at home) were obtained, perhaps after a preliminary ultimatum to Japan to surrender or at least to evacuate certain regions as an alternative to their total destruction."

A brief final section of the Report addressed possible methods of international control, centering on rationing and careful tracking of raw and processed materials. A Summary section then presented a final recommendation:

> To sum up, we urge that the use of nuclear bombs in this war be considered as a problem of long-range national policy rather than of military expediency, and that this policy be directed primarily to the achievement of an agreement permitting an effective international control of the means of nuclear warfare.

Franck hand-delivered the report to Compton in Washington on June 12, asking that he pass it on to Stimson. The latter was not available, but Compton did pass it to Harrison. Compton added his own cover letter to the report, summarizing its essence:

> The proposal is to make a technical but not military demonstration, preparing the way for a recommendation by the United States that the military use of atomic explosives be outlawed by firm international agreement. It is contended that its military use by us now will prejudice the world against accepting any future recommendations by us that its use be not permitted.

Compton did not offer his own thoughts on this position, but added that the report did not address two important considerations: that failure to make a military demonstration of the new bombs might drag out the war and cost more casualties, and that without a military demonstration, it might be impossible to impress the world with the need for national sacrifices in order to gain lasting security. It is not clear if Stimson ever saw the report.

On June 15, Harrison phoned Compton in Los Alamos to ask the Scientific Panel to also consider the question of the immediate use of nuclear weapons at its meeting scheduled for the following day. The Panel's consequent one-page report made three statements. The first was a rather vague recommendation that, before the weapons were used, countries such as Britain, Russia, France and China be informed of their development and be invited to make suggestions as to how "we can cooperate in making this development contribute to improved international relations." The second and third statements get to the nub of the issue, and are worth reproducing in their entirety:

The opinions of our scientific colleagues on the initial use of these weapons are not unanimous: they range from the proposal of a purely technical demonstration to that of the military application best designed to induce surrender. Those who advocate a purely technical demonstration would wish to outlaw the use of atomic weapons, and have feared that if we use the weapons now our position in future negotiations will be prejudiced. Others emphasize the opportunity of saving American lives by immediate military use, and believe that such use will improve the international prospects, in that they are more concerned with the prevention of war than with the elimination of this specific weapon. We find ourselves closer to these latter views; we can propose no technical demonstration likely to bring an end to the war; we see no acceptable alternative to direct military use.

With regard to these general aspects of the use of atomic energy, it is clear that we, as scientific men, have no proprietary rights. It is true that we are among the few citizens who have had occasion to give thoughtful consideration to these problems during the past few years. We have, however, no claim to special competence in solving the political, social, and military problems which are presented by the advent of atomic power.

The Interim Committee met on June 21; Groves was present, but not Stimson or the members of the Scientific Panel. The morning was spent dealing with draft publicity statements and some legal issues. After lunch, the Scientific Panel's report was taken up. Discussion of future policy was left to an eventual "Post-War Control Commission," but as to use of the weapon,

Mr. Harrison explained that he had recently received through Dr. A. H. Compton a report from a group of the scientists at Chicago recommending, among other things, that the weapon not be used in this war but that a purely technical test be conducted which would be made known to other countries. Mr. Harrison had turned this report over to the Scientific Panel for study and recommendation. Part II of the report of the Scientific Panel stated that they saw no acceptable alternative to direct military use. The Committee reaffirmed the position taken at the 31 May and 1 June meetings that the weapons be used against Japan at the earliest opportunity, that it be used without warning, and that it be used on a dual target, namely, a military installation or war plant surrounded by or adjacent to homes or other buildings most susceptible to damage.

The Interim Committee held a number of subsequent meetings, but never revisited the use-versus-demonstration issue.

Despite the reaffirmation of the May 31 decision (the June 1 meeting dealt largely with post-war industrial issues), members the Committee were not monolithic in their thinking. On June 27, Ralph Bard prepared a brief memorandum:

Ever since I have been in touch with this program I have had a feeling that before the bomb is actually used against Japan that Japan should have some preliminary warning for say 2 or 3 days in advance of use. The position of the United States as a great humanitarian nation and the fair play attitude of our people generally is responsible in the main for this feeling. During recent weeks I have also had the feeling very definitely that the Japanese government may be searching for some opportunity which they could use as a medium of surrender ... emissaries from this country could contact representatives from Japan ... and ... give them some information regarding the proposed use of atomic power, together with whatever assurances the President might care to make with regard to the Emperor ... The stakes are so tremendous that it is my opinion very real consideration should be given to some plan of this kind ...

Harrison passed Bard's memo on to Stimson and Byrnes, and Bard secured an interview with President Truman, during which he tried to argue that naval blockade would make an invasion unnecessary. Truman assured him that the questions of invasion and offering a warning had received careful attention.

On July 2, two weeks before the *Trinity* test, Henry Stimson sent President Truman a three-page memorandum titled "Proposed Program for Japan." Recognizing that an invasion of Japan would almost certainly lead to a costly, drawn-out battle which would leave that country destroyed, Stimson raised the question of whether some alternative could be proposed that would avoid an invasion while securing the equivalent of unconditional surrender. In particular, Stimson suggested that a warning which made clear that the Allies did not desire to destroy Japan as a nation, coupled with a policy of not excluding a constitutional monarchy, might improve the chances of success. Japan's situation was desperate: she had no allies, her Navy was effectively destroyed, she was vulnerable to air attack, the rising force of China was against her, the threat of Russia loomed, and America had the industrial capacity to continue the war and the "moral superiority through being the victim of her first sneak attack." The memorandum made no mention of atomic bombs. Many of Stimson's suggestions would appear in the Potsdam Declaration just over 3 weeks later, but not the clause regarding a constitutional monarchy. It may well be that the Japanese response would have been the same; the faction within the Japanese government that sought peace could not yet point to the specter of further atomic bombings to bolster their position.

If political statements were being formulated, Leo Szilard was certain to be a center of activity. Convinced that Project hierarchy stifled any real avenue for making known his concern that an arms race would be inevitable if no international control agreement was reached, Szilard decided to attempt another direct approach to the President. In early March, 1945, he drafted a memorandum titled "Atomic Bombs and the Postwar Position of the United States in the World," wherein he argued that if a control agreement with Russia could not be achieved, America would be forced to engage in a costly arms race, and that the greatest danger might be the outbreak of a "preventative war." Szilard finished his memo on March 12, and decided to again enlist Albert Einstein to prepare a letter of introduction. Szilard traveled to Princeton, where Einstein obliged him with a one-page letter dated March 25. Secrecy forbade Szilard from disclosing the contents of his memorandum (Einstein knew little of the details of the Project); Einstein summarized the issue by writing that "I understand … he is now greatly concerned about the lack of adequate contact between scientists who are doing this work and those members of your cabinet who are responsible for formulating policy," and asked Roosevelt to give Szilard's presentation his personal attention.

Szilard dispatched a copy of Einstein's letter to Mrs. Roosevelt, who replied in early April with a proposal that Szilard meet with her in New York on May 8. But before that date arrived, President Roosevelt died (April 12), and Szilard found himself in limbo. Ingeniously, he found an employee at the Met Lab, mathematician Albert Cahn, who had some political connections in President Truman's home town of Kansas City. Cahn managed to secure an appointment at the White

House for Friday, May 25. Szilard traveled to Washington with Cahn and University of Chicago Dean of Science Walter Bartky, but they were redirected by the President's Appointments Secretary to meet with James Byrnes, who was then living in South Carolina. Szilard and Bartky, now accompanied by Harold Urey, proceeded by train to South Carolina (tailed by some of Groves' agents), where they met with Byrnes on May 28, the day that the last Target Committee meeting was underway in Washington. The meeting was a disaster: Byrnes was not happy with Szilard's attempt to interfere in policy-making, and Szilard felt that Byrnes completely failed to grasp the significance of atomic energy.

Not to be deterred, Szilard moved on to his next tactic: a direct petition to the new President. The first version of his petition, dated July 3 and signed by Szilard and 58 others, expressed the opinion that atomic bombing of Japan could not be justified in the present circumstances, and that atomic bombs were primarily a means for the "ruthless annihilation" of cities. The signers reminded the President that in his hands lay the fateful decision of whether or not to use these bombs, and argued that "Thus a nation which sets the precedent of using these newly liberated forces of nature for purposes of destruction may have to bear the responsibility of opening the door to an era of devastation on an unimaginable scale." The text closed with a plea that the President exercise his power as Commander-in-Chief to rule that the United States not, "in the present phase of the war," resort to the use of atomic bombs.

Perhaps through Compton, word of Szilard's activity reached Oak Ridge. Kenneth Nichols asked Compton to poll his colleagues' attitudes on use of the bomb. Compton delegated the task to Farrington Daniels, formal Director of the Met Lab. Five options were offered (paraphrased):

1. Use the weapons in the most effective military manner;
2. Give a demonstration in Japan followed by an opportunity to surrender before full use of the weapon is employed;
3. Perform a demonstration within the United States with Japanese representatives present;
4. Withhold military use of the weapons but make public experimental demonstration of their effectiveness;
5. Maintain as secret as possible all developments of the new weapons and refrain from using them in the present war.

Responses were received from 150 of approximately 250 employees; Daniels reported the results on July 13. The distribution of votes was 23, 69, 39, 16, and 3 (15, 46, 26, 11, 2 %). At the level of destruction caused by a nuclear weapon, the distinction between options (1) and (2) is not clear, but it is evident that over half of the respondents felt that some direct use of the bomb against Japan was appropriate. In the meantime, Szilard re-drafted his petition, producing a second version on July 17—the day after *Trinity*—which garnered 69 co-signers. This version dropped the "ruthless annihilation" phrase of the original, but added a moral dimension:

The added material strength which this lead gives to the United States brings with it the obligation of restraint and if we are to violate this obligation our moral position would be weakened in the eyes of the world and in our own eyes. It would then be more difficult for us to live up to our responsibility of bringing the unloosened forces of destruction under control.

Szilard handed the petition to Compton on July 19 with a request that it be forwarded to the President. Compton instead sent the petition and the results of the poll to Nichols, who passed them on to Groves. Groves held on to them until an August 1 meeting with Stimson, after which George Harrison filed them with his papers; the President apparently never saw the petition. Groves' action may seem high-handed, but the scientists had had their chance for input through the Scientific Panel of the Interim Committee. By the beginning of August, the 509th Composite Group's orders had been approved by the President, and the full machinery of preparations for the bombing missions was in motion.

The question of whether a demonstration shot should have been carried out continues to be debated. Rudolf Peierls offered an assessment in his memoirs:

To me the obvious answer would have been to drop a bomb on a sparsely populated area to show its effects, coupled with an ultimatum to the Japanese government to avoid a large-scale nuclear attack. This would have involved killing some people and destroying some buildings, since otherwise the power of the bomb would not have been obvious; the effects visible after the Alamogordo test were frightening to the expert but not impressive to the layman. Of course such an ultimatum might have failed, but at least it would have been an attempt to avoid unnecessary casualties. … My regrets are that we did not insist on more dialogue with the military and political leaders, based on full and clear scientific discussions of the consequences of possible courses of action. It is not clear, of course, that such discussions would have made any differences in the end.

This author of this book has so far refrained from stating opinions on political issues. With the caveat that retrospection is easy, I offer the following purely personal comments on the issues of international control and a demonstration shot. The notion that other nations, Russia in particular, would be happy to settle for America being armed with nuclear weapons while some sort of control system was worked out seems to me untenable; the arms race was born with the creation of the Manhattan District, if not the discovery of nuclear fission itself. The idea that Russia or America (or any other nation, for that matter) would be willing to subject itself to invasive scrutiny from some newly-constituted international "agency" seems equally dubious. The proposal of a demonstration shot, while sincere and humanely conceived, seems to me to be fraught with more problems than advantages. Fissile material, obtained at great expense and effort, was limited; why should a good fraction of it be spent in an effort that might be interpreted by a mortal enemy as a sign of vacillation? I believe that the use of the bomb to hasten the end of the war and establish the strategic position of America in the postwar world was implicit in its development, and fully justified. Had the bombs *not* been used and the consequences of nuclear combat so starkly demonstrated to the world, what much worse horrors might have unfolded in a subsequent war? Finally, the momentum that the Project had acquired by the summer of 1945 was practically unstoppable. President Truman did not make a "hard decision" to use the bomb so much as he elected not to alter a chain of events that was already far along when he inherited the Presidency. Indeed, as described in the following section, the decision of when to end the war lay largely in the hands of the Japanese cabinet.

8.5   The Bombing Missions

When President Truman approved the Handy/Groves orders of July 25 (Sect. 7.14) and replied to Henry Stimson's request for permission to prepare public statements for release, the last formal high-level authorizations for deployment of atomic bombs against Japan were completed. The intricate program that General Groves had developed over the preceding 3 years to design, develop, and deliver a revolutionary new weapon was about to come to fruition.

In the Pacific, August 1, 1945, saw various organizational changes come into effect. General LeMay moved up to become Chief of Staff to General Spaatz, and the Twenty-First Bomber Command and the Twentieth Air Force came under the command of Lieutenant General Nathan Twining. Thus it came to be that Twentieth Air Force Field Order number 13, issued on August 2, was over Twining's signature. The orders specified Hiroshima, Kokura Arsenal, and Nagasaki as the primary, secondary, and tertiary targets. Niigata had been scratched for being too far away from the other targets. Hiroshima had been bombed on May 7 and June 2, but the bombs had fallen ineffectively in the Ota river. Nagasaki had been the target of two bombing raids, on July 22 and August 1.

The weather for the first few days of August was overcast and rainy, but on Saturday, August 4, Commander Parsons was informed that the forecast was improving. At 4:00 p.m. that afternoon, 509th Composite Group flight crews were given their first briefing. Tibbets opened the briefing, telling his men that what they had trained for was at hand, but he did not reveal the nature of their payload. He then introduced Parsons, who attempted to show a film of the *Trinity* test. The projector jammed and chewed up the film, so Parsons could give only a verbal description of the test. He began his comments with "The bomb you are going to drop is something new in the history of warfare. It is the most destructive weapon ever produced. We think it will knock out everything within a three mile area."

General LeMay authorized the mission order on August 5 (Fig. 8.5). At the local level, this took the form of 509th Operations Order number 35, dated the same day. The mission called for sorties by seven aircraft, identified by their "Victor" numbers. V-82, the *Enola Gay*, to be piloted by Paul Tibbets, was the "strike" plane—the one which carried the bomb (notice that the type of bombs is indicated on the order as "Special"). Victors 83, 71, and 85 were weather planes, directed toward Nagasaki, Kokura, and Hiroshima, respectively, and which were to depart an hour before the strike planes. Victors 89 and 91 carried blast-measurement instruments and high-speed cameras. Victor 90 was deployed to Iwo Jima as a backup for the *Enola Gay*. Victors 72 and 88 did not fly the Hiroshima mission.

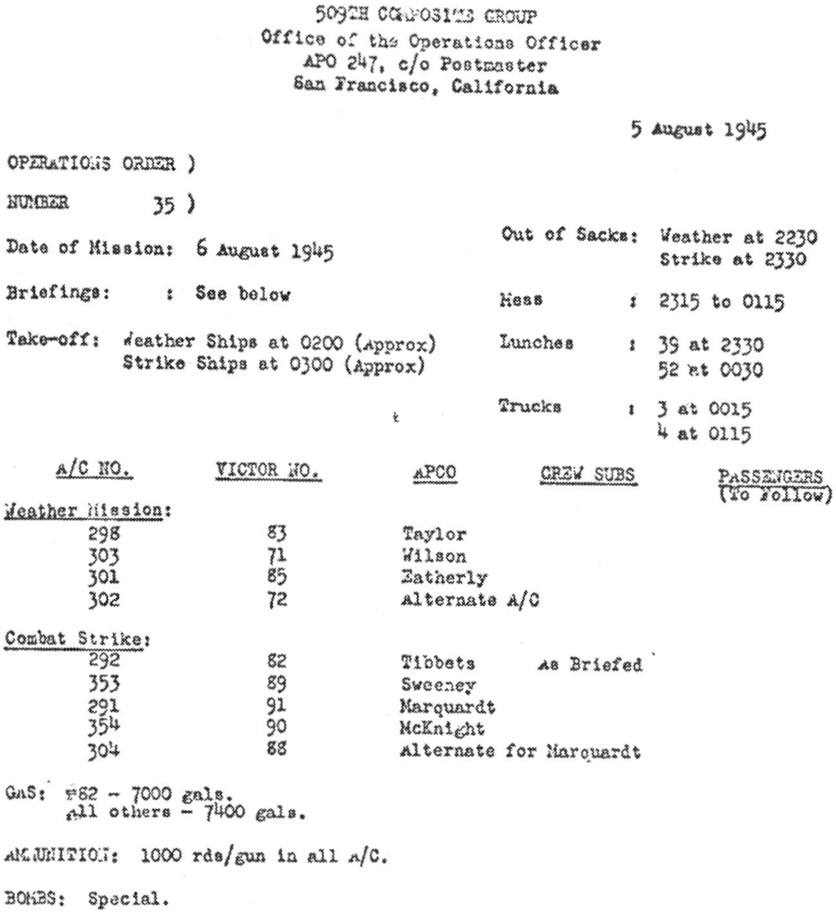

```
                              509TH COMPOSITE GROUP
                           Office of the Operations Officer
                           APO 247, c/o Postmaster
                           San Francisco, California

                                                            5 August 1945

OPERATIONS ORDER )

NUMBER        35 )

Date of Mission:  6 August 1945          Out of Sacks:  Weather at 2230
                                                        Strike at 2330

Briefings:     :  See below             Mess         :  2315 to 0115

Take-off:  Weather Ships at 0200 (Approx)  Lunches    :  39 at 2330
           Strike Ships at 0300 (Approx)                52 at 0030

                                          Trucks       :  3 at 0015
                                                         4 at 0115
```

A/C NO.	VICTOR NO.	APCO	CREW SUBS	PASSENGERS (To Follow)
Weather Mission:				
298	83	Taylor		
303	71	Wilson		
301	85	Eatherly		
302	72	Alternate A/C		
Combat Strike:				
292	82	Tibbets	As Briefed	
353	89	Sweeney		
291	91	Marquardt		
354	90	McKnight		
304	88	Alternate for Marquardt		

```
GAS:   #62 - 7000 gals.
       All others - 7400 gals.

AMMUNITION:  1000 rds/gun in all A/C.

BOMBS:  Special.
```

Fig. 8.5 Partial reproduction of Hiroshima mission operational orders. *Source* http://www.lanl.gov/history/admin/files/509th_Composite_Group.JPG

Hiroshima is located on the delta of the Ota river in the southern part of Honshu, the main island of Japan. The river breaks into channels which divide the city into islands, giving it a distinctive fingered appearance as seen from above (Fig. 8.6). Before the war, Hiroshima was the seventh-largest city in Japan, with a population of about 340,000. Its population in August, 1945, has been estimated at some 280,000 civilians plus approximately 43,000 soldiers, although some estimates have put the number somewhat lower; many civilians had been evacuated, but a number of troops and workers had been brought into the city. Flat and unbroken by hills, Hiroshima was a perfect target for determining the effects of the new weapon.

Little Boy was wheeled out of its assembly building at 2:00 p.m. on Sunday afternoon. By 2:30, it had arrived at the loading pit, into which it was lowered so

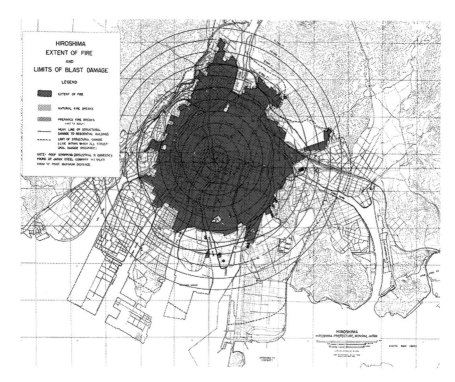

Fig. 8.6 United States strategic bombing survey map of Hiroshima atomic bomb damage. The darkened area shows the extent of fire damage. The *curved solid line* is the mean line of structural damage to residential buildings, and the *dashed line* is the limit of structural damage. The *circles* are in 1000-foot increments from ground zero out to 11,000 feet. *Source* http://commons.wikimedia.org/wiki/File:Hiroshima_Damage_Map.gif

that the *Enola Gay* could be backed over it. The plane was in position by 3:00, and loading was complete by 3:45. Fusing checks were completed by 5:45, and a final inspection made at 6:45. Tibbets had the words *Enola Gay* painted on the left-side nose of the airplane, and guards were posted to prevent any tampering (Fig. 8.7).

On Saturday, General Farrell had informed Groves by cable that the *Enola Gay* should take off at approximately noon on Sunday, Washington time (Washington was 14 hours behind Tinian; this would be equivalent to 2:00 a.m. Monday, Tinian time.) Far from Groves' reach, Parsons decided that he would arm the bomb in flight, and spent Sunday afternoon practicing the procedure. After the bomb had been loaded, he practiced again in the cramped confines of the bomb bay. Not until Sunday evening Tinian time did Farrell cable Groves with the change in plan—too late for Groves to interfere. Final crew briefings began at 11:00 p.m.

The three weather planes began departing at 1:37 a.m., about an hour before the strike and observation planes. The weather crews missed the show back at Tinian that began at 2:00 as *Enola Gay* was floodlit and camera crews began filming; Groves wanted the mission recorded for posterity. Norman Ramsey compared the

Fig. 8.7 *Little boy* in its loading pit. *Source* http://commons.wikimedia.org/wiki/File:Atombombe_Little_Boy_2.jpg

scene to a Hollywood premiere; one scientist allegedly compared it to the opening of a drugstore. Harlow Russ, who had helped engineer the implosion mechanism, estimated the crowd at about 350. Tibbets began the *Enola Gay's* takeoff roll at 2:45 a.m. Tinian time, Monday, August 6, using practically every yard of the two-mile runway to get airborne (Fig. 8.8). The instrument, photo, and backup planes followed at 2 min intervals. In Washington, the time was 12:45 p.m. on Sunday afternoon. Table 8.2 lists the crews of the Hiroshima and Nagasaki strike planes, and Table 8.3 some of the parameters of the missions.

Fifteen minutes after takeoff, Parsons and Second Lieutenant Morris Jeppson crawled into the bomb bay to begin the arming procedure. Jeppson held a flashlight and handed Parsons tools as the latter worked through his 10-step checklist:

Fig. 8.8 *Enola Gay* on Tinian. *Source* http://commons.wikimedia.org/wiki/File:050607-F-1234P-090.jpg

Table 8.2 Hiroshima and Nagasaki Strike Crews

Position	Hiroshima	Nagasaki
Commander	Paul Tibbets 1915–2007	Charles Sweeney 1919–2004
Pilot	Robert Lewis 1917–1983	Don Albury 1920–2009
Co-Pilot		Fred Olivi 1922–2004
Navigator	Theodore Van Kirk 1921–	James Van Pelt 1918–1994
Bombardier	Thomas Ferebee 1918–2000	Kermit Beahan 1918–1989
Bomb commander	William Parsons 1901–1953	Frederick Ashworth 1912-2005
Electronic countermeasures	Jacob Beser 1921–1992	Jacob Beser
Electronics test officer	Morris Jeppson 1922–2010	Philip Barnes 1917–1998
Flight Engineer	Wyatt Duzenbury 1913–1992	John Kuharek 1914–2001
Assistant Engineer	Robert Shumard 1920–1967	Ray Gallagher 1921–1999
Radio operator	Richard Nelson 1925–2003	Abe Spitzer 1912–1984
Radar operator	Joseph Stiborik 1914–1984	Edward Buckley 1913–1981
Tail gunner	George Caron 1919–1995	Albert Dehart 1915–1976

Source Campbell 30, 32

Table 8.3 Some parameters of the Hiroshima and Nagasaki missions

Parameter	Hiroshima	Nagasaki
Strike aircraft	*Enola Gay*	*Bockscar*
Takeoff (Tinian time)	02:45 Aug 6	03:48 Aug 9
Takeoff (Washington time)	12:45 Aug 5	13:48 Aug 8
Bombing (Japan time)	08:15 Aug 6	11:08 Aug 9
Bombing (Washington time)	19:15 Aug 5	22:08 Aug 8
Landing (Tinian time)	14:58 Aug 6	23:06 Aug 9
Landing (Washington time)	00:58 Aug 6	09:06 Aug 9
Mission duration	12 h 13 min	19 h 18 min
Drop height (ft/m)	31,600/9,630	28,900/8,810
Bomb detonation height (ft/m)	1,900/580	1,650/503
Bomb yield (kt)	~ 15	~ 21

Sources Coster-Mullen, 39, 326; Campbell 31–34; Los Alamos report LA-8819. Mission time for Bockscar includes 3-h stop at Okinawa

1. Check that green plugs are installed.
2. Remove rear plate.
3. Remove armor plate.
4. Insert breech wrench in breech plug.
5. Unscrew breech plug, place on rubber pad.
6. Insert charge, 4 sections, red end to breech.
7. Insert breech plug and tighten home.
8. Connect firing line.
9. Install armor plate.
10. Remove and secure catwalk and tools.

In step 1, the "green plugs" were three "safing" plugs that isolated the firing system of the bomb from its batteries; Jeppson would later replace these with red-colored "live" plugs. The entire procedure took about 20 minutes.

At some point not long into the flight, Tibbets went on the plane's intercom system to inform his men that they were carrying the world's first combat atomic bomb. At the request of *New York Times* reporter William Laurence, who was disappointed that he was not allowed to fly as an observer, co-pilot Robert Lewis kept a journal, which, in 1971, would be auctioned for $37,000. Laurence got his wish on the Nagasaki mission, when he flew on the instrument plane.

About 3 hours after takeoff, *Enola Gay* rendezvoused at Iwo Jima with the camera and instrument planes, *Number 91* and *The Great Artiste*. (After the atomic missions, *Number 91* would be dubbed *Necessary Evil*). One of the crew members on *The Great Artiste*, Lawrence Johnston, is believed to have witnessed all three of the *Trinity*, Hiroshima, and Nagasaki explosions. Johnston was a student of Luis Alvarez who had joined Los Alamos in May, 1944, to work on detonators for the implosion device.

Parsons kept a log of the mission. In terse, unadorned words, it narrated the progress of what would prove to be a textbook operation. (Events in brackets were not in Parson's original log, but have been added here for completeness. All times are Tinian time; subtract one hour for Japan time, and subtract 14 hours for Washington time. All events occurred on August 5, Washington time):

02:45   Take off
03:00   Started final loading of gun
03:15   Finished loading
05:52   (Approach Iwo Jima. Begin climb to 9,300 feet)
06:05   Headed for Empire from Iwo
07:30   Red plugs in

After Jeppson had installed the red arming plugs, the bomb was "live." In his journal, Robert Lewis wrote "The bomb is now independent of the plane. It was a peculiar sensation. I had a feeling the bomb had a life of its own now that had nothing to do with us." Jeppson kept one of the green safing plugs and a spare red live plug as souvenirs; they sold at auction in 2002 for $167,000.

As the *Enola Gay* approached Hiroshima, Lewis added to his journal: "There'll be a short intermission while we bomb our target." Resuming with Parsons' log:

07:41   Started climb. Weather report received that weather over primary and tertiary targets was good but not over secondary target
08:25   (Weather plane—cloud cover less than 3/10 at all altitudes. Advice: bomb primary)
08:38   Leveled off at 32,700 feet
08:47   All Archies tested to be OK
09:04   Course west
09:09   Target (Hiroshima) in sight
09:12   (Initial point)
09:14   (Glasses on)
09:15 ½   Dropped bomb. Flash followed by two slaps on plane. Huge cloud
10:00   Still in sight of cloud which must be over 40,000 feet high

Fig. 8.9 Hiroshima
mushroom cloud. *Source*
http://commons.wikimedia.
org/wiki/File:Atomic_
cloud_over_Hiroshima.jpg

10:03	Fighter reported
10:41	Lost sight of cloud 363 miles from Hiroshima with aircraft being 26,000 feet high
14:58	Landed at Tinian

Little Boy free-fell for about 43 seconds before detonating (Fig. 8.9). Bombardier Thomas Ferebee's aiming point was the distinctive T-shaped Aioi bridge in the heart of the city; he missed by only a few hundred feet. Van Kirk's navigation had been flawless. The scheduled time for the drop was 09:15; after a flight of eight and one-half hours, *Enola Gay* arrived at its target only seconds behind schedule. Figure 8.10 shows a post-strike photo; the Aioi bridge, which survived, is clearly visible in the center of the left image.

Tibbets executed his escape maneuver, and then turned south to permit the crew to observe the city for a couple minutes before setting course back to for Tinian. As thousands suffered below, Robert Lewis wrote "My God, what have we done?" He was later quoted as saying "If I live a hundred years, I'll never quite get these few minutes out of my mind."

Fig. 8.10 *Left* Aerial view of Hiroshima, post-bombing. The Aioi bridge is in the center of the image. *Source* http://commons.wikimedia.org/wiki/File:AtomicEffects-p7a.jpg. *Right* General view of damage at Hiroshima. Source: http://commons.wikimedia.org/wiki/File:AtomicEffects-Hiroshima.jpg

In Washington, Groves expected to hear by about 2:00 p.m. that the *Enola Gay* had taken off, but communications were delayed. To work off his nervous energy he went for a game of tennis, and then had dinner with his family. Finally at about 6:45 p.m. the call came through that the plane had taken off; by that time *Enola Gay* had climbed to her bombing altitude and was approaching Hiroshima. Groves returned to his office, where he intended to spend the night. In his memoirs, Groves described how he abandoned his usual formality: "In order to ease the growing tension in the office, I made a point of taking off my tie, opening up my collar and rolling up my sleeves."

Immediately after the drop, Parsons sent Groves a brief coded message, which finally arrived about 11:30 p.m. Washington time, more than four hours after the bombing:

> Results clearcut, successful in all respects. Visible effects greater than New Mexico tests. Conditions normal in airplane following delivery.
> Target at Hiroshima attacked visually. One-tenth cloud at 052315Z. No fighters and no flak.

By the time Groves received Parsons' message, *Enola Gay* was only ninety minutes from returning to Tinian. The 052315Z in Parsons' message means August 5, 23:15 Greenwich time, or 7:15 p.m. Sunday evening in Washington. Groves promptly informed General Marshall of the message, and before going to bed on a cot in his office prepared a rough draft report to be delivered to Marshall in the morning.

Enola Gay landed at Tinian at about 1:00 a.m., Washington time. Tibbets was immediately decorated with a Distinguished Service Cross by General Spaatz; Parsons was later awarded a Silver Star. Farrell sent Groves a lengthier cable:

Following additional information furnished by Parsons, crews, and observers on return to Tinian at 060500Z. Report delayed until information could be assembled at interrogation of crews and observers. Present at interrogation were Spaatz, Giles, Twining, and Davies.

Confirmed neither fighter or flak attack and one tenth cloud cover with large open hole directly over target. High speed camera reports excellent record obtained. Other observing aircraft also anticipates good records although films not yet processed. Reconnaissance aircraft taking post-strike photographs have not yet returned.

Sound—None appreciable observed.

Flash—Not so blinding as New Mexico test because of bright sunlight. First there was a ball of fire changing in a few seconds to purple clouds and flames boiling and swirling upward. Flash observed just after airplane rolled out of turn. All agreed light was intensely bright and white cloud rose faster than New Mexico test, reaching thirty thousand feet in minutes it was one-third greater in diameter.

It mushroomed at the top, broke away from column and the column mushroomed again. Cloud was most turbulent. It went at least to forty thousand feet. Flattening across its top at this level. It was observed from combat airplanes three hundred sixty-three nautical miles away with airplane at twenty-five thousand feet. Observation was then limited by haze and not curvature of the earth.

Blast—There were two distinct shocks felt in combat airplane similar in intensity to close flak bursts. Entire city except outermost ends of dock areas was covered with a dark grey dust layer which joined the cloud column. It was extremely turbulent with flashes of fire visible in the dust. Estimated diameter of this dust layer is at least three miles. One observer stated it looked as though whole town was being torn apart with columns of dust rising out of valleys approaching the town. Due to dust visual observation of structural damage could not be made.

Parsons and other observers felt this strike was tremendous and awesome even in comparison with New Mexico test. Its effects may be attributed by the Japanese to a huge meteor.

Farrell's message reached Groves about 4:30 a.m. The two shocks felt in the plane were due to the direct shock wave of the explosion, and the reflection of the shock wave from the ground. Groves revised his report to Marshall, and was at the latter's office by 7:00 a.m.

Unfortunately, when film from the camera plane came back from being developed, half of the emulsion was gone; it was never determined whether any images had been recorded. The brief films one sees of these explosions were taken by crew-members with hand-held cameras; in the case of the Hiroshima mission, Los Alamos scientist Harold Agnew, riding aboard *The Great Artiste*, filmed the explosion. Reconnaissance planes found that Hiroshima was still mostly covered by the cloud created by the explosion, although fires could be seen around the edges; clearer images would have to wait until the next day.

The world learned of the bombing when President Truman's pre-authorized statement was released in Washington at 11:00 a.m.; Truman was still at sea on his way home from Potsdam, and would not arrive until the evening of the seventh. The text of the release read as

Sixteen hours ago an American airplane dropped one bomb on Hiroshima, an important Japanese Army base. That bomb had more power than 20,000 t of T.N.T. It had more than two thousand times the blast power of the British "Grand Slam" which is the largest bomb ever yet used in the history of warfare.

The Japanese began the war from the air at Pearl Harbor. They have been repaid many fold. And the end is not yet. With this bomb we have now added a new and revolutionary increase in destruction to supplement the growing power of our armed forces. In their present form these bombs are now in production and even more powerful forms are in development.

It is an atomic bomb. It is a harnessing of the basic power of the universe. The force from which the sun draws its power has been loosed against those who brought war to the Far East.

Before 1939, it was the accepted belief of scientists that it was theoretically possible to release atomic energy. But no one knew any practical method of doing it. By 1942, however, we knew that the Germans were working feverishly to find a way to add atomic energy to the other engines of war with which they hoped to enslave the world. But they failed. We may be grateful to Providence that the Germans got the V-1's and V-2's late and in limited quantities and even more grateful that they did not get the atomic bomb at all.

The battle of the laboratories held fateful risks for us as well as the battles of the air, land and sea, and we have now won the battle of the laboratories as we have won the other battles.

Beginning in 1940, before Pearl Harbor, scientific knowledge useful in war was pooled between the United States and Great Britain, and many priceless helps to our victories have come from that arrangement. Under that general policy the research on the atomic bomb was begun. With American and British scientists working together we entered the race of discovery against the Germans.

The United States had available the large number of scientists of distinction in the many needed areas of knowledge. It had the tremendous industrial and financial resources necessary for the project and they could be devoted to it without undue impairment of other vital war work. In the United States the laboratory work and the production plants, on which a substantial start had already been made, would be out of reach of enemy bombing, while at that time Britain was exposed to constant air attack and was still threatened with the possibility of invasion. For these reasons Prime Minister Churchill and President Roosevelt agreed that it was wise to carry on the project here. We now have two great plants and many lesser works devoted to the production of atomic power. Employment during peak construction numbered 125,000 and over 65,000 individuals are even now engaged in operating the plants. Many have worked there for two and a half years. Few know what they have been producing. They see great quantities of material going in and they see nothing coming out of these plants, for the physical size of the explosive charge is exceedingly small. We have spent two billion dollars on the greatest scientific gamble in history—and won.

But the greatest marvel is not the size of the enterprise, its secrecy, nor its cost, but the achievement of scientific brains in putting together infinitely complex pieces of knowledge held by many men in different fields of science into a workable plan. And hardly less marvelous has been the capacity of industry to design, and of labor to operate, the machines and methods to do things never done before so that the brain child of many minds came forth in physical shape and performed as it was supposed to do. Both science and industry worked under the direction of the United States Army, which achieved a unique success in managing so diverse a problem in the advancement of knowledge in an amazingly short time. It is doubtful if such another combination could be got together in the world. What has been done is the greatest achievement of organized science in history. It was done under high pressure and without failure.

We are now prepared to obliterate more rapidly and completely every productive enterprise the Japanese have above ground in any city. We shall destroy their docks, their factories, and their communications. Let there be no mistake; we shall completely destroy Japan's power to make war.

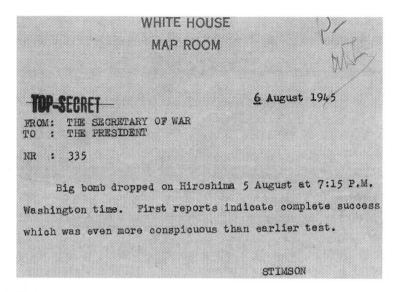

Fig. 8.11 President Truman is informed of the Hiroshima bombing. *Source* http://presidential libraries.c-span.org/Content/Truman/bomb.pdf

It was to spare the Japanese people from utter destruction that the ultimatum of July 26 was issued at Potsdam. Their leaders promptly rejected that ultimatum. If they do not now accept our terms they may expect a rain of ruin from the air, the like of which has never been seen on this earth. Behind this air attack will follow sea and land forces in such numbers and power as they have not yet seen and with the fighting skill of which they are already well aware.

The Secretary of War, who has kept in personal touch with all phases of the project, will immediately make public a statement giving further details.

His statement will give facts concerning the sites at Oak Ridge near Knoxville, Tennessee, and at Richland near Pasco, Washington, and an installation near Santa Fe, New Mexico. Although the workers at the sites have been making materials to be used in producing the greatest destructive force in history they have not themselves been in danger beyond that of many other occupations, for the utmost care has been taken of their safety.

The fact that we can release atomic energy ushers in a new era in man's understanding of nature's forces. Atomic energy may in the future supplement the power that now comes from coal, oil, and falling water, but at present it cannot be produced on a basis to compete with them commercially. Before that comes there must be a long period of intensive research.

It has never been the habit of the scientists of this country or the policy of this Government to withhold from the world scientific knowledge. Normally, therefore, everything about the work with atomic energy would be made public.

But under present circumstances it is not intended to divulge the technical processes of production or all the military applications, pending further examination of possible methods of protecting us and the rest of the world from the danger of sudden destruction.

I shall recommend that the Congress of the United States consider promptly the establishment of an appropriate commission to control the production and use of atomic power within the United States. I shall give further consideration and make further recommendations to the Congress as to how atomic power can become a powerful and forceful influence towards the maintenance of world peace.

Truman's 20,000 t was an overestimate, probably caused by confusing *Little Boy* with the *Trinity* test. The War Department release referred to was considerably longer, and included details regarding the manufacturing plants, some of the contractors and universities involved, the cost of the project, and the existence of the Interim Committee.

Henry Stimson dispatched a message to President Truman, who received it while he was having lunch aboard the *USS Augusta* (Fig. 8.11).

At 2:00 p.m. Washington time, Groves telephoned Oppenheimer to extend his congratulations. A partial transcript of their conversation:

Groves: I'm very proud of you and all of your people.
Oppenheimer: It went alright?
Groves: Apparently it went with a tremendous bang.
Oppenheimer: When was this, was it after sundown?
Groves: No, unfortunately it had to be in the daytime on account of security of the plane and that was left in the hands of the Commanding General over there and he knew what the advantages were of doing it after sundown and he was told just all about that and I said it was up to him; that it was not paramount but that it was very desirable.
Oppenheimer: Right. Everybody is feeling reasonably good about it and I extend my heartiest congratulations. It's been a long road.
Groves: Yes, it has been a long road and I think one of the wisest things I ever did was when I selected the director of Los Alamos.
Oppenheimer: Well, I have my doubts, General Groves.
Groves: Well, you know I've never concurred with those doubts at any time.

At Los Alamos that evening, a crowd gathered in the auditorium. As related by physicist Sam Cohen:

That evening we gathered, long before the appointed time of Oppenheimer's appearance .... Normally at one of these colloquia Oppenheimer, more or less punctual, would walk unobtrusively onstage from a wing, quiet down the audience, make a few remarks in his low-key manner and introduce the speaker. But that was not to be the case on this heroic day: He was late, very late. He did not casually slip onstage from a wing. He came in from the rear of the theatre, strode down the aisle and up the stairs onto the stage, and he made no effort to quiet a yelling, clapping, foot-stomping bunch of scientists who began to cheer him when he entered and continued to do so long after he got onstage.

Now, keep in mind that while this pandemonium was going on, about seventy thousand Japanese civilians lay dead in Hiroshima, with an equal number injured. About 30 % of the victims had received lethal or injurious does of nuclear radiation .... Most of the scientists were, or should have been, very much aware that radiation would take a terrible toll, but at this moment of triumph they couldn't have cared less about any particular moral transgression associated with it. They were flushed with their success and they showed it. And I was one of them.

Finally Oppenheimer was able to quiet the howling crowd and he began to speak, hardly in low key. It was too early to determine what the results of the bombing might have been, but he was sure that the Japanese didn't like it. More cheering. He was proud, and he showed it, of what we had accomplished. Even more cheering. And his only regret was that we hadn't developed the bomb in time to have used it against the Germans. This practically raised the roof.

As the implications seeped in over the following hours and days, the reaction at Los Alamos was by no means one of unrestrained celebration. Alice Smith, wife of metallurgist Cyril Smith, described the atmosphere:

> As the days passed the revulsion grew, bringing with it—even for those who believed that the end of the war justified the bombing—an intensely personal experience of the reality of evil. It was this, and not a feeling of guilt in the ordinary sense, that Oppenheimer meant by his much quoted, and often misunderstood, remark that scientists had known sin.

McAllister Hull, who cast implosion lenses (Sect. 7.11):

> I do not fault Truman's decision to use the bombs, for he was accountable for every Allied casualty he had a means to prevent. I had no such responsibility. I just wish he—or we—had found a way to use them to stop the war immediately without making those of us who had worked on them accessory to several hundred thousand deaths—and scarring wounds to thousands more—in Hiroshima and Nagasaki. I do not know about my friends, but I have never for a moment forgotten that responsibility.

Following the bombing, some six million leaflets were dropped over 47 Japanese cities, encouraging ordinary citizen to pressure the Emperor and ruling militarists to end the war. Ironically, Nagasaki did not receive its quota of leaflets until after it was bombed. The text read

> To the Japanese People: America asks that you take immediate heed of what we say on this leaflet.
>
> We are in possession of the most destructive explosive ever devised by man. A single one of our newly developed atomic bombs is actually the equivalent in explosive power to what 2000 of our giant B-29s can carry on a single mission. This awful fact is one for you to ponder and we solemnly assure you it is grimly accurate.
>
> We have just begun to use this weapon against your homeland. If you still have any doubt, make inquiry as to what happened to Hiroshima when just one atomic bomb fell on that city.
>
> Before using this bomb to destroy every resource of the military by which they are prolonging this useless war, we ask that you now petition the Emperor to end the war. Our president has outlined for you the thirteen consequences of an honorable surrender. We urge that you accept these consequences and begin the work of building a new, better and peace-loving Japan.
>
> You should take steps now to cease military resistance. Otherwise, we shall resolutely employ this bomb and all our other superior weapons to promptly and forcefully end the war.

The Japanese government was not yet ready to quit, but its situation was becoming more perilous by the hour. At 5:00 p.m. local time on the afternoon of August 8, the Japanese ambassador in Moscow was informed that as of August 9, the Soviet Union would consider itself in a state of war with Japan. Five time zones to the east, it was already 10:00 p.m., and Russian forces were advancing in Manchuria. The Japanese government had been hoping to use Russia as a go-between in surrender negotiations, but that hope was now dashed. When President Truman announced the news at 3:00 p.m. in Washington, *Fat Man* was already airborne over the Pacific.

The second nuclear strike was originally scheduled for August 20, but by late July enough time had been made up to permit advancing the date to August 11. By August 7, the day after the Hiroshima mission, it appeared that the schedule could be further tightened to August 10. Good weather was forecast for the 9th, but bad weather for the 5 days thereafter; Groves wanted the second atomic blow to follow the first as quickly as possible. Project Alberta staff set to work to try to have the first live *Fat Man* ready by the evening of the August 8. From its start, however, the Nagasaki mission suffered almost every possible misfortune that the Hiroshima mission had avoided. The front and rear halves of F31's protective armor-plate ballistic casing were out of round, with the result that bolt holes for attaching the casing segments to an equatorial flange on the spherical high-explosive case did not align properly. No other armor-plate casings were available, so an attempt was made to hammer the parts into shape. When that failed, an effort was made to enlarge the bolt-holes with a two-man drill, but it jammed and gashed the leg of one of the workers. Desperate and running short of time, the assemblers substituted an ordinary steel casing; *Fat Man* would have to take its chances against Japanese machine-gun fire. After receiving a coat of pumpkin-colored paint and sealant to close off cracks which might result in erroneous barometric readings, the assembly crew made a small profile-view stencil of *Fat Man* and applied it to the nose of the bomb (Fig. 7.16), along with the letters JANCFU. The first four stood for "Joint Army Navy Civilian"; the meaning of the last two can be extrapolated from popular vernacular. Before it was rolled out for loading, a number of people autographed the bomb, including Purnell, Farrell, Parsons, and Ramsey; the bomb ended up carrying some 60 signatures in total.

The casing was not the only problem. On the night of August 7, Bernard O'Keefe, one of the members of the assembly team, was responsible for carrying out a last check of *Fat Man's* firing unit before it was encased:

By ten o'clock on the night of August 7, the sphere was complete, the radars installed, and the firing set bolted onto the front end of the sphere. I broke out for some sleep while others did final checkup and the mechanical assembly crew put the final touches on the casing. I was to come back at midnight for final checkout and to connect the two ends of the cable between the firing set and the radars; the cable had been installed the day before. Then I would turn the device over to the mechanical crew for installation of the fin and the nose cap.

When I returned at midnight, the others in my group left to get some sleep; I was alone in the assembly room with a single Army technician to make the final connection …

I did my final checkout and reached for the cable to plug it into the firing set. It wouldn't fit!

"I must be doing something wrong," I thought. "Go slowly; you're tired and not thinking straight."

I looked again. To my horror, there was a female plug on the firing set and a female plug on the cable. I walked around the weapon and looked at the radars and the other end of the cable. Two male plugs. The cable had been put in backward. I checked and double-checked. I had the technician check; he verified my findings. I felt a chill and started to sweat in the air-conditioned room.

What had happened was obvious. In the rush to take advantage of good weather, someone had gotten careless and put the cable in backward. Worse still, the checklist had been bypassed so that it was not double-checked before assembling the casing.

Fixing the problem would mean unsoldering the connectors from the two ends of the cable and reversing them. But to follow orders that no source of heat was to be allowed in the explosives assembly room would mean partially disassembling the bomb, which would take time. O'Keefe decided to proceed on his own:

> My mind was made up. I was going to change the plugs without talking to anyone, rules or no rules. I called in the technician. There were no electrical outlets in the assembly room. We went out to the electronics lab and found two long extension cords and a soldering iron. We ... propped the door open (another safety violation) so it wouldn't pinch the extension cords. I carefully removed the backs of the connectors and unsoldered the wires. I resoldered the plugs onto the other ends of the cable, keeping as much distance between the soldering iron and the detonators as I could as I walked around the weapon ... We must have checked the cable continuity five times before plugging the connectors into the radars and the firing set and tightening up the joints.

Field Order number 17 and Operations order number 39 detailed primary and secondary targets: Kokura Arsenal and City, and the Nagasaki Urban Area; there was no tertiary target for this mission. Located about 100 miles apart on the southernmost main island of Kyushu, both areas were rich in targets. Kokura, a city of about 168,000, was home to Kokura Arsenal, a large armaments complex where vehicles, machine guns, and anti-aircraft guns were manufactured. Nagasaki, with a population was estimated to be about 250,000 at the time, is located at what has been described as the best natural harbor of Kyushu. A shipbuilding center and military port, major targets there included the Mitsubishi Heavy Industries shipbuilding complex and the adjacent Mitsubishi Steel and Arms Works. The latter was where torpedoes used at Pearl Harbor had been manufactured. Unlike Hiroshima and Kokura, Nagasaki is a somewhat constricted city, surrounded by hills.

Fat Man was ready by 10:00 p.m. on the evening of August 8, and loaded into *Bockscar* (Fig. 8.12). Major Charles Sweeney was assigned to pilot the strike plane; its usual commander, Captain Frederick Bock, would pilot *The Great Artiste*. The final crew briefing took place at 00:30 on the 9th (Fig. 8.13).

As *Bockscar* was prepared for takeoff, another problem arose. As ballast to compensate for the weight of the bomb, the rear bomb-bay of the aircraft had been fitted with two 320-gallon fuel tanks. Flight Engineer John Kuharek discovered that a pump for transferring fuel from the tanks appeared to be malfunctioning. The fuel would not only be inaccessible, but at about six pounds per gallon would represent almost 2 t of dead weight to be carried through the mission. To empty the tanks, replace the pump, or transfer the bomb to another plane would be too time-consuming; the window of good weather was narrowing. Sweeney decided to proceed with the mission. Bockscar departed at 03:48 Tinian time, Thursday, August 9; in Washington, it was 1:48 p.m. on Wednesday afternoon, August 8.

Fig. 8.12 The author with a
full-scale model of *Fat Man*
at the national Museum of
Nuclear Science and History,
Albuquerque, NM, 2004

Fig. 8.13 Partial *Bockscar*
crew. Standing (l-r): Kermit
Beahan, James Van Pelt, Don
Albury, Fred Olivi, Charles
Sweeney; kneeling (l-r):
Edward Buckley, John
Kuharek, Ray Gallagher,
Albert Dehart, Abe Spitzer.
Not present: Frederick
Ashworth, Philip Barnes.
Photo courtesy John Coster-
Mullen

The rendezvous point for *Bockscar* and the camera and instrument planes was
at the island of Yakushima, immediately off the southern coast of Kyushu
(Fig. 8.14). After flying through a storm, *Bockscar* arrived at about 09:00 and was
promptly joined by *The Great Artiste*, but the camera plane, *Big Stink*, piloted by
Captain James Hopkins, was nowhere to be seen. Hopkins was there, but for some
reason was flying at 39,000 feet versus *Bockscar's* 30,000. In his memoirs,
Sweeney claims that he was told later that Hopkins began making 50-mile dog-leg
sweeps in the area of Yakushima, as opposed to circling as he should have been.
Although Tibbetts had instructed Sweeney to wait for no more than 15 minutes at

Fig. 8.14 The Hiroshima
and Nagasaki bombing
missions. The distance from
Tinian to Hiroshima is about
2740 km (1700 miles).
Source http://
commons.wikimedia.org/
wiki/File:Atomic_bomb_
1945_mission_map.svg

the rendezvous point, he waited about 45 minutes before deciding to strike out for Kokura.

Another element of confusion seems to have been that Commander Ashworth, who was overseeing the bomb, wanted to be sure that at least the instrument plane accompanied *Bockscar* on the strike mission. Ashworth claims that Sweeney never informed him which other plane they had rendezvoused with, and *The Great Artiste* remained too distant for Ashworth to get a visual identification. Sweeney did not address this issue in his own memoirs except to say that he felt that it was vital to have the photographic plane along to fulfill the mission plan. Ashworth claims to have stuck his head up into the flight deck to recommended that they proceed to their primary target; Sweeney implied that it was his decision to do so. The positive news, however, was the both weather planes were reporting good conditions at the targets.

Hopkins' incorrect altitude was not *Big Stink*'s only problem. Robert Serber was to fly on Hopkins' plane for the specific purpose of operating a high-speed camera to record the explosion. As Hopkins taxied to the end of the runway at

Tinian in preparation for take-off, he called for a parachute check. Serber had not been issued one, and was forced off the plane, which then departed without him. After walking back to base (and fearing the presence of Japanese snipers), Serber was authorized to break radio silence in an attempt to transmit instructions to the plane, but this proved to be for naught. At one point, Hopkins, speaking in the clear, radioed "Has Sweeney aborted?" At Tinian this was heard as "Sweeney aborted," which caused General Farrell to run outside and throw up.

Bockscar's flight to Kokura from the rendezvous point took about 50 minutes, but by the time it arrived at its Aiming Point at about 10:44 (Tinian time), the city was obscured by smoke and industrial haze. The nearby city of Yawata had been firebombed the previous day, and smoke was drifting over Kokura. The Japanese started sending up flak, so Sweeney rose to 31,000 feet. The smoke and haze made visual bombing runs impossible; after three attempts from different directions at different altitudes, Sweeney decided to head for Nagasaki. By this time, *Bockscar's* fuel supply was getting low. Sweeney estimated that they would have enough fuel for one run over Nagasaki, but that they would likely have to ditch in the ocean some fifty miles from Okinawa, the nearest friendly base (Fig. 8.14). *Bockscar* departed Kokura about 11:30 a.m. (10:30 Japan time). The term "Kokura luck" is sometimes used by Japanese as a euphemism for the unknown avoidance of a horrible misfortune.

The flight to Nagasaki from Kokura took only about 20 minutes; *Bockscar* arrived at about 11:50 a.m., Tinian time. But the weather had changed there as well, with the city now obscured by 80–90 % cumulus clouds between 6,000 and 8,000 feet. The fuel situation was becoming critical. Some accounts have Ashworth directing bombardier Kermit Beahan to make a radar-based bomb run, for which Ashworth would take responsibility. Sweeney claims in his memoirs that he gave the same order. But about 30–45 seconds before the drop, a hole opened in the clouds, and Beahan shouted something to the effect of "I see it! I see it! I've got it!" They had already passed the original Aiming Point in the dock area of the city, so Beahan chose a new one in the industrial area. Control of the aircraft was relinquished to him, and he released *Fat Man* from an altitude of about 29,000 feet at 11:08 a.m. Nagasaki time (10:08 p.m. Washington time, August 8). The bomb detonated over the Mitsubishi complex; because of the reflective hilly geography, the crew felt five shock waves (Fig. 8.15).

Sweeney ordered radio operator Abe Spitzer to transmit a strike report:

> Bombed Nagasaki 090158Z visually. No opposition. Results technically successful. Visible effects about equal to Hiroshima. Proceeding to Okinawa. Fuel problem.

(The time given in Spitzer's report differs by 10 min from that listed in Table 8.2; slightly different times have been reported by various sources.) Eighty miles away, the crew of *Big Stink* noticed the explosion. As related by Group Captain Leonard Cheshire, a British observer aboard Hopkins' plane:

Fig. 8.15 *Left Bockscar* nose art, added after the Nagasaki mission. *Source* http://commons.wikimedia.org/wiki/File:Bockscar.jpg *Right* The Nagasaki mushroom cloud. *Source* http://commons.wikimedia.org/wiki/File:Atomic_cloud_over_Nagasaki_from_B-29.jpg

> We reached the target some 10 min after the explosion at a height of 39,000 feet. At this time the cloud had become detached from the column and extended up to a height of approximately 60,000 feet. From the bomb aimer's compartment I had an excellent view of the ground and could see that the center of the impact was some four miles north-east of the aiming point and that the city proper was untouched. Fortunately however the bomb had accidentally hit the industrial center north of town and had caused considerable damage.

After lingering only briefly to view the results of his morning's work, Sweeney set course for Okinawa. Spitzer sent a Mayday call, but received no reply. "Fuel problem" was an understatement; Sweeney estimated that they had one hour of flying time available, but Okinawa was about 75 minutes away. By utilizing a technique known as "flying on the step" where he would leave power settings steady but put the plane into a very gradual descent, Sweeney was able to pick up a bit more airspeed without using additional fuel. Alternating descents and level-offs allowed him to stretch the fuel supply to Okinawa.

But *Bockscar* was not yet out of the woods. As they approached Yontan Field on Okinawa, Spitzer was unable to raise the tower. The nearest American base to Japan, Okinawa was always busy with incoming and outgoing traffic. Sweeney ordered Fred Olivi to fire emergency flares. Different-colored flares were used to indicate different emergency conditions (low fuel, damage, prepare for crash, dead and wounded aboard, fire, etc.). Olivi fired all of them, and the field began to clear of aircraft and vehicles. Cutting into the active traffic pattern, Sweeney brought his plane in directly behind a B-24 that was taking off. *Bockscar* bounced into the air and slammed back down just as its number two (left inboard) engine cut out; only by using the reversible propellers were Sweeney and co-pilot Don Albury able to bring the craft to a stop before running out of runway. As Sweeney described it, "I was so mentally and physically exhausted at that point that I just let the airplane roll to the side of the runway and onto a taxiway. Another engine quit." According to various accounts, they arrived with only 7 or 35 gallons of fuel remaining— exclusive of the trapped fuel. After the crew had a meal and *Bockscar* was refueled, they made their way back to Tinian, arriving about 11:00 p.m. to no fanfare after a mission of over 19 hours. Some sources state that *Bockscar* spent more time over enemy territory than any other plane on a single mission in all of World War II.

Wars are full of inhumane and indiscriminate cruelties, but random occurrences of astonishing survivals also occur. In the history of Hiroshima and Nagasaki, one of these improbable stories involves what the Japanese came to call the "nijyuu hibakusha," which translates roughly as "twice bombed." After the bombing of Hiroshima, a number of survivors were relocated or moved of their own accord to Nagasaki, where, three days later, they experienced the *Fat Man* explosion. While it is estimated that some 165 people survived both bombings, the Japanese government officially recognized only one: Mr. Tsutomu Yamaguchi. A Mitsubishi engineer, Yamaguchi was in Hiroshima on a business trip on the morning of August 6, and was stepping off a streetcar less than two miles from ground zero when the *Little Boy* blast struck. His eardrums were ruptured and he sustained some burns, but was able to return to Nagasaki after spending the night in a bomb shelter. On the morning of the 9th, he was in his office telling his boss about what he had witnessed at Hiroshima, when "suddenly the same white light filled the room." Mr. Yamaguchi died of stomach cancer in early 2010 at the age of 93; his daughter has been reported as stating that he remained in good health for most of his life.

In the weeks and days before the bombings, American intelligence services had been intercepting and decrypting Japanese messages; it was known that many elements in the Japanese government wished to find a way toward what they considered to be an honorable surrender. The sticking point was the fate of Emperor Hirohito in the context of the "unconditional surrender" sought by the Allies. In Tokyo on August 9, high-level conferences ran on through the day. At a morning meeting of the Supreme War Council, it was decided that an absolute condition of accepting the Potsdam terms would have to be retention of the imperial house. A militarist faction demanded that if occupation of Japan could not be avoided, then the Japanese should at least be responsible for their own disarmament and dealing with any war criminals. As the meeting progressed, word was received of the strike on Nagasaki. The meeting continued into the late evening with no consensus being reached. At about midnight, the Council met with the Emperor himself, who made it known that he was in favor of ending the war.

At 8:47 a.m. Tokyo time on the 10th (7:47 p.m. on the 9th in Washington), a deliberately low-security message went out from the Foreign Ministry to legations in Switzerland and Sweden. The text included a statement that the Japanese were ready to accept the Potsdam conditions so long as they were understood to not include "any demand for modification of the prerogatives of His Majesty as a sovereign ruler." Intercepted and decrypted, the message was on President Truman's desk early on the morning of the 10th. By noon, a response had been developed that stipulated that "the authority of the Emperor and the Japanese Government to rule the state shall be subject to the Supreme Commander of the

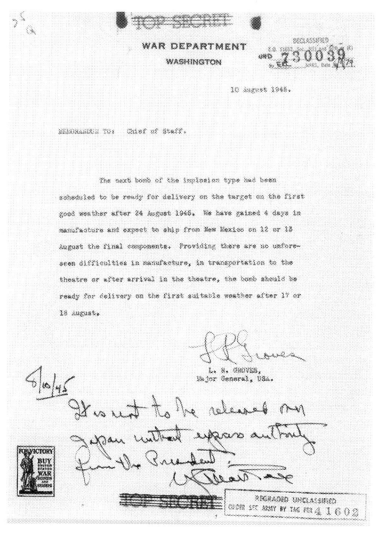

Fig. 8.16 General Groves to Army Chief of Staff General George C. Marshall, August 10, 1945. The handwritten note at the bottom is Marshall's, and reads "8/10/45 It is not to be released upon Japan without express authority from the President." Source: M1109(3), 0653-0654

Allied Powers who will take such steps as he deems proper to effectuate the surrender terms".

Also on August 10, Groves informed General Marshall as to the delivery schedule of the next bomb (Fig. 8.16):

> The next bomb of the implosion type had been scheduled to be ready for delivery on the target on the first good weather after 24 August 1945. We have gained 4 days in manufacture and expect to ship from New Mexico on 12 or 13 August the final components. Providing there are no unforeseen difficulties in manufacture, in transportation to the theatre or after arrival in the theatre, the bomb should be ready for delivery on the first suitable weather after 17 or 18 August.

But the President had exercised his prerogative as Commander-in-Chief, ordering a halt to any more atomic strikes. Henry Wallace, who had preceded Truman as Vice-President and was serving as Secretary of Commerce, recorded in his diary that afternoon that

> The President, who usually comes to cabinet not later than 2:05, came in about 2:25 saying he was sorry to be late but that he and Jimmie [Byrnes] had been busy working on a reply to Japanese proposals … Truman said he had given orders to stop atomic bombing. He said the thought of wiping out another 100,000 people was too horrible. He didn't like the idea of killing, as he said, "all those kids."

The Allied reply to the Japanese proposal began to be picked up by radio intercepts in Tokyo in early hours of August 12. Japanese officials debated through the day and into the evening. The continue-the-war faction favored holding out, with some speaking of mounting a coup. On the morning of the 14th, the Emperor himself called for an Imperial Conference at 10:30 a.m. (9:30 p.m. on the 13th in Washington). Again making clear to the gathered ministers his desire for peace, Hirohito directed that an Imperial Rescript (public statement) be prepared, which he would record for broadcast over national radio; this would be the first time many Japanese would hear their Emperor's voice. That evening, a formal statement accepting the proposed compromise on the status of the Emperor was drafted. But the national Japanese news agency was already broadcasting a message indicating that an Imperial message accepting the Potsdam conditions was expected soon. At 11:48 p.m. (10:48 a.m. on the 14th in Washington), the Foreign Ministry began sending the appropriate coded messages to Switzerland and Sweden.

With negotiations dragging on, General Arnold felt that the Japanese needed more motivation, and decided to mount one last punch: 449 B-29's carried out daylight strikes on the 14th. Raids continued into the night, with the last bombs of the war falling on the city of Tsuchizaki at 3:39 a.m. on the 15th, Japan time (2:39 p.m. on the 14th in Washington). The official surrender note was received at the State Department at 6:10 p.m., three and a half hours later. President Truman announced the surrender to reporters in the Oval Office at 7:00 p.m., and then publicly from the portico of the White House. In Tokyo, Emperor Hirohito's statement was broadcast at noon on the 15th, just four hours after Truman's statement. Hirohito's public statement did not include the word "surrender," referring instead to effecting "a settlement of the present situation by resorting to an extraordinary measure. We have ordered our Government to communicate ... that our Empire accepts the provisions of the Joint Declaration. ... Our one hundred million people, the war situation has developed not necessarily to Japan's advantage, while the general trends of the world have all turned against her interest. Moreover, the enemy has begun to employ a new and most cruel bomb". Formal surrender documents were signed aboard the battleship *USS Missouri* in Tokyo Bay on September 2.

8.6 Effects of the Bombs

With bombs delivered and surrender in the offing, Leslie Groves moved to his next task: assessing the effects of his creations. On August 11, he directed Colonel Nichols to begin organizing teams to carry out on-site investigations in Japan; General Farrell would be in charge of organization in the Pacific. The resulting Manhattan Project Atomic Bomb Investigating Group consisted of three teams: one for Hiroshima, one for Nagasaki, and one to investigate Japanese activities in the field of atomic bombs. Nichols quickly brought together a group of 27, including Los Alamos physicists Robert Serber, Philip Morrison, and William Penney.

The results of the surveys were published in June, 1946, in a Manhattan Engineer District (MED) report titled "The Atomic Bombings of Hiroshima and Nagasaki." The group carried out preliminary inspections in Hiroshima on September 8 and 9, and in Nagasaki on September 13 and 14; these were to ensure that occupying forces would not be exposed to any excessive lingering radiation. In total, the Manhattan teams spent sixteen days in Nagasaki and four in Hiroshima. At the same time, the United States Strategic Bombing Survey (USSBS) also conducted its own analysis of the bombings, with a particular emphasis on surveying their effects on Japanese morale. A selection of statistics drawn from the MED and USSBS reports testify to the power of the bombs. "Point X" is ground zero, the location on the ground below the point of explosion of the bomb:

At Hiroshima:

- Estimated 66,000 dead and 69,000 injured of estimated pre-raid population of 255,000; a Japanese survey indicated some 71,000 dead and 68,000 injured. 60 % of deaths were attributed to burns, and 30 % to falling debris.
- Of over 200 doctors in the city before the attack, over 90 % were casualties, with only about 30 able to perform their normal duties a month after the bombing.
- Of 1,780 nurses, 1,654 were killed or injured.
- Only three of 45 civilian hospitals could be used after the bombing.
- 60,000 of 90,000 buildings destroyed or severely damaged.

- 70,000 breaks in water pipes.
- Heavy fire damage in a circular area of about 6,000 feet radius and a maximum radius of about 11,000 feet.
- Almost everything up to about one mile from X was completely destroyed except for about 50 heavily-reinforced concrete buildings, most of which had been designed to withstand earthquakes. Multistory brick buildings were completely demolished to 4,400 feet from X, and suffered structural damage to 6,600 feet. Steel-framed buildings destroyed to 4,200 feet, and suffered severe structural damage to 5,700 feet. Light concrete buildings in both cities collapsed out to 4,700 feet.
- Firestorm burnt out about 4.4 square miles around X.
- People suffer burns to 7,500 feet.
- Roof tiles were melted out to 4,000 feet.
- In both cities, trolley cars were destroyed up to 5,500 feet and damaged to 10,500 feet.
- Flash ignition of dry combustible material observed to 6,400 feet.
- All homes seriously damaged to 6,500 feet; most to 8,000 feet.
- Flash charring of telephone poles to 9,500 feet.
- Fires started by primary heat radiation in both cities to about 15,000 feet.

At Nagasaki:

- Estimated 39,000 dead and 25,000 injured of estimated pre-raid population of 195,000.
- 95 % of deaths attributed to burns.
- About 20,000 of 50,000 buildings and houses destroyed. Total destruction area about 3 square miles.
- Nearly everything was destroyed within 0.5 miles of X, including heavy structures.
- At 1,500 feet from X, high-quality steel buildings were not collapsed, but suffered mass distortion, and all panels and roofs were blown in. At 2,000 feet, reinforced concrete buildings with 10-inch walls were collapsed; buildings with 4-inch walls were badly damaged. At 3,500 feet, church buildings with 18-inch walls were completely destroyed. Multistory brick buildings were destroyed to 5,300 feet, and suffered structural damage to 6,500 feet. Steel-framed buildings destroyed to 4,800 feet and suffered severe structural damage to 6,000 feet. The extreme range of building collapse was 23,000 feet.
- Twelve-inch brick walls were severely cracked as far as 5,000 feet.
- Roof tiles were melted out to 6,500 feet.
- People suffered burns to almost 14,000 feet.
- Flash ignition of dry combustible material observed to 10,000 feet.
- About 27 % of 52,000 residential units completely destroyed, and a further 10 % half-burned or destroyed. All homes seriously damaged to 8,000 feet; most to 10,500 feet.

- Hillsides scorched to 8,000 feet.
- Foliage turned yellow to about 1.5 miles.
- Flash charring of telephone poles to 11,000 feet.
- Heavy fire damage south of X up to 10,000 feet, stopped by a river.

At Nagasaki, mortality was estimated at 93 % within 1,000 feet of X, falling to 49 % at 5,000 feet. By far, blast and burn effects were the greatest causes of mortality and injury. The Manhattan Project's medical director, Dr. Stafford Warren, estimated that some 7 % of deaths resulted primarily from radiation, although some estimates of radiation-caused deaths ran as high as 15–20 %. Radiation effects included depressed blood counts, loss of hair, bleeding into the skin, inflammations of the mouth and throat, vomiting, diarrhea, and fever. Deaths from radiation began about a week after exposure, peaked in about 3–4 weeks, and ceased by 7–8 weeks. A person who survived but remained continuously in the city for 6 weeks after the explosion could expect to receive a dosage estimated at 6–25 rems (Hiroshima) or 30–110 rems (Nagasaki), with the latter figure referring to a localized area (For a summary of the damage caused by various rem doses, review Sect. 7.13). The USSBS report states that of women in Hiroshima in various stages of pregnancy who were known to be within 3,000 feet of ground zero, all suffered miscarriages, and some miscarriages and premature births where the infant died shortly after birth were recorded up to 6,500 feet. Two months after the bombing, the city's total incidence of miscarriages, abortions and premature births ran to 27 %, as opposed to a normal rate of 6 %.

The USSBS report offered a comparison of the atomic bombings with the March 9/10 firebombing raid on Tokyo:

By November 1, 1945, the USSBS estimated that the population of Hiroshima was back to 137,000, although the city required complete rebuilding. The population of Nagasaki had come back to 143,000.

The survey teams used a number of methods to determine parameters such as blast pressure and the detonation heights of the bombs. Concrete from the remains of buildings could be tested for breaking strength. William Penney sought out gas cans at various distances that had been more or less crushed. After taking them back to England, he had similar cans made up and measured the pressure necessary to crush them. At the Post Office Building in Hiroshima just a mile from Ground Zero, Robert Serber found a room facing the explosion where the glass had been blown out of a large window, but the frames of the windowpanes had remained intact and had cast shadows on an adjacent wall. By measuring the angles of the shadows, he determined that the bomb had detonated at an altitude of 1,900 feet, and by measuring the penumbra of the shadow he could get an idea of how big the fireball had been. In a more humorous vein, William Penney found an unusual situation in Nagasaki: a door with paper panels where half were broken and half were intact. On asking the woman who lived in the house "Atomic bomb?", her reply was "No. Small boy." In 1970, Penney and some collaborators published an extensive paper on results of measurements of the yields of the explosions, determining 11–13 kilotons for Hiroshima and 20–24 kilotons for Nagasaki.

Scores of accounts of the horrifying deaths and injuries suffered by the people of Hiroshima and Nagasaki have been published. While such accounts may seem out of place in a physics text, it would be unconscionable not to include a few. Psychiatrist and writer Robert Jay Lifton interviewed a number of Hiroshima survivors in the 1960s for his book *Death in Life: Survivors of Hiroshima*. A few excerpts will make the point:

A grocer who was severely burned:

> The appearance of people was … well, they all had skin blackened by burns. … They had no hair because their hair was burned, and at a glance you couldn't tell whether you were looking at them from in front or back. … They held their arms bent … and their skin—not only on their hands, but on their faces and bodies too—hung down. … If there had been only one or two such people … perhaps I would not have had such a strong impression. But wherever I walked I met these people. … many of them died along the road—I can still picture them in my mind …

A sociologist at twenty-five at hundred meters from ground zero:

> Everything I saw made a deep impression—a park nearby covered with dead bodies waiting to be cremated … The most impressive thing I saw was some girls, very young girls, not only with their clothes torn off but their skin peeled off as well … I thought that should there be a hell, this was it … And I imagined that all of these people I was seeing were in the hell I had read about.

A thirteen-year-old trying to save his mother from the debris of their house:

> The fire was all around us so I thought I had to hurry. … I was suffocating from the smoke and I thought if we stayed like this, then both of us would be killed. I thought if I could reach the wider road, I could get some help, so I left my mother there and went off. … I was later told by a neighbor that my mother had been found dead, face down in a water tank. … If I had been a little older or stronger I could have rescued her. … Even now I still hear my mother's voice calling me to help her …

A seventeen-year-old, looking for her parents:

> I walked past Hiroshima station … and saw people with their bowels and brains coming out. … I saw an old lady carrying a suckling infant in her arms. … I just cannot put into words the horror I felt …

A professional cremator who suffered radiation sickness:

> I was all right for three days … but then I became sick with fever and bloody diarrhea. … After a few days I vomited blood also. … There was a very bad burn on my hand, and when I put my hand in water something strange and bluish came out if it, like smoke. After that my body swelled up and worms crawled on the outside of my body.

In 1946, President Truman directed the National Academy of Sciences to conduct investigations of the effects of radiation among survivors of Hiroshima and Nagasaki. The resulting Atomic Bomb Casualty Commission (ABCC) functioned until 1975, when it was replaced by the Radiation Effects Research Foundation, a nonprofit Japanese foundation binationally managed and supported with equal funding by the governments of Japan and the United States. Most notable of the Commission's work was a long-term genetic study on the effects of ionizing radiation and its effects on pregnant women and their children. No widespread evidence of genetic damage was found, although some instances of microcephaly and mental retardation in children exposed in utero did turn up.

What can be said about the effect of the bombs on the Japanese decision to surrender? The USSBS report considered this matter in detail, and came to mixed conclusions. As far as public morale went, it was apparent that there was a substantial effect only within about 40 miles of Hiroshima and Nagasaki, likely a result of censorship and lack of mass communication. While the bombs had more effect on the thinking of government leaders, the report concluded that (excerpted)

> It cannot be said, however, that the atomic bomb convinced the leaders who effected the peace of the necessity of surrender. The decision to surrender, influenced in part by knowledge of the low state of popular morale, had been taken at least as early as 26 June at a meeting of the Supreme War Guidance Council in the presence of the Emperor. ... The atomic bombings considerably speeded up these political maneuverings within the government. ... The bombs did not convince the military that defense of the home islands was impossible, if their behavior in government councils is adequate testimony. It did permit the government to say, however, that no army without the weapon could possibly resist an enemy who had it, thus saving "face" for the Army leaders ... There seems little doubt, however, that the bombing of Hiroshima and Nagasaki weakened their inclination to oppose the peace group. ... It is apparent that in the atomic bomb the Japanese found the opportunity which they had been seeking, to break the existing deadlock within the government over acceptance of the Potsdam terms.

8.7   The Aftermath

In the United States, demand for information on the Manhattan Project by media outlets and the public following the bombings was voracious. Not surprisingly, Groves had anticipated this, and had been laying groundwork to deal with the onslaught. In early 1944, he had discussed with James Conant the necessity of having some account of the Project ready for release upon the successful use of an atomic bomb, and in April of that year he asked Henry Smyth of Princeton University to take on the task of preparing a report. The purpose of the report was not only to satisfy the public's demand for information, but also to make clear what information Project employees could disclose. Groves exempted Smyth from his usual compartmentalization rules in order that he could gather information from all parts of the Project, and Richard Tolman was appointed to review the report to ensure that no security protocols were breached. Smyth completed the report on July 28, 1945. Before the Hiroshima bombing, Groves had a thousand copies printed up using top-secret reproduction facilities at the Pentagon. Despite some misgivings that it might help the Russians, Stimson recommended release of the report on August 2, and President Truman gave his own clearance on the 9th. The report was released for use by radio broadcasters after 9:00 p.m. on August 11, and for the Sunday-morning newspapers of August 12.

The formal title of Smyth's report is *Atomic Energy for Military Purposes: The Official Report on the Development of the Atomic Bomb under the Auspices of the United States Government, 1940–1945*. The original public version was published by Princeton University Press; it is now available online and has come to be known as the *Smyth Report*. While the report does not reveal any information regarding the actual construction of a nuclear weapon, what it did disclose was remarkable given the secrecy with which the Project was pursued. After chapters dealing with background physics, readers were informed of general ideas of critical size and the use of a tamper, how to separate isotopes and produce plutonium, and the idea of assembly via a target/projectile arrangement. Implosion was not discussed. Smyth's report was not intended for broad public consumption, but rather, as stated in its Preface, "to be intelligible to scientists and engineers generally and to other college graduates with a good grounding in physics and chemistry". In a summary section which alludes to the political and social questions raised by the development of the bomb, its appeal to public education is still worth contemplating:

> In a free country like ours, such questions should be debated by the people and decisions must be made by the people through their representatives. This is one reason for the release of this report. It is a semi-technical report which it is hoped men of science in this country can use to help their fellow citizens in reaching wise decisions. The people of the country must be informed if they are to discharge their responsibilities wisely.

Given Groves' obsession with secrecy, release of such an extensive report seems out of character. His own attitude, as expressed in a memo he later wrote for his own files, was surprisingly liberal:

> Maintaining security is always a losing battle in the end. ... No one can predict exactly the scientific developments of the next decade or two, but it can be assumed that most of them will come from the minds of young men working untrammeled and undirected, with full access to information, in an atmosphere of freedom. ... America's capacity to win wars with new weapons ... depends on the general scientific, technical, and industrial strength of the country, not on secret researches in either private or government laboratories. ... Therefore we should put our trust in continued scientific progress rather than solely in the keeping of a secret already attained.

In the days immediately following Hiroshima and Nagasaki, public opinion in the United States was strongly in favor of the bombings. A Gallup poll taken between August 10 and 15, 1945, showed 85 % of respondents approving use of the bomb, 10 % disapproving, and 5 % having no opinion. The next Gallup poll on the issue, taken in 1990 (there were apparently none conducted between 1946 and 1989—the period of the Cold War), had approval at 53 % and disapproval at 41 %; in 2005 the numbers were 57 and 38 %. The 2005 poll indicated that 80 % of respondents felt that dropping the bombs saved American lives by shortening the war, but, curiously, 47 % felt that dropping the bombs ultimately cost *more* Japanese lives than would have been lost had the war continued.

As the immediacy of the war faded and the implications of the bombs began to become more deeply appreciated, second-guessing as to the necessity of using them began to arise. A significant factor in this evolution was the publication of an article titled *Hiroshima* in *The New Yorker* magazine in August, 1946, by journalist John Hersey; it soon became a best-selling book. In direct, understated prose, Hersey described the stories of six survivors of the bombing of that city. For many Americans, this was their first exposure to the human costs of nuclear warfare.

In response to concern that the United States had callously deployed an inhumane weapon, individuals involved in the Manhattan Project soon began telling their own side of the story. In the December, 1946, edition of *The Atlantic Monthly*, Karl Compton published a three-page article aimed at refuting what he described as the "wishful thinking among those after-the-event strategists who now deplore the use of the atomic bomb on the ground that its use was inhuman or that it was unnecessary because Japan was already beaten." Some excerpts:

> It is easy now, after the event, to look back and say that Japan was already a beaten nation, and to ask what therefore was the justification for the use of the atomic bomb to kill so many thousands of helpless Japanese in this inhuman way; furthermore, should we not better have kept it to ourselves as a secret weapon for future use, if necessary? This argument has been advanced often, but it seems to me utterly fallacious. ... I believe, with complete conviction, that the use of the atomic bomb saved hundreds of thousands— perhaps several millions—of lives, both American and Japanese; that without its use the war would have continued for many months; that no one of good conscience knowing, as Secretary Stimson and the Chiefs of Staff did, what was probably ahead and what the atomic bomb might accomplish could have made any difference decision.

Compton offered arguments as to the role of the bomb in accelerating the Japanese surrender:

> (1) Some of the more informed and intelligent elements in Japanese official circles realized that they were fighting a losing battle ... These elements, however, were not powerful enough to sway the situation against the dominating Army organization ... (2) The atomic bomb introduced a dramatic new element into the situation, which strengthened the hands of those who sought peace ... (3) When the second atomic bomb was dropped, it became clear that this was not an isolated weapon, but that there were others to follow. With dread prospect of a deluge of these terrible bombs and no possibility of preventing them, the argument for surrender was convincing.

By far the most influential such article was one which appeared in the February, 1947, edition of *Harper's Magazine* under Henry Stimson's name, although it was actually written by Stimson and a number of others. Stimson opened by describing his April 25 meeting with Truman and Groves, the work of the Interim Committee and the Scientific Panel, estimates of Japanese force levels in the summer of 1945, his July 2 "Proposed Program for Japan," and, like Compton, details of the surrender process which were theretofore largely unknown to the public. He then offered some reflections (excerpted):

> But the atomic bomb was more than a weapon of terrible destruction; it was a psychological weapon. ... The bomb thus served exactly the purpose we intended. The peace party was able to take the path of surrender, and the whole weight of the Emperor's prestige was exerted in favor of peace. ... I cannot see how any person vested with such responsibilities as mine could have taken any other course or given any other advice to his chiefs. ... My chief purpose was to end the war in victory with the least possible cost in the lives of men in the armies which I had helped to raise. In light of the alternatives which, on a fair estimate, were

open to us I believe that no man, in our position and subject to our responsibilities, holding in his hands a weapon of such possibilities for accomplishing this purpose and saving those lives, could have failed to use it and afterwards looked his countrymen in the face.

Stimson came in for no small amount of criticism over the fact that a number of American government officials felt that a surrender might have been possible as early as June had America been willing to clarify its position on the fate of the Emperor. General Groves' opinion, expressed shortly after the end of the war, was not surprising:

> I have no qualms of conscience about the making or using of it. It has been responsible for saving perhaps thousands of lives. ... From an official standpoint I knew its success would be greatly to our advantage and from a personal standpoint it might save my own son.

In the words of Groves' biographer, Robert Norris:

> The bomb was not necessary to end the war, but it was critical in ending it when it did. Had the bombs taken longer to prepare, history might have turned out quite differently. ... What we do know is that Groves succeeded in building atomic bombs by July 1945; that the two dropped on Japan concentrated certain tendencies and forces at work within the ruling circles of Japan; and that the war ended on August 14. All the rest is speculation.

Questions of policy and morality lie outside the laws of physics; they remain for readers to contemplate for themselves. As a final thought along these lines, however, it is perhaps appropriate to quote some words from Robert Oppenheimer. Oppenheimer resigned as Director of Los Alamos on October 16, 1945, at which time General Groves presented the Laboratory with a Certificate of Appreciation from the Secretary of War. Oppenheimer's remarks on that occasion:

> It is with appreciation and gratitude that I accept from you this scroll for the Los Alamos Laboratory, for the men and women whose work and whose hearts have made it. It is our hope that in years to come we may look at this scroll, and all that it signifies, with pride.
>
> Today that pride must be tempered with a profound concern. If atomic bombs are to be added as new weapons to the arsenals of a warring world, or to the arsenals of nations preparing for war, then the time will come when mankind will curse the names of Los Alamos and Hiroshima.
>
> The peoples of this world must unite or they will perish. This war, that has ravaged so much of the earth, has written these words. This atomic bomb has spelled them out for all men to understand. Other men have spoken them, in other times, of other wars, of other weapons. They have not prevailed. There are some, misled by a false sense of human history, who hold that they will not prevail today. It is not for us to believe that. By our works we are committed to a world united, before this common peril, in law, and in humanity.

Exercises

8.1 The Franck Report (Sect. 8.4) estimated that an atomic bomb containing 20 kg of fissile material detonating at 6 % efficiency would have an effect equal to 20,000 t of TNT. Look back to Chap. 3 for the energy released in fission of 1 kg of uranium. Are the figures given in the Report internally consistent?

Further Reading

Books, Journal Articles, and Reports

K. Bird, M.J. Sherwin, *American Prometheus: The Triumph and Tragedy of J* (Robert Oppenheimer. Knopf, New York, 2005)

J.M. Blum (ed.), *The Price of Vision: The Diary of Henry A. Wallace, 1942–1946* (Houghton Mifflin, Boston, 1973)

R.H. Campbell, *The Silverplate Bombers* (McFarland & Co., Jefferson, 2005)

A. Christman, *Target Hiroshima: Deak Parsons and the Creation of the Atomic Bomb* (Naval Institute Press, Annapolis, 1998)

S. Cohen, *The Truth About the Neutron Bomb: The Inventor of the Bomb Speaks Out* (William Morrow, New York, 1983)

A.H. Compton, *Atomic Quest* (Oxford University Press, New York, 1956)

K.T. Compton, If the Atomic Bomb Had Not been Used. Atlantic Mon. **178**(6), 54–56 (1946)

J. Coster-Mullen, *Atom Bombs: The Top Secret Inside Story of Little Boy and Fat Man* (Coster-Mullen, Waukesha, 2010)

R.H. Ferrell, *Harry S. Truman and the Bomb* (High Plains Publishing Co., Worland, 1996)

L.R. Groves, *Now It Can be Told: The Story of the Manhattan Project* (Da Capo Press, New York, 1983)

P. Guinnessy, Components of 'Little Boy' sold at auction. Phys. Today **55**(8), 23 (2002)

D. Hawkins, *Manhattan District History. Project Y: The Los Alamos Project.* Volume I: Inception until August 1945 (Los Alamos Scientific Laboratory, Los Alamos, 1947) http://library.lanl.gov/cgi-bin/getfile?LAMS-2532.htm

J. Hersey, *Hiroshima* (Vintage, New York, 1989)

R.G. Hewlett, O. E. Anderson Jr., *A History of the United States Atomic Energy Commission, Vol. 1: The New World, 1939–1946* (Pennsylvania State University Press, University Park, 1962)

L. Hoddeson, P.W. Henriksen, R.A. Meade, C. Westfall, *Critical Assembly: A Technical History of Los Alamos during the Oppenheimer Years* (Cambridge University Press, Cambridge, 1993)

M. Hull, A. Bianco, *Rider of the Pale Horse: A Memoir of Los Alamos and Beyond* (University of New Mexico Press, Albuquerque, 2005)

V.C. Jones, *United States Army in World War II: Special Studies—Manhattan: The Army and the Atomic Bomb* (Center of Military History, United States Army, Washington, 1985)

C.C. Kelly (ed.), *The Manhattan Project: The Birth of the Atomic Bomb in the Words of Its Creators, Eyewitnesses, and Historians* (Black Dog & Leventhal Press, New York, 2007)

W. Lanouette, B. Silard, *Genius in the Shadows: A Biography of Leo Szilard, the Man Behind the Bomb* (University of Chicago Press, Chicago, 1994)

R.J. Lifton, *Death in Life: Survivors of Hiroshima* (Vintage, New York, 1969)

J. Malik, The Yields of the Hiroshima and Nagasaki Explosions. Los Alamos report 8819 (1985) http://library.lanl.gov/cgi-bin/getfile?00313791.pdf

S.L. Malloy, *Atomic Tragedy: Henry L. Stimson and the Decision to Use the Bomb Against Japan* (Cornell University Press, Ithaca, 2008)

Manhattan Engineer District, The Atomic Bombings of Hiroshima and Nagasaki. (1946) http://www.atomicarchive.com/Docs/MED/index.shtml

D. McCullough, *Truman* (Simon and Schuster, New York, 1992)

M. McDonald, Survivor of 2 Atomic Bombs Dies at 93. New York Times, 6 Jan 2010

R. S. Norris, *Racing for the Bomb: General Leslie R. Groves, the Manhattan Project's Indispensable Man* (Steerforth Press, South Royalton, 2002)

B.J. O'Keefe, *Nuclear Hostages* (Houghton Mifflin, Boston, 1983)

A. Pais, R.P. Crease, *J. Robert Oppenheimer: A Life* (Oxford University Press, Oxford, 2006)

R. Peierls, *Bird of Passage: Recollections of a Physicist* (Princeton University Press, Princeton, 1985)

W.G. Penney, D.E.J. Samuels, G.C. Scorgie, The Nuclear explosive yields at Hiroshima and Nagasaki. Phil. Trans. Roy. Soc. London **A266**(1177), 357–424 (1970)

N. Polmar, *The Enola Gay: The B-29 That Dropped the Atomic Bomb on Hiroshima* (Brassey's, Inc., Dulles, 2004)

M. Price, Roots of Dissent: The Chicago met lab and the origins of the Franck report. Isis **86**(2), 222–244 (1995)

N. Ramsey, History of Project A: http://www.alternatewars.com/WW2/WW2_Documents/War_Department/MED/History_of_Project_A.htm

R. Rhodes, *The Making of the Atomic Bomb* (Simon and Schuster, New York, 1986)

H. Russ, *Project Alberta: The Preparation of Atomic Bombs for use in World War II* (Exceptional Books, Los Alamos, 1984)

R. Serber, R.P. Crease, *Peace and War: Reminiscences of a Life on the Frontiers of Science* (Columbia University Press, New York, 1998)

M.J. Sherwin, *A World Destroyed: Hiroshima and the Origins of the Arms Race* (Vintage, New York, 1987)

A.K. Smith, *A Peril and a Hope: the Scientists' Movement in America, 1945-47* (University of Chicago Press, Chicago, 1965)

H.D. Smyth, *Atomic Energy for Military Purposes: The Official Report on the Development of the Atomic Bomb under the Auspices of the United States Government, 1940–1945* (Princeton University Press, Princeton, 1945). http://www.atomicarchive.com/Docs/SmythReport/index.shtml

H. Stimson, The Decision to Use the Atomic Bomb. Harper's Mag. **194**(1161), 97–107 (1947)

M.B. Stoff, J.F. Fanton, R.H. Williams, *The Manhattan Project: A Documentary Introduction to the Atomic Age* (McGraw-Hill, New York, 1991)

C.W. Sweeney, J.A. Antonucci, M.K. Antonucci, *War's End: An Eyewitness Account of America's Last Atomic Mission* (Avon, New York, 1997)

G. Thomas, M. Morgan-Witts, *Enola Gay—Mission to Hiroshima* (White Owl Press, Loughborough, 1995)

United States Strategic Bombing Survey, The Effects of the Atomic Bombings of Hiroshima and Nagasaki. Washington (1946). http://www.trumanlibrary.org/whistlestop/study_collections/bomb/large/documents/pdfs/65.pdf#zoom=100

S. Weintraub, *The Last Great Victory: The End of World War II July/August 1945* (Dutton, New York, 1995)

Websites and Web-based Documents

Atomic Bomb casualty Commission: http://en.wikipedia.org/wiki/Atomic_Bomb_Casualty_Commission; http://www7.nationalacademies.org/archives/ABCC_1945-1982.html

Field orders for 509th Composite Group http://www.beserfoundation.org/FO13.pdf http://www.beserfoundation.org/Field%20orders%2017.pdf

Franck Report: http://www.atomicarchive.com/Docs/ManhattanProject/FranckReport.shtml

Gallup poll on use of atomic bombs: http://www.gallup.com/poll/17677/majority-supports-use-atomic-bomb-japan-wwii.aspx

Groves' April 23, 1945, memorandum to Stimson: http://www.gwu.edu/%7Ensarchiv/NSAEBB/NSAEBB162/3a.pdf

Hiroshima mission operations order: http://www.lanl.gov/history/admin/files/509th_Composite_Group.JPG

Interim Committee meetings: http://www.nuclearfiles.org/menu/key-issues/nuclear-weapons/history/pre-cold-war/interim-committee/index.htm

Mount Holyoke College site on documents relating to Hiroshima: http://www.mtholyoke.edu/acad/intrel/hiroshim.htm

Robert Lewis log: http://en.wikipedia.org/wiki/Robert_A._Lewis

Smyth Report: http://www.atomicarchive.com/Docs/SmythReport/index.shtml

Truman library documents on decision to drop the bomb: http://www.trumanlibrary.org/whistlestop/study_collections/bomb/large/index.php

Website on Twenty-First Bomber Command and bombing of Tokyo: http://en.wikipedia.org/wiki/XXI_Bomber_Command, http://en.wikipedia.org/wiki/Bombing_of_Tokyo

Chapter 9
The Legacy of Manhattan

The bombings of Hiroshima and Nagasaki and the end of the war brought the formal work of the Manhattan Project to a close, and simultaneously brought to the fore all of the postwar-planning issues described in Chap. 8. This chapter offers a low-resolution survey of postwar nuclear weapons developments. Volumes have been written about this period, and many of the issues involved remain active today. The goal here is to touch upon the main points to provide guideposts to orient readers who might be interested in pursuing more detailed study on their own.

9.1 The AEC and the Fate of International Control

At Los Alamos, Norris Bradbury succeeded Robert Oppenheimer as Director of the Laboratory in October, 1945. Many scientists left to go back to academic positions at the start of the fall school term, and there was naturally a great relaxation of efforts as the staff had their first real chance to rest after over 2 years of intense work. Work on weapons production and some theoretical studies of the "super" fusion bomb continued, but for many a sense of purposelessness pervaded the Laboratory. As Robert Christy described it, "When it finally was done, suddenly everyone stopped working. No one could push papers around anymore or do anything.... Basically, work stopped. I believe it was a mass reaction.... No one had the mental energy to push forward with anything for quite some time". It was not long, however, before many Project scientists began to take up activity in a very new world for them: the political arena.

In the summer of 1945, the Interim Committee (Sect. 8.4) appointed two War department lawyers, Kenneth Royall and William Marbury, to draft an atomic energy bill. With considerable input from General Groves, they drafted a proposal for a nine-person commission comprising five civilians and two representatives from each of the Army and Navy, although in the version eventually put to Congress the Chairman and any or all of the commissioners could be military

B. C. Reed, *The History and Science of the Manhattan Project*,
Undergraduate Lecture Notes in Physics, DOI: 10.1007/978-3-642-40297-5_9,
© Springer-Verlag Berlin Heidelberg 2014

officers. The commissioners would be supported by four advisory boards, which would concern themselves with military applications, industrial uses, research, and medical applications of atomic energy.

The powers to be granted the commission were sweeping: custody of raw materials, facilities, and equipment; technical information and patents; all contracts relating to production of fissionable materials; authority to carry out research in commission-owned facilities or to contract with other institutions; and authority to direct, supervise, and regulate all atomic activities, even those pursued by outside organizations. Vannevar Bush and James Conant were concerned that such powers could greatly interfere with university-based research, and also felt that for the commission to conduct its own research would conflict with its regulatory responsibilities. Particularly disturbing was a security provision that would enable the commission to jail an individual for 10 years and levy a fine of $10,000 for disclosing information designated as sensitive to national security; a professor lecturing on cross-sections could unwittingly find himself in very deep trouble. The War Department was fundamentally unwilling to budge, however, and the Royall-Marbury draft was sent to President Truman soon after the end of the war. On October 3, the President addressed Congress to emphasize the need for prompt action, and the Royall-Marbury text was introduced the same day as the May-Johnson bill, named after its sponsors, Congressman Andrew May and Senator Edwin Johnson. Remarkably, May scheduled only one day of hearings on the proposed bill.

The bill's draconian provisions spurred scientists to action. At Los Alamos, Oak Ridge, and Chicago, groups formed to oppose the legislation. Los Alamos saw the formation of ALAS, the Association of Los Alamos Scientists. In Chicago arose the Atomic Scientists of Chicago, and from Oak Ridge came the Association of Oak Ridge Scientists. On November 1, these groups merged as the Federation of Atomic Scientists, which in December became the still-extant Federation of American Scientists (FAS).

Despite support for the May-Johnson bill by Arthur Compton and Robert Oppenheimer, rank-and-file scientists proved effective at raising awareness of how the bill's provisions could throttle research and hamper prospects for international control of atomic energy. Many scientists found themselves the center of attention of the press and politicians, a new experience. May reluctantly scheduled a second day of hearings by the House Military Affairs Committee for October 18, but a number of witnesses, including many scientists, were treated with overt hostility. The situation became further confounded by introduction of competing bills, and maneuvering over which congressional committees held jurisdiction over the issue. By late October, President Truman's support for the measure had waned, and it was effectively dead.

Even as the May-Johnson bill was being defeated, an ambitious young senator from Connecticut, Brien McMahon, was developing his own draft legislation. On October 10, McMahon introduced a resolution to create a special committee to study atomic energy and all bills and resolutions related to it. A committee of 11 members, with McMahon as Chair, was established on October 26, and hearings took place from November 27 until December 20. On the latter date, McMahon introduced a bill to establish a civilian agency, the Atomic Energy Commission (AEC).

Hearings on McMahon's bill were held between January 27 and April 8, 1946, with a number of scientists testifying. McMahon proposed to establish a commission of five civilian members appointed by the President, but there would be no single executive Commissioner. Groves felt that such a committee could accomplish little, and testified against the proposal; he also saw it as having weak security provisions, and was particularly annoyed that the bill failed to provide that active military officers could serve either as commissioners or as the Commission's General Manager (Despite this, three of the first five General Managers of the AEC were military officers, albeit who had to retire from active service to take the position.). Deep animosity arose between Groves and McMahon, but Groves was becoming steadily more isolated by shifting post-war political sentiments. In a remarkable display of public interest for a scientific issue, McMahon's committee received over 75,000 letters.

McMahon's bill placed much less emphasis on military control, and much more on supporting research and clarifying what information could be freely exchanged versus what would be regarded as restricted. The Commission would be the exclusive owner of fissionable-material production plants, but would contract out their operation; it would also be authorized to conduct research and development in the military application of atomic power; to take custody of all assembled or unassembled atomic bombs and bomb parts; and, as authorized by the President, engage in the production of atomic bombs. Weapons could be delivered to the armed forces, but again only under the authority of the President. Basic scientific information could be freely disseminated, as could "related technical information" that was not considered sensitive to national defense. The Commission was also authorized to issue licenses for the operation of equipment or devices utilizing fissionable materials, including reactors. To respect the role of the military, an amendment established a Military Liaison Committee, which would advise the Commission on matters that related to military applications of atomic energy; in case of a dispute between the Liaison Committee and the Commission, the President would be the court of final decision. Also established was a General Advisory Committee (GAC), which was to offer advice on technical matters; Robert Oppenheimer would become that body's first Chair. Similar to the provisions of the May-Johnson bill, the work of the Commission would be supported by four Divisions: research, production, engineering, and military applications. The Senate approved the McMahon bill on June 1, 1946, followed by the House of Representatives on July 20. President Truman signed it into law on August 1, and the Commission formally came into existence on January 1, 1947. The AEC remained in existence until 1974, when a reorganization split its responsibilities between the Energy Research and Development Administration (which later became part of the Department of Energy), and the Nuclear Regulatory Commission. Despite Groves' misgivings with the AEC laying formally in civilian hands, subsequent development of improved nuclear weapons seems to have in no way been impeded.

In parallel with domestic developments, international control of atomic energy also came under consideration, although efforts in this area would ultimately come to nothing. At age 78, Henry Stimson, exhausted and in failing health, was about to resign as Secretary of War, but gave the issue one last shot in a memo to President Truman written on September 11, 1945. While admitting that trying to demand change within Russia to make that nation a more open society as a condition of sharing the atomic bomb would be so resented as to be hopeless, Stimson felt that some trust had to be extended to the Soviets to prevent "a secret armament race of a rather desperate character". Considering the problem of satisfactory relations with Russia as not merely connected with but rather virtually dominated by the "problem of the atomic bomb," Stimson offered some homespun advice: "The chief lesson I have learned in a long life is that the only way you can make a man trustworthy is to trust him; and the surest way to make him untrustworthy is to distrust him and show your distrust".

Skeptical of the possibility of achieving any results by way of international debate, the crux of Stimson's proposal was for America to make a direct approach to the Soviets (after discussions with the British) to develop an arrangement to control and limit the use of atomic bombs as instruments of war, and to encourage the development of atomic power for humanitarian purposes. Specifically, he suggested that it might be proposed to stop work on improving and manufacturing bombs, and for America to impound what bombs it had in hand, provided that an agreement could be reached with Britain and Russia to never use a bomb as an instrument of war unless all agreed to do so. However, Secretary of State James Byrnes was opposed to attempting to cooperate with Russia, and Truman's cabinet divided on the issue. In a time-honored bureaucratic maneuver, Byrnes appointed, in January, 1946, a special committee to formulate American policy on international control of atomic energy. The United Nations was about to establish the United Nations Atomic Energy Commission (UNAEC), and America would have to appoint a representative and offer policy initiatives.

Byrnes' committee was chaired by Undersecretary of State Dean Acheson; the other members were Bush, Conant, Groves, and recently retired Assistant Secretary of War John J. McCloy. The committee held its first meeting on January 14, and Acheson promptly proposed that they appoint a panel of scientific experts to advise them about nuclear energy. Groves objected on the grounds that he, Conant, and Bush already knew more about the issues than any other group that could be assembled, but was outvoted. Acheson appointed David E. Lilienthal, Chairman of the Tennessee Valley Authority and a long-time government administrator, as head of the panel (Fig. 9.1).

To fill out his group, Lilienthal chose Oppenheimer, Charles Thomas of the Monsanto Chemical Company (Sect. 7.7.1), Chester Barnard (President of the New Jersey Telephone Company), and Harry Winne, a Vice-President of General Electric who had been involved with the electromagnetic plant at Oak Ridge. The panel opened its work on January 28, with Oppenheimer giving the other members a two-day crash course on nuclear physics. Consulting with other experts as needed, they worked through February to develop a four-volume draft report for

Fig. 9.1 *Left* David E. Lilienthal (1899–1981), ca. 1947. *Right* Winston Churchill (1874–1965) and Bernard Baruch (1870–1965) in a car outside Baruch's home, 1961. *Sources* http://commons.wikimedia.org/wiki/File:David_E_Lilienthal_c1947.jpg , http://commons.wikimedia.org/wiki/File:Winston_Churchill_and_Bernard_Baruch_talk_in_car_in_front_of_Baruch%27s_home,_14_April_1961.jpg

Acheson's committee. The essential ideas were largely Oppenheimer's. At the core of the panel's proposal was the establishment of an international Atomic Development Authority, which would control, mine, and refine world supplies of uranium and thorium; operate separation plants and piles for breeding plutonium; conduct its own research; license and inspect reactor operators; and distribute "denatured" uranium that could be used for generating power but not bombs. All countries were to renounce ownership of nuclear weapons, but the plan was silent regarding sanctions for countries which violated the terms of the proposed Authority.

In an often-quoted phrase from the introduction to the report, the panel saw their work as "not as a final plan, but as a place to begin, a foundation on which to build." The draft was presented to the Acheson committee on March 7, and quickly became known as the Acheson-Lilienthal report. Skepticism began to arise almost immediately. General Groves doubted that raw materials could be effectively controlled, and thought that the proposal needed to be much more explicit in spelling out transition steps. The idea of "denaturing" was criticized as illusory: uranium useable in a reactor would not be immune from being illicitly enriched. Vannevar Bush argued that the bomb represented a means for the United States to offset the much larger army of the Soviet Union, and thought that America should give up its nuclear weapons only after steps to moderate and liberalize the Soviet Union had been effected. The meeting adjourned with a request to the scientific panel to prepare a section on implementation steps. Over the following week, the panel did add a section to this effect, but it was short on details. The main idea proposed was for the United States to offer to disclose purely theoretical knowledge and an assessment of raw materials, but disclosure of more sensitive information and transfer of physical facilities would have to await a time when the Authority was ready to begin operations. The final version of the report was transmitted to Byrnes on March 17. Soon leaked (Groves blamed the State Department), it became interpreted as the United States' official policy.

President Truman had to appoint a representative to the UNAEC, and here again Byrnes' hand seems to have been at work. On March 16, the President appointed Bernard Baruch, a wealthy financier and government advisor known to be vain, conservative, and hostile to the Soviet form of government; he had no particular technical background (Fig. 9.1). Lilienthal recorded in his journal that he was "quite sick" at the news of Baruch's appointment, and Oppenheimer later said that "That was the day I gave up hope, but that was not the day for me to say so publicly."

Baruch was given latitude to inject his own ideas into America's position, and began working major revisions into the Acheson-Lilienthal document. In his version, violation of the Authority's provisions would be regarded as an international crime, which could be punishable by declaration of war against the offending party. But most controversial was the idea that in the event of a violation, no power on the United Nations Security Council could veto punishment of the offending nation(s). Baruch presented his proposal—now known as the Baruch Plan—at the opening session of the UNAEC on June 14, 1946. On the 19th, Soviet ambassador Andrei Gromyko presented the Soviet response. The Soviet Union rejected any change in the veto procedure, and proposed that a total prohibition on the production, possession, and use of atomic weapons had to preceded establishment of any international authority. In effect, Gromyko called for America to destroy its supply of weapons before any system of controls or inspections had been established. Debates dragged on, but both sides dug in their heels. On December 31, UNAEC delegates voted 10-0 on the Baruch Plan, but the result was meaningless: Russia and Poland abstained, and the Plan was effectively dead. In view of Russian intransigence, Oppenheimer himself turned strongly against any form of international control by early 1947. Discussions continued pointlessly until the UNAEC recommended suspension of its own activities on May 17, 1948. The present International Atomic Energy Agency, which is autonomous of but reports to the UN General Assembly and Security Council, was not established until 1957.

In 1980, Norris Bradbury offered the opinion that the Baruch plan was "Far-seeing, amazing in its general concept," but that it was far ahead of its time and that nobody was willing to subscribe to it. However far-seeing the plan was, it is hard to avoid the conclusion that hypocrisy was evident on both sides. As negotiations dragged on, the Russians were busy constructing their own first graphite reactor, code-named F-1 (Physics-1). Essentially a copy of a reactor built at Hanford for testing fuel slugs, F-1 went critical for the first time on the evening of Christmas Day, 1946, at a power of 10 W. In America, plans had been underway since soon after the end of the war for a series of tests to determine the effects of atomic bombs on naval vessels, and the *Crossroads* tests were conducted on the Pacific island of Bikini on July 1 and 25, 1946, during the first phase of the UNAEC negotiations. By the summer of 1946, the Cold War was settling in.

At Bikini, two *Fat Man* bombs were detonated, one air-dropped and one suspended underwater. *Crossroads Able* was detonated 500 feet above a fleet of

American and Japanese vessels, but was somewhat of a disappointment in that it sunk only five ships. The bomb fell some 1,800 feet horizontally from its intended aiming point, apparently a consequence of incorrect ballistic data. *Crossroads Baker*, detonated at a depth of 90 feet, spectacularly lofted a shaft of water half a mile in diameter a full mile into the air, sank ten ships, and exposed many men who later boarded surviving vessels to radioactivity. A 1996 government-sponsored mortality study of *Crossroads* veterans showed that 46 years after the tests, those veterans had experienced 4.6 % higher mortality than a control group of non-veterans. Glenn Seaborg called *Baker* "the world's first nuclear disaster." A third proposed test, *Charlie*, was to have been detonated even deeper, but was scrubbed due to inability to decontaminate the target fleet following the *Baker* test. These tests, however, gave a somewhat illusory view of America's "nuclear stockpile" at the time. President Truman was shocked to learn in March, 1947, that the country held *no* operable weapons at all, although the number grew to 13 by the end of that year and to 50 by the end of 1948.

9.2   Joe-1, the Super, and the P-5

Between the first criticality of Enrico Fermi's CP-1 reactor and the *Trinity* test, 938 days elapsed. The Russians essentially duplicated this feat, detonating their first atomic bomb on August 29, 1949, exactly 978 days after the startup of their F-1 reactor. Known to the Soviets as RDS-1 (in the West as Joe-1), this device was a plutonium bomb identical to *Fat Man*; the design was based on information transmitted by Klaus Fuchs. Fission products from the test were picked up by B-29 bombers equipped with air-sampling devices which flew weather reconnaissance missions over Japan, Alaska, and the North Pole; they were also detected in rainwater collected in Alaska. President Truman announced the test on September 23. General Groves had estimated that it would probably take the Russians 10–20 years to catch up to the United States; ironically, he underestimated the effort which a command economy can bring to bear on a desired objective.

In both the United States and the Soviet Union, attention also turned to the possibility of the more powerful fusion weapons which had so captivated the attention of Edward Teller since 1942. On October 29 and 30, 1949, the AEC's General Advisory Committee met to discuss whether or not the United States should pursue an all-out effort to develop a hydrogen bomb. It was not at all clear whether technical difficulties in producing such a weapon could be overcome, or even if there was any sensible military use for a weapon 1,000 times as powerful as a fission bomb. In its report to AEC Commissioner Lilienthal, the Committee recommended unanimously against pursuing such development. The GAC had actually split into two groups, each of which appended an Annex to the report to Lilienthal. In recognizing that a super-bomb was essentially a weapon of genocide, the majority group, which included Oppenheimer and James Conant, offered the following commentary (excerpted paragraphs):

The existence of such a weapon in our armory would have far-reaching effects on world opinion; reasonable people the world over would realize that the existence of a weapon of this type whose power of destruction is essentially unlimited represents a threat to the future of the human race which is intolerable. Thus we believe that the psychological effect of the weapon in our hands would be adverse to our interests.

We believe a super bomb should never be produced. Mankind would be far better off not to have a demonstration of the feasibility of such a weapon, until the present climate of world opinion changes.

It is by no means certain that the weapon can be developed at all and by no means certain that the Russians will produce one within a decade. To the argument that the Russians may succeed in developing this weapon, we would reply that our undertaking it will not prove a deterrent to them. Should they use the weapon against us, reprisals by our large stock of atomic bombs would be comparably effective to the use of a super.

In determining not to proceed to develop the super bomb, we see a unique opportunity of providing by example some limitations on the totality of war and thus of limiting the fear and arousing the hopes of mankind.

The minority statement, signed by I. I. Rabi and Enrico Fermi, was even stronger in its opposition to such a development (excerpted paragraphs):

Necessarily such a weapon goes far beyond any military objective and enters the range of very great natural catastrophes. By its very nature it cannot be confined to a military objective but becomes a weapon which in practical effect is almost one of genocide.

It is clear that the use of such a weapon cannot be justified on any ethical ground which gives a human being a certain individuality and dignity even if he happens to be a resident of an enemy country. It is evident to us that this would be the view of peoples in other countries. Its use would put the United States in a bad moral position relative to the peoples of the world.

The fact that no limits exist to the destructiveness of this weapon makes its very existence and the knowledge of its construction a danger to humanity as a whole. It is necessarily an evil thing considered in any light.

For these reasons we believe it important for the President of the United States to tell the American public, and the world, that we think it wrong on fundamental ethical principles to initiate a program of development of such a weapon. At the same time it would be appropriate to invite the nations of the world to join us in a solemn pledge not to proceed in the development or construction of weapons of this category. If such a pledge were accepted even without control machinery, it appears highly probable that an advanced stage of development leading to a test by another power could be detected by available physical means. Furthermore, we have our possession, in our stockpile of atomic bombs, the means for adequate "military" retaliation for the production or use of a "super".

With the Soviet Joe-1 test and the 1948/1949 Berlin blockade, anti-Soviet political pressure on President Truman was intense, and on January 31, 1950, he announced that he was ordering the AEC to "continue work on all forms of atomic weapons, including the so-called hydrogen or super bomb." Isidor Rabi was horrified at the announcement: "For him to have alerted the world that we were going to make a hydrogen bomb at a time when we didn't even know how to make one was one of the worst things he could have done".

A detailed description of the development of fusion bombs lies outside the scope of this book; interested readers are urged to consult Richard Rhodes' *Dark Sun* for an excellent survey of this very complex history. Here I will only briefly describe the physics underlying these weapons.

The first step in the development of fusion weapons was the notion of *boosting* a fission weapon. In a boosted fission weapon, a gas of deuterium and tritium is introduced into the fission core, where temperatures and pressures are great enough to initiate the so-called D-T fusion reaction:

$$_1^2H + {}_1^3H \rightarrow {}_0^1n + {}_2^4He. \tag{9.1}$$

The Q-value of this reaction is 17.6 MeV. The "boosting" comes not from the 17.6 MeV (which is small compared to a typical fission release of ~ 200 MeV), but from the fact that the neutrons created carry off about 14 MeV of kinetic energy and can induce extra fissions in surrounding fissile material. Jacketing the fission–fusion core in a casing of natural uranium will make for a yet more powerful "fission-fusion-fission" device, since these neutrons are energetic enough to induce fissions in U-238. The first test of the boosting principle was carried out in the United States' *Greenhouse Item* test of May, 1951. This device achieved a yield of about 45 kt, about twice as much than if it had been unboosted. Tritium has a beta-decay half-life of about 12 years, and so needs to be periodically replaced in such weapons; it is synthesized by neutron bombardment of lithium in a reactor (below).

What is considered a true "thermonuclear" weapon involves yet another stage to provide a large amount of fusion-liberated energy. In such a device, the X-rays and gamma-rays created by a boosted fission core are energetic enough to compress a secondary device containing deuterium, usually in the form of solid lithium deuteride. Figure 9.2 shows a highly idealized sketch of such a device.

This *radiation compression* initiates a deuterium–deuterium (D–D) reaction, which has two channels of about equal probability of occurrence:

$$_1^2H + {}_1^2H \rightarrow \begin{cases} {}_1^3H + {}_1^1H & (Q = 4.03 \text{ MeV}) \\ \\ {}_0^1n + {}_2^3He & (Q = 3.27 \text{ MeV}). \end{cases} \tag{9.2}$$

Fig. 9.2 Sketch of a fission–fusion–fission thermonuclear device

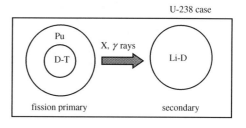

The first branch of this reaction produces tritium, which helps to further boost the D-T reaction in the primary. The neutron created in the second branch creates yet more tritium by reacting with lithium:

$$\,_0^1 n + \,_3^6 Li \rightarrow \,_1^3 H + \,_2^4 He \quad (Q = 4.78 \quad \text{MeV}). \tag{9.3}$$

In such a device, fission and fusion each produce about 50 % of the overall energy release. In the boosted primary, the D–T reaction is used instead of the D–D reaction because, at the temperatures created in these devices, the D–T reaction proceeds at a rate about 100 times that of the D–D reaction, and liberates four times as much energy per reaction. The D–D reaction in the secondary not only produces more tritium, but is "cheaper" in the sense that deuterium occurs naturally in heavy water. The United States' *Greenhouse George* test of May, 1951, which yielded 225 kt (Fig. 9.3), was an experimental test of whether a thermonuclear reaction could be initiated.

On October 31, 1952, the *Ivy Mike* test saw the detonation of America's first full-scale thermonuclear weapon. This achieved a yield of 10.4 megatons (MT), over 400 times as much as *Fat Man*. This device was a test of the so-called Teller-Ulam design, which is the basis for all modern fusion weapons. With a weight of 60 t, however, this was by no means a deliverable weapon. In this sense, the Soviet Union leapt ahead of the United States when it tested a deliverable fusion weapon on August 12, 1953. The first deliverable American thermonuclear device was detonated in the *Castle Bravo* test of February 28, 1954, and yielded 15 MT, about three times what was expected. Fallout from *Castle Bravo* covered some 7,000 square miles, and contaminated the crew of a Japanese fishing vessel, the *Lucky Dragon 5*. One of the crew members died, and many tons of their cargo entered the Japanese market. The highest-yield thermonuclear device ever detonated was the Soviet Union's "Tsar Bomba," in October, 1961. This device yielded almost 60 MT, a remarkable 97 % of which resulted from thermonuclear

Fig. 9.3 *Left* Greenhouse George test, May 9, 1951; *Right* Ivy Mike test, October 31, 1952. *Sources* http://commons.wikimedia.org/wiki/File:Greenhouse_George.jpg, http://commons.wikimedia.org /wiki/File:Ivy_Mike_-_mushroom_cloud.jpg

reactions—a "clean" bomb. Had it been built as designed with uranium surrounding the secondary, it would have achieved 100 MT.

The largest pure *fission* bomb ever detonated by the United States was the *Ivy King* shot of November, 1952, at 500 kt. This raises a question: If fission weapons can be developed to such levels of efficiency, why go to the complex task of developing a fusion weapon? Fundamentally, it is a matter of economics. In their *Megawatts and Megatons*, Richard Garwin and Georges Charpak lay out an example. If one desired to obtain a 10-MT yield from a plutonium fission device, then some 600 kg of material would have to undergo fission (recall from Chap. 3 the rule-of-thumb of 17 kt yield per kilogram in the fission process). If the weapon is 30 % efficient, one would have to provide something like 600/0.3 = 2,000 kg of fissile material. On the other hand, if a fission primary that uses only 6 kg of plutonium (like *Fat Man*) can be used to trigger a thermonuclear explosion, then the same 2,000 kg could fuel 2,000/6 ∼ 333 bombs. Further, by designing bombs wherein a weaponeer can select that amount of D-T gas in the primary or is able to decouple the secondary from the primary before launch, yields can be made "dialable" at the time of bomb delivery according as the needs of the mission. Bombs in the current United States stockpile use a combination of fission and fusion, as described above.

The fusion process depends solely on being able to compress and heat the fusible material; unlike a fission reaction, it does not depend on "catalyzing" particles such as neutrons to keep propagating itself. In principle, the yield of a thermonuclear weapon is unlimited. In practice, as Edward Teller calculated, there is not much destruction gained in going above a few megatons; any additional yield goes largely into blowing away a chunk of the earth's atmosphere. The highest-yield weapon currently deployed by the United States is the B83 warhead, which has a variable yield up to 1.2 MT; these warheads are deployed on B-2 and B-52 bombers.

Between 1949 and 1964, Britain, France, and China also developed nuclear weapons (Table 9.1). The first British test was conducted on the Montebello Islands off Western Australia, and the first French test was conducted in the Sahara Desert in Algeria. The British first tested a boosted device in 1956, and the French and Chinese both in 1966. Britain tested its first true thermonuclear device in 1957; the Chinese followed in 1967, and the French in 1968. The United States, Russia, the United Kingdom, France, and China are now known as the "primary five," or "P5" nuclear weapons states.

In addition to the P5 countries, four other nations are also recognized as nuclear powers: India tested its first weapon in 1974, Pakistan in 1998, and North Korea in 2006, although the latter may have been a fizzle (Some sources claim that the Chinese tested a Pakistani weapon in 1990; the 1998 test was in Pakistan.). It is not clear to what extent the North Koreans have weaponized their devices. Israel is widely regarded as having acquired nuclear weapons in the 1960s and to possess a stockpile on the order of 80 devices, but that country maintains a policy of official ambiguity by never having formally declared itself as a nuclear power (or not).

Table 9.1 Some nuclear milestones for the P-5 nuclear states as of 2013. See also Table 9.2

Parameter	United States	USSR/ Russia	Britain	rance	China
Date of first test	16-Jul-45	29-Aug-49	3-Oct-52	13-Feb-60	16-Oct-64
First test yield (kt)	21	22	25	60–70	20–22
First test name	Trinity	RDS-1/Joe-1	Hurricane	Gerboise Bleue	596
Peak number warheads/ year attained	31,255 1967	45,000 1986	520 1975–1980	540 1993	240 2012?
Number of tests/ Detonations[a]	1,030/1,125	715/969	45/?	210/?	45/?
Total warheads built	66,500	55,000	850	1,260	750
Warheads in stockpile[b]	2,150	4,480	225	300	240
Date of last test	23-Sep-92	24-Oct-90	26-Nov-91	27-Jan-96	29-Jul-96
Largest test, megatons[c]	15/5	50/2.8–4	3/<150 kt	2.6/120 kt	4/420 kt
Total megatonnage expended[b]	141/38	247/38	8/0.9	10/4	21.3/1.3

Source Robert S. Norris and Hans M. Kristensen, Nuclear pursuits, 2012. Bulletin of the Atomic Scientists **68**(1), 94–98 (2012); John R. Walker, British Nuclear Weapons Stockpiles, 1953–78. RUSI Journal **156**(5), 66–72 (2011); Robert S. Norris, private communications; Bulletin of the Atomic Scientists: Russian nuclear forces, 2013: **69**(3), 71–81 (2013); US nuclear forces, 2013: **69**(2), 77–86 (2013)

[a] Some tests involved simultaneous detonation of more than one warhead

[b] United States stockpile number does not include ∼2,650 reserve and 3,000 warheads awaiting dismantlement; Russian stockpile number does not include ∼4,000 warheads awaiting dismantlement

[c] Atmospheric/underground

Other countries such as Iraq, South Africa, and Libya had nuclear weapons development programs, but have abandoned them. At this writing, the situation with Iran is so fluid that no clear assessment can be offered.

9.3 A Brief Survey of Nuclear Tests and Current Deployments

As advances in weapons physics led to the development of lighter and more compact designs with a wide range of yields, the spectrum of missions to which nuclear weapons could be applied grew rapidly. Also, each branch of the armed forces naturally wanted its own piece of the nuclear action. In a 2009 article, Robert Norris and Hans Kristensen estimated that between 1945 and 2009, the United States produced over 66,500 nuclear warheads of 100 different basic types and variants of types. This corresponds to creating on average about 1,000 warheads per year, or almost three per day over 7 decades. These included weapons to be carried on bombers; mounted on land, surface, and submarine-based ballistic missiles; in landmines; on short-range artillery rockets; on ground, air, and submarine-launched cruise missiles; on anti-submarine rockets; in torpedoes; and on air-to-air, air-to-ground, and earth-penetrating missiles. The bomb type of which the most were built was the W68 warhead (40–50 kt); over 5,200 were deployed on submarines between 1970 and 1991. Some tactical (battlefield-scale) nuclear devices were small enough to be carried by a single person.

Such a plethora of designs demanded an extensive testing program. Between 1945 and 1992, the United States conducted 1,030 nuclear tests, plus an additional 24 in conjunction with the United Kingdom. Because some tests involved simultaneous detonation of more than one weapon, the total number of detonations involved in these tests was 1,149. As illustrated in Figs. 9.4 and 9.5, the United States conducted the most tests, but the Soviet Union was not far behind.

Depending on a warhead's anticipated mission, tests were configured to subject a variety of structures, vehicles, vessels, and environments to the effects of nuclear explosions. Detonations were conducted at surface level, underground (mostly), underwater, and at high altitudes via platforms such as airdrops, balloons, barges, rockets, and towers. The 1.4-MT *Starfish Prime* test of July, 1962, was detonated at an altitude of 400 km, and resulted in the discovery of the electromagnetic pulse phenomenon, which caused electrical damage some 900 miles away in Hawaii. Of the United States' 1,030 tests, 210 were atmospheric, 815 were underground, and 5 were underwater. The most frequently-used test location was the Nevada Test Site, which saw 928 tests involving 1,021 detonations.

Such a vast development, testing, and deployment complex involved a correspondingly great budget. A 1995 study by the Brookings Institution of Washington analyzed costs associated with the U. S. nuclear weapons program from 1940 onwards. Including research, development, testing, deployment, command and

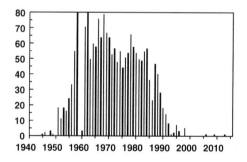

Fig. 9.4 Distribution of 2,054 nuclear tests worldwide by year. The totals for 1958 and 1962 are 116 and 178 tests. For 1958, (US, USSR, UK) = (77, 34, 5); in 1962, (US, USSR, France, UK) = (96, 79, 1, 2). The scale is set to a maximum of 80 to make visible the small numbers of tests in some years. The Soviet Union has not tested since 1990, nor the United States since 1992. The Hiroshima and Nagasaki bombs are not included here as they are considered to be combat weapons, not tests. Individual country scores as of 2013: (US, USSR, France, UK, China, India, Pakistan, North Korea) = (1,030, 715, 210, 45, 45, 4, 2, 3). Data from Natural Resources Defense Council. See also R. S. Norris and W. M. Arkin, Known Nuclear Tests Worldwide, Bulletin of the Atomic Scientists **54**(6), 65–67 (1998)

Fig. 9.5 Distribution of
2,045 "P-5" postwar nuclear
tests 1946–1996. (US, USSR,
France, UK,
China) = (1,030, 715, 210,
45, 45). The UK figure
includes 24 tests conducted
underground in the United
States. Not included here are
one Indian test in 1974, three
Indian tests in 1998
comprising five claimed
detonations, two Pakistani
tests in 1998, and three North
Korean tests (2006, 2009,
2013). Data from Natural
Resources Defense Council

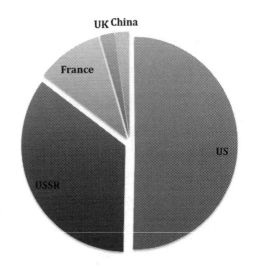

control, defense and dismantling of weapons systems, waste cleanup, compensation for persons harmed by the production and testing of nuclear weapons, estimated future costs for storing and disposing of waste, and dismantling and disposing of surplus materials, the total came to $5.8 trillion in 1996 dollars. This represented 29 % of all military spending between 1940 and 1996, and 11 % of all government expenditures during that period. Spending on the nuclear weapons complex exceeded all other government spending except for non-nuclear defense and social security. At present, the United States still maintains 15 warhead types in its active arsenal. New warhead production ceased in 1992, but existing devices regularly undergo modifications and refurbishments; so called "life-extension programs". Only the Air Force and the Navy currently deploy nuclear weapons, the Army and the Marines have none.

Another legacy of the weapons production and testing program is radioactive contamination. A 1996 publication estimated that the total amount of radioactivity released by the United States and the Soviet Union amounted to a then-current value of 1.7 billion Curies. While this figure is only 0.4 % as much as exists naturally in the world's oceans (the latter is due mostly to potassium-40), the weapons-related radioactivity is concentrated in small areas, which creates substantial local environmental impacts. Of these 1.7 billion Curies, the vast majority was released by the Soviet Union; the United States was responsible for only about 3 million Curies, less than 0.2 % of the total. The greatest concentrations in the United States are at Oak Ridge (about 1 million Curies due to underground injections of cesium and strontium); Savannah River, Georgia (900,000 Curies from release of fission products into streams and seepage basins); and Hanford (700,000 Curies from fission products released into soils and surface ponds). According to a 2012 estimate, the cost of cleanup operations at Hanford alone through the year 2065 is expected to run to $112 billion.

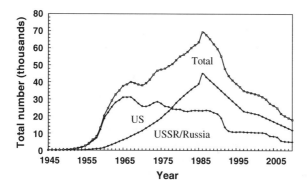

Fig. 9.6 Estimated global nuclear weapons inventories, 1945–2010. Included in the total curve are "smaller" nuclear powers (United Kingdom, France, China, Israel, India, Pakistan) which in 2010 were estimated to hold a total of about 1,000 weapons. Data from Robert S. Norris and Hans M. Kristensen, Global nuclear inventories, 1945–2010 [Bulletin of the Atomic Scientists **66**(4), 77–83 (July 2010)]

The global inventory of deployed and readily-deployable nuclear weapons began to grow dramatically in the 1950s. This growth continued through the first half of the 1980s to the point when, in 1986, just over 69,000 were available for use. Over 98 % of these were in American and Russian hands (Fig. 9.6). Since that time, reductions in numbers due to various arms-control treaties (Sect. 9.4) and unilateral withdrawals from various venues on the parts of both America and Russia have brought the current total inventory down to about 20,000 weapons, not including those awaiting dismantlement.

Many nuclear powers are not overly transparent about the number of weapons that they have on hand at any given time. Table 9.2 shows some estimates for the

Table 9.2 Nuclear weapons deployments for the P-5 nuclear states as of 2011–2013

Parameter	United States	USSR/Russia	Britain	France	China
Strategic weapons					
ICBMs	500	1,050			~138
SLBMs	1,150	620	160	240	
Deployed on bombers & aircraft	300	810		60	~20
Nonstrategic weapons	200	2,000		50	Few
Reserve/maintenance	2,650		65		~62
Awaiting dismantlement	3,000	~4,000			
Grand total	7,800	~8,500	225	350	~240

All numbers should be regarded as approximate. *ICBM* inter-continental ballistic missile; *SLBM* submarine-launched ballistic missile. *Sources* Robert S. Norris and Hans M. Kristensen, "Nuclear notebook" series published in Bulletin of the Atomic Scientists: Russian nuclear forces, 2013: **69**(3), 71–81 (2013); US nuclear forces, 2013: **69**(2), 77–86 (2013); British nuclear forces 2011: **67**(5), 89–97 (2011); French nuclear forces, 2008: **64**(4), 52–54 (2008); Chinese nuclear forces, 2011: **67**(6), 81–87 (2011); Indian nuclear forces, 2012: **68**(4), 96–101 (2012); Pakistan's nuclear forces, 2011: **67**(4), 91–99 (2011); Nonstrategic nuclear weapons, 2012: **68**(5), 96–104 (2012)

P-5 states as of 2011–2013; all numbers should be regarded as approximate. In May 2010, the United States Department of Defense took the unprecedented step of announcing, for the first time ever, the exact number of weapons in the stockpile at that time (deployed and reserve): 5,113. Current United States ICBM warheads have yields of 300 and 335 kt, while SLBM warheads have yields of 100 and 455 kt. Warheads carried on smaller aircraft (so-called "tactical" or "non-strategic" weapons) have yields that can vary from a few tenths of a kiloton up to about 170 kt. Of an estimated 500 United States non-strategic weapons, 200 are actively deployed in Europe, and the balance are in storage in the U. S. Current Russian maximum ICBM and SLBM yields are estimated at 800 and 100 kt, respectively.

Some addenda to Table 9.2: India is estimated to possess some 80–100 warheads; that country's delivery platforms include fighter-bombers and a short-range land-based missile, but longer-range land-based missiles and sea-based (surface and submarine) missiles are under development. Pakistan is thought to have the world's fastest-growing nuclear arsenal, with a current stockpile estimated at 90–110; the Pakistani stockpile could exceed that of Britain by the early 2020s. Pakistani delivery systems include aircraft and ballistic missiles, with cruise missiles under development. Britain deploys nuclear weapons only on submarine-launched missiles; of an estimated stockpile of 225, no more than 160 are thought to be operational at any time. In 2010, the British government announced plans to decrease its stockpile to no more than 180 warheads over the following 15 years. French weapons are deployed on aircraft (both land and carrier-based; yields up to 300 kt) and on submarine-launched missiles (100 kt). Hard numbers for China are difficult to come by, but that country is believed to be the only P-5 state that is increasing its arsenal; the figure of 62 weapons in reserve for China includes warheads awaiting dismantlement plus some produced for SLBMs but which are not yet operational because of difficulties in deploying submarines.

As weapons are retired and dismantled, another issue comes to the fore: secure storage of highly-enriched uranium (HEU) and plutonium. According to the Global Fissile Material Report published by the International Panel on Fissile Materials, the 2011 global stockpile of civilian-plus-military HEU was about 1,440 metric tons; 1 metric ton is equivalent to 1,000 kg. Russia held the greatest amount at about 740 metric tons, and the United States was second at 610. Since the technical definition of HEU is 20 % or greater concentration of U-235, by no means is all of this material of weapons-grade, but it could fairly readily be enriched to be so. Given that a crude Hiroshima-type weapon could be made with about 50 kg of weapons-grade U-235, these tons of HEU potentially represent tens of thousands of weapons. Russia stopped producing HEU in the late 1980s, and the United States in 1992. Both countries are "down-blending" HEU for use as reactor fuel, but at an aggregate rate of only tens of tons per year. The same report estimated the global stockpile of separated plutonium (that not bound up in spent fuel rods) to be about 495 metric tons, of which Russia, the United Kingdom, and the United States respectively hold about 176, 95, and 92 metric tons. These countries have not yet begun to dispose of their excess plutonium.

9.4 Nuclear Treaties and Stockpile Stewardship

As higher-yield bombs were developed and tested through the 1950s, public concern grew over radioactive fallout that was raining out into the food chain. In August, 1957, President Dwight Eisenhower announced that the United States would be willing to suspend testing of nuclear weapons for up to two years if the Soviet Union agreed to a permanent cessation of production of fissionable materials for weapons, and the installation of an inspection system to ensure compliance. The Soviets responded in March, 1958 that they would unilaterally halt all nuclear tests, provided Western nations also stopped testing. Between April and August of 1958, a conference of experts convened in Geneva to study technical issues involved in a test ban, and concluded that a comprehensive ban could be verified through a worldwide network of monitoring stations. On October 31 of that year, the United States, Britain, and the Soviet Union began negotiations on a comprehensive nuclear test ban. Simultaneously, the United States and Britain voluntarily begin a one-year testing moratorium, which the Soviet Union soon joined.

In August, 1959, as negotiations continued, President Eisenhower extended the United States' moratorium on testing to the end of that year; the Soviets stated that they would not resume testing provided that the Western powers continued to observe a moratorium. When the American self-moratorium expired on December 31, President Eisenhower announced that America felt free to resume testing, but would not do so without advance notice. A few weeks later, however, the French carried out their first test. Citing the French test along with rising international tensions, the Soviet Union resumed atmospheric testing on September 1, 1961, and carried out 59 tests over the remainder of that year. The United States reciprocated with a series of underground tests beginning on September 15, and resumed atmospheric tests in the spring of 1962. Despite these setbacks, however, progress was being made on a proposal to ban atmospheric testing altogether.

The result of this first round of negotiations was the Limited Test Ban Treaty (LTBT) of 1963. This treaty prohibits nuclear weapons tests or any other nuclear explosions in the atmosphere, in outer space, under water, and in any other environment if such explosion causes radioactive debris to be present outside the territorial limits of the State under whose jurisdiction or control such explosion is conducted; it does not prohibit underground tests. The LTBT was signed in Moscow on August 5, 1963; the U. S. Senate ratified it on September 24, and it came into force on October 10. The LTBT has been signed by 108 countries, but not France or China.

Over the following years, a number of subsequent treaties concerning nuclear weapons came into effect. Some of these are described briefly in the following paragraphs.

Perhaps the most significant nuclear agreement is the Treaty on the Non-Proliferation of Nuclear Weapons, which is known as the NPT. Signed on July 1, 1968, the NPT entered into force in March, 1970. The NPT recognizes two classes of countries: so-called Nuclear Weapons States (NWS), which at that time comprised the P-5 countries, and Non-Nuclear Weapons States (NNWS). A total of 189 nations are party to the NPT, but four are not: India, Israel, Pakistan and North Korea (North Korea acceded to the NPT, but withdrew in 1993 and again in 2003. Iran is a party to the treaty). The NPT comprises three so-called "pillars": nonproliferation, disarmament, and peaceful use of nuclear energy. The P-5 states agree to not transfer "nuclear weapons or other nuclear explosive devices" and "not in any way to assist, encourage, or induce" NNWS to acquire nuclear weapons." NNWS states agree not to "receive, manufacture or acquire" nuclear weapons or to "seek or receive any assistance in the manufacture of nuclear weapons." NNWS parties also agree to accept safeguards by the IAEA to verify that they are not diverting nuclear research from peaceful uses to nuclear weapons or other nuclear explosive devices. While Article VI of the treaty imposes a very vague, non-binding, obligation on all signatories to move in the general direction of nuclear and total disarmament, the NPT imposed no restrictions on the number of warheads that NWS could possess. The third pillar of the treaty allows for transfer of nuclear technology and materials to signatory countries for the development of civilian nuclear energy programs, as long as they can demonstrate that those nuclear programs are not being used for the development of nuclear weapons. All such treaties have "escape clauses," and Article X of the NPT allows signatories the right to withdraw upon 3 months notice. North Korea is the only nation to have withdrawn from the treaty.

The 1972 Anti-Ballistic Missile (ABM) Treaty between the United States and the Soviet Union concerned limitations on anti-ballistic missile systems used in defending areas against nuclear weapons delivered on missiles. This treaty was in force until the U.S. unilaterally withdrew from it in 2002. Under the terms of this treaty, each country was allowed two sites at which it could base defensive systems: one for the capital city and one for ICBM silos. A later amendment (1974) reduced the number of sites to one per country, largely because neither country had developed a second site. The sole United States system, termed Safeguard, was located in North Dakota, but was deactivated after being operational for less than 4 months. The Russian A-135 ABM system protecting Moscow remains the only operational system deployed to this writing. On December 13, 2001, President George W. Bush gave Russia notice of the United States' withdrawal from the treaty; this was the first time in recent history that the U.S. has withdrawn from a major international arms treaty.

The first treaty to address numbers of warheads was the 1991 Strategic Arms Reduction Treaty between the U.S. and the U.S.S.R. (START I). This treaty prohibited its signatories from deploying more than 6,000 warheads atop a total of 1,600 ICBMs, SLBMs, and bombers. Its final implementation in late 2001 resulted in the removal of about 80 % of all strategic nuclear weapons then in existence. While the treaty was signed on July 31, 1991, its entry into force was delayed by the collapse of the Soviet Union a few months later; the treaty had to be extended to include the newly-independent states of Russia, Belarus, Kazakhstan, and Ukraine, all of which "inherited" a number of Soviet weapons. The latter three countries agreed to transport their nuclear arms to Russia for disposal.

In January 1993, the START II treaty was signed by President George H. W. Bush and Russian President Boris Yeltsin. This treaty banned the use of multiple independently targetable re-entry vehicles (MIRVs) on land-based Inter-Continental Ballistic Missiles. Although ratified, START II never entered into force because Russia withdrew from the it on June 14, 2002, 1 day after the U.S. withdrew from the 1972 ABM treaty. As time passed, START II became less relevant; it was effectively bypassed by the Strategic Offensive Reductions Treaty (SORT) of November 2001, which called for both sides to reduce operationally deployed strategic nuclear warheads to 1,700–2,200 by 2012.

The most ambitious effort to limit nuclear testing is the Comprehensive Nuclear-Test-Ban Treaty (CTBT), which would ban all nuclear explosions in all environments for any purposes. It was adopted by the United Nations General Assembly in September, 1996, but it has not yet entered into force. As of February 2012, a total of 157 states have ratified the CTBT, and another 25 have signed but not yet ratified it. To enter into force, 44 states listed in "Annex 2" of the treaty must ratify it. Annex 2 states are defined as those that participated in CTBT negotiations between 1994 and 1996, and which possessed power or research reactors at that time. As of December 2011, five Annex 2 states have signed but not ratified the treaty (China, Egypt, Iran, Israel, United States), while three have not signed it (India, Pakistan, North Korea); Russia ratified the treaty in 2004. The U.S. Senate rejected ratification of the CTBT in October, 1999, over concerns that other countries could easily cheat. However, the argument that violations could go undetected is becoming harder to sustain. The Preparatory Commission for the Comprehensive Test Ban Treaty Organization, an international organization headquartered in Vienna, was created to build a verification regime which includes establishment and operation of a worldwide network of 337 detection and analysis facilities. These include seismological, hydroacoustical, infrasound, and radionu-clide monitors which transmit their data back to Vienna for analysis and distribution to signatory countries. The sensitivity of the system is evidenced by the fact that fission products from the very low-yield underground North Korean test of 2006 were readily detected at a monitoring station in northern Canada. While it is conceivable that a cheater might get away with a extremely low-yield explosion, such a weapon would be of little practical use.

Although the United States has not ratified the CTBT, it has abided by its provisions. This, however, raises another question: If weapons cannot be tested, how can one be sure of their safety and reliability as they age? At this writing, the United States has not tested a nuclear weapon in over 20 years, yet they must be ready for use on potentially short notice. As weapons age, a number of possible degrading effects can crop up: chemical changes in the high-explosives in the primaries could affect their performance; alpha-decay in plutonium can affect its crystalline structure; hydrogen gas in the fusion-based components can cause corrosion. To deal with this, the National Nuclear Security Administration has established an extensive Science-Based Stockpile Stewardship program. In this program, weapon components are routinely subject to analyses to monitor their aging processes, and they can be refurbished or replaced as needed. These analyses are supported by study of historic test data and computer simulations of how variations in the properties of a material might affect weapon performance. The work of this program can be likened to maintaining a car in a condition to be ready to be started and driven at a moment's notice. You can change the oil, replace the battery, and pull out replace any component that you desire, but the car cannot be started. At least one new weapon in the current U. S. arsenal, the earth-penetrating B61 Mod-11, was deployed "live" without testing in 1996.

The most recent nuclear arms agreement is the "New START" treaty, which was signed by the U. S. and Russia in April, 2010. In brief, this treaty requires the United States and Russia to reduce their arsenals to 700 deployed ICBMs, SLBMs, and heavy bombers, with a total of 1,550 warheads on deployed ICBMs, SLBMs, and "warheads counted" for deployed heavy bombers. Further, the total number of deployed and non-deployed ICBM launchers, deployed and non-deployed SLBM launchers, and deployed and non-deployed heavy bombers is limited to 800. Table 9.3 shows some estimated eventual warhead numbers. The New START numbers for the U. S. are predicated on assuming 420 Minuteman III missiles, each of which carries a single warhead; 240 SLBMs, each of which carries four warheads; and 60 bombers. According to the treaty's counting rules, each bomber is counted as carrying only one warhead, even if it is physically capable of carrying more; although the United States would have 1,440 "accounted" warheads in this scheme, the actual number could be up to about 1,800. These limits are to be accomplished within 7 years after the treaty enters into force, which occurred in February, 2011. The treaty is enforced by a system of mutual data sharing, telemetry, and on-site inspections.

The numbers in Table 9.3 appear to give the United States an advantage, but it must be remembered that Russia possesses many more tactical nuclear weapons than does the U. S. (Table 9.2). No treaties have yet addressed that class of weapons, although successive Russian and American administrations have significantly drawn down tactical weapons deployments over the last 20 years. All American tactical nuclear weapons are deployed to Europe.

As the numbers of American and Russian weapons continues to decline, new questions will come to the fore. At what point should other countries be brought into the negotiations, and with whom will America and Russia be willing to accept parity? As numbers decline, each weapon becomes relatively more important, so

Table 9.3 Estimated nuclear arsenals in 2010 and after the New START reductions. Numbers are deployed warheads under New START counting rules

	United States		Russia	
	2010	New START	2010	New START
ICBMs	550	420	1,250	550
SLBMs	1,100	960	448	640
Strategic bombers	60	60	76	76
Total Warheads	2,000	1,440	1,774	1,266

Source P. Podvig, Physics & Society **39**(3), 12–15 (2010)

even a numerically modest amount of cheating could be significant. Will countries be willing to accept more intrusive inspections as a result? Finally, what do military strategists now see as the role of nuclear weapons? If the value of such weapons lies largely in their deterrent effect, could yields be reduced to well below what current weapons are capable of? That nuclear weapons are viewed by high-level military officers as being of declining importance in the post-Cold-War world is evidenced by a May, 2012, statement by General James E. Cartwright. Now retired, Cartwright served as a Vice-Chairman of the Joint Chiefs of Staff and as commander of the United States Nuclear Forces. In an article published in the *New York Times*, Cartwright said that the United States' nuclear deterrence could be guaranteed with a total arsenal of 900 warheads, with only half of them deployed at any one time: "The world has changed, but the current arsenal carries the baggage of the cold war. There is the baggage of significant numbers in reserve. There is the baggage of a nuclear stockpile beyond our needs. What is it we're really trying to deter? Our current arsenal does not address the threats of the 21st century."

While the extent of the nuclear stockpile may still be inconsistent with credible military needs, it is encouraging to know that reductions in the stockpile should continue over the next several years.

9.5 Epilogue

For many of the scientists involved with the Manhattan Project, that work represented the most dramatic time of their careers. As chemist Joseph Hirschfelder put it:

> I believe in scientific-technological miracles since I saw one performed at Los Alamos during World War II. The very best scientists and engineers were enlisted in the Manhattan Project. They were given overriding priorities. They got everything which they deemed essential to their program; the cost was unimportant. They had the full cooperation of everyone and they, themselves, devoted long hours in mixing together their ingenuity and technical skills. ... In a period of two-and-a-half years, they produced a miracle – an atomic bomb which creates temperatures of the order of 50,000,000 °C ... pressures of the order of 20,000,000 atmospheres ... while unleashing the tremendous energy stored in the atomic nuclei. ... At Los Alamos during World War II there was no moral issue with respect to working on the atom bomb. ... The whole fate of the civilized world depended upon our succeeding before the Germans! ... It is an open question as to whether the world is better or worse for our having made the atom bomb. ... After Otto Hahn's and Fritz Strassmann's discovery it became evident that sooner or later some country would make an atom bomb. If an atom bomb had not been made and detonated in World War II, the world would be unprepared to cope with the tremendous threat of nuclear warfare. ... warfare is no longer a rational means of settling differences between nations.

As the world's first large-scale, government-funded, science-based initiative, Manhattan established the template for such endeavors for decades to come.

During the war, federal funding for research in the United States grew from $50 million to $500 million per year, and currently exceeds $100 billion per year. In the immediate postwar years, three separate federal research agencies were established: the Office of Naval Research, the National Science Foundation, and the Atomic Energy Commission; a host of others would follow. Several national laboratories, including Los Alamos, are now distributed around the country. These organizations often adopted the pattern of large-scale, cooperative, hierarchically-organized research and development pioneered in the Project. Technological developments pioneered at these organizations underpin many of today's medical treatments, electronic consumer goods, and the instantaneous worldwide communications that we now take for granted. The first components of the internet, for example, were developed by the Defense Advanced Research Projects Agency.

The legacy of nuclear energy will forever be a mixed one. Thousands of nuclear weapons still exist. On the other hand, thousands of people benefit daily from radioisotope-based medical treatments, and some 20 % of the electricity generated in the United States comes from non-carbon-emitting nuclear reactors that do not contribute to global warming. Radioactivity, isotopes, and nuclear fission cannot be un-discovered.

The Manhattan Project changed the course of history. Herbert Marks, an aide to Dean Acheson, observed that

> The Manhattan District bore no relation to the industrial or social life of our country; it was a separate state, with its own airplanes and its own factories and its thousands of secrets. It had a peculiar sovereignty, one that could bring about the end, peacefully or violently, of all other sovereignties.

At this writing, the time that has elapsed since the establishment of the Manhattan District is equivalent to about the life expectancy of a person born in America in 1945. The number of living veterans of the Project is steadily dwindling, and many of the physical structures associated with it have been torn down or fallen into disrepair. In the last few years, however, a number of efforts to preserve at least some components of remaining facilities have begun to gain traction. In July 2011, the United States Departments of Interior and Energy transmitted to Congress recommendations for a Manhattan Project National Historical Park, which would involve sites at Los Alamos, Oak Ridge, and Hanford. In June, 2012, bills to establish the Park were introduced in both the House of Representatives and the Senate. In April, 2013, the House Committee on Natural Resources approved the House version of the bill, and in May, 2013, the Senate Committee on Energy Natural Resources approved that body's version of the bill. The bills must be considered by each full body; at this writing, these votes had yet to be scheduled. If this initiative is successful, future generations will be able to view, touch, and reflect upon artifacts from one of the most singular eras of human history.

Further Reading

Books, Journal Articles, and Reports

L. Badash, J.O. Hirschfelder, H.P. Broida (eds.), *Reminiscences of Los Alamos 1943–1945* (Reidel, Dordrecht, 1980)

S. Biegalski, International monitoring system of the comprehensive test-ban treaty. Phys. Soc. **39**(1), 9–12 (2010)

D.J. Bradley, C.W. Frank, Y. Mikerin, Nuclear contamination from weapons complexes in the Former Soviet Union and the United States. Phys. Today **49**(4), 40–45 (1996)

R.P. Carlisle, J.M. Zenzen, *Supplying the Nuclear Arsenal: American Production Reactors, 1942–1992* (Johns Hopkins University Press, Baltimore, 1996)

C. Carson, D.A. Hollinger (eds.), *Remembering Oppenheimer: Centennial Studies and Reflections* (University of California, Berkeley, 2005)

J. Cirincione, *Bomb Scare: The History and Future of Nuclear Weapons* (Columbia University Press, New York, 2007)

J. Davis, Technical and policy issues for nuclear weapons reductions. Phys. Soc. **41**(2), 6–9 (2012)

Department of Energy Nevada Operations Office: United States Nuclear Tests July 1945 through September 1992 (Report DOE/NV-209 REV 15). http://www.nv.doe.gov/library/publications/historical/DOENV_209_REV15.pdf

H. Friedman, L.B. Lockhart, I.H. Blifford, Detecting the Soviet bomb: Joe-1 in a rain barrel. Phys. Today **49**(11), 38–41 (1996)

R.L. Garwin, G. Charpak, *Megawatts and Megatons: A Turning Point in the Nuclear Age?* (Knopf, New York, 2001)

F.G. Gosling, *The Manhattan Project: Making the Atomic Bomb*. U. S. Department of Energy, Washington (2010). http://energy.gov/management/downloads/gosling-manhattan-project-making-atomic-bomb

L.R. Groves, *Now It Can be Told: The Story of the Manhattan Project* (Da Capo Press, New York, 1983)

D. Hawkins, *Manhattan District History. Project Y: The Los Alamos Project. Volume I: Inception until August 1945*. Los Alamos Scientific Laboratory, Los Alamos, NM (1947). Los Alamos publication LAMS-2532. Available online at http://library.lanl.gov/cgi-bin/getfile?LAMS-2532.htm

R.G. Hewlett, O.E. Anderson, Jr., *A History of the United States Atomic Energy Commission, Vol. 1: The New World, 1939/1946* (Pennsylvania State University Press, University Park, PA, 1962)

L. Hoddeson, P.W. Henriksen, R.A. Meade, C. Westfall, *Critical Assembly: A Technical History of Los Alamos during the Oppenheimer Years* (Cambridge University Press, Cambridge, 1993)

R. Jeanloz, Science-based stockpile stewardship. Phys. Today **53**(12), 44–50 (2000)

V.C. Jones, *United States Army in World War II: Special Studies—Manhattan: The Army and the Atomic Bomb* (Center of Military History, United States Army, Washington, 1985)

Y. Khariton, Y. Smirnov, The Khariton version. Bull. At. Sci. **49**(4), 20–31 (1993)

D. Lang, *From Hiroshima to the Moon: Chronicles of Life in the Atomic Age* (Simon and Schuster, New York, 1959)

S.L. Lippincott, A conversation with Robert F. Christy—Part I. Physics in Perspective **8**(3), 282–317 (2006); A conversation with Robert F. Christy—Part II. Physics in Perspective **8**(4), 408–450 (2006)

National Nuclear Security Administration: The United States Plutonium Balance, 1944–2009. Report DOE/DP-0137 (1996; updated 2012). http://nnsa.energy.gov/sites/default/files/nnsa/06-12-inlinefiles/PU%20Report%20Revised%2006-26-2012%20%28UNC%29.pdf

R.S. Norris, *Racing for the Bomb: General Leslie R. Groves, the Manhattan Project's Indispensable Man* (Steerforth Press, South Royalton, VT, 2002)

R.S. Norris, H.M. Kristensen, U.S. Nuclear warheads, 1945–2009. Bull. At. Sci. **65**(4), 72–81 (2009)

A. Pais, R.P. Crease, *J. Robert Oppenheimer: A Life* (Oxford University Press, Oxford, 2006)

P. Podvig, New START treaty and beyond. Phys. Soc. **39**(3), 12–15 (2010)

R. Rhodes, *Dark Sun: The Making of the Hydrogen Bomb* (Simon and Schuster, New York, 1995)

S.I. Schwartz, (ed.), *Atomic Audit: The Costs and Consequences of U. S. Nuclear Weapons Since 1940* (Brookings Institution Press, Washington 1995)

E. Teller, J.L. Shoolery, *Memoirs: A Twentieth-Century Journey in Science and Politics* (Perseus, Cambridge, MA, 2001)

H. Thayer, *Management of the Hanford Engineer Works in World War II* (American Society of Civil Engineers, New York, 1996)

E.C. Truslow, R.C. Smith, Manhattan District History: Project Y—The Los Alamos Project. Vol. II: August 1945 through December 1946. Los Alamos report LAMS-2532 (vol. II). http://www.fas.org/sgp/othergov/doe/lanl/docs1/00103804.pdf

J. Weisgall, *Operation Crossroads: The Atomic Tests at Bikini Atoll* (Naval Institute Press, Annapolis, MD, 1994)

Websites and Web-Based Documents

Acheson-Lilienthal report: http://www.learnworld.com/ZNW/LWText.Acheson-Lilienthal.html#nuclear

Atomic Heritage Foundation: http://www.atomicheritage.org/

Comprehensive Test-Ban Treaty chronology: http://www.fas.org/nuke/control/ctbt/chron1.htm

General Advisory Committee report on super bomb: http://www.pbs.org/wgbh/amex/bomb/filmmore/reference/primary/extractsofgeneral.html

General Cartwright on the nuclear arsenal and tactical nuclear weapons in Europe: http://www.cfr.org/proliferation/nuclear-posture-review/p21861, http://www.nytimes.com/2012/05/16/world/cartwright-key-retired-general-backs-large-us-nuclear-reduction.html?_r=1

Hanford cleanup: http://www.oregonlive.com/pacific-northwest-news/index.ssf/2012/02/hanford_cleanup_oversight_to_c.html

International Panel on Fissile Materials report on highly-enriched uranium and plutonium: http://fissilematerials.org/library/gfmr11.pdf

Nuclear Weapon Archive: http://nuclearweaponarchive.org/

Nuclear test-ban and arms-limitation treaties: http://en.wikipedia.org/wiki/Nuclear_non_proliferation_treaty; http://en.wikipedia.org/wiki/START_I; http://en.wikipedia.org/wiki/START_II; http://en.wikipedia.org/wiki/START_III; http://en.wikipedia.org/wiki/SORT; http://en.wikipedia.org/wiki/Comprehensive_Nuclear-Test-Ban_Treaty; http://en.wikipedia.org/wiki/Anti-Ballistic_Missile_Treaty

Operation Crossroads: http://en.wikipedia.org/wiki/Operation_Crossroads

Glossary

Cross-References Appear in *Italics*

25 Manhattan Engineer District code for uranium-235; from 9<u>2</u>-U-23<u>5</u>

49 Manhattan Engineer District code for plutonium-239; from 9<u>4</u>-Pu-239

Activation energy Generic term for energy that must be supplied to cause a reaction to happen; see also *Fission barrier* and *Coulomb barrier*. In nuclear reactions, activation energies are usually expressed in millions of electron volts (*MeV*).

AEC Atomic Energy Commission (United States). Succeeded by the Nuclear Regulatory Commission (*NRC*).

ALAS Association of Los Alamos Scientists. Superseded by Federation of American Scientists (*FAS*).

Alpha (α) decay Natural radioactive decay mechanism characteristic of heavy elements such as radium and uranium in which a nucleus ejects an alpha-particle, which is a nucleus of helium-4. Notationally designated by $^A_Z X \rightarrow \ ^{A-4}_{Z-2} Y + \ ^4_2 He$, or $^A_Z X \rightarrow \ ^{A-4}_{Z-2} Y + \alpha$, where X and Y designate so-called parent and daughter nuclei.

Ångstrom Unit of length equivalent to 10^{-10} meters; one ten-billionth of a meter. Characteristic of the effective sizes of atoms.

Atomic number (Z) Number of protons in the nucleus of an atom. Identifies the chemical element to which the atom belongs.

Atomic weight (A) The weight of an atom in atomic mass units; see Sects. 2.1.4 and 2.5. The symbol A is also used to designate the *nucleon number,* the total number of protons plus neutrons within a nucleus.

Barn (bn) Unit of reaction cross-section equivalent to 10^{-24} cm^2 = 10^{-28} m^2.

Baruch plan A plan for control of nuclear materials and weapons submitted by the United States to the United Nations in June 1946. Named after Bernard Baruch, U. S. representative to the United Nations Atomic Energy Commission. Despite months of debate, the plan was never implemented; Chap. 9.

B. C. Reed, *The History and Science of the Manhattan Project,*
Undergraduate Lecture Notes in Physics, DOI: 10.1007/978-3-642-40297-5,
© Springer-Verlag Berlin Heidelberg 2014

Becquerel (Bq) A unit of rate of radioactive decay; 1 Bq = 1 decay per second. See also *Curie*.

Beta (β) decay Natural radioactive decay mechanism of nuclei that are neutron or proton-rich. If a nucleus is neutron-rich, a neutron spontaneously transmutes into a proton plus an electron, ejecting the latter to the outside world: $_Z^A X \rightarrow {}_{Z+1}^A Y + {}_{-1}^0 e^-$, where X and Y designate parent and daughter nuclei. In this case, known as β^- decay (with the electron known as a β^- particle), the daughter nucleus is one element heavier in the Periodic Table than the parent nucleus. Conversely, if a nucleus is proton-rich, a proton spontaneously decays into a neutron and an *positron*, ejecting the latter to the outside world: $_Z^A X \rightarrow {}_{Z-1}^A Y + {}_1^0 e^+$; in this case ($\beta^+$ decay) the daughter nucleus is one element lighter in the Periodic Table than the parent nucleus. A sequence of such decays may follow until the nucleus achieves stability.

Binding Energy A form of energy which is created from mass, and which can be transformed back into mass; Sects. 2.1.4 and 2.5. In reactions where the mass of the output product(s) is less than that of the input reactants, binding energy is said to be liberated ($E = mc^2$), and the energy appears in the form of kinetic energy of the products and/or one or more of the products being in an "internally excited" energy state. If the mass of the output products is greater than that of the input reactants, kinetic energy from the input reactants is transmuted into mass. See also *Mass defect* and *Q-value*.

Bockscar Name of the B-29 bomber which carried the Nagasaki *Fat Man* nuclear weapon.

B-Pile First large-scale (250 MW) nuclear reactor constructed at the Hanford Engineer Works (*HEW*, Washington) for the purpose of breeding plutonium. B-pile began operation in late 1944, and was soon followed by the D and F piles at the same site; Chap. 6.

Calutron A device based on a *Cyclotron* which is used for separating isotopes of different atomic weights by ionizing them and passing them through a strong magnetic field; Sect. 5.3. A contraction of *Cal*ifornia *U*niversity cyclo*tron*. See also *Mass spectroscopy*.

CEW Clinton Engineer Works, Tennessee. Location of Manhattan Project uranium enrichment facilities; Chap. 5.

CIW Carnegie Institution of Washington.

Combined Policy Committee (CPC) American-British-Canadian committee established in August, 1943, to coordinate nuclear research and to serve as the focal point for interchanging information; Sect. 7.4.

Control rod Device made of a neutron-absorbing material that is used in a nuclear reactor to control the reaction rate. Cadmium and boron are excellent neutron absorbers.

Coulomb barrier Amount of kinetic energy that an "incoming" nucleus which is approaching a "target" nucleus must possess in order to overcome the repulsive electrical force between protons within the two nuclei in order to collide and induce a nuclear reaction with the target nucleus. Typically measured in millions of electron volts (*MeV*); Sect. 2.1.8.

CP-1 Critical (or Chicago) Pile number 1, the first nuclear reactor to achieve a self-sustaining nuclear chain reaction. This uncooled, graphite-moderated device operated for the first time on December 2, 1942; Sect. 5.2.

Critical mass Minimum mass of a fissile material necessary to achieve a self-sustaining fission chain reaction, taking into account loss of neutrons through the surface of the material. If the material is not surrounded by a neutron-reflecting tamper, the term "bare" critical mass is used. For uranium-235 and plutonium-239, the bare critical masses are respectively about 45 and 17 kg; Sect. 7.5.

Cross-section A quantity which measures the probability that a given *nuclide* will undergo a particular type of reaction (fission, scattering, absorption ...) when struck by an incoming particle. Cross-sections are expressed as areas in *barns*, where 1 barn $= 10^{-24}$ cm^2, and are usually designated by the symbol σ along with a subscript designating the type of reaction involved. Cross sections depend on the type of particle being struck, the type of striking particle, and the energy of the striking particle; Sect. 2.4.

Curie (Ci) A unit of rate of radioactive decay; 1 Ci $= 3.7 \times 10^{10}$ decays per second. This is the alpha-decay rate of one gram of freshly-isolated radium-226. See also *Becquerel*.

Cyclotron A modified mass spectrometer (see *Mass spectroscopy*) used for accelerating electrically charge particles to very great energies by the use of electric and magnetic fields; Sect. 2.1.8. See also *Calutron*.

Diffusion Generic term for the passage of particles through space. The speed of the particles depends on their mass and the temperature of the environment. In the Manhattan Project, uranium was enriched by both gaseous and thermal diffusion processes; Sects. 5.4 and 5.5.

Dragon machine Colloquial name for an experimental device developed at Los Alamos wherein a slug of uranium-235 would be dropped through a hole in a plate of uranium-235, momentarily creating a fast-neutron fission chain reaction; Sect. 7.11.

D-T Reaction Fusion of deuterium and tritium to produce helium and a neutron: $^2_1H + ^3_1H \rightarrow ^1_0n + ^4_2He$; Sect. 9.2.

Electron capture A decay mechanism wherein an inner-orbital electron is captured by a nucleus. The captured electron combines with a proton to form a neutron, rendering the process as a reverse β^- decay, equivalent to a β^+ decay.

Enola Gay Name of the B-29 bomber which carried the Hiroshima *Little Boy* nuclear weapon.

Enrichment Generic term for any process which alters the abundance ratio of isotopes in a sample of some input feed material. Usually used in the sense of a process which increases the number of fissile uranium-235 nuclei in comparison to the number of non-fissile uranium-238 nuclei. In the Manhattan Project, both electromagnetic and diffusion enrichment techniques were employed; Chap. 5.

eV Electron-volt. A unit of energy equivalent to 1.602×10^{-19} Joules. Chemical reactions typically involve energy exchanges of a few eV. See also *MeV*.

FAS Federation of American Scientists.

Fat Man Code name for the Nagasaki implosion-type plutonium bomb, which achieved an explosive *yield* of about 22 kt.

First criticality Moment in the detonation of a nuclear weapon when the core first achieves conditions necessary for a self-sustaining chain reaction.

Fissile A fissile material is one whose nuclei will undergo fission when struck by bombarding neutrons of any energy. Uranium-235 and plutonium-239 are both fissile. Fissile is a subset of *Fissionable*. See also *Fission barrier*.

Fission Nuclear reaction wherein a nucleus splits into two roughly equal fragments, typically accompanied by a significant release of energy (~ 200 MeV). Fission may be induced by striking the nucleus with an outside particle (usually a neutron), but also happens spontaneously in some heavy elements. Compare *Fusion* below.

Fission barrier Minimum amount of kinetic energy a bombarding neutron must possess in order to induce fission in a target nucleus. Typically measured in millions of electron volts (MeV); Sect. 3.3. For nuclei of elements in the middle of the Periodic Table the fission barrier can be as high as ~ 55 MeV, but for heavy nuclei such as those of uranium atoms is on the order of 5–6 MeV, depending on the isotope involved. In these latter cases the barrier may be low enough to be exceeded by the *binding energy* liberated upon neutron absorption, rendering a nuclide *fissile*.

Fissionable A fissionable material is one whose nuclei can be made to fission when struck by bombarding neutrons. In practice, the term is usually reserved for materials that fission only under bombardment by "fast" neutrons, typically of kinetic energy ~ 1 MeV or greater. Compare to *Fissile* above. Uranium-238 is fissionable, but not fissile.

Franck report Document prepared by University of Chicago scientists in June, 1945, addressing political and social problems associated with nuclear weapons; Sect. 8.4. Now considered a founding document of the nuclear non-proliferation movement. See also *Jeffries report*.

Frisch-Peierls memorandum Memorandum prepared in early 1940 by Otto Frisch and Rudolf Peierls at Birmingham University, which alerted British government authorities to the possibility of fission bombs. Sect. 3.7.

Fusion Nuclear reaction wherein two nuclei "fuse" to form a heavier nucleus, typically accompanied by an energy release of a few or few tens of MeV. Used in fusion weapons, which are known colloquially as "hydrogen bombs." Fusion reactions liberate less energy than fission reactions, but liberate more energy per mass of reactant nuclei, and often generate particles which can catalyze further fission and fusion reactions; Sect. 9.2. Compare to *Fission* above.

General Advisory Committee (GAC) An advisory committee to the Atomic Energy Commission, established to provide advice on technical issues; Sect. 9.1.

Greenhouse George First United States test of a radiation implosion weapon, May 1951. Yield 225 kt; Sect. 9.2.

Half-life Characteristic time required for one-half of the nuclei of a naturally-decaying isotope to undergo a specified decay process. Half-lives vary from tiny fractions of a second to billions of years.

Heavy water A form of water in which the hydrogen atoms are replaced with deuterium, an isotopic form of hydrogen. Chemical symbol D_2O. D designates a deuterium, or "heavy hydrogen" nucleus, ${}^2_1H$. Heavy water occurs naturally, and can be extracted from ordinary water. Heavy water is of interest in nuclear power and research as it makes an excellent neutron *moderator*.

HEW Hanford Engineer Works, Washington state. Location of Manhattan Project plutonium production facilities; Chap. 6.

Hex Colloquial term for uranium hexafluoride, UF_6.

Hibakusha Japanese term for people who survived both the Hiroshima and Nagasaki bombings.

IAEA International Atomic Energy Agency.

ICBM Inter-Continental Ballistic Missile.

Implosion A chemical explosion which is directed "inwards". In the context of nuclear weapons, used to crush an initially sub-critical mass to critical density; Sect. 7.11.

Initiator Device at the core of a nuclear weapon that releases neutrons to initiate the chain reaction. In the Manhattan Project, initiators were also known as Urchins.

Interim Committee Advisory group established by Secretary of War Henry Stimson in May, 1945, to advise on postwar atomic-energy planning; Sect. 8.4.

Isotope See also *Nuclide*. Nucleus or atom of an element that has the number of protons characteristic of the element (*Atomic number*), and some specific number of neutrons. All nuclei of a given element have the same number of protons, but different isotopes of an element have different numbers of neutrons. Different isotopes of a given element consequently have different *Atomic weights*.

Ivy King Largest pure fission weapon ever detonated by the United States, November, 1952. Yield $\sim$500 kt; Sect. 9.2.

Ivy Mike First true American thermonuclear (fusion) weapon, detonated November 1952. Yield $\sim$10.4 Mt; Sect. 9.2.

Jeffries report A document prepared by University of Chicago scientists in late 1944 describing anticipated postwar research and industrial applications in the area of nuclear energy; Sect. 8.2. Also known as the "Prospectus on Nucleonics." See also *Franck report*.

Joe-1 Western term for the first test of a Soviet nuclear weapon, 1949; Sect. 9.2.

Jumbo Name of a 200-ton steel vessel that was intended to be used to contain the first test explosion of a nuclear weapon. Jumbo was never used, and parts of it still remain at the *Trinity* site; Sect. 7.12.

K-25 Code name for the gaseous diffusion plant at the Clinton Engineer Works (*CEW*); Sect. 5.4.

Kiloton (kt) A unit of energy equivalent to that released by the explosion of 1,000 metric tons of conventional explosive (1 metric ton = 1,000 kg), commonly used to quantify the energy *yield* of nuclear weapons; 1 kt = 4.2×10^{12} Joules = 1.17 million kWh. World War II-era nuclear weapons had yields in the 10–20 kt range.

kWh kilowatt-hour, a unit of energy corresponding to a power consumption (or generation) of 1,000 Watts (= 1,000 Joule/sec) over a time of 1 h (3,600 s). 1 kWh = 3.6×10^6 Joule.

Lewis Committee There were various Lewis Committees during the Manhattan Project, all involving MIT chemical engineer Warren Lewis. The most important ones reviewed the entire atomic-energy program at the time the CP-1 reactor went critical in late 1942 (Sect. 4.10), and the proposed research program at Los Alamos in March/April 1943 (Sect. 7.2).

Little Boy Code name for the Hiroshima gun-type uranium fission bomb, which achieved a *yield* of about 13 kt.

LTBT Limited Test-Ban Treaty. 1963 treaty which prohibits nuclear weapons tests or any other nuclear explosions in the atmosphere, outer space, or under water. Does not prohibit underground tests; Sect. 9.4.

Mass defect Difference in mass between an "assembled" nucleus and the sum of the masses of the individual protons and neutrons that comprise it; usually expressed in equivalent energy units. All stable nuclei have masses less than the sum of the masses of their constituent *nucleons;* Sects. 2.14 and 2.5.

Mass spectroscopy An experimental technique for determining masses of atoms to high precision. Ionized atoms or molecules are directed into a region of space containing a magnetic field. The trajectories of the particles consequently depend on their mass; by noting where particles "land", masses can be accurately measured; Sect. 2.1.4. See also *Cyclotron* and *Calutron*.

MAUD committee British government committee established in response to the *Frisch-Peierls memorandum* to investigate possible military uses of nuclear fission; Sects. 3.7 and 4.4. In a July, 1941, report (Sect. 4.4) the committee analyzed the possible use of uranium in a fission bomb.

May-Johnson bill Legislation concerning atomic energy introduced to the United States Congress in October, 1945; Sect. 9.1. The bill's harsh control and security provisions generated considerable criticism within the scientific community, which led to its being abandoned in favor of the *McMahon bill*.

McMahon bill Legislation which established the United States Atomic Energy Commission; Sect. 9.1.

Mean Free Path (MFP) Average distance that a particle will travel through some material before striking another particle and possibly inducing some reaction. In the context of nuclear weapons, usually applied to the passage of neutrons through a sample of fissile material; Sect. 7.5. Commonly designated by the symbol λ.

MED Manhattan Engineer District of the United States Army; Sect. 4.9.

Megaton (Mt) A unit of energy equivalent to that released by the explosion of one million metric tons of conventional explosive, commonly used to quantify the energy release of extremely powerful nuclear weapons. 1 Mt $= 4.2 \times 10^{15}$ Joules $= 1.17$ billion kWh.

Metallurgical Laboratory Code name for the atomic research laboratory at the University of Chicago, directed by Arthur Compton. This laboratory had particular responsibility for development of nuclear reactors and plutonium-separation chemistry.

MeV Mega electron-volt; one million electron-volts. A unit of energy equivalent to 1.602×10^{-13} Joules. Nuclear reactions typically involve energy exchanges of a few MeV. See also eV.

Military Policy Committee (MPC) Established in September, 1943, by Secretary of War Henry Stimson to advise on development and use of nuclear weapons. The MPC acted as a sort of Board of Directors of the Manhattan Project; Sect. 4.10.

Moderator Material within a nuclear reactor which slows high-energy neutrons to "thermal" velocities (Sect. 2.4) to increase their chance of fissioning U-235 nuclei. Graphite and heavy water make excellent moderators. Ordinary water can also be used, but requires a reactor fueled with enriched uranium.

MW Megawatt (one million Watts). A unit of power for quantifying the rate of consumption of energy. 1 Watt = 1 Joule/sec.

NAS National Academy of Sciences (United States).

NDRC National Defense Research Committee. Established by President Roosevelt in June, 1940, to support and coordinate research conducted by civilian scientists which might have military applications. The Uranium Committee was absorbed into the NDRC when the latter was established (Sect. 4.2). Absorbed into the *OSRD* in June, 1941.

Neutron Electrically neutral constituent particle of atomic nuclei. Given the number of protons in the nucleus (*Atomic number*), the number of neutrons in a nucleus dictates the *isotope* of the element involved. Neutrons can be thought of as a form of "nuclear glue" that holds nuclei together against repulsive electrostatic forces that protons exert on each other.

Neutron number (*N*) Number of neutrons within a nucleus. The number of neutrons N plus the number of protons Z (*Atomic number*) totals to the *Nucleon number A*. See also *Atomic weight*.

NBS National Bureau of Standards (United States).

NPT Acronym for the Treaty on the Non-Proliferation of Nuclear Weapons (1968); Sect. 9.4.

NRC National Research Council; Nuclear Regulatory Commission (United States).

NRL Naval Research Laboratory (United States).

Nucleon Collective term for neutrons and protons.

Nucleon number (*A*) Total number of protons plus neutrons within a nucleus, always an integer number. See *atomic number* and *neutron number*.

Nuclide Generic term for a nucleus of a given number of protons and neutrons. Notation: $^{A}_{Z}X$, where X is the symbol for the element involved, Z is the number of protons (*Atomic number*), and A is the total number of protons plus neutrons (*Atomic weight*). Essentially synonymous with *Isotope*, except that use of the latter term is usually in the context of referring to nuclides of a given element, which will all have the Z same value but different atomic weights.

Nucleus Positively-charged core of an atom, comprising protons and neutrons.

OSRD Office of Scientific Research and Development. Established by President Roosevelt in June, 1941, to coordinate research and development of devices that might be of military value (e.g., radar, proximity fuses, fission weapons).

Overpressure Condition of atmospheric pressure above "normal" atmospheric pressure, caused by the detonation of a nuclear weapon, usually measured in pounds per square inch (psi); Sect. 7.13.

P-5 The "primary five" nuclear weapons states: United States, Russia, Britain, France, China.

Parity Oddness or evenness of the number of protons and neutrons in a nucleus; Sect. 3.2. In non-proliferation parlance, the relative evenness of numbers of nuclear weapons held by various countries.

Pile Historic term for a nuclear reactor.

Planning Board The Manhattan Project involved two Planning Boards. The first was established in November, 1941, to develop recommendations concerning plans for production of fissile materials and contracts for engineering studies; Sect. 4.6. The second was at Los Alamos, organized to coordinate technical work at the laboratory; Sect. 7.2.

Positron A positively charged electron, also known as a beta-positive (β^+) particle.

Predetonation Detonation of a nuclear explosive before the bomb core is fully assembled, resulting in an explosive *yield* less than intended. May be caused by neutron-emitting impurities or spontaneous fissions; Sect. 7.7.

Project Alberta Code name for Los Alamos program to prepare bombs for combat.

Proton Constituent positively-charged particle of atomic nuclei. The number of protons in a nucleus is equal to the *Atomic number* of the nucleus.

Q-value Amount of energy liberated or consumed in a nuclear reaction, typically measured in millions of electron volts (MeV); Sect. 2.1.6.

Queen Marys Colloquial name for plutonium-processing facilities at the Hanford Engineer Works (*HEW*); Sect. 6.5. These 800-foot-long buildings rivaled the ocean liner Queen Mary in length (1,020 feet).

RaLa Abbreviation for the "radiolanthanum" implosion diagnostic technique developed at Los Alamos; Sect. 7.11.

Reaction channel One of a number of possible outcomes in a reaction involving two (or more) input particles. With neutron-induced reactions involving light elements, a number of possible channels can occur; Sect. 2.4.

Rem Unit of radiation exposure; "Radiation Equivalent in Man." Synonymous with *Roentgen;* Sect. 7.13. For humans, a single-shot dose on the order of 500 rems will often result in death.

Reproduction factor Measure of the net number of neutrons generated per each consumed in a nuclear reactor, designated by the symbol k. If $k \geq 1$, a self-sustaining reaction is in progress.

Roentgen See *Rem.*

S-1 Committee; S-1 Section New name acquired by the *Uranium Committee* after it was absorbed into the Office of Scientific Research and Development (*OSRD*) when the latter was established in July, 1941 (Sects. 4.4, 4.5).

S-1 Executive Committee Successor to the S-1 Committee established June, 1942, within the *OSRD* to coordinate research into various methods of fissile-material production; Sect. 4.9. Chaired by James Conant, the other members were Lyman Briggs, Ernest Lawrence, Harold Urey, Arthur Compton, and Eger Murphree.

S-50 Code name for the thermal diffusion plant at the Clinton Engineer Works (*CEW*); Sect. 5.5.

Scientific Panel A subcommittee of the *Interim Committee* (1945) established to provide advice on technical issues related to the use and future development of nuclear weapons; Sect. 8.4. Members were Robert Oppenheimer, Arthur Compton, Enrico Fermi, and Ernest Lawrence. Another Scientific Panel was that appointed to advise on postwar atomic policies; Sect. 9.1.

Second criticality Moment in the course of the detonation of a nuclear weapon where the core has expanded to the point where conditions necessary for a self-sustaining chain reaction no longer hold.

SED Special Engineer Detachment; a group of military personnel with technical and scientific training; Sect. 7.3.

SF Spontaneous fission.

SLBM Submarine-Launched Ballistic Missile.

Smyth Report Colloquial title of a report authored by Henry Smyth and issued by the United States government just after the bombings of Hiroshima and Nagasaki in August, 1945; Sect. 8.7. This document was the first public description of the Manhattan Project; its full tile was "Atomic Energy for Military Purposes: The Official Report on the Development of the Atomic Bomb under the Auspices of the United States Government, 1940-1945".

SODC Standard Oil Development Company.

SORT Strategic Offensive Reductions Treaty (2001); Sect. 9.4.

START Strategic Arms Reduction Treaty (1991, 1993 and 2010); Sect. 9.4. There are multiple START treaties between the United States and Russia.

Tamper A heavy (usually metallic) structure that surrounds the core of a nuclear weapon, designed to reflect escaping neutrons back into the core and briefly retard expansion of the core while it explodes. Both effects act to increase weapon efficiency.

Target Committee Group of military officers and scientists established April, 1945, to advise on targeting of nuclear weapons against Japanese cities; Sect. 8.1.

Top Policy Group Committee of government, military, and scientific personnel established by President Roosevelt, October, 1941, to advise on policy considerations raised by nuclear issues; Sect. 4.5.

Trinity First test of a nuclear weapon, July 16, 1945, in southern New Mexico. This implosion device achieved a yield of about 22 kt.

TVA Tennessee Valley Authority, an agency of the United States government.

Uranium Committee Formally, the Advisory Committee on Uranium, established October, 1939, to investigate possible military applications of nuclear fission; Sect. 4.1. This was the first United States government group convened to consider the possibility of fission weapons and nuclear power. The Uranium Committee was absorbed into the *NDRC* in June, 1940, and became known as Section S-1 of the Office of Scientific Research and Development (*OSRD*) when the latter was established in July, 1941 (Sect. 4.4).

USSBS United States Strategic Bombing Survey; Sect. 8.6.

X-10 Code name for the graphite reactor at the Clinton Engineer Works (*CEW*); Sect. 5.2.

Xenon poisoning Xenon is a product of nuclear fissions; as it accumulates within a reactor, it "poisons" the reaction due to its tendency to absorb neutrons; Sect. 6.5. If not for the short half-life involved (9 h), the responsible isotope, Xe-135, would continue to accumulate until the reaction could not longer proceed.

Y-12 Code name for the electromagnetic separation complex at the Clinton Engineer Works (*CEW*); Sect. 5.3.

Yield Energy released by a nuclear weapon, usually measured in *kilotons* (kt) or *megatons* (Mt)

Index

Printed in the United States
By Bookmasters